INTERNATIONAL SERIES OF MONOGRAPHS ON PHYSICS

INTERNATIONAL SERIES OF MONOGRAPHS ON PHYSICS

109. A. Aharoni: *Introduction to the theory of ferromagnetism 2/e*
108. R. Dobbs: *Helium three*
107. R. Wigmans: *Calorimetry: energy measurement in particle physics*
106. J. Kübler: *Theory of itinerant electron magnetism*
105. Y. Kuramoto, Y. Kitaoka: *Dynamics of heavy electrons*
104. D. Bardin, G. Passarino: *The Standard Model in the making*
103. G. C. Branco, L. Lavoura, J. P. Silva: *CP Violation*
102. T. C. Choy: *Effective medium theory*
101. H. Araki: *Mathematical theory of quantum fields*
100. L. M. Pismen: *Vortices in nonlinear fields*
99. L. Mestel: *Stellar magnetism*
98. K. H. Bennemann: *Nonlinear optics in metals*
97. D. Salzmann: *Atomic physics in hot plasmas*
96. M. Brambilla: *Kinetic theory of plasma waves*
95. M. Wakatani: *Stellarator and heliotron devices*
94. S. Chikazumi: *Physics of ferromagnetism*
92. J. Zinn-Justin: *Quantum field theory and critical phenomena*
91. R. A. Bertlmann: *Anomalies in quantum field theory*
90. P. K. Gosh: *Ion traps*
89. E. Simánek: *Inhomogeneous superconductors*
88. S. L. Adler: *Quaternionic quantum mechanics and quantum fields*
87. P. S. Joshi: *Global aspects in gravitation and cosmology*
86. E. R. Pike, S. Sarkar: *The quantum theory of radiation*
84. V. Z. Kresin, H. Morawitz, S. A. Wolf: *Mechanisms of conventional and high-T_c superconductivity*
83. P. G. de Gennes, J. Prost: *The physics of liquid crystals*
82. B. H. Bransden, M. R. C. McDowell: *Charge exchange and the theory of ion–atom collision*
81. J. Jensen, A. R. Mackintosh: *Rare earth magnetism*
80. R. Gastmans, T. T. Wu: *The ubiquitous photon*
79. P. Luchini, H. Motz: *Undulators and free-electron lasers*
78. P. Weinberger: *Electron scattering theory*
76. H. Aoki, H. Kamimura: *The physics of interacting electrons in disordered systems*
75. J. D. Lawson: *The physics of charged particle beams*
73. M. Doi, S. F. Edwards: *The theory of polymer dynamics*
71. E. L. Wolf: *Principles of electron tunneling spectroscopy*
70. H. K. Henisch: *Semiconductor contacts*
69. S. Chandrasekhar: *The mathematical theory of black holes*
68. G. R. Satchler: *Direct nuclear reactions*
51. C. Møller: *The theory of relativity*
46. H. E. Stanley: *Introduction to phase transitions and critical phenomena*
32. A. Abragam: *Principles of nuclear magnetism*
27. P. A. M. Dirac: *Principles of quantum mechanics*
23. R. E. Peierls: *Quantum theory of solids*

Anomalies in Quantum Field Theory

REINHOLD A. BERTLMANN
Institute for Theoretical Physics
University of Vienna

OXFORD
UNIVERSITY PRESS

OXFORD
UNIVERSITY PRESS

Great Clarendon Street, Oxford OX2 6DP

Oxford University Press is a department of the University of Oxford.
It furthers the University's objective of excellence in research, scholarship,
and education by publishing worldwide in

Oxford New York

Athens Auckland Bangkok Bogotá Buenos Aires Calcutta
Cape Town Chennai Dar es Salaam Delhi Florence Hong Kong Istanbul
Karachi Kuala Lumpur Madrid Melbourne Mexico City Mumbai
Nairobi Paris São Paulo Singapore Taipei Tokyo Toronto Warsaw

with associated companies in Berlin Ibadan

Published in the United States
by Oxford University Press Inc., New York

First published in hardback 1996
Paperback edition (with corrections) 2000

A catalogue record for this book is available from the British Library

Library of Congress Cataloging-in-Publication Data

ISBN 0 19 850762 3

Typeset by the author using LaTeX
Printed in Great Britain
on acid-free paper by
Bookcraft (Bath), Midsomer Norton, Avon

FOR RENATA

Preface

When I was a fellow at CERN (1978–1980) I had the great fortune to meet John Bell. In the course of my stay I had the chance to become acquainted with his deep understanding of (nonrelativistic) quantum mechanics and of (relativistic) quantum field theory (see [Bell 1987]). This led to an intense and wonderful collaboration in the field [Bell, Bertlmann 1980–1984].

In the mid-eighties the anomalies of quantum field theory were a very exciting subject of research. During this time I visited John regularly. As one of the fathers of the anomaly he had a great interest in these current developments. The main subject of our discussions was the work of Stora [Stora 1984] and Zumino [Zumino 1984], who found within an elegant mathematical formalism the so-called chain of descent equations. Subsequently we discussed the differential-geometric and topological aspects of quantum field theory, especially the important Atiyah–Singer index theorems. Returning to Vienna I gave a series of lectures on this topic; these lectures form the basis of this book.

Anomalies are the key to a deeper understanding of quantum field theory. The aim of this book is, therefore, to present all the different aspects of anomalies in quantum field theory. Much emphasis is placed on a modern mathematical formulation using differential geometry and topology, a formulation which will prove to have great importance in the future. The different aspects of anomalies are introduced step by step to make the whole field accessible to a broad audience. The derivations and calculations are given explicitly to make the subject more attractive to students entering this field.

All the mathematics needed is not presupposed but explained at the very beginning. I also give an introduction to the path integral formalism which will be needed for studying anomalies. The reader should be familiar with the fundamentals of quantum field theory. Finally I consider gravitation and study gravitational anomalies because I believe that gravitation has to be included in a (unified) quantum field theory.

Although this field is still evolving and many gaps remain to be filled, recent years have nevertheless been a period of consolidation, so that I have been able to present a rather coherent and self-contained description of the anomalous nature of quantum field theory. It is my sincere hope that, at the end, the reader will have a deeper understanding of the subject as a whole.

There are many people whose assistance and criticism have been of great value to me while I was writing and revising the manuscript of this book.

I would like to recognize in a special way the contribution of Raymond Stora, who has taken the time and trouble to read parts of the book, pointing out errors and making critical comments and helpful suggestions on several subjects. I am very grateful to Gérard Emch for his valuable and encouraging advice on early drafts of some chapters and for his fruitful suggestions for improvements. I have benefited very much from stimulating discussions with Helmuth Urbantke and I would like to thank him for his numerous detailed critical comments.

I would also like to thank Jiři Hořejší for his careful reading of some sections. I would like to thank Herbert Pietschmann for his interest and Gerhard Ecker, Walter Grimus, Helmut Kühnelt, Hanns Stremnitzer, Helmut Neufeld and Franz Schöberl for fruitful discussions.

My great and especial thanks go to Franziska Wagner for her masterly work transforming a long and complex manuscript into LaTeX and for her great patience and endurance, which made it possible to complete this book on time. I would also like to single out for special thanks Christoph Adam for his great help and assistance over a long time until the book reached its final form, for reading the manuscript, for making the index and for the final LaTeX corrections.

I also wish to thank Khosrow Chadan and Hellmuth Satke for private communications and their kind support.

Finally, and most of all, I would like to thank my wife very warmly for her unfailing encouragement and support over a long period.

Vienna
August 1995 R. A. B.

Preface to the paperback edition

This production of a paperback edition gave me the chance to correct some misprints and errors. In this connection I am thankful to Walter Thirring, Manfred Breitenecker, and Michael Moser for their comments. I am also grateful to Stephan Adler for his correspondence concerning the discovery of the anomaly. Finally, I also would like to thank the editorial staff of Oxford University Press for their cooperation.

Vienna
May 2000 R. A. B.

Contents

1
Introduction

Symmetries and their corresponding conservation laws play an important role in describing the fundamental forces of nature. However, it might turn out that a certain conservation law, valid in the classical theory, is violated in the quantized version. Then we speak of an anomaly. Actually, this is not so surprising; we know by now that 'naïve' classical conceptions are often demolished by quantum effects. Familiar examples are the uncertainty relation of space and momentum, or the nonvanishing ground state energy of the oscillator, etc. Why then are the anomalies so extraordinary in quantum field theories?

The basis of modern quantum field theory (QFT)—gauge theory—is the principle of gauge symmetry. There an anomaly—the violation of a classically conserved current—signals the breakdown of the gauge symmetry and, in consequence, the ruin of the consistency of the theory. Avoiding, on one hand, the anomaly—which may be possible—leads to severe constraints on the physical content of the theory; for example, to the prediction of the top quark. But, on the other hand, anomalies are also needed to describe certain experimental facts. It is this double-feature which makes anomalies so important for physics.

The anomalies which we consider in our book are the axial- or the chiral anomaly corresponding to an axial- or a chiral fermion current. The discovery of the axial anomaly has quite a long history. It began in 1949 with Jack Steinberger [Steinberger 1949], who calculated in his Ph.D. a Feynman diagram in an at that time fashionable pion-nucleon (π–N) model (which, as we know now, represents the anomaly) in order to describe the decay $\pi^0 \to \gamma\gamma$. Independently H. Fukuda and Y. Miyamoto [Fukuda, Miyamoto 1949] performed similar calculations. Steinberger's π–N model containing a γ_5-vertex was in excellent agreement with experiment. However, his comparison with a π–N amplitude containing a $\gamma_\mu\gamma_5$-vertex (which, as we know now, corresponds to the axial Ward identity), of course, failed. Noting this puzzle, Steinberger left theory and became a Nobel Laureate in experimental physics.

Two years later, in 1951, Julian Schwinger [Schwinger 1951] pointed out that the conservation of the axial current in QED—an immediate consequence of the axial symmetry—is violated when the current operator is appropriately regularized. After a pause of 12 years it was Ken Johnson [Johnson 1963] who remarked that in a massless 2-dimensional QED one cannot have both the conservation of the gauge current *and* the conservation of the axial current. However, little attention was paid to the importance of these results in the subsequent years.

In the sixties Gell-Mann's current algebra became rather popular, and within this framework, relying on PCAC (partial conservation of the axial current), Sutherland and Veltman [Sutherland 1967], [Veltman 1967] proved the theorem that the neutral pion *cannot* decay into two photons, in contradiction to experiment! Impressed by the analysis of his friend Martinus Veltman, John S. Bell from CERN stressed that 'the subject of current algebra must not be closed until this puzzle is resolved' (citation from Roman Jackiw's anomaly recollections [Jackiw 1991]) and urged the study of it. In 1969 Bell and Jackiw [Bell, Jackiw 1967] (Jackiw was a CERN fellow in 1967 when their collaboration began) studied this puzzle by calculating $\pi^0 \longrightarrow \gamma\gamma$ in the σ-model. Finding that the σ-model does not satisfy the requirements of PCAC—the effective coupling constant does not vanish in the σ-model—they discovered the role which the anomaly plays in this problem. Independently, in the same year, Stephan L. Adler [Adler 1969] from the Institute of Advanced Study, Princeton, investigated the axial-vector vertex in spinor electrodynamics. He found that the axial Ward identity fails in case of the triangle diagram which led him to modify PCAC by an extra term—the *anomaly* (as he phrased it). When applying the anomaly modified PCAC equation to calculate the pion decay rate Adler obtained a result in excellent agreement with experiment.

By calculating the now famous triangle Feynman diagram made up of one axial- and two vector currents, containing an UV divergency, the following result is found after regularization: while conservation of the vector current can be maintained, the conservation law of the axial current is broken (we consider the massless case, $m = 0$, for simplicity)

$$\partial^\mu j^5_\mu = \mathcal{A}, \tag{1.1}$$

where $\mathcal{A}$ represents the celebrated ABJ anomaly

$$\mathcal{A} = \frac{e^2}{16\pi^2}\varepsilon^{\mu\nu\alpha\beta}F_{\mu\nu}F_{\alpha\beta} \tag{1.2}$$

(ABJ in honour of the three discoverers).

Now the ice was broken, the ABJ anomaly opened the door to a deeper understanding of QFT, a new era for field theory research began.

It is a remarkable fact that the anomaly is already totally given by the 1-loop calculation, the anomaly receives *no* contributions from radiative corrections. This is the content of the important Adler–Bardeen theorem [Adler, Bardeen 1969].

The extension of the theory to non-Abelian fields $A_\mu = A^a_\mu T^a$, $F_{\mu\nu} = F^a_{\mu\nu} T^a$, leads on one hand to the singlet anomaly

$$\mathcal{A} = \partial^\lambda j^5_\lambda = \frac{1}{16\pi^2} \varepsilon^{\mu\nu\alpha\beta} \mathrm{tr}\, F_{\mu\nu} F_{\alpha\beta}, \tag{1.3}$$

and on the other to Bardeen's non-Abelian anomaly result [Bardeen 1969]

$$-G^a[A_\mu] = (D_\mu j^\mu)^a = \pm \frac{1}{24\pi^2} \varepsilon^{\mu\nu\rho\sigma} \mathrm{tr}\, T^a \partial_\mu (A_\nu \partial_\rho A_\sigma + \frac{1}{2} A_\nu A_\rho A_\sigma), \tag{1.4}$$

where $\pm$ corresponds to positive or negative chiral fields.

Soon after these perturbation calculations, in the year 1971, Dolgov and Zakharov [Dolgov, Zakharov 1971] proposed a different method, with certain merits, to calculate the anomaly: the technique of dispersion relations. There the source of the anomaly is the existence of a superconvergence sum rule for an infrared divergent amplitude.

However, the anomaly is not just a perturbation effect, resulting from the regularization of some divergent diagrams, but reflects the deep laws of quantum physics. So the anomaly must be seen from a different point of view as well.

In the years 1976–77 there occurred the first topological investigations of the anomaly. Several authors like Jackiw, Rebbi, Nielsen, Römer, Schroer [Jackiw, Rebbi 1976, 1977], [Nielsen, Schroer 1977], [Nielsen, Römer, Schroer 1977] discovered that the singlet anomaly is determined by an index theorem. The reason is that the anomaly can be expressed by a sum of eigenfunctions of the Dirac operator, where only the zero-modes of a given chirality (n_+, n_-) survive

$$\frac{1}{2i} \int dx \mathcal{A}(x) = \int dx \sum_n \varphi^\dagger_n(x) \gamma_5 \varphi_n(x) = n_+ - n_- = \text{index } D_+. \tag{1.5}$$

The difference in the chirality zero-modes represents the index of the Weyl operator $D_+ = \not{D} P_+$. Using now the distinguished Atiyah–Singer (AS) index theorem [Atiyah, Singer 1968a,b,c, 1971a,b] one can express the index in terms of characteristic classes, here in terms of the Chern character in 4 dimensions

$$\text{index } D_+ = -\frac{1}{8\pi^2} \int \mathrm{tr}\, FF, \tag{1.6}$$

that establishes the anomaly on pure topological grounds.

Another important step towards understanding anomalies happened in 1979. Kazuo Fujikawa worked with the path integral of quantized fermions in an external gauge field (in Euclidean space)

$$Z[A_\mu] = e^{-W[A_\mu]} = \int d\psi d\bar{\psi} e^{S[A_\mu]} \tag{1.7}$$

with the classical action

$$S = \int dx \bar{\psi}(i\not{D} - m)\psi \quad , \quad \not{D} = \gamma^\mu(\partial_\mu + A_\mu). \tag{1.8}$$

Naïvely, no anomaly would appear since under a chiral transformation

$$\psi \to e^{i\beta(x)\gamma_5}\psi \quad , \quad \bar{\psi} \to \bar{\psi} e^{i\beta(x)\gamma_5} \tag{1.9}$$

the classical action clearly remains invariant. However, Fujikawa [Fujikawa 1979, 1980] discovered that the path integral measure does transform too

$$d\psi d\bar{\psi} \to d\psi d\bar{\psi} \exp[-2i \int dx \beta(x) \sum_n \varphi_n^\dagger(x)\gamma_5\varphi_n(x)] \tag{1.10}$$

(φ_n are the eigenfunctions of the Dirac operator $\not{D}$), and the Jacobian contains—after appropriate regularization—precisely the anomaly

$$2\sum_n \varphi_n^\dagger(x)\gamma_5\varphi_n(x) = -\frac{1}{16\pi^2}\varepsilon^{\mu\nu\alpha\beta}\mathrm{tr}\, F_{\mu\nu}F_{\alpha\beta}. \tag{1.11}$$

So this procedure correponds to a local evaluation of the index (described before). Nowadays we know that Fujikawa's procedure—also called the non-perturbative approach—is closely related to the heat kernel method and to the zeta function regularization procedure.

In the eighties the field of anomalies began to flourish again. Modern mathematical techniques—differential geometry, cohomology and topology—became very popular and helped to shed more light onto the anomalous phenomena of QFT. The anomalies are now rewritten in terms of differential forms
singlet anomaly

$$\mathcal{A} = d * j^5 = \frac{1}{4\pi^2}\mathrm{tr}\, FF = \frac{1}{4\pi^2} d\, \mathrm{tr}\,(AdA + \frac{2}{3}A^3) \tag{1.12}$$

non-Abelian anomaly

$$G^a[A] = -(D * j)^a = \pm\frac{1}{24\pi^2}\mathrm{tr}\, T^a d(AdA + \frac{1}{2}A^3). \tag{1.13}$$

The equation which defines the anomaly is the Wess–Zumino (WZ) consistency condition; on a BRS level it can be formulated in a very compact way

$$sG(v, A) = s \int v^a G^a[A] = 0 \tag{1.14}$$

($v = v^a T^a$ denotes the Faddeev–Popov ghost, s the BRS operator with $s^2 = 0$). Generally, any nontrivial solution represents a consistent anomaly; of course, the trivial solution is

$$G_{\text{triv}} = s\widehat{G}[A]. \tag{1.15}$$

Mathematically, the WZ consistency condition corresponds to cocycles and cohomologies.

The modern, differential geometric treatment of anomalies started with the lectures of Stora [Stora 1984] and Zumino [Zumino 1984]. They found a chain of descent equations where several polynomials in v, A, F—so-called chain terms—are linked together in different dimensions

$$\begin{aligned} P(F^n) - dQ^0_{2n-1} &= 0 \\ sQ^0_{2n-1} + dQ^1_{2n-2} &= 0 \\ sQ^1_{2n-2} + dQ^2_{2n-3} &= 0 \\ &\dots \\ sQ^{2n-1}_0 &= 0. \end{aligned} \tag{1.16}$$

$P(F^n)$ is a symmetric, gauge invariant polynomial in F. The chain terms Q^k_{2n-1-k}, where the lower index denotes the form degree, the upper index the power in v, can be solved explicitly in a rather easy way. (Actually, the above chain can be traced back to Stora's work in 1977 [Stora 1977].)

The amazing feature of this pure mathematical chain is that it has counterparts in physics. For example, the third equation of the Stora–Zumino (SZ) chain represents a local version of the WZ consistency condition, so that we can identify the chain term Q^1_{2n-2} (mathematics) with the anomaly $G(v, A)$ (physics) in $(2n-2)$ dimensions

$$G(v, A) = N \int_{M_{2n-2}} Q^1_{2n-2}(v, A) \tag{1.17}$$

(N is some normalization not determined by the chain).

Choosing as the invariant polynomial the symmetrized trace str F, we recover the singlet anomaly in $2n$ dimensions. So the two anomalies, the singlet- and the non-Abelian anomaly, although different in their nature, are linked together in different dimensions.

A second boom of topological activities started with the work of Atiyah and Singer [Atiyah, Singer 1984] and Alvarez, Singer and Zumino [Alvarez, Singer, Zumino 1984] who discovered that the non-Abelian anomaly is related to a more refined index theorem—the family index theorem. Equivalently, Alvarez-Gaumé and Ginsparg [Alvarez-Gaumé, Ginsparg 1984] related the anomaly in $2n$ dimensions to an AS index theorem in $(2n+2)$ dimensions. The extra 2 dimensions arise from considering two-parameter families of gauge potentials. The result is

$$\begin{aligned} -G(v, A) &= 2\pi i \text{ Index } i \not{D}_{2n+2} \\ &= 2\pi i \frac{i^{n+1}}{(2\pi)^{n+1}(n+1)!} \int_{S^{2n}} Q^1_{2n}. \end{aligned} \tag{1.18}$$

The anomaly is given by the SZ chain term Q^1_{2n} with the correct normalization.

Parallel to the activities in gauge theories there was rapid development in gravitation. Gravitation is an important part of field theory which cannot be left aside. It began with the pioneering work of Alvarez-Gaumé and Witten [Alvarez-Gaumé, Witten 1983] on gravitational anomalies, and the enthusiasm culminated in the discovery of Green and Schwarz [Green, Schwarz 1984] that gauge *and* gravitational anomalies may cancel each other, however, in a supersymmetric theory in 10 dimensions.

Here gravitation is regarded as a gauge theory where the gauges are the general coordinate transformations—the diffeomorphisms—or the rotations in the tangent frame—the Lorentz transformations. Then the classical conservation law of the energy-momentum tensor can be broken in the quantum case—an Einstein anomaly occurs

$$\delta^c_\xi W[g_{\mu\nu}] = -\int dx \sqrt{|g|} \xi_\nu \nabla_\mu \langle T^{\mu\nu} \rangle =: G^{\mathrm{E}}(\xi) \tag{1.19}$$

(ξ is an infinitesimal coordinate shift). On the other hand, a Lorentz anomaly may also occur

$$\delta^{\mathrm{L}}_\alpha W[e^a{}_\mu] = \int dx e \alpha_{ab} \langle T^{ab} \rangle =: G^{\mathrm{L}}(\alpha) \tag{1.20}$$

(α is the parameter of the infinitesimal Lorentz transformation), which is equivalent to the existence of an antisymmetric part of the energy-momentum tensor. Bardeen and Zumino [Bardeen, Zumino 1984] discovered that both types of gravitational anomaly are not independent, but can be shifted into each other with help of a counterterm.

The organization of the whole material is as follows:

In Chapter 2 we introduce all necessary mathematical concepts like homotopy, manifolds and differential forms, homology and cohomology, Lie derivative and Lie groups, and finally the fibre bundle set-up. In Chapter 3 we explain the path integral formalism, specifically the Faddeev–Popov method for quantizing gauge theories and the BRS transformations.

We spend the whole of Chapter 4 on a thorough investigation of the anomaly within perturbation theory. The anomaly is calculated within all commonly used regularization techniques (e.g. Pauli–Villars and 't Hooft–Veltman regularization, point splitting method), and we also work out the method of dispersion relations. The famous $\pi^0 \to \gamma\gamma$ decay is described in detail. We establish the anomalies for non-Abelian fields—the singlet- and the non-Abelian anomalies, which are the topic throughout the book—and we close this chapter with a discussion about their importance for physics.

In the subsequent Chapter 5 we present the nonperturbative view. We work in path integral formalism and explain Fujikawa's method for calculating the anomaly. We demonstrate the case of the singlet anomaly as well as the covariant non-Abelian anomaly. (The consistent anomaly calculation is preferably shifted into Chapter 11.5.2.) The connection to the heat kernel method and to the zeta function regularization is carried out in detail.

A little practice in differential forms and fibre-bundle concepts for our geometric point of view of field theories like electrodynamics, Yang–Mills theory, Dirac monopole, the Aharonov–Bohm effect and instantons is given in Chapter 6; while Chapter 7 is committed to invariant polynomials, to the Chern–Simons form and 'transgression', and to the use of the homotopy operator.

The modern treatment of anomalies begins in Chapter 8, where we introduce the gauge operator and the BRS machinery. The WZ consistency condition is established and illuminated from an algebraic, cocycle and cohomological point of view. In Chapter 9 we derive the SZ chain of descent equations, for both, for trivial and nontrivial gauge bundles. We find explicit solutions and formulae for the chain terms and we discuss the identifications with physics. The covariant type of anomaly is studied separately in Chapter 10; its relation to the consistent type is analysed in terms of differential forms on the space–time manifold M as well as in terms of differential forms on Sp $\mathcal{A}$, the space of all gauge connections.

The topological investigations of anomalies are performed in Chapter 11. We begin with the relation of the singlet anomaly to the AS index theorem, and we present a bit of index theory afterwards, emphasizing the relation of the index to the heat kernel. Introducing a modified Dirac operator we calculate the consistent non-Abelian anomaly first à la Fujikawa and then à la Alvarez-Gaumé and Ginsparg, with help of an AS index theorem in $(2n+2)$ dimensions.

The final Chapter, 12, is devoted to gravitation. We first review some basic concepts of Riemannian geometry and establish our notation. Then we discuss the gravitational action, specifically the fermionic action in detail. We introduce the Einstein-, Lorentz- and Weyl anomalies by violating the corresponding Einstein-, Lorentz- and Weyl symmetries and we establish the consistency conditions. The equivalence of the Einstein- and Lorentz anomaly is demonstrated, the covariant gravitational anomaly type is also discussed. Finally we treat gravitation on a BRS level, deriving the SZ chain of descent equations, and we use the index theorems to carry out explicit anomaly examples.

Summarizing, we would like to show these many different aspects of the anomalies, which make them so fascinating, in the following picture:

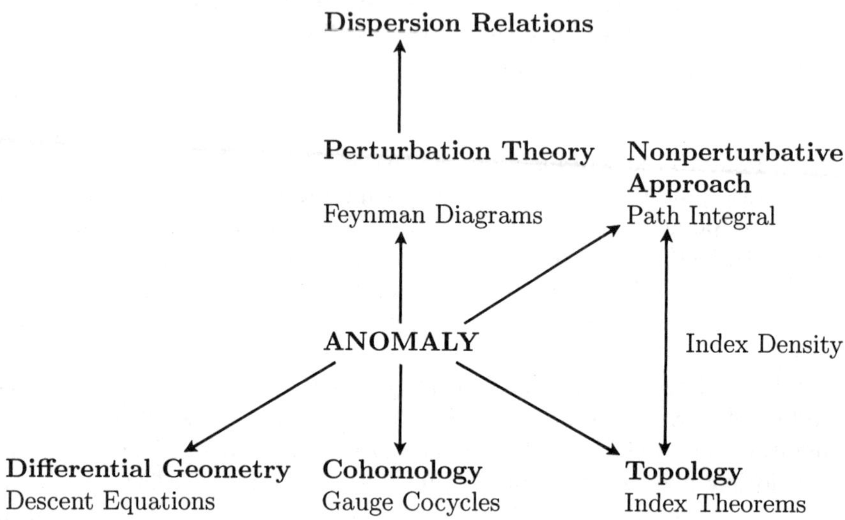

2

Differential geometry, topology and fibre bundles

In the last decade theoretical physicists have made extensive use of differential geometry, topology and fibre bundles. Particularly in quantum field theory—which is our main interest—these modern mathematical techniques prove to be fruitful. In fact, it is our aim to formulate physics in a differential geometric and topological way, which is of increasing importance in the future. We do not assume the knowledge of these techniques, but introduce the basic ideas, especially those concepts we use continuously in the book.

In Section 2.1 we briefly summarize some topological definitions; in Section 2.2 we explain the homotopy of maps and the homotopy of groups. In Section 2.3 we introduce the concept of differentiable manifolds and in Section 2.4 we present the differential forms together with their Hodge duals, along with the differentiation and integration. This is one of the basic sections since we will formulate all laws of physics in terms of differential forms. In Section 2.5 we speak about homology and de Rham cohomology and in Section 2.6 we explain important concepts like the pullback of a differential form, the Lie derivative, the Lie group and the Lie algebra. Finally, Section 2.7 is devoted to the construction of fibre bundles including connection and curvature. They turn out to be a suitable mathematical concept to describe the physics of gauge theories.

The mathematics is sometimes presented in an intuitive way avoiding such a rigorous derivation. For more rigor and a broader presentation the reader should consult the classical mathematical literature: [Flanders 1963], [Kobayashi, Nomizu 1963, 1969], [Choquet-Bruhat, DeWitt-Morette 1982], [Singer, Thorpe 1967], [Greub, Halperin, Vanstone 1972], [Bott, Tu 1982], [Warner 1983], or the books which have recently appeared and are close to our main interest: [Nash, Sen 1983], [Göckeler, Schücker 1987], [Isham 1989], [Nakahara 1990]. We refer to the appropriate literature in each section.

2.1 Topology

A topological space is a set with a structure such that we can define a neighbourhood and continuous functions in a general way.

Topological space: Let U be a system of subsets of a set X. U defines a **topology** on X if

i) $\emptyset \in U$ and $X \in U$,
ii) for any finite or infinite subsets $\{U_{i_k}\}$ the union satisfies $\bigcup_k U_{i_k} \in U$,
iii) for any *finite* subsets $\{U_{i_1}, \dots, U_{i_n}\}$ the intersection satisfies $\bigcap_{k=1}^{n} U_{i_k} \in U$.

Then X or the pair (X, U) is called a **topological space** and the U_{i_k} are **open sets**.

For example, let $X \equiv \mathbf{R}$, then all open intervals (a, b) and their unions define the **usual topology**.

Closed set: A set $A \subset X$ is **closed** if $A^c = X \setminus A$ is open. $A^c = \{x \in X | x \notin A\}$ is the **complement** of A.

Neighbourhood: Let X be a topological space. $N(x)$ is a **neighbourhood** of a point $x \in X$ if $N(x)$ contains an open set $U(x)$ which contains $x : N(x) \supset U(x)$.

For example, the closed interval $[a, b]$ is a neighbourhood for any point $x \in (a, b)$ of the open interval.

Theorem: A subset $A \subset X$ is open iff it is a neighbourhood for each of its points.

Limit point or accumulation point: A point $x \in X$ is a **limit point** of the set $A \subseteq X$ if each neighbourhood $N(x)$ contains at least one point $a \in A$ (different from x): $(N(x) - \{x\}) \cap A \neq \emptyset$, $\forall\, N(x)$.

Theorem: A is closed iff it contains all its limit points.

Closure: The **closure** $\bar{A}$ of A in X is the union of A with all its limit points. It is the smallest closed set containing A.

Interior: The **interior** A^0 of A is the largest open set contained in A.

Dense: The set A is **dense** in X if $\bar{A} \equiv X$.

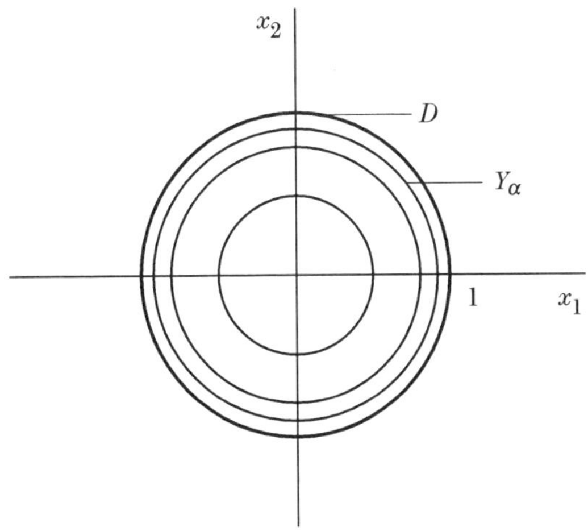

Fig. 2.1. Open disk D with open covering $\{Y_\alpha\}$

Boundary: A **boundary** ∂A of a set A is the complement of the interior in the closure of $A : \partial A = \bar{A} \setminus A^0$. A closed set contains its boundary while an open set is disjoint.

For example, all sets (a, b), $(a, b]$, $[a, b)$, $[a, b]$ have the same boundary $\partial A = \bar{A} \setminus A^0 = \{a, b\}$ since $\bar{A} = [a, b]$ and $A^0 = (a, b)$.

For example, the disk $A = \{(x, y)|x^2 + y^2 \leq a^2\}$ has as its boundary the circle $\partial A = \{(x, y)|x^2 + y^2 = a^2\}$.

Remark: Let A be a set, then

$$\begin{aligned} A \cap \partial A = \emptyset \quad &\Longleftrightarrow \quad A \text{ is open} \\ \partial A \subset A \quad &\Longleftrightarrow \quad A \text{ is closed.} \end{aligned}$$

Metric: A **metric** is a map $d : X \times X \to \mathbf{R}$ such that

i) $d(x, y) = d(y, x)$, symmetric,
ii) $d(x, y) \geq 0$, equality iff $x = y$,
iii) $d(x, y) + d(y, z) \geq d(x, z)$, triangle inequality, $\forall\, x, y, z \in X$.

X becomes a topological space via the metric d whose open sets are **open balls** $U_\varepsilon(x) = \{y \in X | d(x, y) < \varepsilon\}$ and all their unions. In this way we get a **metric topology** and X is called a **metric space**.

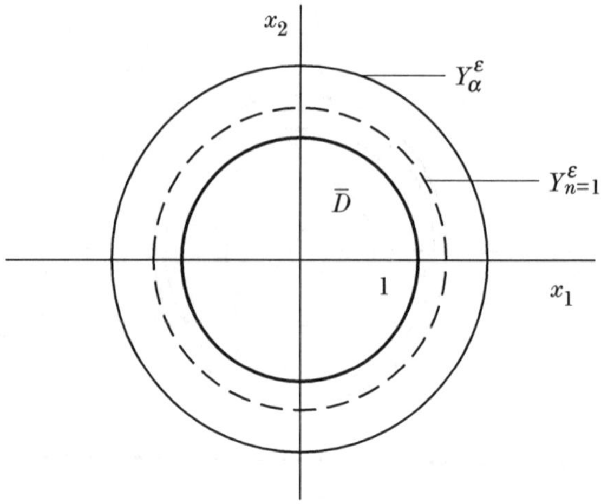

Fig. 2.2. Closed disk $\bar{D}$ with a finite open subcovering $Y_{n=1}^\varepsilon$

Compactness: Given a topological space X and a family of sets $\{Y_\alpha\} = Y$. Then Y is a **covering** of X if $\bigcup_\alpha Y_\alpha \supset X$.

If all Y_α are open sets, the covering Y is called an **open covering**.

The set X is **compact** if for *every* open covering $\{Y_\alpha\}$ there exists a *finite* subcovering $\{Y_1, \ldots, Y_n\}$ of X.

Example: The open disk $D = \{(x_1, x_2)|x_1^2 + x_2^2 < 1\}$ is not compact. Choose an open covering of D, namely concentric disks

$$Y_\alpha = \left\{(x_1, x_2)|x_1^2 + x_2^2 < 1 - \frac{1}{\alpha + 1}\right\} \quad \text{with } \alpha = 1, 2, \ldots. \tag{2.1}$$

Obviously $\bigcup_\alpha Y_\alpha = D$ but $\{Y_\alpha\}$ has *no* finite subcovering (see Figure 2.1), hence D is not compact.

The closed disk $\bar{D} = \{(x_1, x_2)|x_1^2 + x_2^2 \leq 1\}$ is compact. All open coverings

$$Y_\alpha^\varepsilon = \left\{(x_1, x_2)|x_1^2 + x_1^2 < 1 - \frac{1}{\alpha + 1} + \varepsilon\right\} \quad \text{with } \alpha = 1, 2, \ldots \text{ and } \varepsilon > 0 \tag{2.2}$$

now have a finite subcovering Y_n^ε with $n > 1/\varepsilon - 1$, the condition that the radius is larger than 1 or $\varepsilon > 1/(n+1)$ (see Figure 2.2). Hence $\bar{D}$ is compact.

Theorem: The subset $X \subset \mathbf{R}^n$ is compact iff it is closed and bounded.

Thus the n-dimensional spheres S^n are compact since they are closed and bounded in $\mathbf{R}^{n+1}$.

Compactification: The **compactification** of a topological space X is the pair (φ, Y) where Y is a compact space and φ is a homeomorphism: $X \to$ dense subspace of Y.

The compactness is a very important notion in topology. The mathematical analysis and therefore also the physics is much simpler on a compact space. In fact, we only work on compact sets—compact manifolds—like spheres.

Continuous: Let X, Y be two topological spaces. The map $\varphi : X \to Y$ is **continuous** iff for every open set $V \subset Y$ the set $U = \varphi^{-1}(V) \subset X$ is open. Alternatively, φ is **continuous** at $x \in X$ iff $\varphi(x_i) \to \varphi(x)$ whenever $x_i \to x$ (convergence of sequences).

Homeomorphism—diffeomorphism: Let X, Y be two topological spaces. The map φ is a **homeomorphism** if $\varphi : X \to Y$ is a bijection ($1-1$ and onto) and φ and φ^{-1} are continuous. Then X and Y are homeomorphic to each other.

If φ is a bijection and φ and φ^{-1} are continuously differentiable $\varphi, \varphi^{-1} \in C^\infty$ then φ is called a **diffeomorphism**.

2.2 Homotopy

The main aim in topology is to classify spaces with certain properties. For instance, two spaces are equivalent if they can be continuously deformed into each other. More precisely, two topological spaces belong to the same equivalence class if they are homotopic. Such equivalence classes are characterized by **topological invariants**. They are numbers like the number of connected components of a space, or they typify algebraic structures like groups constructed on a space (winding numbers), etc. A complete characterization in terms of invariants is rather difficult; instead what we know is:

Proposition:

- *If two topological spaces have different topological invariants they are not homeomorphic!*

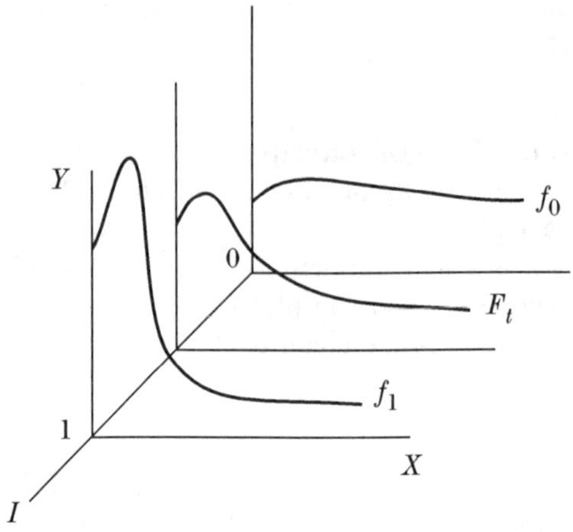

Fig. 2.3. The homotopy F_t: a continuous deformation of f_0 to f_1

In search of invariants another notion—a bit coarser but still powerful—is **homotopy**. There the continuous function mapping one space into another need not have an inverse. Homotopy creates equivalence classes of continuous maps. As we shall see it is also very helpful to visualize the objects we investigate—the topological spaces, the homotopic maps, the loops—by an ideal rubber which we can squeeze and deform smoothly in any direction.

2.2.1 Homotopy of maps

Homotopy: Let X, Y be two topological spaces. Consider two continuous maps $f_0, f_1 : X \to Y$, then f_0 is **homotopic** to f_1 (f_0 can be continuously deformed into f_1):

$$f_0 \sim f_1 \tag{2.3}$$

(the symbol $\sim$ means homotopic throughout the ensuing discussion) if there exists a **continuous family of functions** F_t parametrized over the product space $X \times I$:

$$X \times I \xrightarrow{F_t} Y \quad \text{with } I = [0,1], \tag{2.4}$$

such that

$$F_{t=0}|_{X\times\{0\}} = f_0 \quad \text{and} \quad F_{t=1}|_{X\times\{1\}} = f_1. \tag{2.5}$$

When $t \in [0,1]$ varies from $0 \to 1$ then the map f_0 is deformed continuously into f_1 like a rubber band (see Figure 2.3).

F_t is called a **homotopy**. It defines homotopy classes for functions. Two functions lie in the same class if they are homotopic.

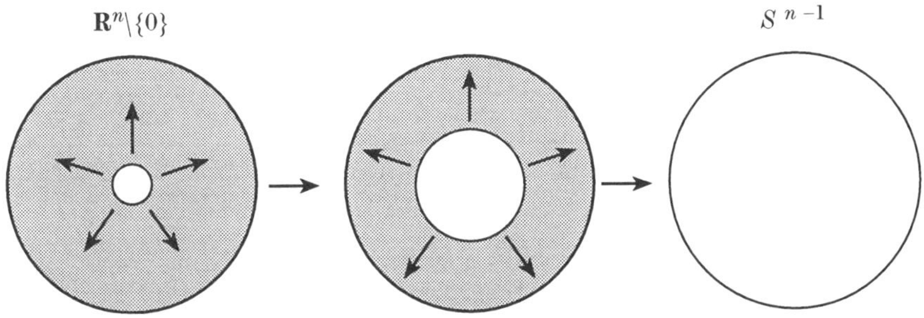

Fig. 2.4. The space $\mathbf{R}^n \setminus \{0\}$ is squeezed to a sphere S^{n-1}

Transitivity: Let f, g, h be continuous maps in topological spaces. Then **transitivity** is satisfied:

$$\text{if } f \sim g \text{ and } g \sim h, \text{ then } f \sim h. \tag{2.6}$$

Homotopic equivalent spaces: Two topological spaces X and Y are homotopic equivalent if there exist continuous maps f, g:

$$X \xrightarrow{f} Y \xrightarrow{g} X, \tag{2.7}$$

such that

$$\begin{aligned} f \cdot g &\sim \mathbf{1}_Y \qquad \text{identity on } Y \\ g \cdot f &\sim \mathbf{1}_X \qquad \text{identity on } X. \end{aligned} \tag{2.8}$$

Example of homotopic spaces: Choose the $\mathbf{R}^n$ space with origin extracted $X = \mathbf{R}^n \setminus \{0\}$ and the $(n-1)$-dimensional sphere $Y = S^{n-1}$, then

$$\mathbf{R}^n \setminus \{0\} \sim S^{n-1}. \tag{2.9}$$

A homotopic space is like a rubber block which can be smoothly squeezed everywhere (see Figure 2.4).

Proof. It is illustrative to prove the above example. We construct the two maps f, g:

$$\begin{aligned} f : X = \mathbf{R}^n \setminus \{0\} \quad &\rightarrow \quad S^{n-1} = Y \\ \text{via } x \quad &\mapsto \quad \frac{x}{|x|} \end{aligned} \tag{2.10}$$

and

$$\begin{aligned} g : Y = S^{n-1} &\rightarrow \mathbf{R}^n \setminus \{0\} = X \\ \text{via } x &\mapsto x \text{ with } |x| = 1. \end{aligned} \tag{2.11}$$

There exists a continuous map F_t:

$$X \times [0,1] \xrightarrow{F_t} X, \tag{2.12}$$

such that

$$F_{t=0}(x) = \mathbf{1}(x) = x \quad \text{and} \quad F_{t=1}(x) = g \cdot f(x) = \frac{x}{|x|}. \tag{2.13}$$

Explicitly it is

$$F_t(x) = (1-t)x + t\frac{x}{|x|}; \tag{2.14}$$

therefore $g \cdot f \sim \mathbf{1}_X$. This construction is typical for homotopic families and we shall use it frequently in this book. The second part $f \cdot g \sim \mathbf{1}_Y$ is trivial. Q.E.D.

Example of homotopic maps: Let $X = \{\theta\}$ be the unit circle described by the angle θ where θ and $\theta + 2\pi$ are identified and let $Y = \{e^{i\varphi}\}$ be the set of unimodular complex numbers, the group $U(1)$. Consider the map $f : X \rightarrow Y$; then the continuous functions

$$f(\theta) = e^{i(n\theta+\alpha)}, \tag{2.15}$$

which are such mappings, generate a **homotopy class** for different α and fixed integer n; explicitly

$$f_0(\theta) = e^{i(n\theta+\alpha_0)} \quad \text{and} \quad f_1(\theta) = e^{i(n\theta+\alpha_1)}. \tag{2.16}$$

Then f_0 is homotopic to f_1, $f_0 \sim f_1$ for fixed n, since there exists a continuous family of functions $F_t(\theta)$:

$$X \times I \xrightarrow{F_t} Y \quad \text{with } I = [0,1], \tag{2.17}$$

such that

$$F_{t=0}(\theta) = f_0(\theta) \quad \text{and} \quad F_{t=1}(\theta) = f_1(\theta). \tag{2.18}$$

It is the homotopy

$$F_t(\theta) = \exp\{i[n\theta + (1-t)\alpha_0 + t\alpha_1]\}. \tag{2.19}$$

Next we consider the **covering** of this map. The function $f(\theta)$: $S^1 \rightarrow U(1) \simeq S^1$ maps a circle—a 1-dimensional sphere—onto a circle since $U(1)$

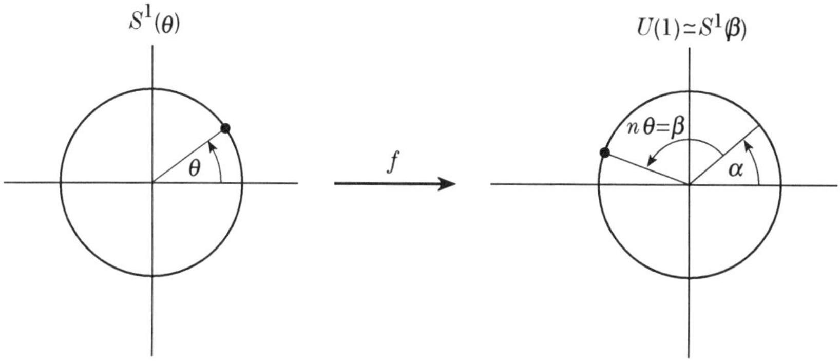

Fig. 2.5. While $S^1(\theta)$ is covered once the group $U(1)$ or the sphere $S^1(\beta)$ is wound around n times

is isomorphic to S^1. When θ varies from $0 \to 2\pi$ the sphere $X = S^1(\theta)$ is covered once. But then $\beta = n\theta$ varies from $0 \to 2\pi n$ and the other sphere $Y = S^1(\beta)$, the group $U(1)$, will be covered n times or wound around n times (see Figure 2.5). Therefore n is named the **winding number**.

Proposition:

- *Each homotopy class is characterized by its* **winding number** n.

For a given mapping $f(\theta)$ the **winding number** can be expressed by an integral

$$n = \frac{1}{2\pi i} \int_0^{2\pi} d\theta \, f^{-1}(\theta) \frac{d}{d\theta} f(\theta). \tag{2.20}$$

If we have a map of winding number 1, say $f(\theta) = e^{i\theta}$, then a mapping of a higher winding number k is obtained by taking k powers of this map: $g(\theta) = [f(\theta)]^k$.

We can also express the map in terms of Cartesian coordinates

$$f(x, y) = x + iy \qquad \text{with } x^2 + y^2 = 1. \tag{2.21}$$

We may expand the domain of the map from the sphere S^1 to the whole $\mathbf{R}^1 : -\infty \leq x \leq \infty$ by identifying the endpoints $x = -\infty$ and $x = \infty$ to be the same point; thus the map satisfies $f(x = \infty) = f(x = -\infty)$. In this way we achieve a **1-point compactification** of $\mathbf{R}^1$ which is topologically equivalent to the S^1.

For example, the map

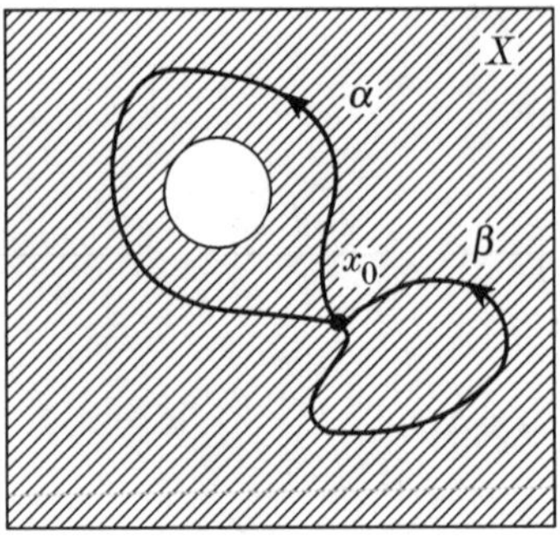

Fig. 2.6. Space X with a hole. Loop α surrounds the hole, loop β does not. Loop α cannot be shrunk to a point, whereas loop β can be shrunk

$$f(x) = \exp\left[i\pi\frac{x}{\sqrt{x^2+\lambda^2}}\right] \tag{2.22}$$

is of this type and has winding number 1. Generally the **winding number** is given by

$$n = \frac{1}{2\pi i}\int_{-\infty}^{\infty} dx\, f^{-1}(x)\frac{d}{dx}f(x). \tag{2.23}$$

The important applications in physics are the Dirac monopole and the instantons which we shall discuss in Sections 6.4 and 6.6.

2.2.2 Homotopy groups

Introducing loops in a topological space it turns out that the equivalence classes of loops reveal a group structure.

Loop: Let X be a topological space and $I = [0,1]$ an interval. The continuous map $\alpha(t) : I \to X$ with $\alpha(0) = \alpha(1) = x_0 \in X$ and $t \in I$ is called a **loop** with base point x_0.

There exist two kinds of loop (see Figure 2.6):

i) a loop containing a hole—it cannot be shrunk to a point,
ii) a loop not containing a hole—it can be shrunk.

Connectedness: A topological space X is **disconnected** if there exist two disjoint open subsets X_1, X_2: $X_1 \cap X_2 = \emptyset$, such that $X_1 \cup X_2 = X$; otherwise the space X is **connected**.

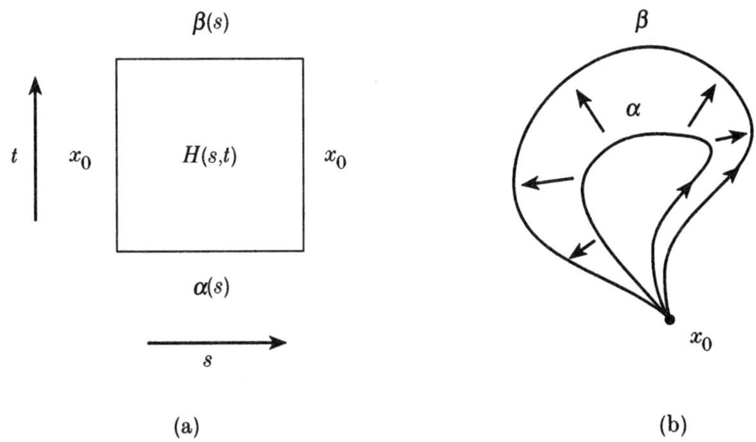

Fig. 2.7. a) The homotopy $H(s,t)$ between $\alpha(s)$ and $\beta(s)$. b) The loop α is continuously deformed into the loop β by the homotopy H

If any loop in X can be continuously shrunk to a point then X is **simply connected.** X is **arcwise connected** (or path connected) if there exists a path between any pair of points $x_0, x_1 \in X$.

Note: arcwise connected $\underset{\not\Leftarrow}{\Rightarrow}$ connected.

Again we imagine the loop to be an ideal rubber ring which can be smoothly stretched or shrunk. Then we consider two loops as **equivalent or homotopic** if they can be continuously deformed into each other. They form an **equivalence or homotopy class** which is characterized by an integer n, the number of times a loop surrounds a hole ($n > 0$ for clockwise). For example, in Figure 2.6 the loop α is not homotopic to the loop β.

Homotopic loops: Two loops α and β based at x_0 are **homotopic** to each other

$$\alpha \sim \beta \tag{2.24}$$

if there exists a **continuous map** $H(s,t)$:

$$I \times I \overset{H(s,t)}{\longrightarrow} X, \qquad \text{with } s,t \in I = [0,1] \tag{2.25}$$

such that

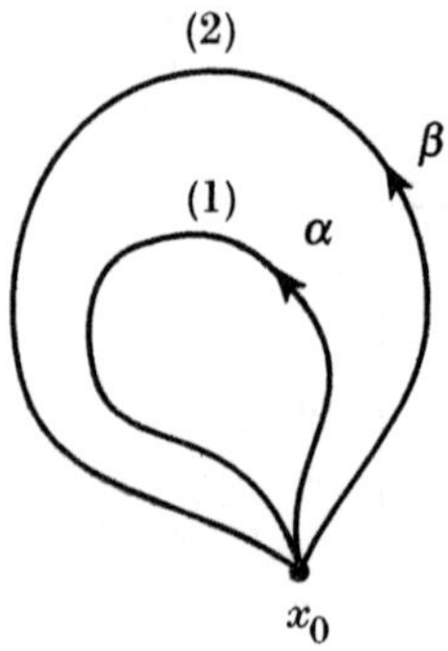

Fig. 2.8. For the product $\alpha \circ \beta$ we first pass along α then along β

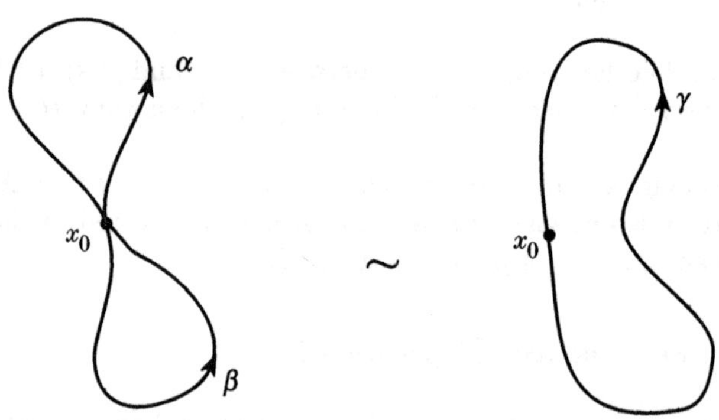

Fig. 2.9. The loop γ is homotopic to the product $\alpha \circ \beta$: $\gamma \sim \alpha \circ \beta$

$$\begin{aligned} H(s,0) &= \alpha(s) \text{ and } H(s,1) = \beta(s), \quad \forall\, s \in I, \\ H(0,t) &= H(1,t) = x_0, \qquad\qquad\quad \forall\, t \in I. \end{aligned} \tag{2.26}$$

$H(s,t)$ is called a **homotopy** between the loops α and β (see Figure 2.7).

Fundamental group or first homotopy group $\Pi_1(X, x_0)$: To equivalence classes of loops one can assign a group structure. For that we first have to define a product of loops, the inverse loop and the unit.

Product of loops: The **product** of two loops $\alpha(t)$ and $\beta(t)$ based at $x_0 \in X$, $\alpha(0) = \alpha(1) = \beta(0) = \beta(1) = x_0 \in X$, is defined by

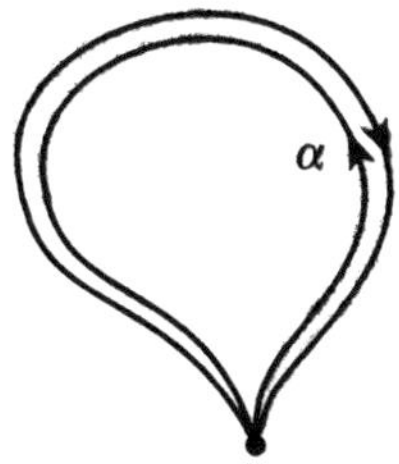

Fig. 2.10. Loop α and its inverse α^{-1}

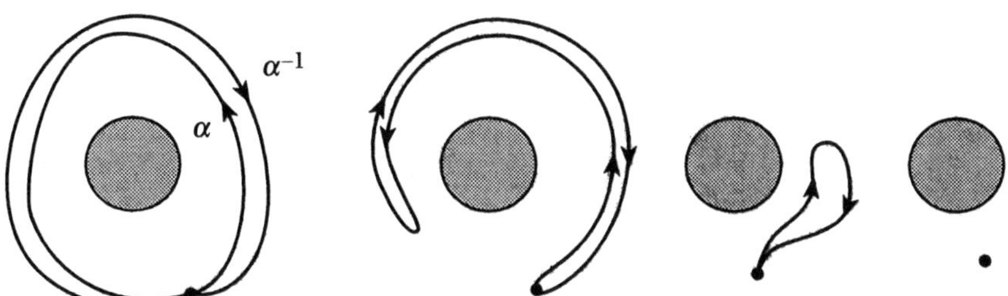

Fig. 2.11. A loop α times its inverse α^{-1} is homotopic to the identity: $\alpha \circ \alpha^{-1} \sim \mathbf{1}$

$$\alpha \circ \beta = \begin{cases} \alpha(2t) & \text{for } 0 \leq t \leq 1/2 \\ \beta(2t-1) & \text{for } 1/2 \leq t \leq 1. \end{cases} \tag{2.27}$$

So starting at x_0 we first pass along the loop α and then along the second loop β (see Figure 2.8).

Remark: Loops homotopic to $\alpha \circ \beta$ start and end at the base point x_0 but need not pass through x_0 in the midjourney; for instance, $\gamma \sim \alpha \circ \beta$ in Figure 2.9.

Inverse loop: The **inverse** loop $\alpha^{-1}(t)$ of α based at $x_0 \in X$ is defined by

$$\alpha^{-1}(t) = \alpha(1-t), \qquad \forall t \in I. \tag{2.28}$$

So for the inverse we pass along the loop in the opposite direction (see Figure 2.10).

Identity or unit element:

$$\mathbf{1}(t) = x_0, \qquad \forall t \in I. \tag{2.29}$$

The identity is just the 'constant' loop at $x_0 \in X$.

However, the product of a loop with its inverse is not the identity: $\alpha \circ \alpha^{-1} \neq \mathbf{1}$! But it is homotopic to the identity: $\alpha \circ \alpha^{-1} \sim \mathbf{1}$ as we demonstrate in Figure 2.11 where the loops surround a hole.

Thus the loops themselves cannot form a group but it is the equivalence class of loops which admits a group structure. Therefore we consider the **homotopy class of loops** denoted by $[\alpha]$ and we finally define the **product** and **inverse** of these equivalence classes

$$[\alpha] \circ [\beta] := [\alpha \circ \beta] \tag{2.30}$$

$$[\alpha]^{-1} := [\alpha^{-1}]. \tag{2.31}$$

Then the group properties are satisfied and the group is named the **fundamental group or first homotopy group** $\Pi_1(X, x_0)$:

$$\begin{array}{lll} & \forall\, [\alpha], [\beta] \in \Pi_1(X, x_0) & \\ \text{i)} & [\alpha] \circ [\beta] = [\gamma] \in \Pi_1(X, x_0) & \text{closed} \\ \text{ii)} & ([\alpha] \circ [b]) \circ [c] = [\alpha] \circ ([b] \circ [c]) & \text{associative} \\ \text{iii)} & \mathbf{1} \circ [\alpha] = [\alpha] & \exists \text{ unit element} \\ \text{iv)} & [\alpha]^{-1} \circ [\alpha] = [\mathbf{1}] & \exists \text{ inverse element} \\ \text{v)} & \text{if } [\alpha] \circ [\beta] = [\beta] \circ [\alpha] & \text{Abelian.} \end{array} \tag{2.32}$$

Examples:

i) Consider the maps into the circle; there the homotopy classes are typified by the winding number n. A product of classes has winding number $n + m$, therefore the group is Abelian and it is isomorphic to the group of integers $\mathbf{Z}$

$$\Pi_1(U(1)) \simeq \Pi_1(S^1) \simeq \mathbf{Z} \tag{2.33}$$

(the symbol $\simeq$ denotes isomorphic). In physics the Dirac monopole is characterized by the fundamental group $\Pi_1(U(1))$ whose integers—the winding numbers—represent the magnetic monopole charge, see Section 6.4.

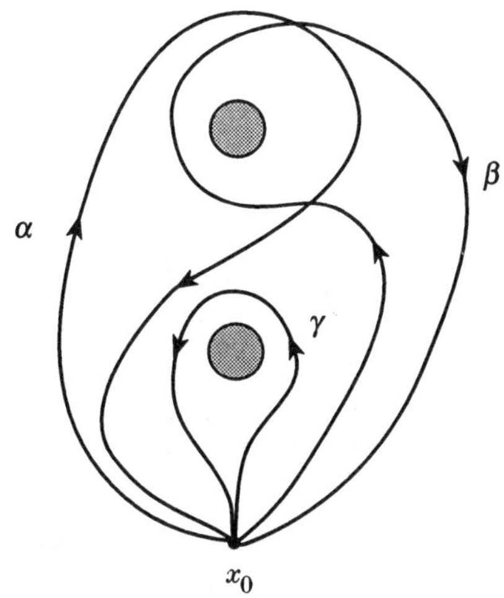

Fig. 2.12. Plane with two holes; $\alpha \not\sim \beta$ but with the help of γ we convert α into β, see Figure 2.13

ii) The mapping of a loop into the sphere S^2 can be shrunk to a point. Therefore the fundamental group of the S^2 contains only the identity and we have (in an additive group notation)

$$\Pi_1(S^2) = 0 \tag{2.34}$$

and furthermore

$$\Pi_1(S^n) = 0 \qquad \text{for } n \geq 2. \tag{2.35}$$

iii) Similarly the fundamental group of the n-dimensional Euclidean space is trivial

$$\Pi_1(\mathbf{R}^n) = 0. \tag{2.36}$$

iv) The first homotopy group Π_1 need not necessarily be Abelian. A **non-Abelian homotopy group** is, for example,

$$\Pi_1(D^2 - \{x_1\} - \{x_2\}), \tag{2.37}$$

where the space is a disk (or plane) with two holes or with two points extracted. This can be readily understood from Figures 2.12 and 2.13 or 2.14. The loop α is not homotopic to the loop β: $\alpha \not\sim \beta$; α cannot be continuously deformed into β because of the lower hole. But with the help of loop γ we can convert α into β (see Figure 2.13)

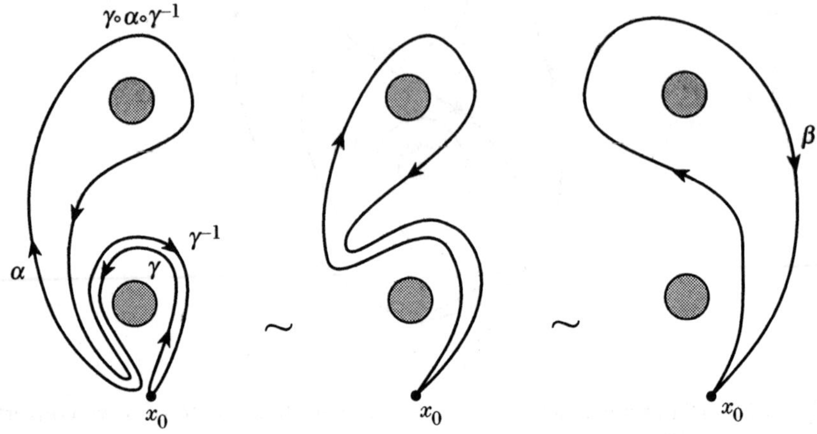

Fig. 2.13. The product of loops $\gamma \circ \alpha \circ \gamma^{-1}$ is homotopic to β

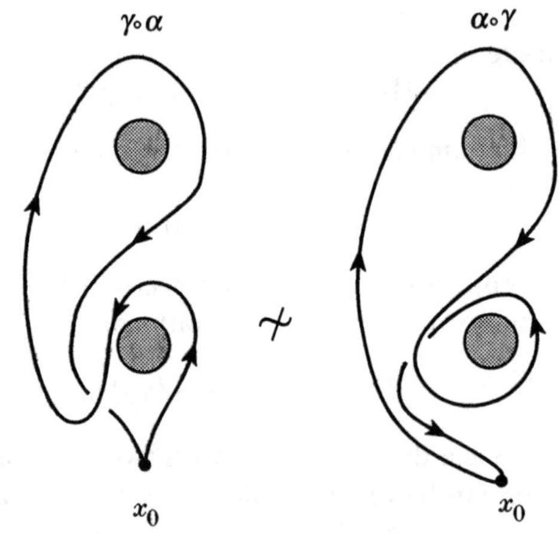

Fig. 2.14. The loop products $\gamma \circ \alpha \not\sim \alpha \circ \gamma$ are not homotopic

$$\begin{aligned} & \gamma \circ \alpha \circ \gamma^{-1} \sim \beta \not\sim \alpha \\ \Rightarrow \quad & \gamma \circ \alpha \not\sim \alpha \circ \gamma \\ \Rightarrow \quad & [\gamma] \circ [\alpha] \not\sim [\alpha] \circ [\gamma]. \end{aligned} \tag{2.38}$$

Alternatively we could directly consider the loop products $\gamma \circ \alpha$ and $\alpha \circ \gamma$. Figure 2.14 illustrates that there exists no homotopy between them: $\gamma \circ \alpha \not\sim \alpha \circ \gamma$.

Considering finally the fundamental groups based at different points x_0, x_1, one finds the following theorems.

Theorem: Let X be an arcwise connected topological space and $x_0, x_1 \in X$, then

$$\Pi_1(X, x_0) \simeq \Pi_1(X, x_1). \tag{2.39}$$

Theorem: Let X and Y be arcwise connected topological spaces, then

$$\Pi_1(X \times Y, (x_0, y_0)) \simeq \Pi_1(X, x_0) \oplus \Pi_1(Y, y_0). \tag{2.40}$$

Both theorems can be generalized to the higher homotopy groups Π_n.

Examples: Choose the torus $T^2 = S^1 \times S^1$, then

$$\Pi_1(T^2) = \Pi_1(S^1 \times S^1) \simeq \Pi_1(S^1) \oplus \Pi_1(S^1) \simeq \mathbf{Z} \oplus \mathbf{Z}, \tag{2.41}$$

whereas for the cylinder $C = S^1 \times \mathbf{R}$ we have

$$\Pi_1(C) = \Pi_1(S^1 \times \mathbf{R}) \simeq \Pi_1(S^1) \oplus \Pi_1(\mathbf{R}) \simeq \mathbf{Z} \oplus \{1\} \simeq \mathbf{Z}. \tag{2.42}$$

Higher homotopy groups: Our homotopy discussion with 1-dimensional loops can now be generalized. We introduce loops in n dimensions.

***n*-loop:** Let X be a topological space and $I^n = I \times \ldots \times I$ the unit n-cube where $I = [0, 1]$ and let ∂I^n be its geometric boundary. The continuous map

$$\alpha_n : I^n \longrightarrow X, \tag{2.43}$$

such that

$$\alpha_n : \partial I^n \longrightarrow x_0 \in X \tag{2.44}$$

is also called an ***n*-loop** in X based at the point x_0.

So an n-loop α_n is a continuous map of an n-dimensional cube I^n into a topological space X such that the boundary ∂I^n is mapped into a single point x_0 of that space X. Effectively we deform the cube I^n—by identifying

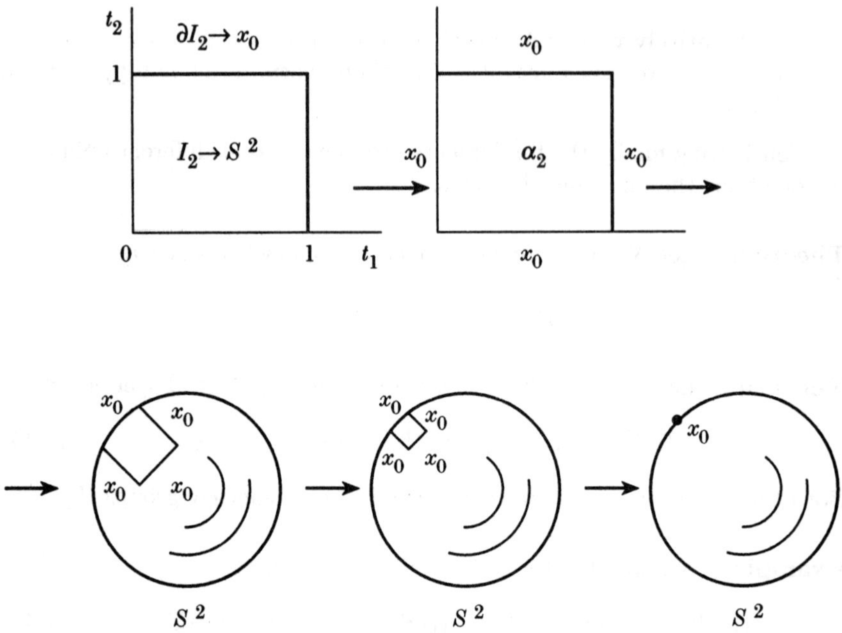

Fig. 2.15. A 2-loop: the square is 'blown up like chewing gum'

its boundary ∂I^n with a single point x_0—into a sphere S^n analogous to the previous 1-dimensional case where we have bent the interval I to a 1-dimensional loop α.

Illustration in two dimensions: The square I_2 is the source of a 2-loop α_2 when the boundary ∂I_2 is identified with the point $x_0 \in S^2$ of the sphere S^2. Then the square is homotopic (even homeomorphic) to the sphere which represents the 2-loop. This we have illustrated in Figure 2.15 by 'blowing up chewing gum'!

Product of n-loops: Be α_n and β_n two n-loops based at $x_0 \in X$ then the **product** is defined by

$$\alpha_n \circ \beta_n = \begin{cases} \alpha_n(2t_1, t_2, \ldots, t_n) & \text{for } 0 \le t_1 \le 1/2 \\ \beta_n(2t_1 - 1, t_2, \ldots, t_n) & \text{for } 1/2 \le t_1 \le 1. \end{cases} \tag{2.45}$$

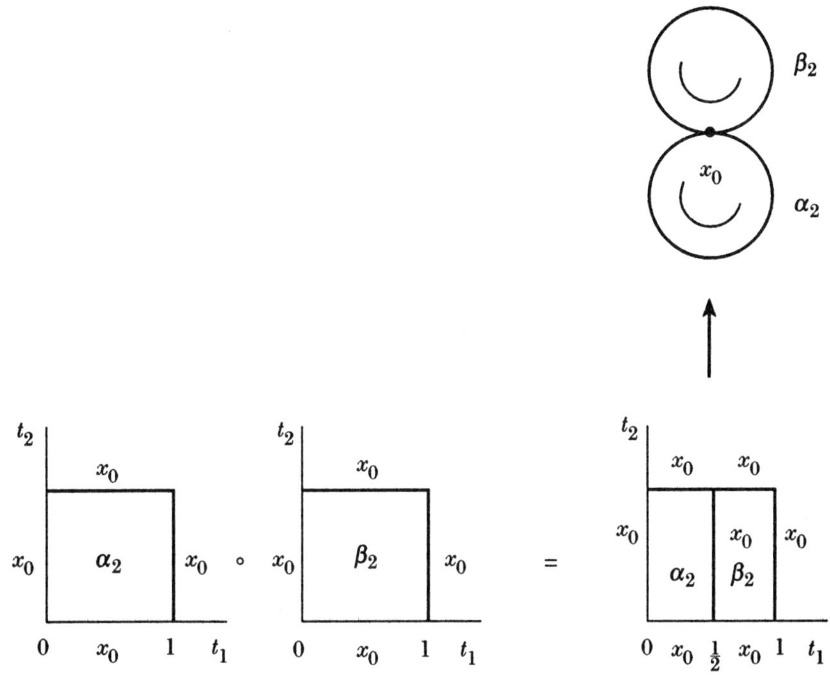

Fig. 2.16. The product of two 2-loops $\alpha_2 \circ \beta_2$

Again in two dimensions, we glue together the two blades α_2 and β_2 at an edge x_0 (see Figure 2.16).

Inverse n-loop:

$$\alpha_n^{-1}(t_1, t_2, \ldots, t_n) = \alpha_n(1 - t_1, t_2, \ldots, t_n), \quad \forall\,(t_1, t_2, \ldots, t_n) \in I^n. \quad (2.46)$$

Identity or unit element:

$$\mathbf{1}(t_1, \ldots, t_n) = x_0, \quad \forall\,(t_1, \ldots, t_n) \in I^n. \quad (2.47)$$

Homotopic n-loops: Two n-loops α_n and β_n based at x_0 are **homotopic** to each other

$$\alpha_n \sim \beta_n \quad (2.48)$$

if there exists a **continuous map**—a **homotopy**—$H_n(s_1, \ldots, s_n; t)$:

$$I^n \times I \xrightarrow{H_n} X \quad \text{with } s_i, t \in I = [0, 1] \quad (2.49)$$

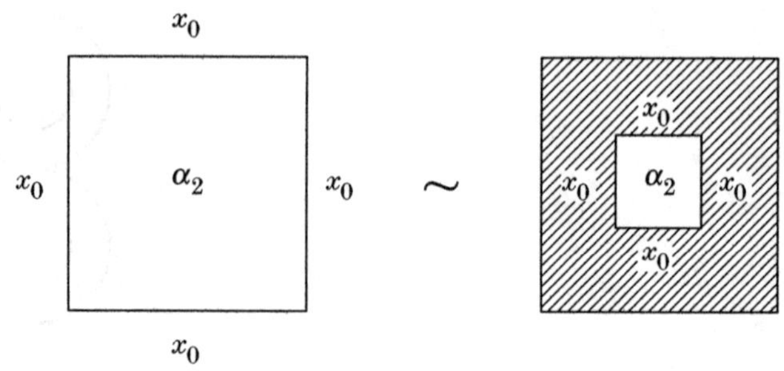

Fig. 2.17. The two loops with a thin and thickened boundary are homotopic to each other

such that

$$\begin{aligned} H_n(s_1, \ldots, s_n; 0) &= \alpha_n(s_1, \ldots, s_n) \\ H_n(s_1, \ldots, s_n; 1) &= \beta_n(s_1, \ldots, s_n) \\ H_n(s_1, \ldots, s_n, t) &= x_0 \in X \quad \text{if } (s_1, \ldots, s_n) \in \partial I^n, \ \forall t. \end{aligned} \tag{2.50}$$

This homotopy generates **equivalence classes of n-loops** denoted by $[\alpha_n]$ and, defining again a **product** and **inverse**,

$$[\alpha_n] \circ [\beta_n] = [\alpha_n \circ \beta_n] \tag{2.51}$$

$$[\alpha_n]^{-1} = [\alpha_n^{-1}] \tag{2.52}$$

then these equivalence classes of n-loops admit a group structure—the **n-th homotopy group** $\Pi_n(X, x_0)$ of the topological space X with base point x_0.

Examples: Recall the isomorphy between $SU(2) \simeq S^3$, then the mapping of a S^3 into a S^3 gives

$$\Pi_3(SU(2)) \simeq \Pi_3(S^3) \simeq \mathbf{Z}. \tag{2.53}$$

This is the homotopic formulation of the physical example of instantons, see Section 6.6, where the integers $\mathbf{Z}$ represent the instanton numbers.

Generally we have

$$\Pi_n(S^n) \simeq \mathbf{Z}. \tag{2.54}$$

The higher homotopy groups are trivial if the sphere dimension—the space where the loop 'lives'—is larger than that of the loop

$$\Pi_n(S^k) = 0 \qquad \text{for } k > n. \tag{2.55}$$

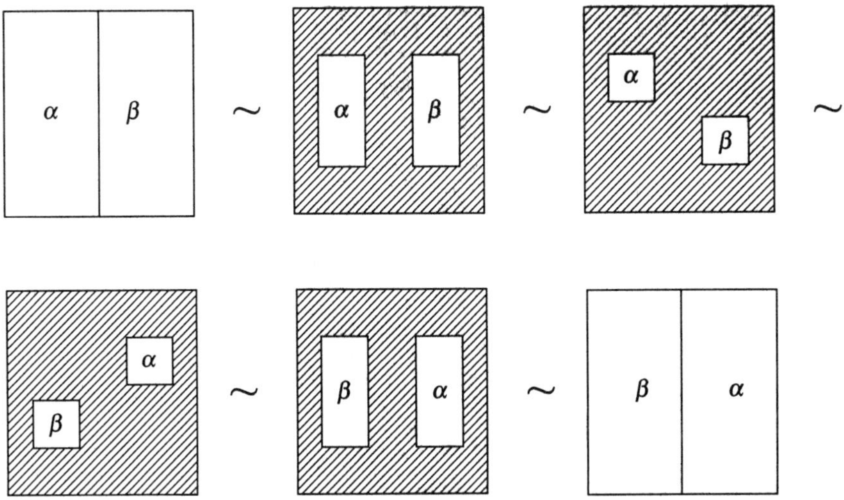

Fig. 2.18. Higher homotopy groups are Abelian. The product of n-loops is commutative $\alpha_n \circ \beta_n \sim \beta_n \circ \alpha_n$

But this is not necessarily so for the reversed $k < n$; for example,

$$\Pi_3(S^2) = \mathbf{Z}. \tag{2.56}$$

Finally we show that in contrast to Π_1 the Π_n is Abelian.

Proposition:

- *The n-th homotopy group Π_n $(n > 1)$ is Abelian!*

This is easily understood by viewing the n-loop as a rubber which can be squeezed around. The n-loop α_n based at x_0 is homotopic to the loop we obtain by thickening the boundary ∂I^n and mapping this thickened boundary into x_0, see Figure 2.17 in the case of $n = 2$.

Then we can immediately deform the loop product $\alpha_n \circ \beta_n$ into $\beta_n \circ \alpha_n$ as in Figure 2.18.

2.3 Differentiable manifolds

Now we turn to differential geometry where the basic concepts are the differentiable manifolds and the differential forms.

What is a manifold?

For example, a so-called 'Schnuller' is such an object (see Figure 2.19 which has been borrowed with permission from Thirring's book [Thirring

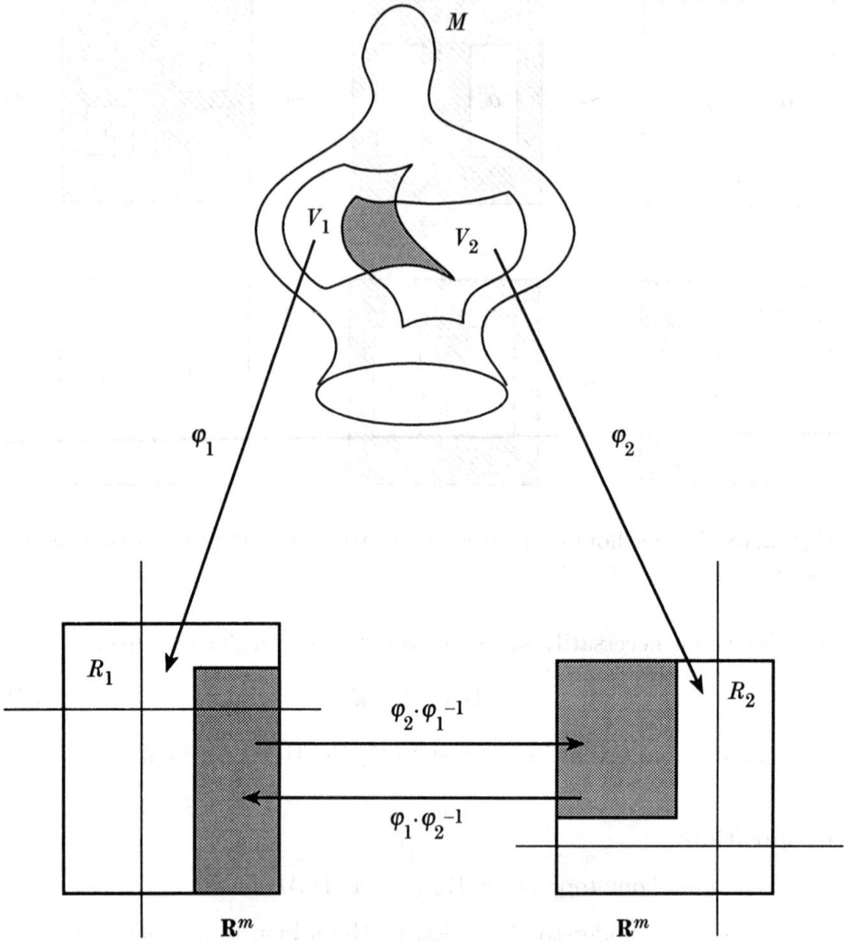

Fig. 2.19. A so-called 'Schnuller' represents a manifold

1992]). Generally, a differentiable manifold M is some smooth surface—a topological space—which locally looks Euclidean but not necessarily in its global extent. Via a homeomorphism to $\mathbf{R}^m$ we attach to each point of M a set of coordinates and require the transition from one set to another to be smooth.

In addition to the literature quoted at the beginning the reader will find the following literature helpful: [Thirring 1992], [Sexl, Urbantke 1983], [Visconti 1992], [Curtis, Miller 1985].

Chart: Let M be a topological space. A **chart** $(V_\alpha, \varphi_\alpha)$ is a homeomorphism φ_α from an open set $V_\alpha \subset M$ into an open set $R_\alpha \subset \mathbf{R}^m$ (see Figure 2.19)

$$V_\alpha \xrightarrow{\varphi_\alpha} R_\alpha. \tag{2.57}$$

Two charts are **compatible** if the overlap maps are continuously differentiable (diffeomorphisms)

$$\varphi_1 \cdot \varphi_2^{-1} \in C^\infty \qquad \text{and} \qquad \varphi_2 \cdot \varphi_1^{-1} \in C^\infty \tag{2.58}$$

or if $V_1 \cap V_2 = \emptyset$ (see Figure 2.19).

Atlas: The set of compatible charts $\{(V_\alpha, \varphi_\alpha)\}$ covering M is named, for obvious reasons, an **atlas**. Two atlases are **compatible** if all their charts are compatible.

Differentiable manifold: We speak of a **differentiable manifold** if the following four conditions are satisfied:

i) M is a topological space.
ii) M is equipped with a family of pairs $\{(V_\alpha, \varphi_\alpha)\}$ (charts).
iii) $\{V_\alpha\}$ is a family of open sets which cover $M : \bigcup_\alpha V_\alpha = M$ and φ_α is a homeomorphism from V_α onto an open subset $R_\alpha \subset \mathbf{R}^m$: $V_\alpha \xrightarrow{\varphi_\alpha} R_\alpha$ (see Figure 2.19).
iv) Given two open sets V_α, V_β with $V_\alpha \cap V_\beta \neq \emptyset$, the overlap maps (see Figure 2.19) $\varphi_\beta \cdot \varphi_\alpha^{-1}$ from subset $\varphi_\alpha(V_\alpha \cap V_\beta)$ to $\varphi_\beta(V_\alpha \cap V_\beta)$ or $\varphi_\alpha \cdot \varphi_\beta^{-1}$ from subset $\varphi_\beta(V_\alpha \cap V_\beta)$ to $\varphi_\alpha(V_\alpha \cap V_\beta)$ are infinitely differentiable (C^∞ functions, diffeomorphisms).

Hence a differentiable manifold is given by a topological space M and the equivalence class of atlases (which is determined by the compatibility). The features ii) and iii) imply that the topological space—the manifold M—is locally Euclidean. M is covered with patches V_α and via the homeomorphisms φ_α we attach coordinates in $\mathbf{R}^m$ to these patches. So within one patch the manifold looks like the Euclidean $\mathbf{R}^m$ but not necessarily globally—we have to know how the several patches fit together.

Feature iv) means that in the overlap region of two patches V_α, V_β we establish two coordinates in $\mathbf{R}^m$, $\varphi_\alpha(V_\alpha \cap V_\beta)$ and $\varphi_\beta(V_\alpha \cap V_\beta)$ and we can change these coordinates smoothly via $\varphi_\beta \cdot \varphi_\alpha^{-1}$ and $\varphi_\alpha \cdot \varphi_\beta^{-1}$. More explicitly:

$$x := \varphi_\alpha(p) \in \mathbf{R}^m \qquad \text{with } p \in V_\alpha \tag{2.59}$$

and

$$y := \varphi_\beta(p) \in \mathbf{R}^m \qquad \text{with } p \in V_\beta \tag{2.60}$$

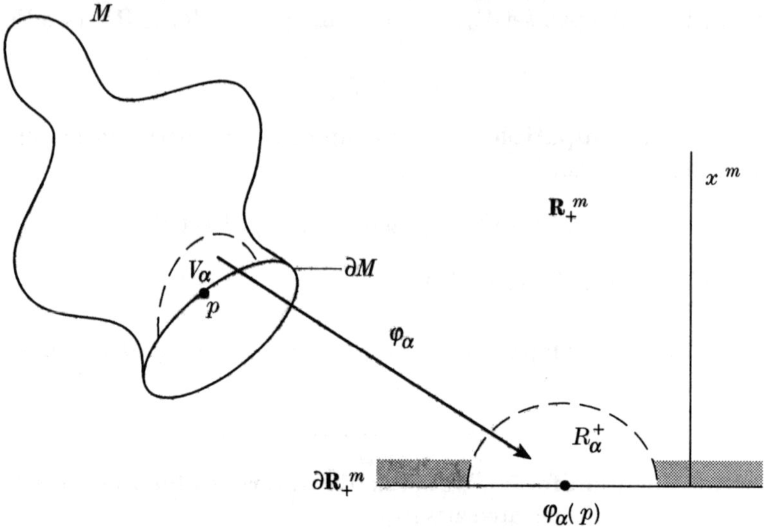

Fig. 2.20. A 'Schnuller' manifold with boundary

then the coordinate transformation is determined by the map $\varphi_\beta \cdot \varphi_\alpha^{-1}$

$$y = \varphi_\beta(p) = \varphi_\beta \cdot \varphi_\alpha^{-1}(x) \equiv y(x). \tag{2.61}$$

The **dimension of the manifold** is given by the dimension of the Euclidean vector space

$$\dim M \equiv \dim \mathbf{R}^m = m. \tag{2.62}$$

If we replace the real space $\mathbf{R}^m \to \mathbf{C}^m$ by a complex space we obtain a **complex analytic manifold**.

Manifold with boundary: The manifold M is defined as a space which locally looks like the Euclidean $\mathbf{R}^m$. If we now replace $\mathbf{R}^m \to \mathbf{R}_+^m$ by the positive half space

$$\mathbf{R}_+^m = \{(x^1, \ldots, x^m) \in \mathbf{R}^m | x^m \geq 0\} \tag{2.63}$$

with the boundary

$$\partial\mathbf{R}_+^m = \{(x^1, \ldots, x^m) \in \mathbf{R}^m | x^m = 0\} \tag{2.64}$$

and x^m denotes the m-th coordinate, then we achieve the notion of a manifold with boundary. In this case an atlas consists of charts $\{(V_\alpha, \varphi_\alpha)\}$

where $\varphi_\alpha : V_\alpha \to R_\alpha^+$ represents a homeomorphism onto the open subset $R_\alpha^+ \subset \mathbf{R}_+^m$ (see Figure 2.20). The overlap maps are again infinitely differentiable $\varphi_\beta \cdot \varphi_\alpha^{-1}, \varphi_\alpha \cdot \varphi_\beta^{-1} \in C^\infty$. Then ∂M, the **boundary of a manifold**, is the set of all points which are mapped into $\partial \mathbf{R}_+^m$ (see Figure 2.20)

$$\partial M = \bigcup_\alpha \varphi_\alpha^{-1}(\varphi_\alpha(V_\alpha) \cap \partial \mathbf{R}_+^m). \tag{2.65}$$

The dimension of ∂M is clearly less by 1

$$\dim \partial M = m - 1. \tag{2.66}$$

For a manifold without boundary we have $\partial M = \emptyset$. The boundary ∂M itself represents a $(m-1)$-dimensional manifold without boundary: $\partial\partial M = \emptyset$.

Remarks:

i) If a manifold M is connected then its boundary ∂M is not necessarily connected. For example, $M = [a, b]$ is connected but $\partial M = \{a\} \cup \{b\}$ is not.
Vice versa:
If ∂M is connected then M is not necessarily connected. For example, the point a is the (connected) boundary $\partial M = \{a\}$ of the disconnected manifold $M = [a, b] \cup (b, c)$.

ii) A manifold with boundary is not necessarily compact. For example, the manifold $M = (0, 1]$ has a boundary $\partial M = \{1\}$ but M is not compact.
Vice versa:
A compact manifold has not necessarily a boundary. For example, the circle $M = S^1$ is compact but boundaryless $\partial M = \partial S^1 = \emptyset$.

Examples:

i) The $\mathbf{R}^m$ is clearly a manifold where just one chart is needed and the homeomorphism φ may be the identity map.

ii) The closed interval $M = [a, b]$ is a manifold with boundary $\partial M = \{a\} \cup \{b\}$. This is intuitively obvious but let us apply our definitions. We find two charts:

$$\begin{aligned}
&(V_1, \varphi_1) \text{ with} \\
&V_1 = [a, b) \\
&\varphi_1 : x \stackrel{\varphi_1}{\longmapsto} x - a = \varphi_1(x) \in R_1^+ \subset \mathbf{R}_+
\end{aligned} \tag{2.67}$$

and

Fig. 2.21. Spaces which are not manifolds

$$\begin{aligned}&(V_2, \varphi_2) \text{ with}\\ &V_2 = (a, b] \\ &\varphi_2 : x \overset{\varphi_2}{\longmapsto} b - x = \varphi_2(x) \in R_2^+ \subset \mathbf{R}_+ .\end{aligned} \tag{2.68}$$

The boundary is then defined by

$$\begin{aligned}\partial M &= \varphi_1^{-1}(\varphi_1(V_1) \cap \partial \mathbf{R}_+) \cup \varphi_2^{-1}(\varphi_2(V_2) \cap \partial \mathbf{R}_+) \\ &= \varphi_1^{-1}(0) \cup \varphi_2^{-1}(0) = \{a\} \cup \{b\}.\end{aligned} \tag{2.69}$$

iii) The closed ball $M = B^m = \{(x^1, \ldots, x^m) \in \mathbf{R}^m | \sum_{i=1}^m (x^i)^2 \leq 1\}$ is a manifold with m dimensions whose boundary is the $(m-1)$-dimensional sphere $\partial B^m = S^{m-1}$.

iv) Note!
The closed cube $M = C^m = \{(x^1, \ldots, x^m) \in \mathbf{R}^m || x^i| \leq 1, \forall i = 1, \ldots, m\}$ does *not* represent a manifold with boundary since there are edges and corners which violate the differentiability of the overlap maps. But the interior is again a manifold.
Similarly, spaces containing junctions (see Figure 2.21) do *not* represent manifolds since at the junctions the space does not look Euclidean.

v) The n-dimensional sphere is a compact manifold without boundary. To get used to our manifold notations it is worthwhile demonstrating the proof in one dimension:

$$S^1 = \{(x_1, x_2) \in \mathbf{R}^2 | x_1^2 + x_2^2 = 1\} \text{ is the 1-dimensional sphere.}$$

Searching for an atlas we find the two charts (at least two are necessary, see Figure 2.22) (V_1, φ_1) with $V_1 = S^1 \setminus \{(1, 0)\}$, and defining the inverse map for convenience

$$\begin{aligned}\varphi_1^{-1} : \theta &\overset{\varphi_1^{-1}}{\longmapsto} (\cos\theta, \sin\theta) \\ (0, 2\pi) &\overset{\varphi_1^{-1}}{\longrightarrow} S^1\end{aligned} \tag{2.70}$$

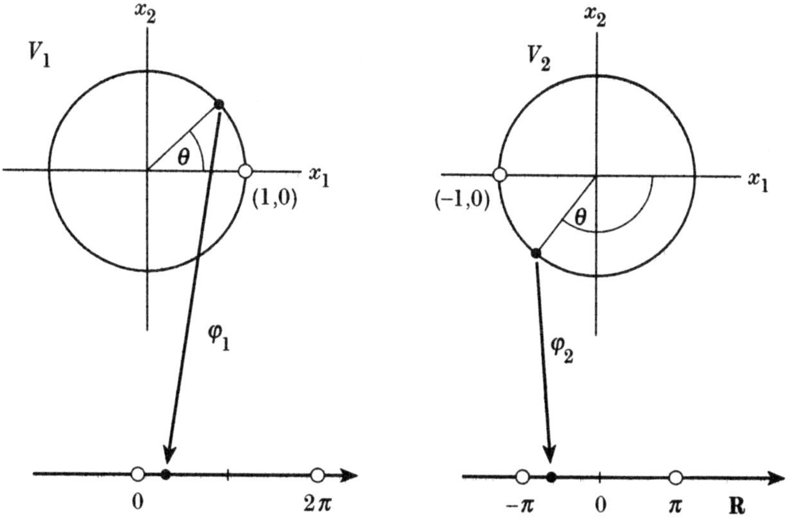

Fig. 2.22. An atlas for S^1 containing at least two charts

and (V_2, φ_2) with $V_2 = S^1 \setminus \{(-1,0)\}$ and

$$\begin{aligned} \varphi_2^{-1} : \theta \quad &\overset{\varphi_2^{-1}}{\longmapsto} \quad (\cos\theta, \sin\theta) \\ (-\pi, \pi) \quad &\overset{\varphi_2^{-1}}{\longrightarrow} \quad S^1. \end{aligned} \tag{2.71}$$

The maps φ_1^{-1}, φ_2^{-1} are invertible. All maps φ_1, φ_2, φ_1^{-1}, φ_2^{-1} are continuous, thus φ_1, φ_2 represent homeomorphisms and the overlap maps $\varphi_1 \cdot \varphi_2^{-1}, \varphi_2 \cdot \varphi_1^{-1} \in C^\infty$ are smooth. So all conditions for a manifold are fulfilled.

Product of manifolds: The Cartesian product $M_1 \times M_2$ of two manifolds M_1 and M_2 with atlases $\{(V_\alpha, \varphi_\alpha)\}$ and $\{(V_\beta, \varphi_\beta)\}$ is defined by the product of the charts

$$(V_\alpha, \varphi_\alpha) \times (V_\beta, \varphi_\beta) = (V_\alpha \times V_\beta, \varphi_\alpha \times \varphi_\beta), \tag{2.72}$$

with the homeomorphic map

$$\begin{aligned} V_\alpha \times V_\beta \quad &\overset{\varphi_\alpha \times \varphi_\beta}{\longrightarrow} \quad \mathbf{R}^{m_1} \times \mathbf{R}^{m_2} = \mathbf{R}^{m_1+m_2} \\ (p_1, p_2) \quad &\overset{\varphi_\alpha \times \varphi_\beta}{\longmapsto} \quad (\varphi_\alpha(p_1), \varphi_\beta(p_2)) \in \mathbf{R}^{m_1+m_2} \end{aligned} \tag{2.73}$$

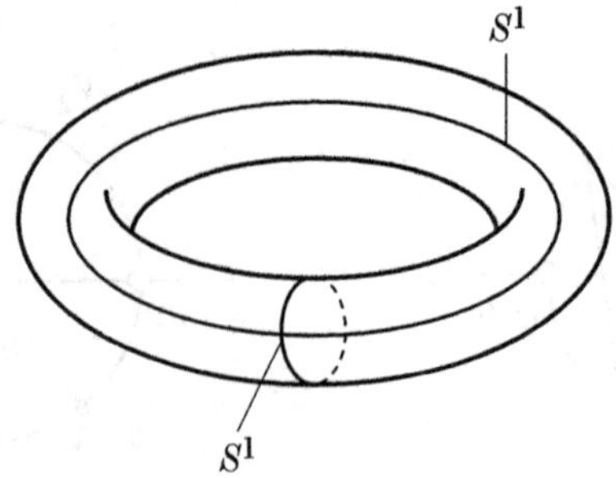

Fig. 2.23. The torus $T^2 = S^1 \times S^1$ as a product manifold

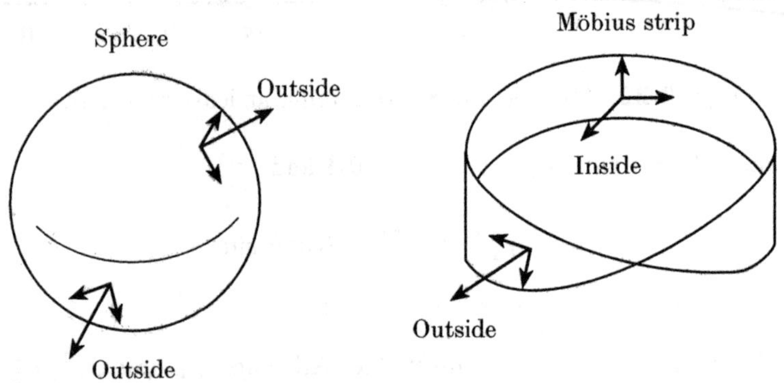

Fig. 2.24. A sphere is orientable, the Möbius strip is not

and $p_1 \in M_1$, $p_2 \in M_2$. So the atlas $\{(V_\alpha \times V_\beta, \varphi_\alpha \times \varphi_\beta)\}$ determines a differentiable structure on the $(m_1 + m_2)$-dimensional manifold $M_1 \times M_2$.

For instance, the 2-dimensional torus $T^2 = S^1 \times S^1$ is such a product manifold with the coordinates (θ_1, θ_2) and $\theta_1, \theta_2 \in [0, 2\pi)$ (see Figure 2.23).

Orientability of a manifold: Let us consider a disk, it has a top- and bottom side; a sphere, it has an in- and outside. These 2-sided surfaces are orientable since we can use this 2-sidedness to define an orientation in $\mathbf{R}^3$. However, if we study the Möbius strip we do not find a definite in- and outside. Moving around the normal, it changes its direction (see Figure 2.24). So the Möbius strip is not orientable.

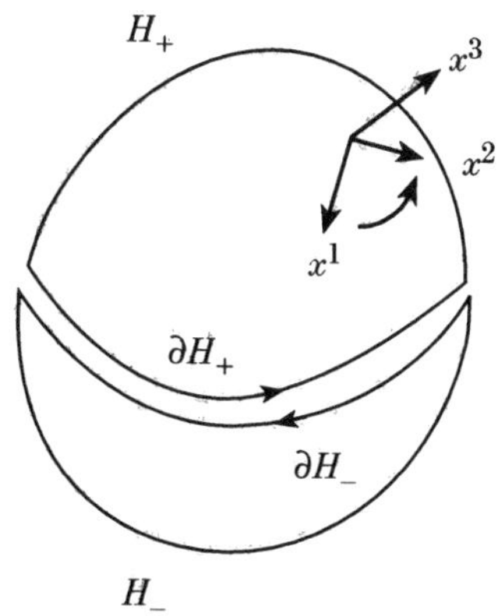

Fig. 2.25. The orientation of the boundaries ∂H_+ and ∂H_- is reversed

Mathematically, the key object for the orientability is the Jacobian determinant. Recall a vector space and choose two different bases $\{e_i\}$ and $\{e'_i\}$. A basis change is given by a matrix $A : e'_i = A_i{}^j e_j$. Then the determinant controls the orientation of the two bases. If $\det A > 0$ ($\det A < 0$) the bases $\{e_i\}$ and $\{e'_i\}$ have the same (opposite) orientation.

Analogously we have to consider a manifold which is covered by charts $\{(V_\alpha, \varphi_\alpha)\}$. There the coordinates are given by the homeomorphisms φ_α and we have to study the coordinate change $\varphi_\beta \cdot \varphi_\alpha^{-1}$. Then the **manifold is orientable** if there exists an atlas such that

$$\det(\varphi_\beta \cdot \varphi_\alpha^{-1}) > 0, \quad \forall V_\alpha, V_\beta \text{ with } V_\alpha \cap V_\beta \neq \emptyset. \tag{2.74}$$

In the following we always work with compact and orientable manifolds unless we specify it differently.

The orientation for the boundary ∂M of a manifold M is induced in a quite natural way which we illustrate with upper and lower hemispheres in Figure 2.25.

Tangent space to a manifold: Vectors on a manifold M always describe tangent vectors to a curve in M. Let $p(t)$, $t \in I$ be some curve passing through a chart. The coordinates of this curve are $x^i(p(t))$, $i = 1, \ldots, m$ and the tangent vector to the curve is given by

$$\frac{d}{dt} x^i(p(t)). \tag{2.75}$$

Next we consider a function $f(p) : M \to \mathbf{R}^m$ defined on M. The change of the function along the curve is

$$\frac{d}{dt} f(p(t)) \tag{2.76}$$

and in terms of local coordinates

$$\frac{\partial}{\partial x^i} f \frac{dx^i(p(t))}{dt}. \tag{2.77}$$

Note that $\frac{\partial}{\partial x^i} f$ actually means $\frac{\partial}{\partial x^i} f \cdot \varphi^{-1}(x)$.

Defining the differential operator

$$X = X^i \frac{\partial}{\partial x^i} \qquad \text{with} \qquad X^i = \frac{dx^i(p(t))}{dt} \tag{2.78}$$

we obtain

$$\frac{d}{dt} f(p(t)) = Xf. \tag{2.79}$$

We consider the differential operator X as the **tangent vector to the manifold** M at the point $p = p(t_0)$, with t_0 fixed, in the direction of the curve $p(t)$.

Example: Applying, for example, the operator X to the position coordinates we obtain the velocity

$$X[x^i] = \frac{dx^j}{dt} \frac{\partial x^i}{\partial x^j} = \frac{dx^i}{dt}. \tag{2.80}$$

For every differentiable curve through a point $p \in M$ of the manifold M there exists a tangent vector. So we can define the **tangent space** $T_p(M)$ to the manifold M at p to be the space of all possible tangents at the point p. $T_p(M)$ describes a vector space and in terms of local coordinates the set of differential operators

$$\left\{ \frac{\partial}{\partial x^i} \right\}, \qquad i = 1, \dots, m \tag{2.81}$$

forms a **basis** in $T_p(M)$. Clearly the dimension is

$$\dim T_p(M) = \dim M = m. \tag{2.82}$$

This notation of tangent vectors—they represent **contravariant vectors**—is a coordinate independent description. If we choose, for instance, another chart with local coordinates y^i we obtain for the tangent vector

$$\begin{aligned} Y &= Y^i \frac{\partial}{\partial y^i} = \frac{\partial y^i}{\partial x^j} \frac{dx^j}{dt} \frac{\partial x^k}{\partial y^i} \frac{\partial}{\partial x^k} \\ &= \frac{dx^j(p(t))}{dt} \frac{\partial}{\partial x^j} = X^j \frac{\partial}{\partial x^j} = X. \end{aligned} \tag{2.83}$$

Cotangent space to a manifold: To the contravariant vectors, which we have considered up to now, there also exist their duals—the covariant vectors. The dual space to $T_p(M)$ is the **cotangent space** $T_p^*(M)$ where duality is defined via the inner product

$$\left(dx^i, \frac{\partial}{\partial x^j}\right) = \delta^i{}_j. \tag{2.84}$$

So the set of differentials

$$\{dx^i\}, \qquad i = 1, \ldots, m, \tag{2.85}$$

which are dual to equation (2.81) forms a **basis** in $T_p^*(M)$.

For example, the differential

$$df = \frac{\partial}{\partial x^i} f(x) dx^i \in T_p^*(M) \tag{2.86}$$

is such an element of $T_p^*(M)$. It is a **cotangent vector** or already a (special) 1-form. Then the action of $df \in T_p^*(M)$ on $X \in T_p(M)$ is defined by the inner product

$$\begin{aligned} (df, X) &= \left(\frac{\partial}{\partial x^i} f dx^i, X^j \frac{\partial}{\partial x^j}\right) = \frac{\partial}{\partial x^i} f X^j \delta^i{}_j \\ &= X^i \frac{\partial}{\partial x^i} f = Xf \in \mathbf{R}. \end{aligned} \tag{2.87}$$

Generally, an element of $T_p^*(M)$ is given by the so-called **1-form**

$$\omega = \omega_i dx^i \in T_p^*(M) \tag{2.88}$$

where ω_i represents some **covariant vector**. Again, the **action of a 1-form on a vector** is determined by

$$(\omega, X) = \left(\omega_i dx^i, X^j \frac{\partial}{\partial x^j}\right) = \omega_i X^j \delta^i{}_j = \omega_i X^i \tag{2.89}$$

or in a slightly different notation we say: a **1-form evaluated at a vector field**

$$\omega(X) = \omega_i dx^i X^j \frac{\partial}{\partial x^j} = \omega_i X^j \delta^i{}_j = \omega_i X^i \in \mathbf{R}. \tag{2.90}$$

The notion of a 1-form (2.88) is also a coordinate independent description; in the overlap region of two charts we have

$$\bar{\omega} = \bar{\omega}_i(y) dy^i = \omega_j(x) \frac{\partial x^j}{\partial y^i} \frac{\partial y^i}{\partial x^k} dx^k = \omega_j(x) dx^j = \omega. \tag{2.91}$$

Tensors: We can now construct **tensors of type** (a, b) by mapping a elements of $T_p^*(M)$ and b elements of $T_p(M)$ into $\mathbf{R}$. So the space of these tensors is defined by

$$T^a{}_b = \underbrace{T_p(M) \otimes \ldots \otimes T_p(M)}_{a \text{ factors}} \otimes \underbrace{T_p^*(M) \otimes \ldots \otimes T_p^*(M)}_{b \text{ factors}}. \tag{2.92}$$

An element of $T^a{}_b$, a mixed tensor with contravariant rank a and covariant rank b, is given in terms of local coordinates

$$T(x) = T^{i_1 \ldots i_a}_{j_1 \ldots j_b}(x) \frac{\partial}{\partial x^{i_1}} \cdots \frac{\partial}{\partial x^{i_a}} dx^{j_1} \ldots dx^{j_b} \in T^a{}_b. \tag{2.93}$$

The action of T on 1-forms $\omega_1, \ldots, \omega_a$ and vectors $X_1, \ldots, X_b$ gives the number

$$T(\omega_1, \ldots, \omega_a, X_1, \ldots, X_b) = T^{i_1 \ldots i_a}_{j_1 \ldots j_b} \omega_{1 i_1} \ldots \omega_{a i_a} X_1^{j_1} \ldots X_b^{j_b}.$$

Allowing the point p to vary smoothly over the whole manifold M the vectors and tensors also vary smoothly over M and we achieve so-called **vector fields** and **tensor fields** on M.

2.4 Differential forms

Differential forms are very practical objects. First, their notation is independent of the choice of a coordinate system and, second, the whole mathematical formalism becomes very simple.

Wedge product: Let us begin with the antisymmetric tensor product

$$dx^\mu \wedge dx^\nu := dx^\mu \otimes dx^\nu - dx^\nu \otimes dx^\mu \tag{2.94}$$

which is the simplest **wedge product**.

Notation: For convenience from now on we drop the symbol wedge $\wedge$ which is habitually used in the literature

$$dx^\mu dx^\nu \equiv dx^\mu \wedge dx^\nu. \tag{2.95}$$

Then we have

$$\begin{aligned} dx^\mu dx^\nu &= -dx^\nu dx^\mu \\ dx^\mu dx^\mu &= 0 \end{aligned} \tag{2.96}$$

by definition.

Analogously we define the higher wedge products as totally antisymmetric tensor products

$$\begin{aligned} dx^\mu dx^\nu dx^\sigma &= dx^\mu \otimes dx^\nu \otimes dx^\sigma + dx^\nu \otimes dx^\sigma \otimes dx^\mu \\ &\quad + dx^\sigma \otimes dx^\mu \otimes dx^\nu - dx^\mu \otimes dx^\sigma \otimes dx^\nu \\ &\quad - dx^\sigma \otimes dx^\nu \otimes dx^\mu - dx^\nu \otimes dx^\mu \otimes dx^\sigma \\ dx^\mu dx^\nu dx^\sigma dx^\rho &= dx^\mu \otimes dx^\nu \otimes dx^\sigma \otimes dx^\rho \\ &\quad + \text{ all } \begin{pmatrix} +1 & \text{even} \\ -1 & \text{odd} \end{pmatrix} \text{ permutations,} \\ &\quad \text{etc.} \end{aligned} \tag{2.97}$$

In this way we obtain totally antisymmetric tensors of rank: $0, 1, 2, \ldots, m = \dim M$. If the rank $p > m = \dim M$ exceeds the dimension of the manifold the wedge product vanishes.

Differential form: With the help of the wedge products we can define now the several **differential forms**

$$\begin{aligned} &\text{0-form} && \omega = \omega(x) \\ &\text{1-form} && \omega = \omega_\mu(x) dx^\mu \\ &\text{2-form} && \omega = \frac{1}{2!}\omega_{\mu\nu}(x) dx^\mu dx^\nu \\ &\ldots && \ldots \\ &p\text{-form} && \omega = \frac{1}{p!}\omega_{\mu_1\ldots\mu_p}(x) dx^{\mu_1} \ldots dx^{\mu_p}, \end{aligned} \tag{2.98}$$

where we understand summation over all indices. $\omega_{\mu_1\ldots\mu_p}(x)$ expresses a totally antisymmetric covariant tensor field of rank $p \leq m$. ω vanishes for $p > m$, $\omega = 0$.

We denote the set of all p-forms by Λ^p. This is a vector space of dimension

$$\dim \Lambda^p = \begin{pmatrix} m \\ p \end{pmatrix} = \frac{m!}{p!(m-p)!}. \tag{2.99}$$

For example, for a manifold of dimension $m = 3$ we get the following notations

$$\begin{array}{cccc} \text{vector space} & p\text{-form} & \text{basis} & \dim \Lambda^p \\ \Lambda^0 & \omega_0 & \{1\} & 1 \\ \Lambda^1 & \omega_1 & \{dx^1, dx^2, dx^3\} & 3 \\ \Lambda^2 & \omega_2 & \{dx^1dx^2, dx^2dx^3, dx^3dx^1\} & 3 \\ \Lambda^3 & \omega_3 & \{dx^1dx^2dx^3\} & 1 \end{array} \tag{2.100}$$

Note that the vector spaces Λ^p and Λ^{m-p} have the same dimensions.

We can construct $(p+q)$-forms out of p-forms and q-forms in a straightforward manner with help of the wedge product $\alpha_p\beta_q \in \Lambda^{p+q}$.

All forms belong to the space

$$\Lambda^* = \Lambda^0 \oplus \Lambda^1 \oplus \Lambda^2 \oplus \ldots \oplus \Lambda^m, \tag{2.101}$$

which is closed under the wedge product operation (or exterior product). Λ^* is a **graded algebra**, also named **Cartan's exterior algebra** (or Grassmann algebra).

Commuting the forms $\alpha_p \in \Lambda^p$ and $\beta_q \in \Lambda^q$ we obtain

$$\alpha_p\beta_q = (-)^{pq}\beta_q\alpha_p. \tag{2.102}$$

So odd forms always anticommute.

Exterior derivative: We differentiate the forms by introducing the **exterior derivative**

$$d = \frac{\partial}{\partial x^\mu} dx^\mu \tag{2.103}$$

acting on a p-form in the following way

$$d\omega = \frac{1}{p!}\frac{\partial}{\partial x^\nu}\omega_{\mu_1\ldots\mu_p}(x)dx^\nu dx^{\mu_1}\ldots dx^{\mu_p}. \tag{2.104}$$

The exterior derivative is a map $d : \Lambda^p \to \Lambda^{p+1}$ which transforms p-forms into $(p+1)$-forms. It satisfies the important property

$$d^2 = 0 \tag{2.105}$$

—named **nilpotency** by the mathematicians—which can be quickly verified by direct application to a p-form

$$d^2\omega = \frac{1}{p!}\frac{\partial}{\partial x^\sigma}\frac{\partial}{\partial x^\nu}\omega_{\mu_1\ldots\mu_p}(x)dx^\sigma dx^\nu dx^{\mu_1}\ldots dx^{\mu_p} = 0. \tag{2.106}$$

This expression has to vanish since the derivatives $\frac{\partial}{\partial x^\sigma}\frac{\partial}{\partial x^\nu}$ are symmetric while the wedge product $dx^\sigma dx^\nu$ is antisymmetric.

Furthermore d obeys the **antiderivation rule**

$$d(\alpha_p\beta_q) = (d\alpha_p)\beta_q + (-)^p\alpha_p d\beta_q, \tag{2.107}$$

where again we have indicated the degrees of the forms.

There is an important property for the exterior derivative which actually represents the basis for the anomaly equations in physics to be studied later on.

Lemma:

$$d\omega(X,Y) = X\omega(Y) - Y\omega(X) - \omega([X,Y]), \tag{2.108}$$

with $\omega = \omega_\mu dx^\mu \in \Lambda^1$ and $X = X^\mu \frac{\partial}{\partial x^\mu}$, $Y = Y^\mu \frac{\partial}{\partial x^\mu} \in T(M)$.

The commutator, being again a vector field,

$$[X,Y] = [X,Y]^\mu \frac{\partial}{\partial x^\mu} \tag{2.109}$$

represents the **Lie bracket** (see Section 2.6.4)

$$[X,Y]^\mu = X^\nu \frac{\partial}{\partial x^\nu} Y^\mu - Y^\nu \frac{\partial}{\partial x^\nu} X^\mu. \tag{2.110}$$

We evaluate the 2-form $d\omega$ at the vector fields X, Y,

$$dx^\nu dx^\mu(X,Y) = X^\nu Y^\mu - Y^\nu X^\mu, \tag{2.111}$$

which we verify by applying our definitions.

Proof. We prove lemma (2.108)

$$\begin{aligned} &X\omega(Y) - Y\omega(X) - \omega([X,Y]) \\ &= X^\nu \frac{\partial}{\partial x^\nu}(\omega_\mu Y^\mu) - Y^\nu \frac{\partial}{\partial x^\nu}(\omega_\mu X^\mu) - \omega_\mu \left(X^\nu \frac{\partial}{\partial x^\nu} Y^\mu - Y^\nu \frac{\partial}{\partial x^\nu} X^\mu \right) \\ &= \frac{\partial}{\partial x^\nu}\omega_\mu (X^\nu Y^\mu - Y^\nu X^\mu) = \frac{\partial}{\partial x^\nu}\omega_\mu dx^\nu dx^\mu (X,Y) \\ &= d\omega(X,Y). \qquad \text{Q.E.D.} \end{aligned} \tag{2.112}$$

Lemma: Generalization to p-forms $\omega \in \Lambda^p$

$$\begin{aligned} d\omega(X_1,\dots,X_{p+1}) &= \sum_{i=1}^{p+1}(-)^{i+1} X_i \omega(X_1,\dots,\widehat{X}_i,\dots,X_{p+1}) \\ &+ \sum_{i<j}(-)^{i+j}\omega([X_i,X_j],X_1,\dots,\widehat{X}_i,\dots,\widehat{X}_j,\dots,X_{p+1}), \end{aligned} \tag{2.113}$$

where $X_i \in T(M)$ is a vector field and the symbol $\widehat{}$ means that the corresponding vector field has been omitted.

Interior product: We construct a contracted multiplication—**interior product**—of a p-form $\omega \in \Lambda^p$ with a vector field $X \in T(M)$. It is defined by

$$\begin{array}{ll} \text{i)} & i_X \quad \text{antiderivative} \\ \text{ii)} & i_X f = 0 \\ \text{iii)} & i_X dx^\mu = X^\mu, \end{array} \tag{2.114}$$

where f denotes some function. Viewed as a mapping the operation $i_X : \Lambda^p \to \Lambda^{p-1}$ maps opposite to the exterior derivative d. Clearly we have

$$i_X^2 = 0. \tag{2.115}$$

For example, for a 1-form we obtain

$$i_X \omega = \omega_\mu X^\mu = \omega(X) \tag{2.116}$$

for a 2-form

$$i_X \omega = \frac{1}{2!} \omega_{\mu\nu} (X^\mu dx^\nu - dx^\mu X^\nu) \tag{2.117}$$

(note the minus sign stems from the antiderivative property) and finally for a p-form

$$\begin{aligned} i_X \omega &= \frac{1}{p!} \sum_{s=1}^{p} X^{\mu_s} \omega_{\mu_1 \ldots \mu_s \ldots \mu_p} (-)^{s-1} dx^{\mu_1} \ldots \widehat{dx^{\mu_s}} \ldots dx^{\mu_p} \\ &= \frac{1}{(p-1)!} X^\nu \omega_{\nu \mu_2 \ldots \mu_p} dx^{\mu_2} \ldots dx^{\mu_p}. \end{aligned} \tag{2.118}$$

If we apply the interior product twice to a 2-form we get a scalar function

$$\begin{aligned} i_Y i_X \omega &= \frac{1}{2} \omega_{\mu\nu} i_Y i_X dx^\mu dx^\nu \\ &= \frac{1}{2} \omega_{\mu\nu} (X^\mu Y^\nu - Y^\mu X^\nu) \\ &= \omega_{\mu\nu} X^\mu Y^\nu = \omega(X, Y). \end{aligned} \tag{2.119}$$

Thus we find

$$i_Y i_X dx^\mu dx^\nu = X^\mu Y^\nu - Y^\mu X^\nu = dx^\mu dx^\nu (X, Y), \tag{2.120}$$

which we had before (equation (2.111)). Analogously we can contract a p-form

$$i_{X_p} \dots i_{X_1}\omega = \omega(X_1, \dots, X_p) \tag{2.121}$$

to arrive at a 0-form.

Integration of differential forms: We can also integrate differential forms. The integration is the inverse operation to the exterior derivative. The integration on a manifold M requires an integration measure or a volume element. The differential forms automatically provide such an object together with the correct transformation properties.

We always assume compact, orientable manifolds. An **orienting** m-**form** is

$$\begin{aligned} \omega &= \frac{1}{m!}\varepsilon_{\mu_1 \dots \mu_m} dx^{\mu_1} \dots dx^{\mu_m} \\ &= dx^1 dx^2 \dots dx^m \equiv dV_x \end{aligned} \tag{2.122}$$

and provides the volume element dV_x (for a right-handed coordinate system). The totally antisymmetric ε-tensor or **Levi–Civita tensor** is defined by

$$\varepsilon_{\mu_1 \dots \mu_m} = \begin{cases} 1 & \text{for any even permutations of } 1, \dots, m \\ 0 & \text{if two indices are equal} \\ -1 & \text{for any odd permutations of } 1, \dots, m. \end{cases} \tag{2.123}$$

Choosing a different coordinate system the m-form with its antisymmetry generates the Jacobian determinant. For example in two dimensions we have

$$\begin{aligned} dx^1 dx^2 &= \left(\frac{\partial x^1}{\partial y^1}dy^1 + \frac{\partial x^1}{\partial y^2}dy^2\right)\left(\frac{\partial x^2}{\partial y^1}dy^1 + \frac{\partial x^2}{\partial y^2}dy^2\right) \\ &= \left(\frac{\partial x^1}{\partial y^1}\frac{\partial x^2}{\partial y^2} - \frac{\partial x^1}{\partial y^2}\frac{\partial x^2}{\partial y^1}\right) dy^1 dy^2 \\ &= \det\left(\frac{\partial x^\mu}{\partial y^\nu}\right) dy^1 dy^2. \end{aligned} \tag{2.124}$$

Generally we obtain

$$\omega = dx^1 dx^2 \dots dx^m = \det\left(\frac{\partial x^\mu}{\partial y^\nu}\right) dy^1 dy^2 \dots dy^m, \tag{2.125}$$

which is the correct transformation property for a volume element.

Now we take a smooth function f on M then any m-form can be written like $f\omega$. We define the integral of such an m-form in the domain of a chart $(V_\alpha, \varphi_\alpha)$ with coordinates $(x^1, x^2, \dots, x^m)$ by

$$\int_{V_\alpha} f\omega = \int_{\varphi_\alpha(V_\alpha)} f(\varphi_\alpha^{-1}(x))dx^1 \dots dx^m = \int_{R_\alpha} f(x)dV_x. \qquad (2.126)$$

This is an ordinary multiple integration of a function with m variables.

For the integral of the m-form over the whole manifold we have to consider an atlas of charts (a finite number)

$$\int_M f\omega = \sum_\alpha \int_{V_\alpha} f_\alpha \omega, \qquad (2.127)$$

where $f = \sum_\alpha f_\alpha$ and f_α has its support on the chart $(V_\alpha, \varphi_\alpha)$.

This is usually called a **partition of unity** and means that we have functions

$$e_\alpha : f_\alpha = f e_\alpha$$

satisfying

i) $0 < e_\alpha(p) < 1$ for $p \in V_\alpha$,
ii) $e_\alpha(p) = 0$ if $p \notin V_\alpha$,
iii) $\sum_\alpha e_\alpha(p) = 1$ in the overlap region.

Stokes' theorem: Let M be an orientable compact manifold with dimension $\dim M = m$ and with a nonempty boundary ∂M; $\omega \in \Lambda^{m-1}$ be an $(m-1)$-form, then

$$\int_M d\omega = \int_{\partial M} \omega. \qquad (2.128)$$

In fact, it expresses an interesting connection between a local information of a quantity in the manifold M and a global information on the boundary ∂M of the manifold.

Examples: Relation (2.128) generalizes in a compact way theorems we are already familiar with in 1, 2, 3 dimensions.

i) *1-dimensional case*, $\dim M = 1$:

$$M = [a, b] \quad , \quad \partial M = \{a, b\} \qquad (2.129)$$

and

$$\omega = \omega(x) \quad , \quad d\omega = \frac{d}{dx}\omega(x)dx \qquad (2.130)$$

Then Stokes' theorem corresponds to the ordinary integration in $\mathbf{R}$

$$\int_M d\omega = \int_a^b \frac{d}{dx}\omega(x)dx = \omega(b) - \omega(a) = \int_{\partial M} \omega. \qquad (2.131)$$

ii) *2-dimensional case*, $\dim M = 2$:

$$M = \{(x_1, x_2) | x_1^2 + x_2^2 \leq 1\}, \partial M = \{(x_1, x_2) | x_1^2 + x_2^2 = 1\} \quad (2.132)$$

and

$$\omega = \omega_1(x)dx^1 + \omega_2(x)dx^2 \equiv \vec{A}(x)d\vec{x}, \quad (2.133)$$

$$\begin{aligned} d\omega &= \frac{\partial}{\partial x^i}\omega_j(x)dx^i dx^j = \frac{1}{2}(\partial_i A_j - \partial_j A_i)dx^i dx^j \\ &\equiv \frac{1}{2}\varepsilon_{ijk}B_k dx^i dx^j = B_k df_k = \vec{B}d\vec{f}, \end{aligned} \quad (2.134)$$

with the curl $B_k = \varepsilon_{kij}\partial_i A_j$ or $\vec{B} = \vec{\nabla} \times \vec{A}$ multiplied by an area element

$$df_k = \frac{1}{2}\varepsilon_{kij}dx^i dx^j \quad \text{or} \quad d\vec{f} = \begin{pmatrix} 0 \\ 0 \\ df_3 \end{pmatrix}. \quad (2.135)$$

Then we recover **Stokes' theorem** in the familiar notation

$$\int_M d\omega = \int_{\text{disk}} (\vec{\nabla} \times \vec{A})d\vec{f} = \int_{\text{circle}} \vec{A}d\vec{x} = \int_{\partial M} \omega. \quad (2.136)$$

In terms of physics, $\vec{A}$ represents the vector potential and $\vec{B}$ the magnetic field. The integral of the magnetic flux through the disk equals the integral of the vector potential along the boundary, the circle.

iii) *3-dimensional case*, $\dim M = 3$:

$$\begin{aligned} M &= \{(x_1, x_2, x_3) | x_1^2 + x_2^2 + x_3^2 \leq 1\} \qquad \text{ball} \\ \partial M &= \{(x_1, x_2, x_3) | x_1^2 + x_2^2 + x_3^2 = 1\} \qquad \text{sphere} \end{aligned} \quad (2.137)$$

and

$$\begin{aligned} \omega &= \frac{1}{2}\omega_{ij}(x)dx^i dx^j \\ &\equiv \frac{1}{2}\varepsilon_{ijk}E_k(x)dx^i dx^j = E_k df_k = \vec{E}d\vec{f} \end{aligned} \quad (2.138)$$

$$\begin{aligned} d\omega &= \frac{1}{2}\partial_\ell \omega_{ij} dx^\ell dx^i dx^j = \frac{1}{2}\varepsilon_{ijk}\partial_\ell E_k dx^\ell dx^i dx^j \\ &= \frac{1}{2}\partial_\ell E_k \varepsilon_{ijk}\varepsilon^{\ell ij} dx^1 dx^2 dx^3 = \partial_k E_k dV = \vec{\nabla}\vec{E}dV. \end{aligned} \quad (2.139)$$

In 3 dimensions Stokes' theorem (2.128) represents what we usually call the **Gauss law**

$$\int_M d\omega = \int_{\text{ball}} \vec{\nabla}\vec{E}dV = \int_{\text{sphere}} \vec{E}d\vec{f} = \int_{\partial M} \omega. \tag{2.140}$$

Again in terms of physics, $\vec{E}$ describes the electric field.

Hodge star operation: Let us recall the space of all p-forms Λ^p and the space Λ^{m-p}. Both vector spaces have the same dimension $\dim \Lambda^{m-p} = \dim \Lambda^p$. There is a duality between these 2 spaces, an isomorphism given by the **Hodge * operation**:

$$\Lambda^p \xrightarrow{\;*\;} \Lambda^{m-p}. \tag{2.141}$$

The star transforms p-forms into $(m-p)$-forms and its action is defined by

$$*dx^{\mu_1}\dots dx^{\mu_p} = \frac{1}{(m-p)!}\varepsilon^{\mu_1\dots\mu_p}{}_{\mu_{p+1}\dots\mu_m} dx^{\mu_{p+1}}\dots dx^{\mu_m}, \tag{2.142}$$

where

$$\varepsilon^{\mu_1\dots\mu_p}{}_{\mu_{p+1}\dots\mu_m} = g^{\mu_1\nu_1}\dots g^{\mu_p\nu_p}\varepsilon_{\nu_1\dots\nu_p\mu_{p+1}\dots\mu_m}. \tag{2.143}$$

$g_{\mu\nu}$ is the metric tensor and $g = \det g_{\mu\nu}$. In a curved space with metric $g_{\mu\nu}$ we have

$$\varepsilon_{1\dots m} = \sqrt{|g|} \to 1 \qquad \text{in flat space}$$

and

$$\varepsilon^{1\dots m} = \begin{cases} \dfrac{1}{\sqrt{|g|}} & \text{Riemannian} \quad \to \quad 1 \quad \text{Euclidean} \\[2ex] -\dfrac{1}{\sqrt{|g|}} & \text{Lorentzian} \quad \to \quad -1 \quad \text{Minkowskian.} \end{cases} \tag{2.144}$$

Note that we raise and lower the indices by

$$\begin{aligned} \varepsilon^{\mu_1\dots\mu_m} &= g^{\mu_1\nu_1}\dots g^{\mu_m\nu_m}\varepsilon_{\nu_1\dots\nu_m} \\ &= g^{-1}\varepsilon_{\nu_1\dots\nu_m}. \end{aligned} \tag{2.145}$$

In the following we only work in flat space throughout the book, except in Chapter 12, which is specially devoted to gravitation. So our metric is

$$g_{\mu\nu} = \eta_{\mu\nu} = \begin{pmatrix} 1 & & & \\ & -1 & & \\ & & -1 & \\ & & & -1 \end{pmatrix} \quad \text{Minkowskian}$$

$$g_{\mu\nu} = \delta_{\mu\nu} = \begin{pmatrix} 1 & & & \\ & 1 & & \\ & & 1 & \\ & & & 1 \end{pmatrix} \quad \text{Euclidean.} \tag{2.146}$$

Given a p-form

$$\omega_p = \frac{1}{p!}\omega_{\mu_1\ldots\mu_p}dx^{\mu_1}\ldots dx^{\mu_p} \tag{2.147}$$

then the **dual p-form** denotes

$$\begin{aligned} *\omega_p &= \frac{1}{p!(m-p)!}\omega_{\mu_1\ldots\mu_p}\varepsilon^{\mu_1\ldots\mu_p}{}_{\mu_{p+1}\ldots\mu_m}dx^{\mu_{p+1}}\ldots dx^{\mu_m} \\ &= \frac{1}{p!(m-p)!}\omega^{\mu_1\ldots\mu_p}\varepsilon_{\mu_1\ldots\mu_p\mu_{p+1}\ldots\mu_m}dx^{\mu_{p+1}}\ldots dx^{\mu_m}. \end{aligned} \tag{2.148}$$

Lemma: When applied to a p-form we have

$$** = \begin{cases} (-)^{p(m-p)} \\ (-)^{p(m-p)+1} \end{cases} \quad \text{or} \quad *^{-1} = \begin{cases} (-)^{p(m-p)}* & \text{Euclidean} \\ (-)^{p(m-p)+1}* & \text{Minkowskian.} \end{cases} \tag{2.149}$$

Proof. We need the useful identity

$$\varepsilon^{\mu_1\ldots\mu_p}{}_{\mu_{p+1}\ldots\mu_m}\varepsilon_{\nu_1\ldots\nu_p}{}^{\mu_{p+1}\ldots\mu_m}dx^{\nu_1}\ldots dx^{\nu_p} = p!(m-p)!dx^{\mu_1}\ldots dx^{\mu_p}, \tag{2.150}$$

which can be quickly verified. Then we calculate

$$\begin{aligned} **\omega_p &= \frac{1}{p!(m-p)!}\omega_{\mu_1\ldots\mu_p}\varepsilon^{\mu_1\ldots\mu_p}{}_{\mu_{p+1}\ldots\mu_m} * dx^{\mu_{p+1}}\ldots dx^{\mu_m} \\ &= \frac{1}{p!p!(m-p)!}\omega_{\mu_1\ldots\mu_p}\varepsilon^{\mu_1\ldots\mu_p}{}_{\mu_{p+1}\ldots\mu_m}\varepsilon^{\mu_{p+1}\ldots\mu_m}{}_{\nu_1\ldots\nu_p}dx^{\nu_1}\ldots dx^{\nu_p}. \end{aligned} \tag{2.151}$$

Reshuffling the indices in the second ε-tensor gives a factor $(-)^{p(m-p)}$ and using identity (2.150) yields

$$
\begin{aligned}
**\omega_p &= (-)^{p(m-p)}\frac{1}{p!}\omega_{\mu_1\ldots\mu_p}dx^{\mu_1}\ldots dx^{\mu_p} \\
&= (-)^{p(m-p)}\omega_p. \qquad (2.152)
\end{aligned}
$$

For Minkowski space identity (2.150) and hence relation (2.152) gets an extra minus sign. Q.E.D.

Note that in Euclidean space we have

$$
\begin{aligned}
\frac{1}{m!}\varepsilon_{\mu_1\ldots\mu_m}dx^{\mu_1}\ldots dx^{\mu_m} &= dx^1\ldots dx^m = *\mathbf{1} \\
dx^{\mu_1}\ldots dx^{\mu_m} &= \varepsilon^{\mu_1\ldots\mu_m}dx^1\ldots dx^m \qquad (2.153)
\end{aligned}
$$

and in Minkowski space

$$
\begin{aligned}
\frac{1}{m!}\varepsilon_{\mu_1\ldots\mu_m}dx^{\mu_1}\ldots dx^{\mu_m} &= dx^0 dx^1\ldots dx^{m-1} = *\mathbf{1} \\
dx^{\mu_1}\ldots dx^{\mu_m} &= -\varepsilon^{\mu_1\ldots\mu_m}dx^0 dx^1\ldots dx^{m-1} \qquad (2.154)
\end{aligned}
$$

since

$$
\varepsilon_{\mu_1\ldots\mu_m}\varepsilon^{\mu_1\ldots\mu_m} = \begin{cases} m! & \text{for Euclidean space} \\ -m! & \text{for Minkowski space.} \end{cases} \qquad (2.155)
$$

Examples:

i) *3-dimensional space*, $\dim M = 3$:

$$
\begin{aligned}
*dx^i dx^j dx^k &= \varepsilon^{ijk} \\
*\,dx^1 dx^2 dx^3 &= \mathbf{1} \\
*\,dx^i dx^j &= \varepsilon^{ij}{}_k dx^k \qquad (2.156) \\
*\,dx^i &= \frac{1}{2!}\varepsilon^i{}_{jk}dx^j dx^k \\
*\,\mathbf{1} &= \frac{1}{3!}\varepsilon_{ijk}dx^i dx^j dx^k = dx^1 dx^2 dx^3
\end{aligned}
$$

vector space	dual p-form	basis	$\dim \Lambda^p$
Λ^3	$*\omega_0$	$\{dx^1dx^2dx^3\}$	1
Λ^2	$*\omega_1$	$\{dx^2dx^3, dx^3dx^1, dx^1dx^2\}$	3
Λ^1	$*\omega_2$	$\{dx^3, dx^1, dx^2\}$	3
Λ^0	$*\omega_3$	$\{\mathbf{1}\}$	1

(2.157)

ii) *4-dimensional Minkowski space*, $\dim M = 4$ ($\varepsilon_{0123} = 1$):

$$
*\mathbf{1} = dx^0 dx^1 dx^2 dx^3
$$

$$\begin{aligned}
*\,dx^0 &= \frac{1}{3!} g^{0\nu} \varepsilon_{\nu\mu_1\mu_2\mu_3} dx^{\mu_1} dx^{\mu_2} dx^{\mu_3} \\
&= dx^1 dx^2 dx^3 \\
*\,dx^1 &= \frac{1}{3!} g^{1\nu} \varepsilon_{\nu\mu_1\mu_2\mu_3} dx^{\mu_1} dx^{\mu_2} dx^{\mu_3} \\
&= -dx^2 dx^0 dx^3 = dx^0 dx^2 dx^3 \\
*\,dx^2 &= -dx^0 dx^1 dx^3 \\
*\,dx^3 &= -dx^0 dx^2 dx^1 = dx^0 dx^1 dx^2 \\
*\,dx^1 dx^2 &= dx^0 dx^3 \\
*\,dx^1 dx^3 &= \frac{1}{2!} g^{1\nu_1} g^{3\nu_2} \varepsilon_{\nu_1\nu_2\mu_1\mu_2} dx^{\mu_1} dx^{\mu_2} \\
&= dx^2 dx^0 = -dx^0 dx^2 \\
*\,dx^0 dx^1 dx^3 &= g^{0\nu_1} g^{1\nu_2} g^{3\nu_3} \varepsilon_{\nu_1\nu_2\nu_3\mu} dx^\mu \\
&= -dx^2 \\
*\,dx^0 dx^2 dx^3 &= dx^1 \\
*\,dx^1 dx^2 dx^3 &= dx^0 \\
*\,dx^0 dx^1 dx^2 dx^3 &= **\mathbf{1} = -\mathbf{1}. \qquad (2.158)
\end{aligned}$$

Coderivative or adjoint exterior derivative: For the exterior derivative we can also construct an adjoint via an inner product of p-forms.

The **inner product of two p-forms** $\alpha_p, \beta_p \in \Lambda^p$ is defined by

$$(\alpha_p, \beta_p) = \int_M \alpha_p * \beta_p \in \mathbf{R}. \qquad (2.159)$$

(Note that for a product of forms we always mean a *wedge product* unless we specify it differently.) How does it look in explicit coordinates?

$$(\alpha_p, \beta_p) = \int_M \alpha_p * \beta_p = \frac{1}{p!p!(m-p)!} \int_M \alpha_{\mu_1\ldots\mu_p}(x) \beta_{\nu_1\ldots\nu_p}(x) \cdot \\ \varepsilon^{\nu_1\ldots\nu_p}{}_{\nu_{p+1}\ldots\nu_m} \varepsilon^{\mu_1\ldots\mu_p\nu_{p+1}\ldots\nu_m} dx^1 \ldots dx^m \qquad (2.160)$$

using the identity

$$\beta_{\nu_1\ldots\nu_p} \varepsilon^{\nu_1\ldots\nu_p}{}_{\nu_{p+1}\ldots\nu_m} \varepsilon^{\mu_1\ldots\mu_p\nu_{p+1}\ldots\nu_m} = p!(m-p)!\beta^{\mu_1\ldots\mu_p} \qquad (2.161)$$

we obtain

$$(\alpha_p, \beta_p) = \int_M \alpha_p * \beta_p = \frac{1}{p!} \int dV \alpha_{\mu_1\ldots\mu_p}(x) \beta^{\mu_1\ldots\mu_p}(x). \qquad (2.162)$$

If the forms contain matrices

$$\alpha_p = \alpha_p^a T^a, \qquad \beta_p = \beta_p^a T^a, \tag{2.163}$$

with T^a some matrices, the generators of a group, then we have to take the (normalized) trace

$$(\alpha_p, \beta_p) = \int_M \text{tr}\ \alpha_p * \beta_p = \frac{1}{p!} \int dV \alpha^a_{\mu_1 \ldots \mu_p}(x) \beta_a^{\mu_1 \ldots \mu_p}(x). \tag{2.164}$$

The inner product is certainly symmetric

$$(\alpha_p, \beta_p) = (\beta_p, \alpha_p) \tag{2.165}$$

since $\alpha_p * \beta_p = \beta_p * \alpha_p$ and it is also positive definite if the manifold M is Riemannian

$$(\alpha_p, \alpha_p) \geq 0, \tag{2.166}$$

with equality only for $\alpha_p = 0$.

Now we consider the following inner product

$$\begin{aligned} (d\alpha_{p-1}, \beta_p) &= \int_M d\alpha_{p-1} * \beta_p \\ &= (-)^p \int_M \alpha_{p-1} d * \beta_p + \int_M d(\alpha_{p-1} * \beta_p). \end{aligned} \tag{2.167}$$

The second term vanishes via Stokes' theorem (2.128). It corresponds to an integral over the boundary of the manifold and we assume a *boundaryless* manifold $\partial M = \emptyset$. Inserting the operation $**^{-1}$ we have

$$(d\alpha_{p-1}, \beta_p) = \int_M \alpha_{p-1} * (-)^p *^{-1} d * \beta_p. \tag{2.168}$$

Definition: The **δ-operator, coderivative, or adjoint derivative** applied on p-forms is defined by

$$\begin{aligned} \delta &:= (-)^p *^{-1} d * \\ &= (-)^{mp+m+1} * d*, \end{aligned} \tag{2.169}$$

where the second line follows by respecting the inverse star $*^{-1}$ (recall equation (2.149)). For a compact manifold without boundary this leads to the following theorem.

Theorem:

$$(d\alpha_{p-1}, \beta_p) = \int_M \alpha_{p-1} * \delta\beta_p = (\alpha_{p-1}, \delta\beta_p). \tag{2.170}$$

The coderivative lowers the degree of a form by one unit and $\delta : \Lambda^p \to \Lambda^{p-1}$ operates opposite to the exterior derivative $d : \Lambda^p \to \Lambda^{p+1}$.

Properties: We have

$$\delta = (-)^p * d * \qquad \text{for } m \text{ odd} \tag{2.171}$$
$$\delta = - * d * \qquad \text{for } m \text{ even} \tag{2.172}$$
$$\delta^2 = 0. \tag{2.173}$$

Instead of δ it is also customary to use the symbol $d^\dagger$.

Laplacian: Finally we have to introduce the **Laplacian**, which is defined by

$$\Delta = d\delta + \delta d = (d + \delta)^2. \tag{2.174}$$

It is a map $\Delta : \Lambda^p \to \Lambda^p$ from p-forms to p-forms. Indicating the degrees of the forms we write

$$\Delta_p \omega_p = d_{p-1}\delta_{p-1}\omega_p + \delta_p d_p \omega_p. \tag{2.175}$$

So altogether we have the following operators acting on the vector spaces

$$\Lambda^{p-1} \quad \underset{\delta_{p-1}}{\overset{d_{p-1}}{\rightleftarrows}} \quad \Lambda^p \xrightarrow{\Delta_p} \Lambda^p \quad \underset{\delta_p}{\overset{d_p}{\rightleftarrows}} \quad \Lambda^{p+1}. \tag{2.176}$$

Example: Let us consider the Laplacian when acting on a 1-form in 3-dimensional Euclidean space. We need

$$\begin{aligned} \delta\omega &= (-)^{3\cdot 1+3+1} * d * \omega_i dx^i \\ &= -\frac{1}{2}\frac{\partial}{\partial x^\ell}\omega_i \varepsilon^i{}_{jk}\varepsilon^{\ell jk} \\ &= -\frac{\partial}{\partial x^\ell}\omega_\ell, \end{aligned} \tag{2.177}$$

where we have used $\varepsilon^i{}_{jk}\varepsilon^{\ell jk} = 2g^{i\ell}$; then

$$d\delta\omega = -\frac{\partial}{\partial x^k}\frac{\partial}{\partial x^\ell}\omega_\ell dx^k. \tag{2.178}$$

On the other hand, we calculate

$$\begin{aligned} \delta d\omega &= (-)^{3\cdot 2+3+1} * d * \frac{\partial}{\partial x^k}\omega_i dx^k dx^i \\ &= \frac{\partial}{\partial x^\ell}\frac{\partial}{\partial x^k}\omega_i \varepsilon^{ki}{}_j \varepsilon^{\ell j}{}_s dx^s \\ &= -\frac{\partial}{\partial x^\ell}\frac{\partial}{\partial x^\ell}\omega_i dx^i + \frac{\partial}{\partial x^k}\frac{\partial}{\partial x^\ell}\omega_\ell dx^k, \end{aligned} \tag{2.179}$$

where we have used $\varepsilon^{ki}{}_j\varepsilon^{\ell j}{}_s = (-)\varepsilon^{ki}{}_j\varepsilon^{\ell}{}_s{}^j = (-)(g^{k\ell}g^i{}_s - g^k{}_s g^{i\ell})$. Altogether we find

$$\Delta\omega = (d\delta + \delta d)\omega = -\frac{\partial^2}{\partial x^\ell \partial x^\ell}\omega, \tag{2.180}$$

which is the familiar Laplacian with a negative sign.

Property: The Laplacian commutes with

$$\begin{array}{rl} \text{Hodge star} & *\Delta = \Delta * \\ \text{derivative} & d\Delta = \Delta d \\ \text{coderivative} & \delta\Delta = \Delta\delta. \end{array} \tag{2.181}$$

Proposition:

- *Δ is a positive operator!*

Proof. We take the inner product

$$\begin{aligned} (\omega_p, \Delta\omega_p) &= (\omega_p, d\delta\omega_p) + (\omega_p, \delta d\omega_p) \\ &= (\delta\omega_p, \delta\omega_p) + (d\omega_p, d\omega_p) \geq 0. \quad \text{Q.E.D.} \end{aligned} \tag{2.182}$$

This equation immediately implies the following theorem.

Theorem:

$$\begin{aligned} \Delta\omega_p = 0 \Longleftrightarrow d\omega_p &= 0 \\ \delta\omega_p &= 0. \end{aligned} \tag{2.183}$$

Definitions: We call a p-form ω_p

$$\begin{array}{rl} \textbf{harmonic} & \text{if } \Delta\omega_p = 0 \\ \textbf{closed} & \text{if } d\omega_p = 0 \\ \textbf{coclosed} & \text{if } \delta\omega_p = 0. \end{array} \tag{2.184}$$

Furthermore: if there exists *globally* a form α such that

$$\begin{array}{ll} \omega_p = d\alpha_{p-1} & \text{then } \omega_p \text{ is } \textbf{exact} \\ \omega_p = \delta\alpha_{p+1} & \text{then } \omega_p \text{ is } \textbf{coexact}. \end{array} \tag{2.185}$$

Remark: Note that

$$\text{(co-)exact} \overset{\Rightarrow}{\not\Leftarrow} \text{(co-)closed.} \tag{2.186}$$

Poincaré's lemma:

- *Any closed form, $d\omega = 0$, can be expressed locally as an exact form, $\omega = d\alpha$!*

This is a very important statement which we use frequently in this book. We devote a separate section to its discussion and proof (Section 7.3).

Example: An example of a closed form which is not exact is the following. Let us consider the manifold $M = \mathbf{R}^2 \setminus \{0\}$, then the 1-form

$$\omega = \omega_i dx^i, \tag{2.187}$$

with $\omega_1 = \frac{-y}{r^2}, \quad \omega_2 = \frac{x}{r^2}$ and $x^1 = x, \quad x^2 = y, \quad r^2 = x^2 + y^2$

is well defined on M. It is closed since

$$d\omega = \frac{\partial}{\partial x^i}\frac{x^i}{r^2} dx^1 dx^2 = 0. \tag{2.188}$$

But ω is *not* exact. Exactness would mean that there is a function f such that

$$\omega = df = d(\arctan \frac{y}{x}). \tag{2.189}$$

However, $f(x, y)$ is not continuous (hence not differentiable) on the whole manifold $M = \mathbf{R}^2 \setminus \{0\}$. Therefore $\omega \neq df$ *globally*, ω is *not* exact.

It is instructive to study this example also in a complex notation (see e.g. [Thirring 1992])

$$z = x + iy, \qquad dz = dx + idy, \qquad \frac{dz}{z} = \frac{xdx + ydy + i(-ydx + xdy)}{r^2}. \tag{2.190}$$

Then

$$\begin{aligned} \omega_{\text{Im}} &= \frac{-ydx + xdy}{r^2} = \text{Im}\,\frac{dz}{z} \\ \omega_{\text{Re}} &= \frac{xdx + ydy}{r^2} = \text{Re}\,\frac{dz}{z} \end{aligned} \tag{2.191}$$

and

$$\omega = \omega_{\text{Re}} + i\,\omega_{\text{Im}}. \tag{2.192}$$

We have $d\omega = 0$, ω is closed, since $d\omega_{\text{Re}} = d\omega_{\text{Im}} = 0$. On the other hand,

$$\omega = \omega_{\text{Re}} + i\omega_{\text{Im}} = \text{Re}\ \frac{dz}{z} + i\ \text{Im}\ \frac{dz}{z} = \frac{dz}{z} = d\ln z. \tag{2.193}$$

But the logarithm $\ln z$ is *not* defined on the whole manifold $M = \mathbf{R}^2 \setminus \{0\}$, hence $\omega \neq d\ln z$ *globally* and ω is *not* exact.

On the other hand, if we cut the plane along the negative x-axis, so if we consider the manifold $M = \mathbf{R}^2 \setminus \{(x,y)|x \leq 0, y = 0\}$, there $\omega = d\ln z$ is well defined, ω is exact.

Trace, (anti-)commutator: Finally we define the trace, the commutator and anticommutator of differential forms.

Let $\alpha_p \in \text{Lie}\ G \otimes \Lambda^p$, $\beta_q \in \text{Lie}\ G \otimes \Lambda^q$ be forms

$$\alpha_p = \frac{1}{p!}\alpha_{\mu_1\ldots\mu_p} dx^{\mu_1}\ldots dx^{\mu_p} \qquad p\text{-form} \tag{2.194}$$

$$\beta_q = \frac{1}{q!}\beta_{\nu_1\ldots\nu_q} dx^{\nu_1}\ldots dx^{\nu_q} \qquad q\text{-form,} \tag{2.195}$$

which are matrix valued, we say Lie algebra valued $\alpha_{\mu_1\ldots\mu_p}, \beta_{\nu_1\ldots\nu_q} \in \text{Lie}\ G$

$$\alpha_{\mu_1\ldots\mu_p} = \alpha^a_{\mu_1\ldots\mu_p} T^a \tag{2.196}$$

$$\beta_{\nu_1\ldots\nu_q} = \beta^b_{\nu_1\ldots\nu_q} T^b, \tag{2.197}$$

with T^a being the generators (matrices) of a Lie group G. Equations (2.194), (2.195) and the subsequent equations actually mean the tensor product, like $\alpha = T_a \otimes \alpha^a \in \text{Lie}\ G \otimes \Lambda^p$, between the basis $\{T_a\}$ of the Lie algebra and the wedge products of the differential form. Then we define the **trace of forms** by

$$\text{tr}\ \alpha_p\beta_q = \frac{1}{p!q!}\ \text{tr}\ (\alpha_{\mu_1\ldots\mu_p}\beta_{\nu_1\ldots\nu_q}) dx^{\mu_1}\ldots dx^{\mu_p} dx^{\nu_1}\ldots dx^{\nu_q}, \tag{2.198}$$

where the trace is to be taken over the matrix product T^aT^b.

We define the **commutator and anticommutator of forms** similarly by separating the (anti-)commutator of the matrices and the wedge product of the forms

$$[\alpha_p, \beta_q] = \frac{1}{p!q!}[\alpha_{\mu_1\ldots\mu_p}, \beta_{\nu_1\ldots\nu_q}] dx^{\mu_1}\ldots dx^{\mu_p} dx^{\nu_1}\ldots dx^{\nu_q} \tag{2.199}$$

$$\{\alpha_p, \beta_q\} = \frac{1}{p!q!}\{\alpha_{\mu_1\ldots\mu_p}, \beta_{\nu_1\ldots\nu_q}\} dx^{\mu_1}\ldots dx^{\mu_p} dx^{\nu_1}\ldots dx^{\nu_q}. \tag{2.200}$$

Since the matrices T^a satisfy the Lie algebra

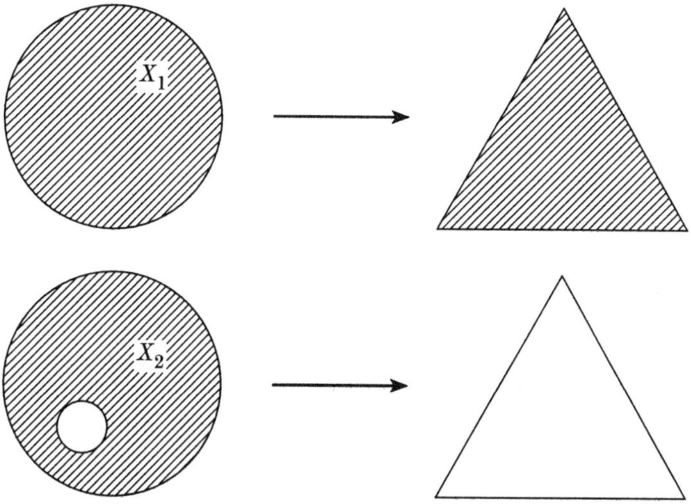

Fig. 2.26. Space X_1 is homeomorphic to a triangle with interior; space X_2, containing a hole, is deformable to a triangle without interior

$$[T^a, T^b] = f^{abc}T^c, \tag{2.201}$$

where f^{abc} are the antisymmetric structure constants, we have

$$\begin{aligned}
[\alpha_p, \beta_q] &= \frac{1}{p!q!} f^{abc} \alpha^a_{\mu_1 \ldots \mu_p} \beta^b_{\nu_1 \ldots \nu_q} T^c dx^{\mu_1} \ldots dx^{\mu_p} dx^{\nu_1} \ldots dx^{\nu_q} \\
&= f^{abc} \alpha^a_p \beta^b_q T^c \\
&= [\alpha_p, \beta_q]^c T^c
\end{aligned} \tag{2.202}$$

and

$$[\alpha_p, \beta_q]^c = f^{abc} \alpha^a_p \beta^b_q. \tag{2.203}$$

Rules:

$$\begin{aligned}
&\text{i)} && [\alpha_p, \beta_q] = \alpha_p \beta_q - (-)^{pq} \beta_q \alpha_p = -(-)^{pq} [\beta_q, \alpha_p] \\
&\text{ii)} && \{\alpha_p, \beta_q\} = \alpha_p \beta_q + (-)^{pq} \beta_q \alpha_p = (-)^{pq} \{\beta_q, \alpha_p\} \\
&\text{iii)} && \alpha_p \beta_q = \frac{1}{2}([\alpha_p, \beta_q] + \{\alpha_p, \beta_q\}) \\
&\text{iv)} && [\alpha_p \beta_q, \omega_r] = \alpha_p [\beta_q, \omega_r] + (-)^{qr} [\alpha_p, \omega_r] \beta_q \\
&\text{v)} && \text{tr } \alpha_p \beta_q = (-)^{pq} \text{ tr } \beta_q \alpha_p \\
&\text{vi)} && \text{tr } [\alpha_p, \beta_q] = 0.
\end{aligned} \tag{2.204}$$

For instance, the commutator of a 1-form $\alpha \in \Lambda^1$ (any odd-form) with itself does *not* vanish but it is

$$[\alpha, \alpha] = 2\alpha\alpha =: 2\alpha^2. \tag{2.205}$$

This is an important result which we use permanently later on when formulating quantum field theory with differential forms.

2.5 Homology and de Rham cohomology

In the homotopy section (Section 2.2) we classified the topological spaces with the help of loops; loops containing a hole or loops not containing a hole. Now we characterize the spaces with their topological invariants in a different way by studying the boundaries of the spaces. This is the topic of homology and cohomology; for literature we refer to [Singer, Thorpe 1967], [Nash, Sen 1983], [Nakahara 1990], [Bott, Tu 1982].

2.5.1 Homology

Let us consider some region X_1 of a topological space, say $\mathbf{R}^2$. It is homeomorphic to a triangle with interior (see Figure 2.26). On the other hand, a region X_2, which contains a hole, can be continuously deformed into a triangle without interior (see Figure 2.26). In the first case the edges of the triangle are the boundary of a connected space, whereas in the second case the closed edges do not represent a boundary of any space. So in classifying spaces we can detect a hole in a space by searching for some closed area which is *not* itself a boundary. Mathematically, we formulate this in a general way by introducing so-called simplexes.

Simplex: Let $v_0, \ldots, v_p$ be $(p+1)$ distinct points in $\mathbf{R}^m$. The set of points—vectors—$\{v_0, \ldots, v_p\}$ is termed **geometrically independent** if the p vectors: $v_1 - v_0$, $v_2 - v_0$, $\ldots$, $v_p - v_0$ are linearly independent.

For example, in $\mathbf{R}^2$ the set of vectors $\{v_0, v_1, v_2\}$ is geometrically independent iff v_0, v_1, v_2 are not collinear.

A ***p*-simplex** σ^p is defined by

$$\sigma^p = \{x \in \mathbf{R}^m | x = \sum_{i=0}^{p} c_i v_i, c_i \geq 0, \sum_{i=0}^{p} c_i = 1\}, \tag{2.206}$$

where the vectors $v_0, \ldots, v_p$ are geometrically independent. The coefficients $(c_0, \ldots, c_p)$ express the **barycentric coordinates** of x; for instance (1/3,1/3,1/3) for $p = 2$, the triangle. The index p represents the dimension of the simplex. It is customary to use also the notation $\sigma^p \equiv [v_0 v_1 v_2]$.

A p-simplex σ^p is compact since it is closed and bounded in $\mathbf{R}^m$.

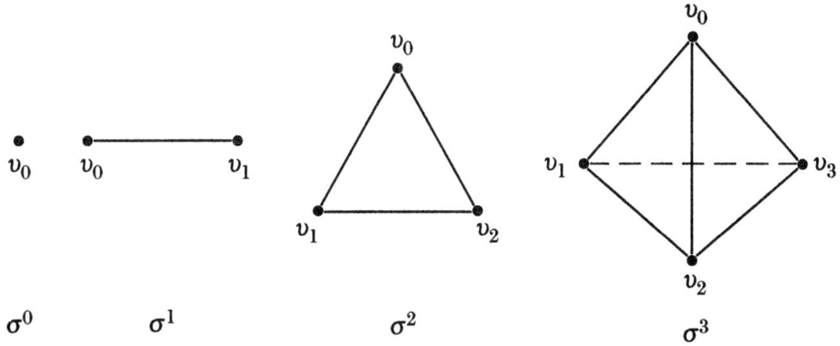

Fig. 2.27. 0-, 1-, 2-, 3-simplexes

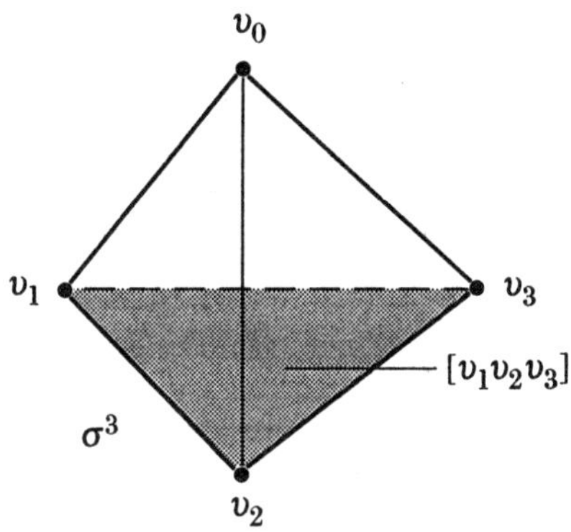

Fig. 2.28. A 2-face $[v_1v_2v_3]$ opposite to the 0-face $[v_0]$ of the simplex $\sigma^3 = [v_0v_1v_2v_3]$

We have drawn examples of simplexes in Figure 2.27: σ^0—a point, σ^1—a line, σ^2—a triangle with interior, σ^3—a tetrahedron with interior, etc.

Face: Recall the barycentric coordinates $c_i \geq 0$. If *all* $c_i > 0$ then the corresponding set of points represents the **interior** of a simplex σ^p—an **open simplex** denoted by $\sigma^p \equiv (v_0v_1v_2)$. If *some* $c_i = 0$ then x describes a **face** of the simplex σ^p opposite to the vertex v_i.

Consider, for instance, the tetrahedron (see Figure 2.28). $[v_0]$ is a 0-face; a 0-simplex has no faces. $[v_1v_2v_3]$ is a 2-face opposite to v_0 of the simplex $[v_0v_1v_2v_3]$. So we have one 3-face, four 2-faces, six 1-faces and four 0-faces.

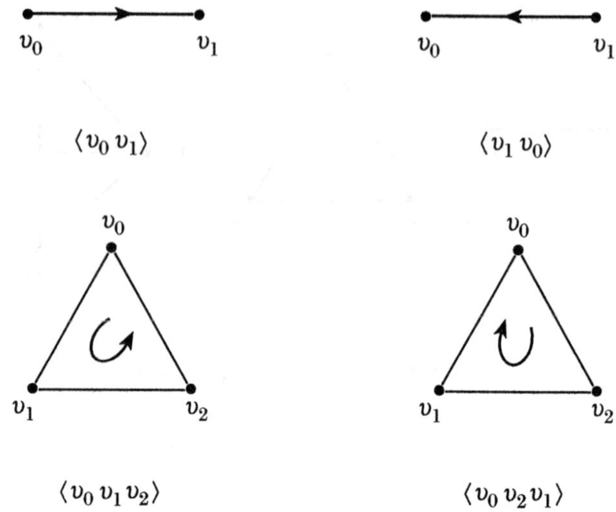

Fig. 2.29. Oriented simplexes

Simplicial complex: A simplicial complex K is a finite set of simplexes in some $\mathbf{R}^m$ satisfying

i) if $\sigma^p \in K$ then all faces of σ^p are elements of K,
ii) if $\sigma^p, \sigma^q \in K$ then either $\sigma^p \cap \sigma^q = \emptyset$ or $\sigma^p \cap \sigma^q$ is a common face of σ^p and σ^q.

The dimension of the complex K is given by the maximum dimension of the simplexes.

Orientation: We can assign an orientation to a p-simplex by directing the line segments. We denote an oriented p-simplex by $\sigma^p \equiv \langle v_0 \ldots v_p \rangle$. In Figure 2.29 we illustrate some examples:

$$\begin{aligned}\langle v_0 v_1 \rangle &= -\langle v_1 v_0 \rangle \\ \langle v_0 v_1 v_2 \rangle &= \langle v_1 v_2 v_0 \rangle = \langle v_2 v_0 v_1 \rangle \\ &= -\langle v_0 v_2 v_1 \rangle = -\langle v_2 v_1 v_0 \rangle = -\langle v_1 v_0 v_2 \rangle. \qquad (2.207)\end{aligned}$$

p-chain: Let K be a k-dimensional simplicial complex containing ℓ oriented p-simplexes (so $p = 0, 1, \ldots, k$). The **p-chain group** of K, denoted by

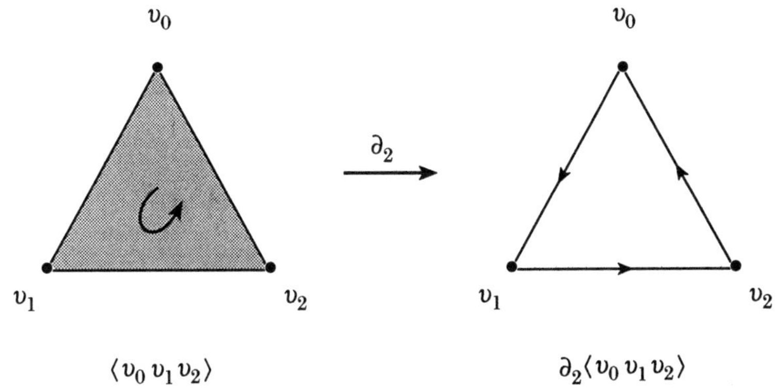

Fig. 2.30. Oriented boundary of the oriented 2-simplex

$C_p(K)$, is a free Abelian group generated by the oriented p-simplexes of K. An element, the **p-chain**, is expressed by

$$c_p = \sum_{i=1}^{\ell} a_i \sigma_i^p \in C_p(K) \qquad \text{with } a_i \in \mathbf{Z}. \tag{2.208}$$

Boundary: The boundary of an oriented simplex can be defined in a very obvious manner. Let us begin with σ^0, which has no boundary (a point is boundaryless)

$$\partial_0 \sigma^0 = \partial_0 \langle v_0 \rangle = 0. \tag{2.209}$$

Next we consider a 1-simplex $\sigma^1 = \langle v_0 v_1 \rangle$, an oriented line segment. Then the boundary is

$$\partial_1 \sigma^1 = \partial_1 \langle v_0 v_1 \rangle = \langle v_1 \rangle - \langle v_0 \rangle. \tag{2.210}$$

The minus sign is determined by the orientation.

Considering a 2-simplex $\sigma^2 = \langle v_0 v_1 v_2 \rangle$, which is an oriented triangle with interior, then its boundary represents the triangle without interior, clearly it is boundaryless, and its orientation has been induced (see Figure 2.30). So we have

$$\begin{aligned} \partial_2 \langle v_0 v_1 v_2 \rangle &= \langle v_1 v_2 \rangle - \langle v_0 v_2 \rangle + \langle v_0 v_1 \rangle \\ &= \langle v_1 v_2 \rangle + \langle v_2 v_0 \rangle + \langle v_0 v_1 \rangle \end{aligned} \tag{2.211}$$

and

$$\begin{aligned}\partial_2\partial_2\langle v_0v_1v_2\rangle &= \partial_2\langle v_1v_2\rangle + \partial_2\langle v_2v_0\rangle + \partial_2\langle v_0v_1\rangle \\ &= \langle v_2\rangle - \langle v_1\rangle + \langle v_0\rangle - \langle v_2\rangle + \langle v_1\rangle - \langle v_0\rangle = 0.\end{aligned} \tag{2.212}$$

As a rule for the boundary we write a plus sign for the first omitted vertex, a minus sign for the second omitted vertex, etc.

Now we generalize to a p-simplex. The **boundary $\partial_p\sigma^b$ of an oriented p-simplex** is a $(p-1)$-chain defined by

$$\partial_p\sigma^p = \sum_{i=0}^{p}(-)^i\langle v_0v_1\dots\widehat{v}_i\dots v_p\rangle, \tag{2.213}$$

where the symbol $\widehat{}$ means that v_i is omitted. Then the **boundary of a p-chain** is

$$\partial_pc_p = \sum_{i=1}^{\ell} a_i\partial_p\sigma_i^p. \tag{2.214}$$

So the **boundary operator** ∂_p represents a map $\partial_p : C_p(K) \to C_{p-1}(K)$ from the p-chains into the $(p-1)$-chains of a complex. The sequence

$$0 \xrightarrow{i} C_m(K) \xrightarrow{\partial_m} C_{m-1}(K) \xrightarrow{\partial_{m-1}} \dots \xrightarrow{\partial_2} C_1(K) \xrightarrow{\partial_1} C_0(K) \xrightarrow{\partial_0} 0 \tag{2.215}$$

is called a **chain complex** and the map i an **inclusion map**.

Lemma: The boundary operator is nilpotent

$$\partial_{p-1}\partial_p = 0. \tag{2.216}$$

This is intuitively clear since the boundary of a simplex—a point, line, triangle, tetrahedron, ...—is boundaryless.

Proof. Generally, from our definition (2.213) we can quickly prove

$$\begin{aligned}\partial_{p-1}\partial_p\sigma^p &= \partial_{p-1}\partial_p\langle v_0\dots v_p\rangle \\ &= \partial_{p-1}\sum_{i=0}^{p}(-)^i\langle v_0\dots\widehat{v}_i\dots v_p\rangle \\ &= \sum_{i=0}^{p}(-)^i\left\{\sum_{j=0}^{i-1}(-)^j\langle v_0\dots\widehat{v}_j\dots\widehat{v}_i\dots v_p\rangle\right. \\ &\quad \left. + \sum_{j=i+1}^{p}(-)^{j-1}\langle v_0\dots\widehat{v}_i\dots\widehat{v}_j\dots v_p\rangle\right\}\end{aligned}$$

$$= \sum_{j<i} \{(-)^{i+j} - (-)^{i+j}\} \langle v_0 \dots \widehat{v}_j \dots \widehat{v}_i \dots v_p \rangle = 0. \quad \text{Q.E.D.} \tag{2.217}$$

Definitions:

***p*-cycle:** If $c_p \in C_p(K)$ satisfies

$$\partial_p c_p = 0 \tag{2.218}$$

c_p is called a ***p*-cycle**. The **set of *p*-cycles**

$$Z_p(K) = \{c_p \in C_p(K) | \partial_p c_p = 0\} = \ker \partial_p \tag{2.219}$$

represents the kernel of the boundary operator ∂_p.

***p*-boundary:** If there exists a $(p+1)$-chain $b_{p+1} \in C_{p+1}(K)$ such that

$$c_p = \partial_{p+1} b_{p+1} \tag{2.220}$$

then c_p is a ***p*-boundary**. The **set of *p*-boundaries**

$$B_p(K) = \{c_p \in C_p(K) | c_p = \partial_{p+1} b_{p+1}\} = \text{image } \partial_{p+1} \tag{2.221}$$

represents the image of the operator ∂_{p+1}.

***p*-homology:** Returning now to our task, mentioned at the beginning, to detect a hole in space we have to eliminate the boundaries $B_p(K)$ contained in $Z_p(K)$. So the elements of $Z_p(K)$ which are *not* boundaries are the desired objects. Clearly we have $B_p(K) \subset Z_p(K)$ since $\partial^2 = 0$. The quotient group

$$H_p(K) = Z_p(K) / B_p(K) \tag{2.222}$$

is called the ***p*-homology group of the complex** K.

Generalization: What we have discussed so far in Euclidean space $\mathbf{R}^m$ and with integer chain coefficients can be generalized to arbitrary topological spaces or manifolds and real chain coefficients. By choosing the vertices

$$\begin{aligned} v_0 &= (0, 0, \dots, 0) \\ v_1 &= (1, 0, \dots, 0) \\ \dots & \quad \dots \\ v_p &= (0, 0, \dots, 1) \end{aligned} \tag{2.223}$$

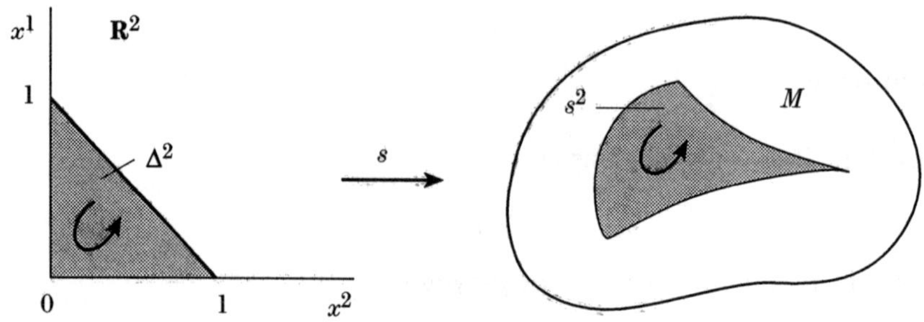

Fig. 2.31. Mapping a standard p-simplex into a manifold

we define the **standard p-simplex in $\mathbf{R}^p$**

$$\Delta^p = \{(x^1, \dots, x^p) \in \mathbf{R}^p | \sum_{i=1}^{p} x^i \leq 1, x^i \geq 0\}. \tag{2.224}$$

Next we map this (oriented) p-simplex into a manifold M, $s : \Delta^p \to M$ (see Figure 2.31), and we obtain the **singular p-simplex in M**

$$s^p = s(\Delta^p). \tag{2.225}$$

The map s is assumed to be smooth but need not have an inverse (for this reason the term singular is used).

A **p-chain in M** is a sum of these p-simplexes with $\mathbf{R}$-coefficients

$$c = \sum_i a_i s_i^p \qquad \text{with } a_i \in \mathbf{R}. \tag{2.226}$$

The **boundary of the p-chain in M** is

$$\partial c = \sum_i a_i \partial s_i^p, \tag{2.227}$$

where $\partial s^p = s(\partial \Delta^p)$ clearly represents the map of the boundary of the standard simplex into the manifold. So ∂s^p describes the geometrical boundary of s^p with an induced orientation. The p-chains in M generate the **chain group in M** denoted by $C_p(M)$. The boundary operator ∂ then defines a map $\partial : C_p(M) \to C_{p-1}(M)$ and is obviously **nilpotent**

$$\partial^2 = 0. \tag{2.228}$$

As before we define the **set of p-cycles**

$$Z_p(M, \mathbf{R}) = \{c \in C_p(M) | \partial c = 0\} = \ker \partial \tag{2.229}$$

and the **set of p-boundaries**

$$B_p(M, \mathbf{R}) = \{c \in C_p(M) | c = \partial b\} = \text{image } \partial \tag{2.230}$$

and by taking the quotient ($B_p(M, \mathbf{R}) \subset Z_p(M, \mathbf{R})$ since $\partial^2 = 0$) we obtain the **singular p-homology group of M**

$$H_p(M, \mathbf{R}) = Z_p(M, \mathbf{R}) / B_p(M, \mathbf{R}). \tag{2.231}$$

Remark:

$$c = \partial b \underset{\not\Leftarrow}{\Rightarrow} \partial c = 0. \tag{2.232}$$

In homology we are interested in those cycles which are *not* themselves boundaries.

Triangulation of space: Take a simplicial complex K. We regard each simplex as a subset of $\mathbf{R}^m$, then the union of all simplexes is again a subset of $\mathbf{R}^m$ and is called a **polyhedron** of K.

A topological space X which is homeomorphic to a polyhedron is said to be **triangulable**. The polyhedron is called a **triangulation** of X which is not at all unique.

Proposition:

- *All compact manifolds are triangulable!*

With the help of triangulations we can calculate the homology groups of manifolds. Let M be a compact manifold (in general a topological space) and K an arbitrary triangulation, then we have for the homology groups

$$H_p(M) \equiv H_p(K), \qquad p = 0, 1, 2, \ldots. \tag{2.233}$$

This is valid independently of the choice of the triangulation by virtue of the following proposition.

Proposition:

- *Homology groups are topological invariants!*

This means the isomorphy

$$H_p(K_1) \simeq H_p(K_2) \tag{2.234}$$

if K_1 and K_2 are two different triangulations of M (in general K_1 and K_2 may be triangulations of the homeomorphic manifolds M_1 and M_2, respectively).

Example: To triangulate the circle S^1, a 1-dimensional space, we have to find a set of suitably joined 1-simplexes which is homeomorphic to S^1. This is the set

$$K = \{\langle v_0\rangle, \langle v_1\rangle, \langle v_2\rangle, \langle v_0v_1\rangle, \langle v_1v_2\rangle, \langle v_2v_0\rangle\}. \tag{2.235}$$

Let us calculate the homology groups $H_p(K)$:

Since there is no 2-simplex in K the 1-simplexes are not boundaries and we have

$$\begin{aligned} B_1(K) &= 0 \\ H_1(K) &= Z_1(K)/B_1(K) = Z_1(K). \end{aligned} \tag{2.236}$$

Considering a 1-cycle

$$z = a\langle v_0v_1\rangle + b\langle v_1v_2\rangle + c\langle v_2v_0\rangle \in Z_1(K) \text{ with } a, b, c \in \mathbf{Z} \tag{2.237}$$

we demand

$$\begin{aligned} \partial_1 z &= a\partial_1\langle v_0v_1\rangle + b\partial_1\langle v_1v_2\rangle + c\partial_1\langle v_2v_0\rangle \\ &= (c-a)\langle v_0\rangle + (a-b)\langle v_1\rangle + (b-c)\langle v_2\rangle \\ &= 0, \end{aligned} \tag{2.238}$$

which is satisfied only for $a = b = c$. So we find

$$Z_1(K) = \{a(\langle v_0v_1\rangle + \langle v_1v_2\rangle + \langle v_2v_0\rangle)|a \in \mathbf{Z}\} \tag{2.239}$$

and

$$H_1(K) = Z_1(K) \simeq \mathbf{Z}. \tag{2.240}$$

Calculating next the 0-homology we clearly have

$$Z_0(K) = C_0(K) = \{a\langle v_0\rangle + b\langle v_1\rangle + c\langle v_2\rangle | a, b, c \in \mathbf{Z}\} \tag{2.241}$$

and

$$\begin{aligned} B_0(K) &= \{a\partial_1\langle v_0v_1\rangle + b\partial_1\langle v_1v_2\rangle + c\partial_1\langle v_2v_0\rangle | a, b, c \in \mathbf{Z}\} \qquad (2.242) \\ &= \{(c-a)\langle v_0\rangle + (a-b)\langle v_1\rangle + (b-c)\langle v_2\rangle | a, b, c \in \mathbf{Z}\}. \end{aligned}$$

Next we define a **surjective** (onto) **homomorphism** (this is a map preserving the group structure, e.g. $f : G_1 \to G_2$ with $f(x+y) = f(x)+f(y)$ $\forall\, x, y \in G_1$):

$$f : Z_0(K) \to \mathbf{Z} \tag{2.243}$$

by

$$f(a\langle v_0\rangle + b\langle v_1\rangle + c\langle v_2\rangle) = a + b + c \tag{2.244}$$

and we observe that

$$\ker f = f^{-1}(0) \equiv B_0(K) \tag{2.245}$$

since

$$f((c-a)\langle v_0\rangle + (a-b)\langle v_1\rangle + (b-c)\langle v_2\rangle) = (c-a) + (a-b) + (b-c) = 0.$$

Then we use the **theorem of homomorphism**:

$$Z_0(K)/\ker f \simeq \text{image } f, \tag{2.246}$$

we recall that

$$\text{image } f = \mathbf{Z} \tag{2.247}$$

and we find

$$H_0(K) = Z_0(K)/B_0(K) \simeq \mathbf{Z}. \tag{2.248}$$

The **homology groups of the circle** S^1 are, finally,

$$\begin{aligned} H_0(S^1) &= H_0(K) \simeq \mathbf{Z} \\ H_1(S^1) &= H_1(K) \simeq \mathbf{Z}. \end{aligned} \tag{2.249}$$

2.5.2 de Rham cohomology

In a certain sense, dual to the chains discussed before are the differential forms that the physicists are more familiar with. So we find a dual group to the homology group, this is the de Rham cohomology group. What is it more precisely?

Definitions: Let us consider the space of all differential forms Λ^* (2.101) and we define the **p-cocycles** (closed forms)

$$Z^p(M, \mathbf{R}) = \{\omega \in \Lambda^p | d\omega = 0\} = \ker d \tag{2.250}$$

and the **p-coboundaries** (exact forms)

$$B^p(M, \mathbf{R}) = \{\omega \in \Lambda^p | \omega = d\beta\} = \text{image } d. \tag{2.251}$$

Clearly we have $B^p(M, \mathbf{R}) \subset Z^p(M, \mathbf{R})$ since $d^2 = 0$, so the quotient of both groups (2.250) and (2.251) makes sense. The **p-de Rham cohomology group of M** is defined by

$$H^p(M, \mathbf{R}) = Z^p(M, \mathbf{R})/B^p(M, \mathbf{R}). \tag{2.252}$$

Let be $\omega \in Z^p(M,\mathbf{R})$ then $[\omega] \in H^p(M,\mathbf{R})$ represents an equivalence class. Two forms $\omega', \omega \in Z^p(M,\mathbf{R})$ are equivalent or **cohomologous** if they differ by an exact form:

$$\omega' \sim \omega \qquad \text{if } \omega' = \omega + d\beta.$$

So in cohomology we are searching all closed forms which are not exact.

The sequence

$$0 \xrightarrow{i} \Lambda^0(M) \xrightarrow{d_0} \Lambda^1(M) \xrightarrow{d_1} \ldots \xrightarrow{d_{p-1}} \Lambda^p(M) \xrightarrow{d_p} \ldots \xrightarrow{d_{m-1}} \Lambda^m(M) \xrightarrow{d_m} 0 \quad (2.253)$$

is called the **de Rham complex** and i is again the **inclusion map**.

Duality: Duality between chains and differential forms is defined via an inner product. Take a p-chain $c \in C_p(M)$ and a p-form $\omega \in \Lambda^p(M)$ where $1 \leq p \leq m = \dim M$, an **inner product of a p-chain and a p-form** is defined by the bilinear map

$$\langle\,,\,\rangle \;:\; C_p(M) \times \Lambda^p(M) \to \mathbf{R} \qquad (2.254)$$

$$c, \omega \mapsto \langle c, \omega\rangle =: \int_c \omega \in \mathbf{R}. \qquad (2.255)$$

Then Stokes' theorem (2.128) introduced in Section 2.4

$$\int_c d\omega = \int_{\partial c} \omega \qquad (2.256)$$

can be written as an inner product

$$\langle c, d\omega\rangle = \langle \partial c, \omega\rangle. \qquad (2.257)$$

In this sense the boundary operator ∂ and the exterior derivative d are adjoint to each other.

Period: An inner product of a cycle ($\partial c = 0$) with a cocycle ($d\omega = 0$) is called a **period** Π

$$\Pi(c,\omega) = \langle c, \omega\rangle = \int_c \omega \in \mathbf{R}, \qquad (2.258)$$

with

$$c \in Z_p(M,\mathbf{R}) \qquad \text{and} \qquad \omega \in Z^p(M,\mathbf{R}). \qquad (2.259)$$

The mapping Π on

$$\Pi : Z_p(M,\mathbf{R}) \times Z^p(M,\mathbf{R}) \to \mathbf{R}, \qquad (2.260)$$

however, may be degenerate since

$$\Pi(c, \omega) = 0, \qquad \forall \omega \in Z^p(M, \mathbf{R}) \tag{2.261}$$

does *not* imply $c = 0$. Identity (2.261) holds whenever $c = \partial b$

$$\Pi(\partial b, \omega) = \Pi(b, d\omega) = 0 \tag{2.262}$$

since $d\omega = 0$. But the map Π is nondegenerate when working on the (co)homologies $H_p(M, \mathbf{R})$, $H^p(M, \mathbf{R})$. This leads us to de Rham's theorem.

de Rham's theorem: The mapping

$$\Pi \; : \; H_p(M, \mathbf{R}) \times H^p(M, \mathbf{R}) \to \mathbf{R} \tag{2.263}$$

$$c, \omega \mapsto \Pi(c, \omega) = \langle c, \omega \rangle = \int_c \omega \in \mathbf{R}, \tag{2.264}$$

with

$$c \in H_p(M, \mathbf{R}) \qquad \text{and} \qquad \omega \in H^p(M, \mathbf{R}) \tag{2.265}$$

is bilinear and nondegenerate. The spaces $H_p(M, \mathbf{R})$ and $H^p(M, \mathbf{R})$ are vector spaces and finite dimensional and

$$\dim H_p(M, \mathbf{R}) = \dim H^p(M, \mathbf{R}). \tag{2.266}$$

Thus $H_p(M, \mathbf{R})$ is dual to $H^p(M, \mathbf{R})$ with respect to the inner product Π. Both spaces are isomorphic

$$H_p(M, \mathbf{R}) \simeq H^p(M, \mathbf{R}) \tag{2.267}$$

and can be identified. Choosing a set of independent p-cycles which form a basis for $H_p(M, \mathbf{R})$

$$\{c_i\} \qquad \text{with } i = 1, \ldots, \dim H_p(M, \mathbf{R}) \tag{2.268}$$

we find the dual basis—a set of independent p-cocycles

$$\{\omega_j\} \qquad \text{with } j = 1, \ldots, \dim H^p(M, \mathbf{R}) \tag{2.269}$$

via the inner product $\Pi(c_i, \omega_i)$, an invertible matrix. The dimension of the (co)homology space is the p**-th Betti number** of the manifold M

$$b_p(M) := \dim H_p(M, \mathbf{R}) = \dim H^p(M, \mathbf{R}) =: b^p(M). \tag{2.270}$$

Euler characteristic: Let us introduce here the **Euler characteristic** $\chi(M)$ of a manifold M

$$\chi(M) = (\text{number of vertices of } K) - (\text{number of edges of } K) + (\text{number of faces of } K), \tag{2.271}$$

where K is a polyhedron homeomorphic to M. $\chi(M)$ represents a pure topological quantity depending only on the kind of manifold.

On the other hand, the Euler characteristic can be expressed by the alternating sum of the Betti numbers

$$\chi(M) = \sum_{p=0}^{m} (-)^p b^p. \tag{2.272}$$

This is a remarkable relation. We can calculate the topological Euler characteristic by studying the analytical de Rham cohomology. We shall meet such connections between topology and analysis again when we discuss the index theorems (see Chapter 11).

Poincaré duality: There exists another type of duality—**Poincaré duality**—when considering the inner product of p-forms introduced in Section 2.4 (equation (2.159))

$$(\alpha_p, \beta_p) = \int_M \alpha_p * \beta_p \in \mathbf{R}. \tag{2.273}$$

The dual form $*\beta_p \in \Lambda^{m-p}$ is clearly a $(m-p)$-form. So it makes sense to define an **inner product of the cohomologies** H^p and H^{m-p} by the bilinear map (note that we consider only compact orientable manifolds without boundaries)

$$[\,,\,] \ : \ H^p(M, \mathbf{R}) \times H^{m-p}(M, \mathbf{R}) \to \mathbf{R} \tag{2.274}$$

$$\alpha, \beta \mapsto [\alpha, \beta] =: \int_M \alpha\beta \in \mathbf{R}, \tag{2.275}$$

where

$$\alpha \in H^p(M, \mathbf{R}) \qquad \text{and} \qquad \beta \in H^{m-p}(M, \mathbf{R}). \tag{2.276}$$

This determines the Poincaré duality between $H^p(M, \mathbf{R})$ and $H^{m-p}(M, \mathbf{R})$. Both spaces are isomorphic

$$H^p(M, \mathbf{R}) \simeq H^{m-p}(M, \mathbf{R}) \tag{2.277}$$

and the Betti numbers are related by

$$b^p = b^{m-p}. \tag{2.278}$$

Consequently the Euler characteristic of an odd-dimensional manifold vanishes, $\chi(M_{\text{odd}}) = 0$.

Künneth formula: If we consider a product manifold $M = M_1 \times M_2$ then the cohomology can be decomposed into a sum of products—the **Künneth formula**

$$H^k(M, \mathbf{R}) = \bigoplus_{p+q=k} [H^p(M_1, \mathbf{R}) \otimes H^q(M_1, \mathbf{R})] \tag{2.279}$$

($\otimes$ denotes the tensor product and $\oplus$ the direct sum). Hence the Betti numbers are related by

$$b^k(M) = \sum_{p+q=k} b^p(M_1) b^q(M_2), \tag{2.280}$$

implying for the Euler characteristics the formula

$$\chi(M) = \chi(M_1)\chi(M_2). \tag{2.281}$$

Harmonic forms and de Rham cohomology: There is an interesting connection between the de Rham cohomology and the harmonic forms. The starting point is a theorem which decomposes any p-form into a closed-, coclosed- and harmonic form.

Hodge decomposition theorem: Let M be a compact orientable manifold without boundary, then any p-form $\omega_p \in \Lambda^p$ can be uniquely decomposed into

$$\omega_p = d\alpha_{p-1} + \delta\beta_{p+1} + \gamma_p, \tag{2.282}$$

with $\alpha_{p-1} \in \Lambda^{p-1}$, $\beta_{p+1} \in \Lambda^{p+1}$ and $\Delta\gamma_p = 0$.

Remark: Let us choose $d\omega_p = 0$, ω_p closed or $\omega_p \in H^p(M, \mathbf{R})$, we have

$$0 = (d\omega_p, \beta_{p+1}) = (d\delta\beta_{p+1}, \beta_{p+1}) = (\delta\beta_{p+1}, \delta\beta_{p+1}), \tag{2.283}$$

which implies $\delta\beta_{p+1} = 0$ and we get

$$\omega_p = d\alpha_{p-1} + \gamma_p. \tag{2.284}$$

If we choose $\delta\omega_p = 0$, ω_p coclosed, we analogously obtain $d\alpha_{p-1} = 0$ and

$$\omega_p = \delta\beta_{p+1} + \gamma_p. \tag{2.285}$$

If the form is harmonic, $\Delta\omega_p = 0$, then clearly $\omega_p = \gamma_p$.

So it is the harmonic component γ_p which determines the cohomology class of the form ω_p. In fact, there exists an isomorphism between the **set of harmonic p-forms**

$$\mathrm{Harm}^p(M) = \{\gamma_p \in \Lambda^p | \Delta\gamma_p = 0\} \tag{2.286}$$

and the p-cohomology

$$\mathrm{Harm}^p(M) \simeq H^p(M, \mathbf{R}). \tag{2.287}$$

In particular for the space dimensions we have

$$\dim \mathrm{Harm}^p(M) = \dim H^p(M, \mathbf{R}) \tag{2.288}$$

or

$$\dim \ker \Delta_p = b^p \tag{2.289}$$

so that the Euler characteristic is given by

$$\chi(M) = \sum_{p=0}^{m} (-)^p \dim \ker \Delta_p. \tag{2.290}$$

So the topological quantity $\chi(M)$ is determined by solving an analytical problem, the eigenvalues of the Laplacian Δ_p. In index theory this connection corresponds to the index of the de Rham complex.

Examples:

i) *de Rham cohomology of the Euclidean space, $M = \mathbf{R}^n$.*
In $\mathbf{R}^n$ all closed forms are also exact except the solutions to $df = 0$, the real constants $f = \text{const.} \in \mathbf{R}$. Therefore we have

$$\begin{aligned} &H^p(\mathbf{R}^n, \mathbf{R}) = 0, && \dim H^p(\mathbf{R}^n, \mathbf{R}) = 0, && 1 \le p \le n \\ &H^0(\mathbf{R}^n, \mathbf{R}) = \mathbf{R}, && \dim H^0(\mathbf{R}^n, \mathbf{R}) = 1. \end{aligned} \tag{2.291}$$

The de Rham cohomology of the Euclidean space is trivial. This is the cohomology version of **Poincaré's lemma**.

ii) *de Rham cohomology of the sphere, $M = S^n$.*
The only nonvanishing cohomologies for S^n are H^0 and H^n. H^0 contains the real constant functions and H^n the constants times the volume element, an n-form. The result is

$$\begin{aligned} &H^0(S^n, \mathbf{R}) = \mathbf{R}, && \dim H^0(S^n, \mathbf{R}) = 1, \\ &H^p(S^n, \mathbf{R}) = 0, && \dim H^p(S^n, \mathbf{R}) = 0, && 1 \le p < n, \\ &H^n(S^n, \mathbf{R}) = \mathbf{R}, && \dim H^n(S^n, \mathbf{R}) = 1. \end{aligned} \tag{2.292}$$

iii) *de Rham cohomology of the torus,* $M = T^2 = S^1 \times S^1$.
With the knowledge of the sphere–cohomology we can calculate the torus–cohomology by using the Künneth formula (2.279)

$$\begin{aligned} H^0(T^2 = S^1 \times S^1, \mathbf{R}) &= \mathbf{R} \otimes \mathbf{R} = \mathbf{R} \\ H^1(T^2 = S^1 \times S^1, \mathbf{R}) &= [\mathbf{R} \otimes \mathbf{R}] \oplus [\mathbf{R} \otimes \mathbf{R}] \\ &= \mathbf{R} \oplus \mathbf{R} \\ H^2(T^2 = S^1 \times S^1, \mathbf{R}) &= \mathbf{R} \otimes \mathbf{R} = \mathbf{R} \end{aligned} \tag{2.293}$$

and

$$\begin{aligned} \dim H^0(T^2 = S^1 \times S^1, \mathbf{R}) &= 1 \\ \dim H^1(T^2 = S^1 \times S^1, \mathbf{R}) &= 2 \\ \dim H^2(T^2 = S^1 \times S^1, \mathbf{R}) &= 1. \end{aligned} \tag{2.294}$$

2.6 Flow, Lie derivative and Lie group

Vector fields on a manifold generate flows. The change of a vector field, generally of a tensor field, along such a flow is described by the Lie derivative. This important mathematical conception is widely used in physics and we shall discuss it now. For further literature we refer to [Choquet-Bruhat, DeWitt-Morette 1982], [Nakahara 1990], [Thirring 1992], [Warner 1983].

2.6.1 Differential map

In the following we deal with mappings from one manifold to another. How do the quantities defined on the manifold, the vectors, transform under such a mapping?

Consider a smooth map $f : M \to N$. It induces the **differential map** f_* **or tangent map**

$$f_* : T_p(M) \to T_{f(p)}(N) \tag{2.295}$$

in the tangent spaces of the manifolds (see Figure 2.32).

Let us find an explicit expression for f_*. We denote by $\mathcal{F}(N)$ the set of smooth functions on the manifold N and we choose some function $g \in \mathcal{F}(N)$; then $gf \in \mathcal{F}(M)$. Concentrating on the chart (V, ψ) in N the coordinate representation of g is determined by the map $g\psi^{-1} : \mathbf{R}^n \to \mathbf{R}$ (see Figure 2.33).

Definition: We choose a vector $X \in T_p(M)$ and we define the **differential map or tangent map** $f_*X \in T_{f(p)}(N)$ by

$$f_*X[g] = X[gf]. \tag{2.296}$$

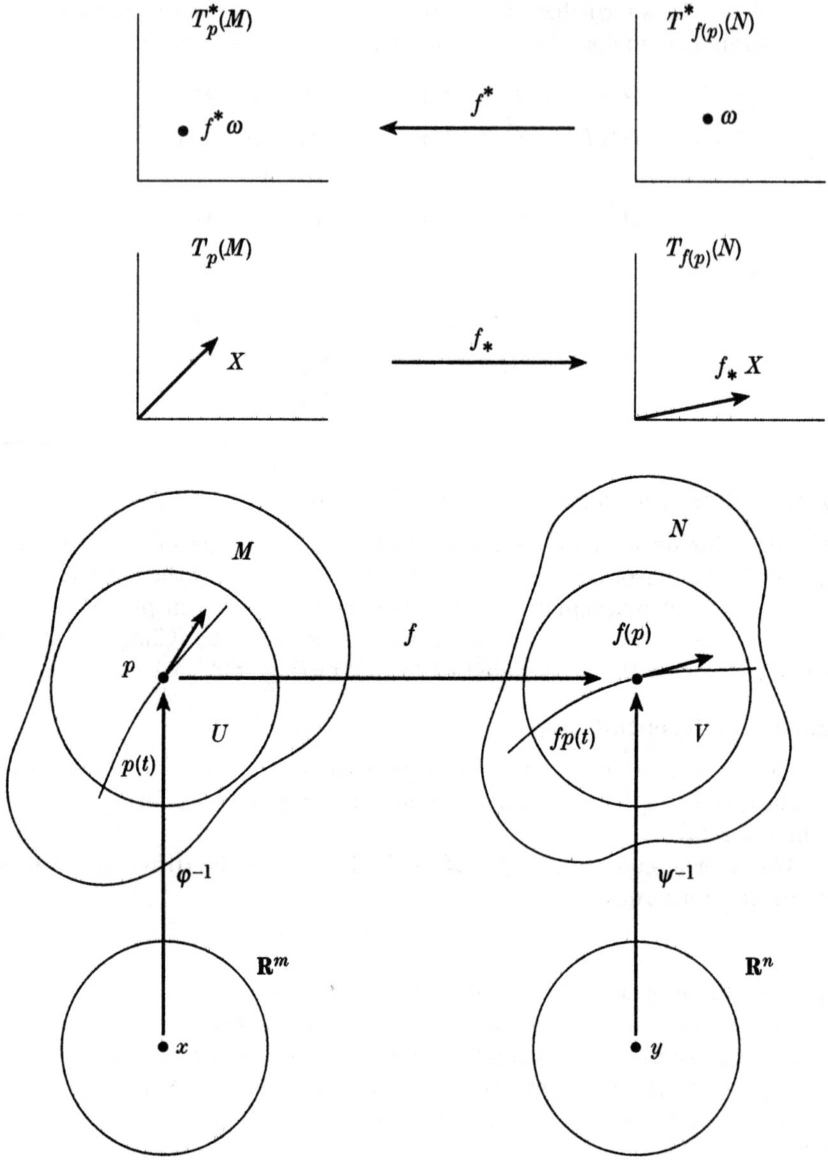

Fig. 2.32. The map f together with the induced differential map f_* and the pullback f^*

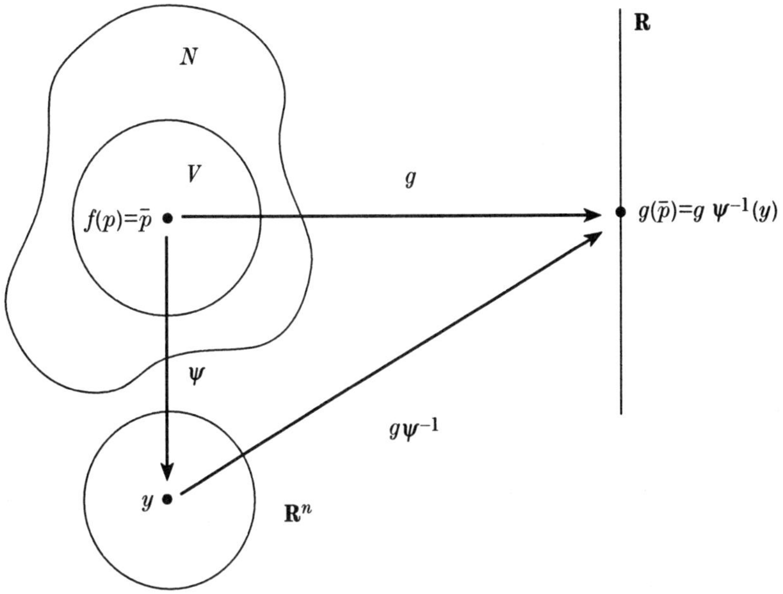

Fig. 2.33. The function $g : N \to \mathbf{R}$ and its coordinate representation $g\psi^{-1}$

In terms of the charts (U, φ) on M and (V, ψ) on N this means

$$f_* X[g\psi^{-1}](y) = X[gf\varphi^{-1}](x), \tag{2.297}$$

where $x = \varphi(p)$ and $y = \psi(f(p))$ are the coordinates (or homeomorphisms) of the charts. Furthermore we have

$$f_* X^\nu \frac{\partial}{\partial y^\nu}[g\psi^{-1}](y) = X^\mu \frac{\partial}{\partial x^\mu}[gf\varphi^{-1}](x) \tag{2.298}$$

and the choice $g = y^\nu$ provides the **components of the tangent map**

$$f_* X^\nu = X^\mu \frac{\partial y^\nu}{\partial x^\mu}. \tag{2.299}$$

Thus the two components X^μ and $f_* X^\nu$ are related by the Jacobian of the map f.

The **composite map** hf with $f : M \to N$ and $h : N \to P$ clearly gives

$$(hf)_* = h_* f_*. \tag{2.300}$$

Example: Let us take two manifolds M and N with coordinates (x^1, x^2) and (y^1, y^2, y^3) and the tangent vector

$$X = 3\,\frac{\partial}{\partial x^1} + 4\,\frac{\partial}{\partial x^2} \tag{2.301}$$

on M. We consider the map $f : M \to N$ with the coordinate representation

$$(y^1, y^2, y^3) = (x^1, x^2, (1 - r^2)^{1/2}) \qquad \text{and} \qquad r^2 = (x^1)^2 + (x^2)^2. \tag{2.302}$$

Then the differential map or tangent map is

$$\begin{aligned} f_* X &= f_* X^\nu \frac{\partial}{\partial y^\nu} = X^\mu \frac{\partial y^1}{\partial x^\mu}\frac{\partial}{\partial y^1} + X^\mu \frac{\partial y^2}{\partial x^\mu}\frac{\partial}{\partial y^2} + X^\mu \frac{\partial y^3}{\partial x^\mu}\frac{\partial}{\partial y^3} \\ &= 3\,\frac{\partial}{\partial y^1} + 4\,\frac{\partial}{\partial y^2} - \frac{3y^1 + 4y^2}{y^3}\frac{\partial}{\partial y^3} \end{aligned} \tag{2.303}$$

according to equation (2.299).

2.6.2 Pullback

Certainly we also may consider the dual space—the cotangent space. Then the smooth map $f : M \to N$ induces a **pullback map**

$$f^* : T^*_{f(p)}(N) \to T^*_p(M) \tag{2.304}$$

in the cotangent spaces of the manifolds (see Figure 2.32). Whereas f_* pushes forward the vector fields (in the direction of f), f^* pulls the differential forms backward (in opposite direction of f).

Definition: Choosing a vector $X = X^\nu \frac{\partial}{\partial x^\nu} \in T_p(M)$ and a 1-form $\omega = \omega_\mu dy^\mu \in T^*_{f(p)}(N)$ we define the **pullback** $f^*\omega = f^*\omega_\mu dx^\mu \in T^*_p(M)$ by

$$(f^*\omega, X) = (\omega, f_* X) \tag{2.305}$$

or in the notation of equations (2.90) and (2.116)

$$f^*\omega(X) = \omega(f_* X). \tag{2.306}$$

We obtain the components by evaluating the inner products

$$(f^*\omega, X) = f^*\omega_\mu(x) X^\mu(x) \tag{2.307}$$

and (recall equation (2.299))

$$(\omega, f_* X) = \omega_\nu(y) f_* X^\nu(y) = \omega_\nu(y) \frac{\partial y^\nu}{\partial x^\mu} X^\mu(x). \tag{2.308}$$

Comparing both results (2.307) and (2.308) implies for the **components of the pullback**

$$f^*\omega_\mu(x) = \omega_\nu(y(x)) \frac{\partial y^\nu}{\partial x^\mu}. \tag{2.309}$$

Again the two components ω_ν and $f^*\omega_\mu$ are related by the Jacobian of the map f. In local coordinates f is defined by $y^\nu(x^\mu)$ with $\mu = 1, \ldots, m = \dim M$ and $\nu = 1, \ldots, n = \dim N$.

It is straightforward to generalize the **pullback f^* for p-forms**

$$\omega = \frac{1}{p!}\omega_{\mu_1 \ldots \mu_p} dy^{\mu_1} \ldots dy^{\mu_p} \in \Lambda^p(N). \tag{2.310}$$

Definition: The map

$$f^* : \Lambda^p_{f(p)}(N) \to \Lambda^p_p(M) \tag{2.311}$$

is defined by (in the notation of equation (2.121))

$$f^*\omega(X_1, \ldots, X_p) = \omega(f_*X_1, \ldots, f_*X_p). \tag{2.312}$$

Explicitly we have

$$f^*\omega = \frac{1}{p!} f^*\omega_{\mu_1 \ldots \mu_p} dx^{\mu} \ldots dx^{\mu_p} \in \Lambda^p(M), \tag{2.313}$$

with

$$f^*\omega_{\mu_1 \ldots \mu_p}(x) = \omega_{\nu_1 \ldots \nu_p}(y(x)) \frac{\partial y^{\nu_1}}{\partial x^{\mu_1}} \cdots \frac{\partial y^{\nu_p}}{\partial x^{\mu_p}}. \tag{2.314}$$

Properties: The pullback has the following properties

$$\begin{aligned} &\text{i)} && d(f^*\omega) = f^* d\omega, \\ &\text{ii)} && (hf)^* = f^* h^* \text{ with } f : M \to N \text{ and } h : N \to P, \\ &\text{iii)} && f^*(\alpha_p \beta_q) = f^*\alpha_p f^*\beta_q \text{ with } \alpha_p \in \Lambda^p(N),\ \beta_q \in \Lambda^q(N). \end{aligned} \tag{2.315}$$

Attention: Consider a smooth map $f : M \to N$ and let X be a vector field on M. It may happen that two distinct points on M get mapped by f to the same point in N. Then f_*X is not a well-defined vector field on N (we would have two different vectors at that point in N). The tangent map f_* does not carry vector fields into vector fields. In contrast, if ω is a covector field (a 1-form) on N then $f^*\omega$ gives always a well-defined covector field on M.

Remark: Let us return to our discussion of the inner product of a p-chain $c = \sum_i a_i s_i^p \in C_p(M)$ with a p-form $\omega \in \Lambda^p(M)$ on a manifold M (recall equations (2.226) and (2.255) of the previous section)

$$\langle c, \omega \rangle = \int_c \omega = \sum_i a_i \int_{s_i^p} \omega. \tag{2.316}$$

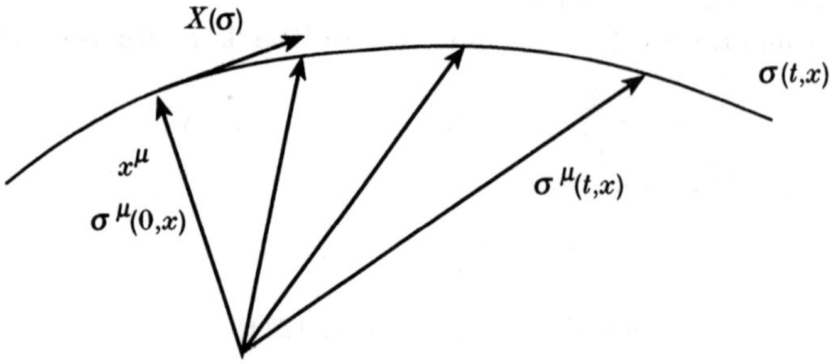

Fig. 2.34. The integral curve $\sigma(t,x)$ of a vector field X

Studying the integration further we consider the integral of a p-form ω over a p-simplex s^p in M. By virtue of the pullback it is

$$\int_{s^p} \omega = \int_{\Delta^p} s^*\omega \in \mathbf{R}, \tag{2.317}$$

where $s : \Delta^p \to M$ describes a smooth map such that $s^p = s(\Delta^p)$ (recall equations (2.224), (2.225)). The pullback $s^*\omega$ then represents a p-form on $\mathbf{R}^p$ and $\Delta^p \subset \mathbf{R}^p$ denotes the p-simplex in $\mathbf{R}^p$—the domain of integration. Thus the r.h.s. of equation (2.317) expresses the usual p-fold integration in $\mathbf{R}^p$ giving a real number.

Example: Choose a map $f : U \to I \times U$ from some area U to a cylinder manifold $I \times U$ given by the coordinates

$$(t,x) = (x^2, x) \qquad \text{with } (t,x) \in I \times U,\ x \in U. \tag{2.318}$$

Consider a 1-form on the cylinder

$$\omega = tdx + xdt \in \Lambda^1(I \times U) \tag{2.319}$$

then its pullback is the following 1-form on the area

$$f^*\omega = x^2 dx + x \cdot 2xdx = 3x^2 dx \in \Lambda^1(U). \tag{2.320}$$

2.6.3 Flow

A vector field on a manifold M describes, quite naturally, a flow in M. We consider the **integral curve** $\sigma(t,x)$ of a vector field $X = X^\mu\ \partial/\partial x^\mu \in T_x(M)$—a curve whose tangent vector is X—passing through x at a time

$t = 0$ (see Figure 2.34). In a given chart (U, φ) we have

$$\frac{d\sigma^\mu(t,x)}{dt} = X^\mu(\sigma(t,x)) \qquad \text{with } \sigma^\mu(0,x) = x^\mu. \tag{2.321}$$

$\sigma^\mu(t,x)$ denote the local coordinates which are actually the components of the homeomorphism $\varphi(x(t))$ and for simplicity we use x to indicate both the point in M and its coordinate in $\mathbf{R}^m$. Such an integral curve $\sigma(t,x)$, representing a map $\sigma : \mathbf{R} \times M \to M$, is called a **flow** generated by the vector field X.

Theorem: Let $X \in T(M)$ be a vector field on the manifold M. For any point $x \in M$ there exists an integral curve of X, a flow $\sigma : \mathbf{R} \times M \to M$, such that $\sigma(t,x)$ is a solution of the differential equation (2.321).

Lemma: A flow satisfies the group property

$$\sigma(t, \sigma(s,x)) = \sigma(t+s, x). \tag{2.322}$$

This follows from the differential equation (2.321).

One-parameter group: When keeping the parameter t fixed we denote the flow by

$$\sigma(t,x) \equiv \sigma_t(x). \tag{2.323}$$

So σ_t is a diffeomorphic map: $M \to M$ and it represents the commutative **one-parameter group**:

$$\begin{aligned} &\text{i)} \quad \sigma_t(\sigma_s(x)) = \sigma_{t+s}(x) \text{ or } \sigma_t \cdot \sigma_s = \sigma_{t+s}, \\ &\text{ii)} \quad \sigma_0 = \mathbf{1} \text{ unit element}, \\ &\text{iii)} \quad \sigma_t^{-1} = \sigma_{-t} \text{ inverse element}. \end{aligned} \tag{2.324}$$

Choosing the parameter t infinitesimal then the differential equation (2.321) determines the **infinitesimal flow**

$$\sigma_t^\mu(x) = x^\mu + tX^\mu(x) \tag{2.325}$$

and X is called the **infinitesimal generator of the flow group** σ_t.

Remark: For a given vector field X the flow σ can be expressed by the **exponentiation** of X

$$\sigma^\mu(t,x) = \exp(tX)\, x^\mu. \tag{2.326}$$

It fulfils the properties:

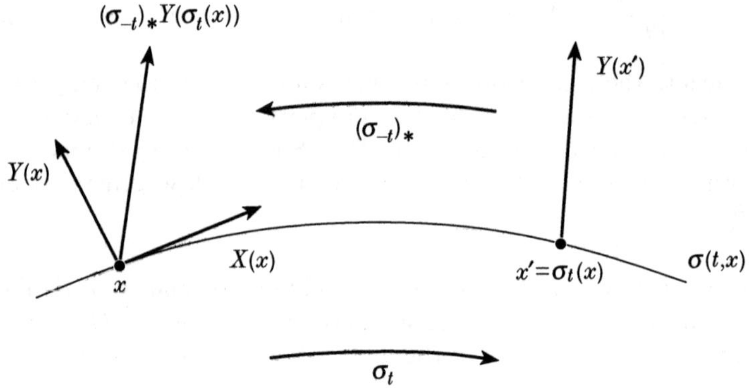

Fig. 2.35. Change of the vector Y along the flow σ of X. The vector $Y(x')$ must be moved back to the point x by the inverse differential map $(\sigma_{-t})_*$ to be compared with the vector $Y(x)$

$$\begin{aligned} &\text{i)} \qquad \sigma(0,x) = x, \\ &\text{ii)} \qquad \frac{d}{dt}\sigma(t,x) = X \exp(tX)\, x, \\ &\text{iii)} \qquad \sigma(t,\sigma(s,x)) = \sigma(t+s,x). \end{aligned} \tag{2.327}$$

We can verify the exponentiation as follows

$$\begin{aligned} \sigma^\mu(t,x) &= \left[1 + t\frac{d}{dt} + \frac{t^2}{2!}\left(\frac{d}{dt}\right)^2 + \ldots\right] \sigma^\mu(t,x)\bigg|_{t=0} \\ &= \exp\left(t\frac{d}{dt}\right) \sigma^\mu(t,x)\bigg|_{t=0} \\ &= \exp(tX)\, x^\mu, \end{aligned} \tag{2.328}$$

where for the last step we used the differential equation (2.321).

Example: On the plane $M = \mathbf{R}^2$ the vector field

$$X(x) = -x^2\frac{\partial}{\partial x^1} + x^1\frac{\partial}{\partial x^2}, \qquad x = (x^1, x^2) \tag{2.329}$$

generates the flow

$$\sigma(t,x) = (\sigma^1, \sigma^2) = (x^1 \cos t - x^2 \sin t, x^1 \sin t + x^2 \cos t) \tag{2.330}$$

which represents a circle through the point x:

$$(\sigma^1)^2 + (\sigma^2)^2 = (x^1)^2 + (x^2)^2 = r^2. \tag{2.331}$$

2.6.4 Lie derivative of a vector field

Let X and Y be two vector fields on a manifold M. We are interested now in the change of the vector field Y along the flow $\sigma(t,x)$ which is generated by the vector field X. So we have to compare the vector Y at the point x with Y at a nearby point $x' = \sigma_t(x)$. However, the vectors $Y(x) \in T_x(M)$ and $Y(x') \in T_{x'}(M)$ belong to different tangent spaces! Therefore we first have to map $Y(x')$ to $T_x(M)$. This can be done by the *inverse differential map*

$$(\sigma_t^{-1})_* = (\sigma_{-t})_* : \ T_{x'=\sigma_t(x)}(M) \to T_{\sigma_t^{-1}\sigma_t(x)}(M) = T_x(M) \tag{2.332}$$

(recall definition 2.295). Then we can take the difference of the two vectors $(\sigma_{-t})_* Y(\sigma_t(x)) \in T_x(M)$ and $Y(x) \in T_x(M)$ which are both lying in the tangent space $T_x(M)$ (see Figure 2.35).

Definition: The **Lie derivative of a vector field** Y **along the flow of** X we define by

$$\mathcal{L}_X\, Y = \lim_{t\to 0} \frac{1}{t}[(\sigma_{-t})_* Y(\sigma_t(x)) - Y(x)]. \tag{2.333}$$

Alternatively we also could consider the limits

$$\begin{aligned} \mathcal{L}_X\, Y &= \lim_{t\to 0} \frac{1}{t}[Y(x) - (\sigma_t)_* Y(\sigma_{-t}(x)] \\ &= \lim_{t\to 0} \frac{1}{t}[Y(\sigma_t(x)) - (\sigma_t)_* Y(x)]. \end{aligned} \tag{2.334}$$

Now we perform the explicit calculation. On a chart (U,φ) with coordinates x we have the two vector fields $X = X^\mu(x)\ \partial/\partial x^\mu$ and $Y = Y^\mu(x)\ \partial/\partial x^\mu$ and the flow $\sigma_t^\mu(x) = x^\mu + tX^\mu(x)$. The vector field Y at the point $x' = \sigma_t(x)$ is

$$\begin{aligned} Y(\sigma_t(x)) &= Y^\mu(\sigma_t(x)) \frac{\partial}{\partial x^\mu}\bigg|_{\sigma_t(x)} = Y^\mu(x^\lambda + tX^\lambda(x)) \frac{\partial}{\partial x^\mu}\bigg|_{\sigma_t(x)} \\ &= \left[Y^\mu(x) + tX^\lambda(x) \frac{\partial}{\partial x^\lambda} Y^\mu(x) \right] \frac{\partial}{\partial x^\mu}\bigg|_{\sigma_t(x)}. \end{aligned} \tag{2.335}$$

Remembering the transformation property (2.299) of a vector under a differential map we obtain

$$\begin{aligned}(\sigma_{-t})_* Y(\sigma_t(x)) &= (\sigma_{-t})_* Y^\nu(\sigma_t(x))\frac{\partial}{\partial x^\nu} \qquad (2.336)\\ &= Y^\mu(\sigma_t(x))\frac{\partial \sigma^\nu_{-t}}{\partial x^\mu}\frac{\partial}{\partial x^\nu}\\ &= Y^\mu(x)\frac{\partial}{\partial x^\mu} + t\left[X^\mu(x)\frac{\partial}{\partial x^\mu}Y^\nu(x) - Y^\mu(x)\frac{\partial}{\partial x^\mu}X^\nu(x)\right]\frac{\partial}{\partial x^\nu}.\end{aligned}$$

Then we find for the **Lie derivative** (2.333)

$$\mathcal{L}_X\, Y = \left[X^\mu\frac{\partial}{\partial x^\mu}Y^\nu - Y^\mu\frac{\partial}{\partial x^\mu}X^\nu\right]\frac{\partial}{\partial x^\nu}. \qquad (2.337)$$

Lie bracket: Next we introduce the **Lie bracket** $[X,Y]$ defined by

$$[X,Y]f = X[Y[f]] - Y[X[f]], \qquad (2.338)$$

with $f \in \mathcal{F}(M)$ some smooth function. Explicitly in components we get

$$\begin{aligned}[X,Y]f &= X^\mu\frac{\partial}{\partial x^\mu}Y^\nu\frac{\partial}{\partial x^\nu}f - Y^\mu\frac{\partial}{\partial x^\mu}X^\nu\frac{\partial}{\partial x^\nu}f\\ &= \left[X^\mu\frac{\partial}{\partial x^\mu}Y^\nu - Y^\mu\frac{\partial}{\partial x^\mu}X^\nu\right]\frac{\partial}{\partial x^\nu}f \qquad (2.339)\end{aligned}$$

or

$$[X,Y] = [X,Y]^\nu\frac{\partial}{\partial x^\nu}, \qquad (2.340)$$

with

$$[X,Y]^\nu = X^\mu\frac{\partial}{\partial x^\mu}Y^\nu - Y^\mu\frac{\partial}{\partial x^\mu}X^\nu, \qquad (2.341)$$

which again expresses a vector field (note: neither XY nor YX alone is a vector field).

So the **Lie derivative of the vector field Y along X**

$$\mathcal{L}_X\, Y = [X,Y] \qquad (2.342)$$

is equal to the Lie bracket of X and Y.

Properties of the Lie bracket: The following properties are easy to verify

i) $[X, aY_1 + bY_2] = a[X,Y_1] + b[X,Y_2]$
   $[aX_1 + bX_2, Y] = a[X_1,Y] + b[X_2,Y]$
   with a, b some constants,

ii) $[X,Y] = -[Y,X]$ antisymmetry,

iii) $[[X,Y],Z] + [[Y,Z],X] + [[Z,X],Y] = 0$ Jacobi identity,

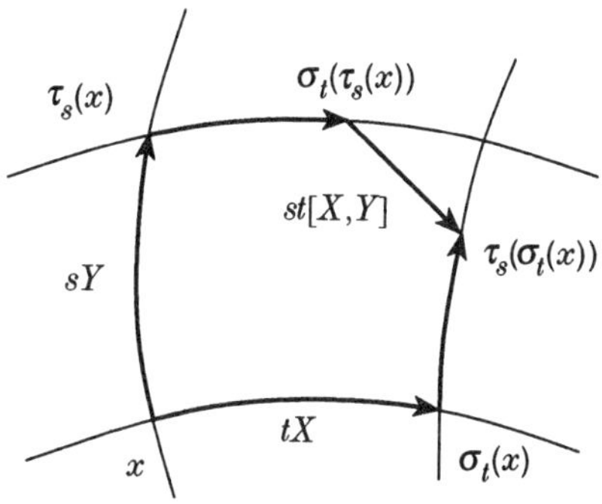

Fig. 2.36. Geometrically the Lie bracket $[X, Y]$ describes the noncommutativity of the two flows σ_t and τ_s

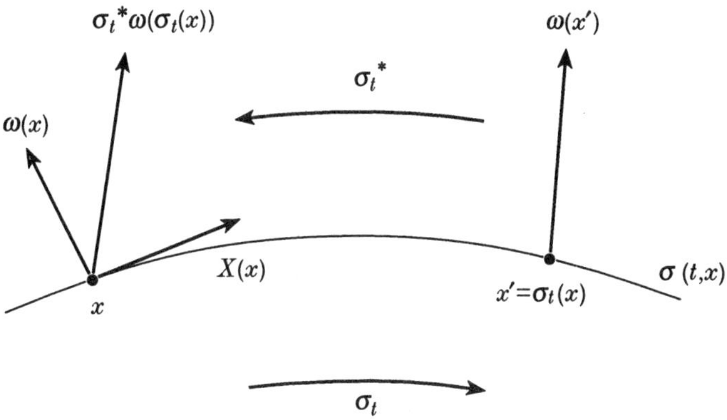

Fig. 2.37. Change of a p-form ω along the flow σ of X. $\omega(x')$ must be pulled back to the point x via σ_t^* to be compared with $\omega(x)$

iv) $\mathcal{L}_X\, fY = [X, fY] = f[X, Y] + X[f]Y$
$\mathcal{L}_{fX}\, Y = [fX, Y] = f[X, Y] - Y[f]X$
with f a smooth function, $f \in \mathcal{F}(M)$,

v) $f_*[X, Y] = [f_*X, f_*Y]$ with map $f : M \to N$. (2.343)

Geometry: The Lie bracket describes geometrically the noncommutativity of two flows. Let us consider the flow $\sigma_t(x)$ generated by the vector field X and the flow $\tau_s(x)$ generated by Y. Moving first along σ_t and then along τ_s and vice versa (see Figure 2.36) we find for the difference of the coordinates the Lie bracket

$$\tau_s^\mu(\sigma_t(x)) - \sigma_t^\mu(\tau_s(x)) = st[X, Y]^\mu. \tag{2.344}$$

So the parallelogram in Figure 2.36 does not close—the two flows σ_t and τ_s do not commute.

2.6.5 Lie derivative of a differential form

Now we consider a differential form $\omega(x) \in \Lambda_x^p(M)$ and study its change along a flow $\sigma(t, x)$. When comparing the differential forms at two nearby points x and $x' = \sigma_t(x)$ they belong to different spaces. Therefore we have to *pull back*

$$\sigma_t^* : \Lambda_{x'=\sigma_t(x)}^p(M) \to \Lambda_x^p(M) \tag{2.345}$$

the flowed differential form $\sigma_t^* \omega(\sigma_t(x))$ in order to be able to take the difference with $\omega(x)$ (see Figure 2.37).

Definition: We define the **Lie derivative of a p-form ω along the flow of a vector field X** by

$$\mathcal{L}_X\, \omega = \lim_{t\to 0} \frac{1}{t}[\sigma_t^* \omega(\sigma_t(x)) - \omega(x)]. \tag{2.346}$$

The explicit calculation gives

$$\begin{aligned}
&\sigma_t^* \omega(\sigma_t(x)) = \\
&= \frac{1}{p!}\omega_{\nu_1\ldots\nu_p}(\sigma_t(x)) \frac{\partial \sigma_t^{\nu_1}(x)}{\partial x^{\mu_1}} \cdots \frac{\partial \sigma_t^{\nu_p}(x)}{\partial x^{\mu_p}} dx^{\mu_1} \ldots dx^{\mu_p} \in \Lambda_x^p \\
&= \omega(x) + t\frac{1}{p!}\Big[X^\nu(x)\frac{\partial}{\partial x_\nu}\omega_{\mu_1\ldots\mu_p}(x) \\
&\quad + \omega_{\nu\mu_2\ldots\mu_p}(x)\frac{\partial}{\partial x^{\mu_1}}X^\nu(x) + \ldots + \omega_{\mu_1\ldots\mu_{p-1}\nu}(x)\frac{\partial}{\partial x^{\mu_p}}X^\nu(x)\Big] \\
&\quad \cdot dx^{\mu_1} \ldots dx^{\mu_p}
\end{aligned} \tag{2.347}$$

so that we find for the **Lie derivative of the p-form ω along the vector field X** the result

$$\begin{aligned}
\mathcal{L}_X\, \omega &= [X^\nu \partial_\nu \omega_{\mu_1\ldots\mu_p} + \omega_{\nu\mu_2\ldots\mu_p}\partial_{\mu_1}X^\nu + \ldots \\
&\quad \ldots + \omega_{\mu_1\ldots\mu_{p-1}\nu}\partial_{\mu_p}X^\nu]\frac{1}{p!}dx^{\mu_1}\ldots dx^{\mu_p} \\
&= [X^\nu \partial_\nu \omega_{\mu_1\ldots\mu_p} + p\, \omega_{\nu\mu_2\ldots\mu_p}\partial_{\mu_1}X^\nu] \\
&\quad \cdot \frac{1}{p!}dx^{\mu_1}\ldots dx^{\mu_p},
\end{aligned} \tag{2.348}$$

with $\partial_\nu = \partial/\partial x^\nu$.

For a 1-form $\omega \in \Lambda^1$ we have

$$\mathcal{L}_X\, \omega = [X^\nu \partial_\nu \omega_\mu + \omega_\nu \partial_\mu X^\nu] dx^\mu. \tag{2.349}$$

For a 0-form $\omega \equiv f \in \Lambda^0$, a function, the **Lie derivative of a function along the vector field X** is

$$\begin{aligned} \mathcal{L}_X\, f &= \lim_{t\to 0} \frac{1}{t}[f(\sigma_t(x)) - f(x)] \\ &= \lim_{t\to 0} \frac{1}{t}[f(x^\mu + tX^\mu(x)) - f(x)] \\ &= X^\mu \frac{\partial}{\partial x^\mu} f = X[f] \end{aligned} \tag{2.350}$$

and corresponds to the usual directional derivative along X.

Tensors: Considering next the tensor product of a covariant and contravariant vector or generally the tensor product of tensor fields t_1, t_2 of arbitrary type, then the Lie derivative satisfies the **Leibniz rule**

$$\mathcal{L}_X(t_1 \otimes t_2) = (\mathcal{L}_X\, t_1) \otimes t_2 + t_1 \otimes \mathcal{L}_X\, t_2. \tag{2.351}$$

From this rule and from the previous results for vector fields and p-forms (2.337), (2.348) we deduce the following formula for the **Lie derivative of a tensor field $T^{\alpha\cdots}_{\beta\ldots}(x)$ along the flow of a vector field X**

$$\begin{aligned} \mathcal{L}_X\, T^{\alpha\ldots}_{\beta\ldots} &= X^\nu \partial_\nu T^{\alpha\ldots}_{\beta\ldots}(x) \\ &\quad - \partial_\nu X^\alpha T^{\nu\ldots}_{\beta\ldots}(x) - \text{all upper indices} \\ &\quad + T^{\alpha\ldots}_{\nu\ldots}(x) \partial_\beta X^\nu + \text{all lower indices}. \end{aligned} \tag{2.352}$$

Introducing the compact **notation**

$$\begin{array}{rcl} \text{matrix} & & \text{components} \\ T & \leftrightarrow & T^{\alpha\ldots}_{\beta\ldots}(x) \\ \partial X & \leftrightarrow & \partial_\nu X^\alpha \\ X\cdot\partial & = & X^\nu \partial_\nu \\ {[\partial X, T]} & = & \partial_\nu X^\alpha T^{\nu\ldots}_{\beta\ldots}(x) + \text{all upper indices} \\ & & - T^{\alpha\ldots}_{\nu\ldots}(x) \partial_\beta X^\nu - \text{all lower indices} \end{array} \tag{2.353}$$

formula (2.352) is re-expressed by

$$\mathcal{L}_X\, T = X \cdot \partial\, T - [\partial X, T]. \tag{2.354}$$

Tensor-valued p-forms: If, on the other hand, the tensor field also represents a differential form

$$\mathcal{T} \to \mathcal{T}^{\alpha\dots}_{\beta\dots} = \frac{1}{p!}\mathcal{T}^{\alpha\dots}_{\beta\dots,\mu_1\dots\mu_p}(x)dx^{\mu_1}\dots dx^{\mu_p} \in \Lambda^p, \tag{2.355}$$

which we need, for instance, in gravitation (see Chapter 12) then we obtain for the **Lie derivative of a tensor-valued p-form $\mathcal{T}^{\alpha\dots}_{\beta\dots}$ along X** the formula

$$\begin{aligned}\mathcal{L}_X\, \mathcal{T}^{\alpha\dots}_{\beta\dots} = \; & [X^\nu\partial_\nu \mathcal{T}^{\alpha\dots}_{\beta\dots,\mu_1\dots\mu_p}(x) \\ & - \partial_\nu X^\alpha \mathcal{T}^{\nu\dots}_{\beta\dots,\mu_1\dots\mu_p}(x) - \text{all upper indices} \\ & + \mathcal{T}^{\alpha\dots}_{\nu\dots,\mu_1\dots\mu_p}(x)\partial_\beta X^\nu + \text{all lower indices} \\ & + \mathcal{T}^{\alpha\dots}_{\beta\dots,\nu\dots\mu_p}(x)\partial_{\mu_1} X^\nu + \text{all form indices}] \\ & \cdot\frac{1}{p!}dx^{\mu_1}\dots dx^{\mu_p}\end{aligned} \tag{2.356}$$

and rewriting the sum of the form index contractions we have

$$\begin{aligned}\mathcal{L}_X\, \mathcal{T}^{\alpha\dots}_{\beta\dots} = \; & [X^\nu\partial_\nu \mathcal{T}^{\alpha\dots}_{\beta\dots,\mu_1\dots\mu_p}(x) + p\, \mathcal{T}^{\alpha\dots}_{\beta\dots,\nu\mu_2\dots\mu_p}(x)\partial_{\mu_1} X^\nu \\ & - \partial_\nu X^\alpha \mathcal{T}^{\nu\dots}_{\beta\dots,\mu_1\dots\mu_p}(x) - \text{all upper indices} \\ & + \mathcal{T}^{\alpha\dots}_{\nu\dots,\mu_1\dots\mu_p}(x)\partial_\beta X^\nu + \text{all lower indices}] \\ & \cdot\frac{1}{p!}dx^{\mu_1}\dots dx^{\mu_p}.\end{aligned} \tag{2.357}$$

When studying the first two terms we find that they correspond to a combination of interior product—exterior derivative for a p-form. We shall prove the result below

$$\begin{aligned}(i_X d + d i_X)\mathcal{T} = \; & [X^\nu\partial_\nu \mathcal{T}^{\alpha\dots}_{\beta\dots,\mu_1\dots\mu_p}(x) + p\, \mathcal{T}^{\alpha\dots}_{\beta\dots,\nu\mu_2\dots\mu_p}(x)\partial_{\mu_1} X^\nu] \\ & \cdot\frac{1}{p!}dx^{\mu_1}\dots dx^{\mu_p}\end{aligned} \tag{2.358}$$

which is valid for a p-form $\mathcal{T} \in \Lambda^p$, $\mathcal{T}$ being tensor-valued or not, so that expression (2.357) can be reformulated in a compact way

$$\mathcal{L}_X\, \mathcal{T} = (i_X d + d i_X)\mathcal{T} - [\partial X, \mathcal{T}]. \tag{2.359}$$

For a scalar valued p-form $\omega \in \Lambda^p$, the case we usually consider, the commutator clearly vanishes and we get the well-known result

$$\mathcal{L}_X\, \omega = (i_X d + d i_X)\omega. \tag{2.360}$$

Formula: The Lie derivative satisfies the formula:

$$\text{a) } \mathcal{L}_X = i_X d + d i_X \tag{2.361}$$

$$\text{b) } \mathcal{L}_X = i_X d + d i_X - [\partial X, \;] \tag{2.362}$$

a) when applied to a scalar valued p-form,
b) when applied to a tensor valued p-form.

Proof. Now we prove equation (2.358).

$$\begin{aligned} d i_X\,\omega &= [(\partial_{\mu_1} X^\nu)\omega_{\nu\mu_2\ldots\mu_p} + X^\nu \partial_{\mu_1}\omega_{\nu\mu_2\ldots\mu_p}]\frac{p}{p!}dx^{\mu_1}\ldots dx^{\mu_p} \\ i_X\,d\omega &= [X^\nu\partial_\nu\omega_{\mu_1\ldots\mu_p} - pX^\nu\partial_{\mu_1}\omega_{\nu\mu_2\ldots\mu_p}]\frac{1}{p!}dx^{\mu_1}\ldots dx^{\mu_p}. \end{aligned} \tag{2.363}$$

Hence we get

$$(i_X d + d i_X)\omega = [X^\nu\partial_\nu\omega_{\mu_1\ldots\mu_p} + p\,\omega_{\nu\mu_2\ldots\mu_p}\partial_{\mu_1}X^\nu]\frac{1}{p!}dx^{\mu_1}\ldots dx^{\mu_p}, \tag{2.364}$$

which is also true for a tensor-valued p-form since i_X and d affect only the form indices; Q.E.D.

Properties: The Lie derivative satisfies the following properties

i) $[\mathcal{L}_X, i_X] = 0$
ii) $[\mathcal{L}_X, d] = 0$
iii) $[\mathcal{L}_X, i_Y] = i_{[X,Y]}$
iv) $[\mathcal{L}_X, \mathcal{L}_Y] = \mathcal{L}_{[X,Y]}$
v) $d\omega(X,Y) = \mathcal{L}_X\,\omega(Y) - \mathcal{L}_Y\,\omega(X) - \omega([X,Y])$
for $\omega \in \Lambda^1$ being a 1-form,
vi)

$$\begin{aligned} d\omega(X_1,\ldots,X_{p+1}) &= \sum_{i=1}^{p+1}(-)^{i+1}\mathcal{L}_{X_i}\omega(X_1,\ldots,\widehat{X}_i,\ldots,X_{p+1}) \\ &+ \sum_{i<j}(-)^{i+j}\omega([X_i,X_j],X_1,\ldots,\widehat{X}_i,\ldots,\widehat{X}_j,\ldots,X_{p+1}) \end{aligned} \tag{2.365}$$

for $\omega \in \Lambda^p$ being a p-form, and the symbol $\widehat{\ }$ means that the corresponding vector field has been omitted.

Proof. Properties i) and ii) are obvious in the light of formula (2.361). Property iii): It is sufficient to check on 0-forms (which is trivial) and on

1-forms; let $\omega = \omega_\mu dx^\mu \in \Lambda^1$ be a 1-form then we have

$$\begin{aligned} [\mathcal{L}_X, i_Y]\omega &= X^\mu \partial_\mu (Y^\nu \omega_\nu) - X^\nu (\partial_\nu \omega_\mu) Y^\mu - \omega_\nu (\partial_\mu X^\nu) Y^\mu \\ &= \omega_\nu (X^\mu \partial_\mu Y^\nu - Y^\mu \partial_\mu X^\nu) \\ &= \omega_\nu [X, Y]^\nu = i_{[X,Y]}\omega. \qquad \text{Q.E.D.} \end{aligned} \tag{2.366}$$

Property iv) follows from property v) which we rewrite

$$i_Y i_X d\omega = \mathcal{L}_X i_Y \omega - \mathcal{L}_Y i_X \omega - i_{[X,Y]}\omega. \tag{2.367}$$

We choose $\omega = df$ and use formulae (2.361) together with $i_X\, f = 0$, then we have

$$\begin{aligned} \mathcal{L}_X i_Y df - \mathcal{L}_Y i_X df - i_{[X,Y]} df = i_Y i_X d^2 f &= 0 \\ (\mathcal{L}_X \mathcal{L}_Y - \mathcal{L}_Y \mathcal{L}_X - \mathcal{L}_{[X,Y]}) f &= 0. \end{aligned} \tag{2.368}$$

Since the Lie derivative commutes with d equation (2.368) is valid for 1-forms and so for all forms. Q.E.D.

Property v):

$$\begin{aligned} &\mathcal{L}_X\, \omega(Y) - \mathcal{L}_Y\, \omega(X) - \omega([X, Y]) \\ &= X^\nu \partial_\nu (Y^\mu \omega_\mu) - Y^\nu \partial_\nu (X^\mu \omega_\mu) - (X^\nu \partial_\nu Y^\mu - Y^\nu \partial_\nu X^\mu)\omega_\mu \\ &= \partial_\nu \omega_\mu (X^\nu Y^\mu - Y^\nu X^\mu) = d\omega(X, Y). \qquad \text{Q.E.D.} \end{aligned} \tag{2.369}$$

Properties v) and vi) are equivalent to equations (2.108) and (2.113) since $\mathcal{L}_X = X$ when applied to a 0-form, a scalar function.

2.6.6 Lie group and Lie algebra

Lie groups and algebras play an important role not only in mathematics—we shall need them in the geometry of fibre bundles—but also in physics. In fact, we mainly work with Lie groups in this book. Let us discuss here some basic features.

Lie group: A **Lie group** G is a differentiable manifold with a group structure such that the following group operations are differentiable (smooth):

i) $G \times G \to G$ given by $(g_1, g_2) \mapsto g_1 g_2 \in G$,
ii) $G \to G$ given by $g \mapsto g^{-1}$.

The dimension of a Lie group G is determined by the dimension of G as a manifold. The **unit element** we denote by e.

An important property of Lie groups is stated in the following proposition.

Proposition:

- *A closed subgroup H of a Lie group G is again a Lie subgroup!*

Examples: $U(N)$, $SU(N)$, $GL(m, \mathbf{R})$, $SO(m)$, ...

Definitions: We need the following **maps** (diffeomorphisms):

$$\begin{aligned} R_g : G &\rightarrow G \\ R_g h &= hg \qquad \textbf{right-translation} \end{aligned} \tag{2.370}$$

$$\begin{aligned} L_g : G &\rightarrow G \\ L_g h &= gh \qquad \textbf{left-translation} \end{aligned} \tag{2.371}$$

with $g, h \in G$.

These maps induce **differential maps in tangent space**

$$R_{g*} : T_h(G) \rightarrow T_{hg}(G) \tag{2.372}$$

$$L_{g*} : T_h(G) \rightarrow T_{gh}(G). \tag{2.373}$$

Let X be a vector field on a Lie group G, then X remains **invariant** if

$$R_{g*}X|_h = X|_{hg} \qquad \textbf{right-invariant vector field} \tag{2.374}$$

$$L_{g*}X|_h = X|_{gh} \qquad \textbf{left-invariant vector field}. \tag{2.375}$$

A vector $V \in T_e(G)$ generates a unique **left-invariant vector field on** G by

$$X_V|_g = L_{g*}V. \tag{2.376}$$

Indeed we have

$$\begin{aligned} X_V|_{gh} &= L_{gh*}V = (L_g L_h)_* V \\ &= L_{g*}L_{h*}V = L_{g*}X_V|_h \,. \end{aligned} \tag{2.377}$$

We denote the set of all left-invariant vector fields on G by $\mathcal{G}$. The map $T_e(G) \rightarrow \mathcal{G}$ given by $V \mapsto X_V$ is an isomorphism, $T_e(G) \simeq \mathcal{G}$, hence $\dim \mathcal{G} = \dim T_e(G) = \dim G$.

We also can define a **Lie bracket** $[\ ,\ ]$ **on** G. Let $X, Y \in \mathcal{G}$ be two left-invariant vector fields on G, then we find

$$L_{g*}[X, Y]|_h = [L_{g*}X|_h \,, L_{g*}Y|_h] = [X, Y]|_{gh} \,. \tag{2.378}$$

So the Lie bracket $[X, Y]$ is again a left-invariant vector field $[X, Y] \in \mathcal{G}$; we say $\mathcal{G}$ *is closed under the Lie bracket.*

Lie algebra: The set of all left-invariant vector fields $\mathcal{G}$ with a Lie bracket $[\ ,\ ] : \mathcal{G} \times \mathcal{G} \to \mathcal{G}$ is called the **Lie algebra** of the Lie group G. From now on we use the notation Lie $G \equiv \mathcal{G}$.

One-parameter subgroup: Let us next consider a map $\phi_V : T_e(G) \to G$ given by

$$\phi_V(t) = \exp(tV), \tag{2.379}$$

with $V \in T_e(G)$ and $t \in \mathbf{R}$. $\phi_V(t)$ represents a commutative **one-parameter subgroup** of G generated by $X_V = L_{g*}V$. This means $\phi_V(t)$ satisfies the properties

$$\begin{aligned} &\text{i)} \qquad \phi(t)\phi(s) = \phi(t+s), \\ &\text{ii)} \qquad \phi(0) = e, \qquad \phi^{-1}(t) = \phi(-t). \end{aligned} \tag{2.380}$$

At $t = 1$ it is conventionally named an **exponential map**

$$\phi_V(1) = \exp V. \tag{2.381}$$

For a matrix group, say $G = GL(m, \mathbf{R})$, we have

$$\phi_V(t) = \exp(tV) = \mathbf{1} + tV + \ldots + \frac{t^n}{n!}V^n + \ldots. \tag{2.382}$$

Flow: The curve

$$\sigma_t(g) = g\exp(tV) \tag{2.383}$$

represents a **flow** through $g \in G$. Thus we have

$$\left.\frac{d\sigma_t(g)}{dt}\right|_{t=0} = \left.\frac{d}{dt}\right|_{t=0} g\exp(tV) = L_{g*}V = X_V|_g\,. \tag{2.384}$$

In the case of a matrix group G we can use expansion (2.382) and we get

$$X_V|_g = L_{g*}V = gV. \tag{2.385}$$

The flow (2.383) can be also interpreted as a **right-translation of the group** G

$$\sigma_t = R_{\exp(tV)} \tag{2.386}$$

which gives $\sigma_t(g) = g\exp(tV)$.

Basis: Let us choose a **basis for the Lie algebra** Lie G

$$\{X_a\} \qquad a = 1, \ldots, k; \qquad k = \dim \text{Lie } G = \dim G,$$

then the commutator of these vector fields satisfies the relation

$$[X_a, X_b] = f_{ab}{}^c X_c, \tag{2.387}$$

where $f_{ab}{}^c$ express the totally antisymmetric **structure constants** of the Lie group/algebra. We say that the commutator closes. The structure constants obey the familiar Jacobi identity (see e.g. equation (3.289) in Section 3.5.1).

We also can introduce a **dual basis** $\{\Theta^a\}$ of the dual Lie algebra $\mathcal{G}^* \equiv {}^*\text{Lie } G$ via the inner product

$$(\Theta^a, X_b) \equiv \Theta^a \cdot X_b = \delta^a{}_b. \tag{2.388}$$

The set $\{\Theta^a\}$ is the *basis for the left-invariant 1-forms on G.*

By defining next **wedge products**

$$\begin{aligned}\Theta^a\Theta^b &\equiv \Theta^a \wedge \Theta^b = \Theta^a \otimes \Theta^b - \Theta^b \otimes \Theta^a \\ \Theta^a\Theta^b\Theta^c &\equiv \Theta^a \wedge \Theta^b \wedge \Theta^c = \Theta^a \otimes \Theta^b \otimes \Theta^c + - \dots\end{aligned} \tag{2.389}$$

(etc.) one can construct ***p*-forms on *G*.**

Analogously to vectors a p-form $\omega \in \Lambda^p(G)$ on the Lie group G remains **invariant** if

$$L_g^*\omega\big|_h = \omega|_{g^{-1}h} \qquad \textbf{left-invariant } p\textbf{-form} \tag{2.390}$$

$$R_g^*\omega\big|_h = \omega|_{hg^{-1}} \qquad \textbf{right-invariant } p\textbf{-form.} \tag{2.391}$$

The dual basis reveals a very peculiar property. It satisfies the following theorem.

Theorem: Maurer–Cartan structure equation on the Lie group

$$d\Theta^a = -\frac{1}{2} f^a{}_{bc}\Theta^b\Theta^c. \tag{2.392}$$

Proof. Using relation (2.108)

$$d\omega(X, Y) = X\omega(Y) - Y\omega(X) - \omega([X, Y]) \tag{2.393}$$

we calculate the l.h.s.

$$\begin{aligned}d\Theta^a(X_b, X_c) &= X_b\ \Theta^a \cdot X_c - X_c\ \Theta^a \cdot X_b - \Theta^a \cdot [X_b, X_c] \\ &= X_b\delta^a{}_c - X_c\delta^a{}_b - f_{bc}{}^d\ \Theta^a \cdot X_d \\ &= -f^a{}_{bc},\end{aligned} \tag{2.394}$$

using (recall equation (2.111))

$$\Theta^{\bar{b}}\Theta^{\bar{c}}(X_b, X_c) = \delta^{\bar{b}}{}_b\delta^{\bar{c}}{}_c - \delta^{\bar{b}}{}_c\delta^{\bar{c}}{}_b \tag{2.395}$$

we calculate the r.h.s.

$$-\frac{1}{2}f^a{}_{\bar{b}\bar{c}}\,\Theta^{\bar{b}}\Theta^{\bar{c}}(X_b,X_c) = -\frac{1}{2}f^a{}_{\bar{b}\bar{c}}\,(\delta^{\bar{b}}{}_b\delta^{\bar{c}}{}_c - \delta^{\bar{b}}{}_c\delta^{\bar{c}}{}_b) = -f^a{}_{bc}. \qquad \text{Q.E.D.} \tag{2.396}$$

Definition: Next we define the **Maurer–Cartan form on the Lie group** G

$$\Theta = V_a \otimes \Theta^a, \tag{2.397}$$

where

$$\begin{aligned} &\{V_a\} && \text{is the basis of } T_e(G) \\ &\{\Theta^a\} && \text{is the dual basis of } T_g^*(G). \end{aligned} \tag{2.398}$$

So the Maurer–Cartan form Θ is a Lie algebra valued 1-form where the tangent vector $X \in T_g(G)$ is moved back to the unit

$$\Theta : X \mapsto L_{g^{-1}*}X = L_{g*}^{-1}X \in T_e(G). \tag{2.399}$$

Then we clearly have for the derivative

$$d\Theta = V_a \otimes d\Theta^a \tag{2.400}$$

and for the commutator

$$[\Theta, \Theta] = [V_a, V_b] \otimes \Theta^a\Theta^b = 2\Theta\Theta. \tag{2.401}$$

Multiplying equation (2.392) by $V_a\otimes$ and using

$$[V_b, V_c] = f^a{}_{bc}V_a \tag{2.402}$$

we obtain the **Maurer–Cartan structure equation in terms of** Θ

$$d\Theta = -\frac{1}{2}[\Theta, \Theta] = -\Theta^2. \tag{2.403}$$

Remark: The Maurer–Cartan form on the Lie group has a very important correspondence in physics. It will be identified with the Faddeev–Popov ghost in a non-Abelian gauge theory. We recover the Maurer–Cartan structure equation in the gauge transformation procedure of Becchi, Rouet and Stora. This we will discuss extensively in Sections 3.6, 8.1 and 8.3.2.

Lie group action on manifold: The **left-action** of a Lie group G on a manifold M is a differentiable map $\sigma : G \times M \to M$ such that

$$
\begin{aligned}
&\text{i)} \quad \sigma(e,p) = p \qquad \forall p \in M,\ e \in G, \\
&\text{ii)} \quad \sigma(g_1, \sigma(g_2, p)) = \sigma(g_1 g_2, p), \qquad g_1 g_2 \in G.
\end{aligned} \tag{2.404}
$$

The usual **notation** will be

$$\sigma(g,p) \equiv gp, \tag{2.405}$$

with

$$
\begin{aligned}
&\text{i)} \quad ep = p, \\
&\text{ii)} \quad g_1(g_2 p) = (g_1 g_2)p.
\end{aligned} \tag{2.406}
$$

Example: Take for example the group $G = GL(m, \mathbf{R})$ and the manifold $M = \mathbf{R}^m$. Let be $A \in GL(m, \mathbf{R})$ and $x \in \mathbf{R}^m$, then the action of $GL(m, \mathbf{R})$ on $\mathbf{R}^m$

$$\sigma(A, x) = Ax \tag{2.407}$$

corresponds to the standard matrix action on a vector.

Definition: Analogously we define the **right-action** of G on M by

$$\sigma(g,p) \equiv pg \tag{2.408}$$

with

$$
\begin{aligned}
&\text{i)} \quad pe = p, \\
&\text{ii)} \quad (pg_1)g_2 = p(g_1 g_2).
\end{aligned} \tag{2.409}
$$

Orbit: The group action shifts a given point $p \in M$ of the manifold to another. The **orbit** through p under the action σ is defined by

$$G_p = \{\sigma(g,p) | g \in G,\ \text{fixed } p \in M\}. \tag{2.410}$$

Induced vector field: Let a Lie group G act on a manifold M from the left: $(g, x) \mapsto gx$. Then a left-invariant vector field X_V generated by $V \in T_e(G)$ induces a vector field in M. The **induced vector field** is

$$
\begin{aligned}
X_V(x) &= \left.\frac{d}{dt}\right|_{t=0} \exp(tV)x \\
&= \left.\frac{d}{dt}\right|_{t=0} \sigma(t,x)
\end{aligned} \tag{2.411}
$$

together with the **flow in** M

$$\sigma(t, x) = \exp(tV)x. \tag{2.412}$$

Example: Discussing again our example from Section 2.6.3 we choose the Lie group $G = SO(2)$ acting on the manifold $M = \mathbf{R}^2$. For the element V of the Lie algebra

$$V = \begin{pmatrix} 0 & -1 \\ 1 & 0 \end{pmatrix} \tag{2.413}$$

we find immediately from the series (2.382)

$$\exp(tV) = \begin{pmatrix} \cos t & -\sin t \\ \sin t & \cos t \end{pmatrix}. \tag{2.414}$$

This gives the flow through $x \in \mathbf{R}^2$

$$\begin{aligned} \sigma(t, x) &= \begin{pmatrix} \cos t & -\sin t \\ \sin t & \cos t \end{pmatrix} \begin{pmatrix} x_1 \\ x_2 \end{pmatrix} \\ &= (x^1 \cos t - x^2 \sin t, x^1 \sin t + x^2 \cos t) \end{aligned} \tag{2.415}$$

together with the induced vector field

$$X_V(x) = \left.\frac{d}{dt}\right|_{t=0} \sigma(t, x) = -x^2 \frac{\partial}{\partial x^1} + x^1 \frac{\partial}{\partial x^2}. \tag{2.416}$$

Adjoint representation: The following homomorphism is called an **adjoint representation of G**

$$\begin{aligned} ad_g : G &\to G \\ h &\mapsto ghg^{-1}, \qquad g, h \in G. \end{aligned} \tag{2.417}$$

The **induced map in tangent space** is

$$ad_{g*} : T_h(G) \to T_{ghg^{-1}}(G). \tag{2.418}$$

If we restrict this map to the unit of $G : h \to e$ we obtain the **adjoint map** (recall the isomorphism $T_e(G) \simeq \text{Lie } G$)

$$\begin{aligned} Ad_g : T_e(G) &\to T_e(G) \quad \text{or} \quad \text{Lie } G \to \text{Lie } G \\ V &\mapsto gVg^{-1}, \qquad g \in G,\ V \in T_e(G). \end{aligned} \tag{2.419}$$

For a matrix group G we just have matrix operations. Let $\phi_V(t) = \exp(tV)$, $V \in T_e(G)$, be the one-parameter subgroup of G, then the adjoint representation is

$$ad_g \phi_V(t) = g \exp(tV) g^{-1} = \exp(tgVg^{-1}) \tag{2.420}$$

and we verify the adjoint map by

$$
\begin{aligned}
Ad_g V &= \left.\frac{d}{dt}\right|_{t=0} ad_g \phi_V(t) = \left.\frac{d}{dt}\right|_{t=0} \exp(tgVg^{-1}) \\
&= gVg^{-1}.
\end{aligned} \tag{2.421}
$$

2.7 Fibre bundles

Fibre bundles are of increasing interest in theoretical physics. They are the proper modern mathematical framework for discussing physical theories such as gauge theories or gravitation. We have already considered a special case—the tangent bundle. The tangent space $T_p(M)$ of a manifold M represents a fibre which—when 'glued' together at all points p—establishes this bundle. Another important bundle we meet in physics is the principal bundle.

The essential quantities of a bundle are the fibres. The way the fibres twist—when moving around in space–time—reveals the topological structure of the physical theory and determines properties like the magnetic monopole charge, the instanton number or, what interests us most, the anomalous content of a theory.

In this section we shall give an introduction into some basic ideas of fibre bundles; for further literature we refer to [Choquet-Bruhat, DeWitt-Morette 1982], [Nash, Sen 1983], [Nakahara 1990], [Isham 1989], [Trautman 1984], [Daniel, Viallet 1980], [Coquereaux, Jadczyk 1988], [Mickelsson 1989].

2.7.1 Bundle set-up

Fibre bundle: We denote a **fibre bundle** by (E, Π, X, F, G). It consists of

i) a topological space E—the **total space**,
ii) a topological space X—the **base space**,
iii) a surjection called the **projection** $\Pi : E \to X$,
iv) a topological space F—the **standard fibre**. It is homeomorphic to all inverse images $\Pi^{-1}(x) = F_x$ with $x \in X$; the F_x are called fibres,
v) a group G—the **structure group**—of homeomorphisms of the fibre F,
vi) a covering of open sets $\{U_\alpha\}$ of X together with **homeomorphisms** ϕ_α

$$\forall U_\alpha \, \exists \phi_\alpha : \Pi^{-1}(U_\alpha) \to U_\alpha \times F$$

such that

$$\Pi\phi_\alpha^{-1}(x, f) = x \qquad \text{with } x \in U_\alpha,\ f \in F.$$

The homeomorphism ϕ_α maps $\Pi^{-1}(U_\alpha) \subset E$—the union of fibres above U_α—onto the direct product space $U_\alpha \times F$ and is called a **local trivialization**.

We have illustrated a fibre bundle in Figure 2.38 with help of a Möbius strip. There the fibre F is a line segment and the Möbius strip represents the total fibre space which is projected down by Π onto the base space X, a circle S^1. The homeomorphism ϕ_α untwists the fibre space $\Pi^{-1}(U_\alpha)$ onto the product space $U_\alpha \times F$, the local trivialization (for recent literature see [Frankel 1997]).

Transition functions and global properties: In order to find the global properties of a fibre bundle we have to study the structure group G. The group G arises by transition from one set of **local bundle coordinates** (U_α, ϕ_α) to another set (U_β, ϕ_β). Suppose U_α and U_β have a nonempty overlap $U_\alpha \cap U_\beta \neq \emptyset$. Let us consider the map $\phi_\alpha \cdot \phi_\beta^{-1}$:

$$(U_\alpha \cap U_\beta) \times F \overset{\phi_\alpha \cdot \phi_\beta^{-1}}{\longrightarrow} (U_\alpha \cap U_\beta) \times F. \tag{2.422}$$

It is continuous and invertible (homeomorphic). Keeping $x \in U_\alpha \cap U_\beta$ fixed and allowing the fibre point $f \in F$ to vary then we have the map $\phi_\alpha \cdot \phi_\beta^{-1} : F \to F$ from one fibre point to another (see Figure 2.39) and for this reason it is named the **transition function**

$$g_{\alpha\beta} := \phi_\alpha \cdot \phi_\beta^{-1}, \tag{2.423}$$

with

$$f' = g_{\beta\alpha} f. \tag{2.424}$$

The letter f denotes a fibre point corresponding to the local trivialization (U_α, ϕ_α) and f' corresponding to (U_β, ϕ_β).

The transition functions tell us how the fibres must be glued together in the overlap of two neighbourhoods. Although the local topology of a bundle is always trivial, its global topology—determined by the transition functions—might be nontrivial and rather complicated due to the twists of the neighbouring fibres.

The transition functions satisfy the **consistency conditions**

$$\begin{aligned} g_{\alpha\alpha}(x) &= \mathbf{1}, & x &\in U_\alpha \\ g_{\alpha\beta}(x) &= g_{\beta\alpha}^{-1}(x), & x &\in U_\alpha \cap U_\beta \\ g_{\alpha\beta}(x) g_{\beta\gamma}(x) &= g_{\alpha\gamma}(x), & x &\in U_\alpha \cap U_\beta \cap U_\gamma. \end{aligned} \tag{2.425}$$

The set of all transition functions generate the **structure group**

$$G = \{g_{\alpha\beta}\}. \tag{2.426}$$

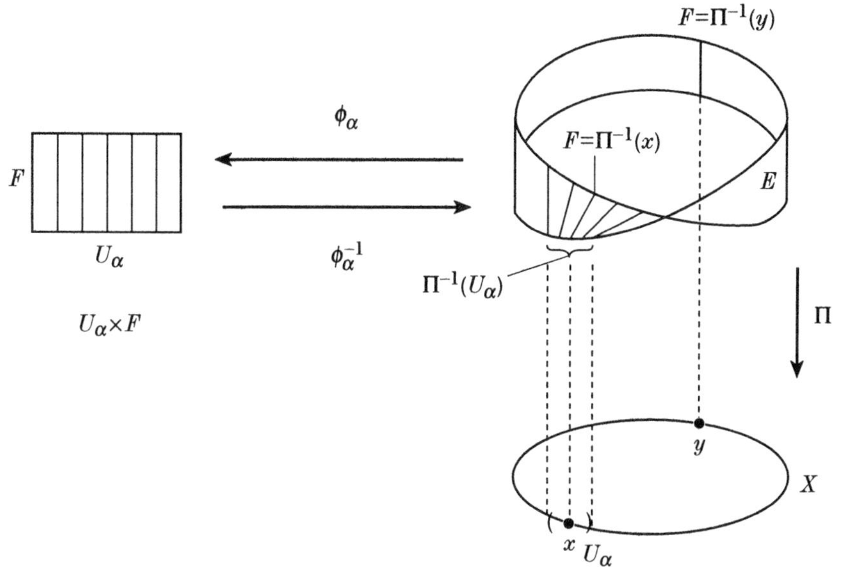

Fig. 2.38. A fibre bundle illustrated by a Möbius strip. F denotes the fibre and E the total space, Π is the projection and X the base space. ϕ_α is a homeomorphic map from the bundle space $\Pi^{-1}(U_\alpha)$ onto the product space $U_\alpha \times F$

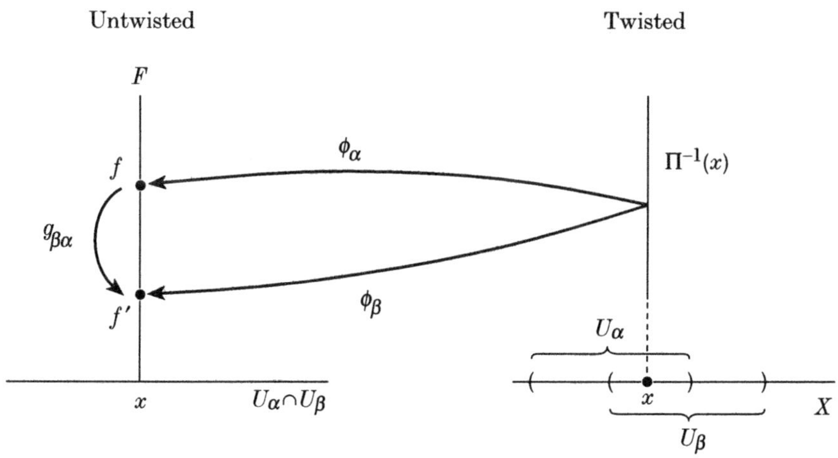

Fig. 2.39. The transition function $g_{\beta\alpha}$ from one fibre point f to another $f' = g_{\beta\alpha} f$

Another important property we have to discuss is expressed in the following proposition.

Proposition:

- *The transition functions are not unique!*

Let us take two fibre bundles E and E' with the *same* base X, fibre F and structure group G. Let (U_α, ϕ_α) be the local coordinates of E and (U_α, ψ_α) those of E'. Next we require that the map

$$h_\alpha(x) := \phi_\alpha \cdot \psi_\alpha^{-1}(x) \tag{2.427}$$

is a homeomorphism $h_\alpha : U_\alpha \times F \to U_\alpha \times F$ *belonging to* the structure group G. Then the transition functions $g_{\alpha\beta}(x)$ and $g'_{\alpha\beta}(x)$ of the two bundles E and E' are related by

$$g'_{\alpha\beta}(x) = h_\alpha^{-1}(x) g_{\alpha\beta}(x) h_\beta(x) \qquad \text{for } x \in U_\alpha \cap U_\beta \tag{2.428}$$

since

$$h_\alpha^{-1} g_{\alpha\beta} h_\beta = \psi_\alpha \cdot \phi_\alpha^{-1} \cdot \phi_\alpha \cdot \phi_\beta^{-1} \cdot \phi_\beta \cdot \psi_\beta^{-1} = \psi_\alpha \cdot \psi_\beta^{-1} = g'_{\alpha\beta}. \tag{2.429}$$

The requirement $h_\alpha \in G$ implies that as $g_{\alpha\beta}$ varies then the related $h_\alpha^{-1} g_{\alpha\beta} h_\beta = g'_{\alpha\beta}$ generates all the elements of the structure group G. So the two bundles E and E' are **topologically equivalent**; they differ only in the choice of the local coordinates ϕ_α and ψ_α. Actually we always have to consider the equivalence class of such bundles, which we usually neglect. For instance, in the case of the Möbius strip the bundles which are twisted once, three times, five times, ... are topologically equivalent.

Remark: When we return to physics and discuss Yang–Mills gauge theories we shall see that the transition functions $g_{\alpha\beta}$ represent the gauge transformations which glue the local coordinates together, whereas the homeomorphism h_α expresses the gauge freedom on the coordinate.

Trivial bundle: If we can choose for *all* transition functions the identity map

$$\{g_{\alpha\beta}(x) = \mathbf{1}\} \tag{2.430}$$

we get a **trivial bundle**, *all* fibres are glued together without twist and the global topology of this bundle corresponds simply to the direct product

$$E = X \times F. \tag{2.431}$$

The transition functions contain the information as to whether a fibre bundle is trivial or not.

Proposition:

- *If a bundle E is trivial then the transition functions factorize according to, and vice versa,*

$$g_{\alpha\beta}(x) = h_\alpha(x)h_\beta^{-1}(x). \tag{2.432}$$

This is in accordance with the choice

$$g'_{\alpha\beta} = h_\alpha^{-1} g_{\alpha\beta} h_\beta = h_\alpha^{-1} h_\alpha h_\beta^{-1} h_\beta = \mathbf{1}. \tag{2.433}$$

Proof. If the bundle is trivial there is always a global homeomorphism $\psi : E \to X \times F$ and together with the usual coordinate choice (U_α, ϕ_α) we take

$$h_\alpha(x) = \phi_\alpha \cdot \psi^{-1}(x) \tag{2.434}$$

implying

$$g_{\alpha\beta}(x) = \phi_\alpha \cdot \phi_\beta^{-1}(x) = \phi_\alpha \cdot \psi^{-1}(x)\, \psi \cdot \phi_\beta^{-1}(x) = h_\alpha(x)h_\beta^{-1}(x). \qquad \text{Q.E.D.}$$

Another way to find the triviality of a bundle is to consider the base space.

Proposition:

- *Any fibre bundle over a contractible base space is trivial!*

Since the space $\mathbf{R}^m$ is contractible the simplest noncontractible spaces we may choose are the spheres S^n.

Reconstruction of a fibre bundle: Let us return to our definition of a fibre bundle. It consists of several components.

The question now is:

Which of the ingredients determine already the structure of the fibre bundle?

The answer is:

Given X, F, G and $g_{\alpha\beta}(x)$ we can reconstruct the whole fibre bundle (E, Π, X, F, G). Thus we can find Π, E and ϕ_α.

The procedure is as follows:

We start with all direct products

$$\widetilde{E} := \bigcup_\alpha U_\alpha \times F \tag{2.435}$$

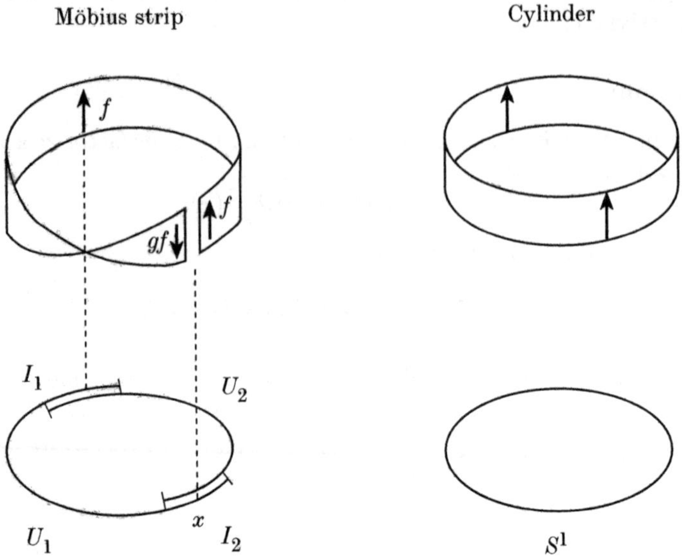

Fig. 2.40. For the Möbius strip—a nontrivial fibre bundle—we have to glue together the twisted fibres. For the cylinder—a trivial fibre bundle—the fibres remain untwisted

and we define for two coordinates $(x, f) \in U_\alpha \times F$ and $(x', f') \in U_\beta \times F$ the **equivalence relation** $\sim$

$$(x, f) \sim (x', f') \tag{2.436}$$

iff $x = x'$ and $g_{\beta\alpha}(x)f = f'$. Then the fibre bundle E is given by the set of all these equivalence classes

$$E = \widetilde{E} \,/ \sim . \tag{2.437}$$

The projection Π is defined by

$$\begin{aligned} \Pi \ : \ & E \to X \\ & [(x, f)] \mapsto x \end{aligned} \tag{2.438}$$

when we denote the equivalence class by $[(x, f)]$. So the projection Π is a map which shrinks each fibre to a point. Finally, we describe the homeomorphism ϕ_α by its inverse

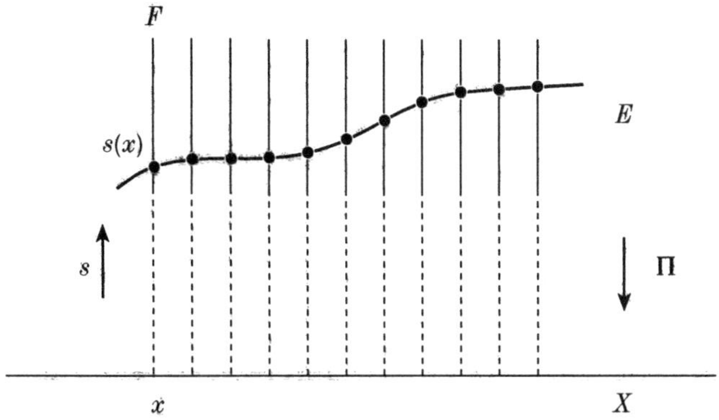

Fig. 2.41. A section $s(x)$ of a fibre bundle E

$$\begin{aligned} \phi_\alpha^{-1} : U_\alpha \times F &\rightarrow \Pi^{-1}(U_\alpha) \\ (x, f) &\mapsto [(x, f)] \in \Pi^{-1}(x) \end{aligned} \tag{2.439}$$

so that

$$\phi_\alpha^{-1}(x, f) \in \Pi^{-1}(x) \Rightarrow \Pi\phi_\alpha^{-1}(x, f) = x \tag{2.440}$$

as required.

Example Möbius strip: Let us come back to the Möbius strip to illustrate the bundle reconstruction. Here the base is $X = S^1$, a circle, and the fibre $F = f$, a line segment. The structure group $G = \{\mathbf{1}, g\}$ contains two elements: $\mathbf{1}$ the identity and g the twist (or reflection of the fibre about its midpoint). The covering of the base S^1 consists of two open arcs U_1 and U_2 with intersection $U_1 \cap U_2 = I_1 \cup I_2$ (see Figure 2.40). The transition functions are

$$\begin{aligned} g_{11}(x) &= g_{22}(x) = \mathbf{1} \\ g_{12}(x) &= \begin{cases} \mathbf{1} & \text{if } x \in I_1 \\ g & \text{if } x \in I_2 \end{cases} \\ g_{21}(x) &= g_{12}^{-1}(x). \end{aligned} \tag{2.441}$$

Now we consider the equivalence classes $[(x, f)]$ in the overlap regions I_1 and I_2. If $x \in I_1$ then $[(x, f)] = (x, f)$ is just one element. If, however, $x \in I_2$ then $[(x, f)] = \{(x, f), (x, gf)\}$ contains two elements—the fibre and the twisted fibre. But the equivalence relation (2.436) identifies these two

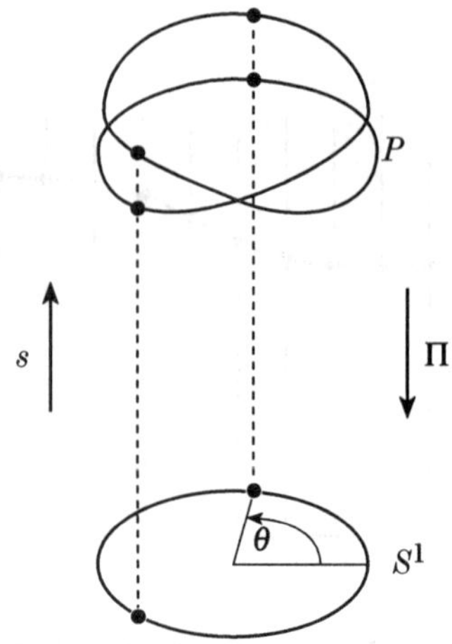

Fig. 2.42. The principal bundle of the Möbius strip

elements; for this reason we have to glue them together and we get the twist for the bundle E, the Möbius strip (see Figure 2.40).

If we consider in constrast the cylinder the structure group is simply $G = \{\mathbf{1}\}$, the fibres remain untwisted, and we obtain a trivial fibre bundle $E = S^1 \times F$.

Section: A **section** of a bundle E is a continuous map

$$s : X \to E \quad \text{with } \Pi s(x) = x, \forall x \in X. \tag{2.442}$$

If the section is defined only over an open subset $U \subset X$ we speak of a **local section** (see Figure 2.41).

Principal bundle: Given a fibre bundle (E, Π, X, F, G) we can always construct a **principal bundle** by taking the same base X and the same transition functions $g_{\alpha\beta}$, but by choosing the fibre to be identical with the structure group itself $F \simeq G$. As before, the transition functions act on the fibre from the left. In addition, we also define an action of G on F from the right, the group G acts freely on the principal bundle P $(ug = u \Rightarrow g = e)$ and is transitive $(ug = u')$, see Section 2.7.2. For the construction we follow the procedure described above. We denote the principal bundle by $P(X, G)$.

The fibre bundles we mainly consider in physics, specifically in gauge theories, are principal bundles. We discuss several examples separately in this book.

There is an important theorem on the triviality of a bundle.

Theorem: The principal bundle $P(X, G)$ and the correlated fibre bundle (E, Π, X, F, G) are trivial iff $P(X, G)$ has a section.

In our example of the Möbius strip we find the following. Choosing the fibre to be the structure group $F \equiv G = \{\mathbf{1}, g\}$, which consists of two elements, we obtain a *double covering* of the base S^1 depicted in Figure 2.42. Choosing a local coordinate $\theta \in [0, 2\pi]$ on S^1 the section $s(\theta)$ is *not* continuous, e.g. $s(\theta) \neq s(\theta + 2\pi)$ due to the double covering. According to the above theorem the principal bundle and the Möbius strip itself are then nontrivial.

Remark:

i) Although the principal bundle $P(X, G)$ has no section, the bundle itself (E, Π, X, F, G) may have a section (as in the case of Möbius)! So the principal bundle is the important one for studying the triviality of a bundle.

ii) Several bundles with different fibres but with common base, transition functions and structure group will have the same principal bundle.

Line bundle: A line bundle is a vector bundle, but with a 1-dimensional vector space as a fibre. For instance, if we replace in the Möbius strip the finite line segment by the real line $\mathbf{R}$ we get a nontrivial real line bundle over S^1. On the other hand, the cylinder $S^1 \times \mathbf{R}$ represents a trivial real line bundle.

Associated bundle: A **vector bundle**—a bundle whose fibre F is a vector space—has a certain representation ρ of the structure group G on the vector space F. It is termed **associated to the principal bundle** if its transition functions are the images under ρ of the corresponding transition functions of the principal bundle. Consider the direct product of a principal bundle and a vector bundle $P \times E$. Let us define the following equivalence relation $(pg, f) \sim (p, \rho(g)f)$, for all $p \in P, g \in G, f \in E$, then the **associated vector bundle** is given by

$$P \times_G E := P \times E / \sim .$$

Tangent bundle: Choosing the fibre F at any point x of some base manifold M to be the tangent space $T_x(M) \equiv F$ we construct the **tangent bundle**

$$E = T(M) = \bigcup_{x \in M} T_x(M). \tag{2.443}$$

The projection Π is defined by

$$\begin{aligned} \Pi \quad : \quad & T(M) \to M \\ & V \in T_x(M) \mapsto x. \end{aligned} \tag{2.444}$$

For the local trivialization we need the homeomorphism ϕ_α on some $U_\alpha \subset M$. U_α is homeomorphic to $\mathbf{R}^m$, $U_\alpha \simeq \mathbf{R}^m$ and also $T_x(M) \simeq \mathbf{R}^m$. In local coordinates the tangent vector V is given by $V = V^i(x)\frac{\partial}{\partial x^i}$ so that the homeomorphism is

$$\begin{aligned} \phi_\alpha : \Pi^{-1}(U_\alpha) &\rightarrow U_\alpha \times T_x(M) \simeq \mathbf{R}^m \times \mathbf{R}^m \\ V &\mapsto (x, V^i(x)) \in \mathbf{R}^m \times \mathbf{R}^m \end{aligned} \tag{2.445}$$

and locally the tangent bundle has the direct product structure

$$T(U_\alpha) \simeq \mathbf{R}^m \times \mathbf{R}^m. \tag{2.446}$$

The structure group of the tangent bundle is $G = GL(m, \mathbf{R})$, the group of all invertible $m \times m$ matrices since the fibres $F = T_x(M)$ are m-dimensional vector spaces and the action on an m-dimensional vector is represented by an $m \times m$ matrix.

Searching for the transition functions from one fibre to another $g_{\alpha\beta} = \phi_\alpha \cdot \phi_\beta^{-1}$ with $\phi_\alpha(V) = (x, V_\alpha^i(x))$ and $\phi_\beta(V) = (x, V_\beta^i(x))$ we find

$$V_\alpha^i \equiv \frac{\partial x_\alpha^i}{\partial x_\beta^j} V_\beta^j = g_{\alpha\beta}^{ij}\, V_\beta^j \tag{2.447}$$

(note: i, j are the vector indices and α, β refer to the charts) the **Jacobian matrix** of vector transformations

$$g_{\alpha\beta} = \left[\frac{\partial x_\alpha^i}{\partial x_\beta^j}\right]_{m\times m}. \tag{2.448}$$

Since the fibre is a vector space the tangent bundle also represents a **vector bundle.** If we consider all ordered bases of $T_x(M)$—a frame—the bundle is called a **frame bundle.** Then the fibre can be identified with $GL(m, \mathbf{R})$, the structure group itself, hence the frame bundle expresses a principal bundle.

Cotangent bundle: Analogous to the tangent bundle we can take as a fibre the cotangent space $F = T_x^*(M)$ and we set up the **cotangent bundle**

$$E = T^*(M) = \bigcup_{x\in M} T_x^*(M). \tag{2.449}$$

$T^*(M)$ is the **dual bundle** to $T(M)$.

The transition functions, a matrix, are determined by the dual vector transformations

$$g_{\alpha\beta} = \left[\frac{\partial x^j_\beta}{\partial x^i_\alpha} \right]_{m \times m} . \tag{2.450}$$

2.7.2 Connection

The 'connection' is of great importance for a fibre bundle (we consider here only principal bundles). It allows for a parallel transport of the fibres. For parallel transport or covariant differentiation we have to compare the points in neighbouring fibres independently of a specific choice of local bundle trivialization. So we need vectors that lead from one fibre to another. We already have vectors at our disposal. They arise from the group action on the bundle—the fundamental vector fields. However, they are tangent to the fibre. Therefore we have to construct vectors that point away from the fibre. This we achieve with help of a 'connection'.

In the interplay between the geometry of fibre bundles and physics the 'connection' takes over an important role. Restricting the 'connection' to a local region represents precisely the Yang–Mills (YM) potential of gauge theories. Furthermore, the 'curvature' of the 'connection' describes the corresponding Yang–Mills field strength.

Vertical subspace: Let us start with a principal bundle $P(M, G)$. Let the fibre $F \equiv G$ be some Lie group G, and the base be some manifold M. The projection is $\Pi : P \to M$. We consider the tangent space $T_u(P)$ of the bundle P with $u \in P$ some fibre point and we now construct the vertical subspace $V_u(P)$ of $T_u(P)$. We take an element $A \in T_e(G) \simeq \text{Lie } G$ of the Lie algebra then we can define a curve in P through u by the right-action of G on P (recall our discussion of Lie groups and algebras in Section 2.6; e.g. $R_g h = hg$; $g, h \in G$)

$$R_{\exp(tA)} u = u \exp(tA) = \sigma(t, u). \tag{2.451}$$

This **flow** $\sigma(t, u)$ lies along the fibre $\Pi^{-1}(p)$ over the base point $p \in M$ since $\Pi(u) = \Pi(u \exp(tA)) = p \in M$. The action of the one-parameter subgroup: $t \to \exp(tA)$ induces a vector field—the **fundamental vector field**—by

$$X_A f(u) := \left. \frac{d}{dt} \right|_{t=0} f(u \exp(tA)) \tag{2.452}$$

with a smooth function $f : P \to \mathbf{R}$. X_A preserves the Lie algebra structure

$$[X_A, X_B] = X_{[A,B]}. \tag{2.453}$$

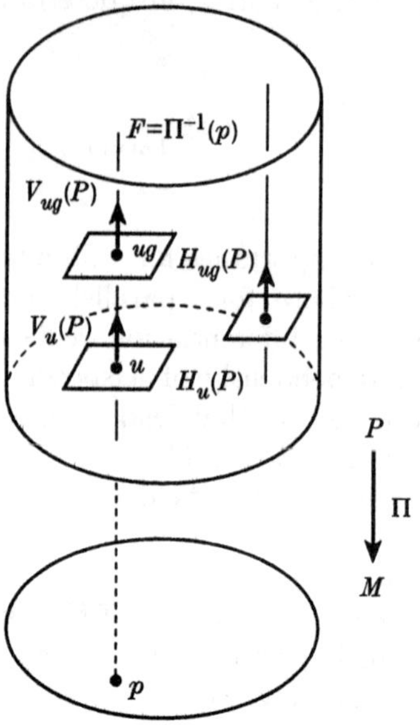

Fig. 2.43. The tangent space $T_u(P)$ of the fibre bundle $P(M,G)$ is decomposed into the vertical subspace $V_u(P)$ and horizontal subspace $H_u(P)$. We move along the fibre via the right-action of the group G on P, $R_g u = ug$

X_A is a tangent to the fibre at u and belongs therefore to a space—the **vertical subspace** $V_u(P)$ of the tangent space $T_u(P)$, $X_A \in V_u(P)$—which is defined by

$$V_u(P) := \{X \in T_u(P) | \Pi_* X = 0\}. \tag{2.454}$$

The map $\iota : A \mapsto X_A$ is an isomorphism Lie $G \to V_u(P)$, hence $\dim V_u(P) = \dim \text{Lie } G = \dim G$.

Horizontal subspace and connection: The complement of $V_u(P)$ in $T_u(P)$—the **horizontal subspace** $H_u(P)$—is described by the so-called connection.

Definition: A **connection** assigns smoothly to each fibre point $u \in P$ a subspace $H_u(P) \subset T_u(P)$:

$$u \in P \mapsto H_u(P) \subset T_u(P) \tag{2.455}$$

such that

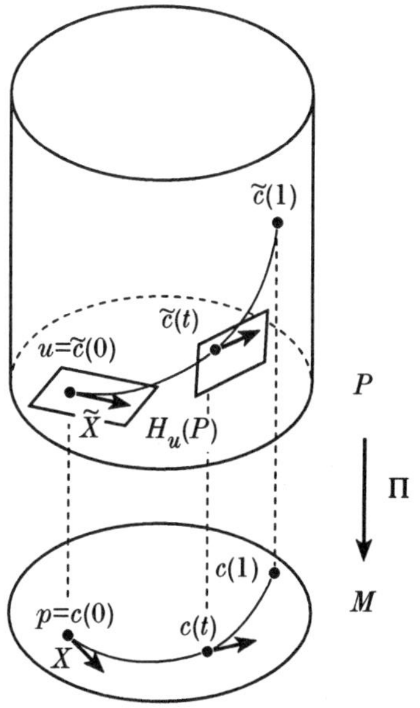

Fig. 2.44. Parallel transport of fibre point u along the base curve c

$$\begin{aligned} &\text{i)} \quad T_u(P) = V_u(P) \oplus H_u(P) \\ &\text{ii)} \quad R_{g*}H_u(P) = H_{ug}(P). \end{aligned} \tag{2.456}$$

We have illustrated this assignment in Figure 2.43. R_{g*} denotes the differential map (recall Section 2.6.1) of the right-action $R_g u = ug$ of G on P; $g \in G$, $u \in P$.

From Property i) it follows that any tangent vector $X \in T_u(P)$ can be decomposed into

$$X = X^V + X^H \qquad \text{with } X^V \in V_u(P),\ X^H \in H_u(P). \tag{2.457}$$

Condition ii) means that the horizontal subspaces $H_u(P)$ and $H_{ug}(P)$ on the *same* fibre are linked together by the differential map R_{g*}. For a given $H_u(P)$ *all* the other $H_{ug}(P)$ on the same fibre are generated by R_{g*} (see Figure 2.43). Hence, if a fibre point u is parallel transported, the fibre point ug is transported too.

Horizontal lift and parallel transport: The projection $\Pi : P \to M$ induces the differential map $\Pi_* : T_u(P) \to T_{\Pi(u)}(M)$ on the tangent spaces with the kernel $\ker \Pi_* = V_u(P)$. Therefore the map from the horizontal subspace into the tangent space $\Pi_* : H_u(P) \to T_{\Pi(u)}(M)$ is an isomorphism.

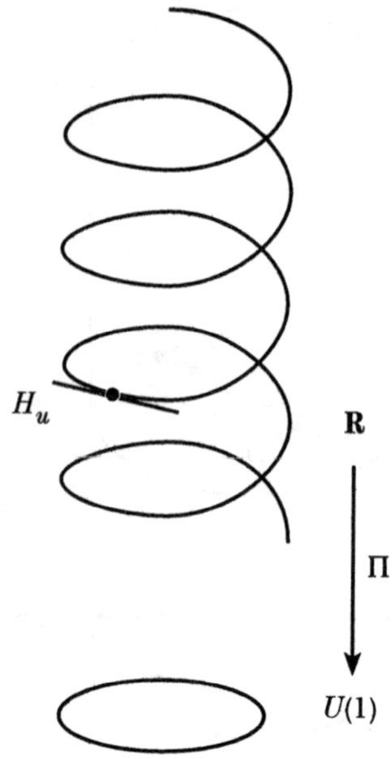

Fig. 2.45. The bundle with projection $\Pi : \mathbf{R} \to U(1)$ has a connection: H_u = tangent space to $\mathbf{R}$ at u. The holonomy group $H = \mathbf{Z} \equiv G$ coincides with the structure group

For this reason any tangent vector $X \in T_p(M)$ from the base manifold M can be **lifted** to a unique vector $\widetilde{X} \in H_u(P)$ in the horizontal subspace of the fibre bundle. We have illustrated this in Figure 2.44.

Definition: The vector $\widetilde{X}$ is a **horizontal lift** of X if

$$\text{i)} \quad \Pi_* \widetilde{X} = X,$$
$$\text{ii)} \quad \widetilde{X}^V = 0 \qquad \text{vertical component.} \tag{2.458}$$

Definition: Let $c : [0,1] \to M$ be a base curve. A curve $\widetilde{c} : [0,1] \to P$ is a **horizontal lift** of c if

$$\text{i)} \quad \Pi\widetilde{c} = c,$$
$$\text{ii)} \quad \text{tangent vector to } \widetilde{c}\text{: } \widetilde{X} \in H_{\widetilde{c(t)}}(P). \tag{2.459}$$

Theorem: Let $c : [0,1] \to M$ be a base curve and let $u \in \Pi^{-1}(c(0))$, then there exists a unique horizontal lift $\widetilde{c} : [0,1] \to P$ such that $\widetilde{c}(0) = u$.

So the parallel transport of a fibre point $u = \tilde{c}(0)$ along the base curve $c(t)$ is provided by the horizontal lift $\tilde{c}(t)$ (see Figure 2.44).

Holonomy: Generally, it occurs that even if the base curve $c(t)$ is closed, $c(1) = c(0)$, the horizontal lift $\tilde{c}(t)$ *need not* be closed: $\tilde{c}(1) = \tilde{c}(0)h$, $h \in G$. The set of such elements is called the **holonomy group** $H = \{h\}$ of the connection at the fibre point $\tilde{c}(0)$. It is a subgroup of the structure group, $H \subset G$.

For example, the bundle with projection $\Pi : \mathbf{R} \to U(1)$ so that

$$\Pi(u) = e^{i2\pi u} \tag{2.460}$$

allows for one connection:

$$H_u = \text{tangent space to } \mathbf{R} \text{ at } u. \tag{2.461}$$

Let $c(t) = e^{i2\pi t}$ be the closed base curve, then the lifted curve is $\tilde{c}(t) = t$, which is not closed, $\tilde{c}(0) = 0$ and $\tilde{c}(1) = 1$ (see Figure 2.45). The holonomy group coincides with the structure group $H = \mathbf{Z} \equiv G$.

Connection 1-form: Next we define the connection in a way which is—as we shall see—more suitable for physics.

Definition: The **connection 1-form** $\omega \in T^*(P) \otimes \text{Lie } G$—a Lie algebra valued 1-form on the bundle P—is defined by a projection of the tangent space $T_u(P)$ onto the vertical subspace $V_u(P) \simeq \text{Lie } G$ satisfying

$$\begin{aligned} &\text{i)} \quad \omega(X_A) = A, \qquad A \in \text{Lie } G, \\ &\text{ii)} \quad {R_g}^*\omega = Ad_{g^{-1}}\omega. \end{aligned} \tag{2.462}$$

${R_g}^*$ is the pullback (recall Section 2.6.2) of the right-action R_g: $u \mapsto ug$ and Ad means the adjoint map $Ad_g : \text{Lie } G \to \text{Lie } G$ given by $A \mapsto gAg^{-1}$, $g \in G$, $A \in \text{Lie } G$ (recall Section 2.6.6.). So for some vector $X \in T_u(P)$ we have

$${R_g}^*\omega_{ug}(X) = \omega_{ug}(R_{g*}X) = g^{-1}\omega_u(X)g. \tag{2.463}$$

With such a defined connection 1-form ω we can now determine the **horizontal subspace** $H_u(P)$ by the kernel of the map $\omega : T_u(P) \to \text{Lie } G$

$$H_u(P) = \{X \in T_u(P) | \omega(X) = 0\}. \tag{2.464}$$

This leads again to the direct sum separation of $T_u(P) = V_u(P) \oplus H_u(P)$. The subspace $H_u(P)$ also satisfies the previous property ii): $R_{g*}H_u(P) = H_{ug}(P)$ (equation (2.456)), therefore both definitions (2.455), (2.456) and (2.462), of the connection are *equivalent*.

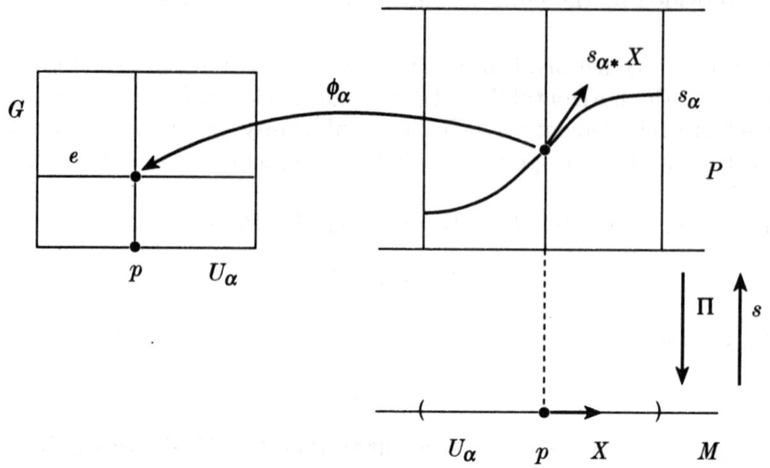

Fig. 2.46. Canonical local trivialization of a fibre bundle

Local connection 1-form and gauge potential: Restricting now to a local region we find the bridge to physics. There we already know a Lie algebra valued 1-form defined on the base M. It is the YM gauge potential. If we pull back the connection ω within a local section down to the base we will find that it satisfies precisely the properties of the YM potential.

Proposition:

- *Locally the connection 1-form ω represents the YM gauge potential $\mathcal{A}$! Be $\{U_\alpha\}$ an open covering of M; choose a local section on U_α*

$$s_\alpha : U_\alpha \to \Pi^{-1}(U_\alpha) \tag{2.465}$$

then

$$\mathcal{A}_\alpha \equiv s_\alpha^* \omega \in \Lambda^1(U_\alpha) \otimes \text{Lie } G. \tag{2.466}$$

Conversely, we have the following theorem.

Theorem: If we consider a Lie G valued 1-form $\mathcal{A}_\alpha$ defined on a patch U_α together with a local section $s_\alpha : U_\alpha \to \Pi^{-1}(U_\alpha)$ then there exists exactly one connection 1-form ω on $\Pi^{-1}(U_\alpha)$ such that $\mathcal{A}_\alpha = s_\alpha^* \omega$. The explicit expression

$$\omega|_{U_\alpha} = g_\alpha^{-1} \Pi^* \mathcal{A}_\alpha g_\alpha + g_\alpha^{-1} d_p g_\alpha \tag{2.467}$$

is found by construction.

The operator d_p means the exterior derivative on the bundle P and g_α expresses a local bundle coordinate given by the **canonical local trivialization** (see Figure 2.46) which is defined by the homeomorphism

$$\begin{aligned} \phi_\alpha : \Pi^{-1}(U_\alpha) &\to U_\alpha \times F \\ u &\mapsto (p, g_\alpha), \end{aligned} \tag{2.468}$$

thus

$$\phi_\alpha(u) = (p, g_\alpha) \qquad \text{with } u = s_\alpha(p) g_\alpha. \tag{2.469}$$

Proof. a) First we have to show that expression (2.467) gives the YM potential (2.466). Take a tangent vector $X \in T_p(M)$ on the base, lift X up to the bundle $s_{\alpha *}X \in T_{s_\alpha}(P)$ and note that $\Pi_* s_{\alpha *} = \mathbf{1}_{T_p(M)}$ and $g_\alpha = e$ on s_α.
Note furthermore that $d_p g_\alpha(s_{\alpha *}X) = 0$ since $g_\alpha = e$ along the section $s_{\alpha *}X$. Then we get

$$\begin{aligned} {s_\alpha}^* \omega(X) &= \omega(s_{\alpha *}X) = \Pi^* \mathcal{A}_\alpha(s_{\alpha *}X) + d_p g_\alpha(s_{\alpha *}X) \\ &= \mathcal{A}_\alpha(\Pi_* s_{\alpha *}X) = \mathcal{A}_\alpha(X). \end{aligned} \tag{2.470}$$

b) Second we demonstrate that expression (2.467) satisfies the axioms (2.462) of a connection 1-form.
Axiom i):
Taking a fundamental vector field $X_A \in V_u(P)$, so $\Pi_* X_A = 0$, we find

$$\begin{aligned} \omega(X_A) &= g_\alpha^{-1} \mathcal{A}_\alpha(\Pi_* X_A) g_\alpha + g_\alpha^{-1} d_p g_\alpha(X_A) \\ &= g_\alpha^{-1}(u) \left.\frac{d}{dt}\right|_{t=0} g_\alpha(u \exp(tA)) \\ &= g_\alpha^{-1}(u) g_\alpha(u) \left.\frac{d}{dt}\right|_{t=0} \exp(tA)) = A. \end{aligned} \tag{2.471}$$

Axiom ii):
Take a tangent vector $X \in T_u(P)$ and a group element $h \in G$, recall that X is a right-invariant vector field $R_{h*}X|_u = X|_{uh}$ (recall Section 2.6.6) and note that $g_\alpha(uh) = g_\alpha(u)h$.
Then we calculate

$$\begin{aligned} {R_h}^* \omega(X) &= \omega(R_{h*}X) \\ &= h^{-1} g_\alpha^{-1}(u) \Pi^* \mathcal{A}_\alpha(X) g_\alpha(u) h \\ &\quad + h^{-1} g_\alpha^{-1}(u) d_p g_\alpha(u)(X) h \\ &= h^{-1} \omega(X) h = Ad_{h^{-1}} \omega(X). \qquad \text{Q.E.D.} \end{aligned} \tag{2.472}$$

Transformation property: The connection 1-form ω is defined uniquely on the bundle; the decomposition of the tangent space $T_u(P) = V_u(P) \oplus H_u(P)$ is unique. Therefore we have in the overlap of two neighbourhoods U_α, U_β the identity

$$\omega|_{U_\alpha} \equiv \omega|_{U_\beta} \qquad \text{on } U_\alpha \cap U_\beta \neq \emptyset. \tag{2.473}$$

As a consequence the YM potential has to obey an appropriate transformation law. We now calculate this explicitly. We start with

$$\begin{aligned} \omega|_{U_\alpha} &= g_\alpha^{-1}\Pi^*\mathcal{A}_\alpha g_\alpha + g_\alpha^{-1}d_p g_\alpha \\ \equiv \omega|_{U_\beta} &= g_\beta^{-1}\Pi^*\mathcal{A}_\beta g_\beta + g_\beta^{-1}d_p g_\beta. \end{aligned} \tag{2.474}$$

In the overlap $U_\alpha \cap U_\beta$ the **transition functions** $h_{\alpha\beta}$ are defined by

$$g_\beta = h_{\beta\alpha}(p)g_\alpha \quad \text{or} \quad s_\beta(p) = s_\alpha(p)h_{\alpha\beta}(p), \quad p \in U_\alpha \cap U_\beta, \tag{2.475}$$

which we insert into equation (2.474) so that we find

$$\Pi^*\mathcal{A}_\beta = h_{\alpha\beta}^{-1}\Pi^*\mathcal{A}_\alpha h_{\alpha\beta} + h_{\alpha\beta}^{-1}d_p h_{\alpha\beta}. \tag{2.476}$$

This is the **compatibility condition** the local YM potentials have to fulfil when pulled back on the bundle P. That means on the base M in the overlap region $U_\alpha \cap U_\beta$ we have the following **transformation law for YM gauge potentials**

$$\mathcal{A}_\beta = h_{\alpha\beta}^{-1}\mathcal{A}_\alpha h_{\alpha\beta} + h_{\alpha\beta}^{-1}dh_{\alpha\beta} \qquad \text{on } U_\alpha \cap U_\beta. \tag{2.477}$$

Remark: A nontrivial principal bundle does not admit a global section, so the pullback $\mathcal{A}_\alpha = s_\alpha{}^*\omega$ exists locally but not to a global extent. Therefore we have to cover the base M with locally trivialized charts $\{U_\alpha\}$ and we get several (at least two) *different* YM potentials. The YM potentials $\mathcal{A}_\alpha$ on U_α and $\mathcal{A}_\beta$ on U_β are then related to each other in the overlap region $U_\alpha \cap U_\beta$ by the above transformation condition (2.477).

So a single YM potential $\mathcal{A}_\alpha$ does *not* carry any global information. It is the connection 1-form ω which contains the global feature of the fibre bundle or equivalently the set of *all* YM potentials $\{A_\alpha\}$ (defined over the covering $\{U_\alpha\}$ of M) satisfying the compatibility condition (2.476), (2.477).

Gauge transformation: Let us finally consider two charts U_1, U_2 and let us choose the local sections on the overlap as

$$s_2(x) = s_1(x)g(x), \qquad x \in U_1 \cap U_2 \tag{2.478}$$

(we identify the point x of the manifold with its coordinate x), then the local YM forms $\mathcal{A}_1$, $\mathcal{A}_2$ are related by

$$\mathcal{A}_2(x) = g^{-1}(x)\mathcal{A}_1(x)g(x) + g^{-1}(x)dg(x). \tag{2.479}$$

In components $\mathcal{A}_1 \to \mathcal{A}$, $\mathcal{A}_2 \to \mathcal{A}^g$ and $\mathcal{A} = \mathcal{A}_\mu dx^\mu$, $d = \partial_\mu dx^\mu$ we get

$$\mathcal{A}^g_\mu(x) = g^{-1}(x)\mathcal{A}_\mu(x)g(x) + g^{-1}(x)\partial_\mu g(x). \tag{2.480}$$

This is the **usual gauge transformation** we are familiar with in physics.

Example: If the structure group is $G = U(1)$—the case of electrodynamics—the transition functions are

$$\begin{aligned} h_{\alpha\beta} : U_\alpha \cap U_\beta &\rightarrow U(1) \\ h_{\alpha\beta}(x) &= e^{i\lambda(x)}, \qquad x \in U_\alpha \cap U_\beta. \end{aligned} \tag{2.481}$$

So $dh_{\alpha\beta} = id\lambda(x)e^{i\lambda(x)}$ and the compatibility condition gives the well-known **gauge transformation law**

$$\mathcal{A}^\lambda(x) = \mathcal{A}(x) + id\lambda(x). \tag{2.482}$$

2.7.3 Curvature

The last concept we need to discuss in our tour of fibre bundles is the curvature.

Covariant derivative: Let us consider a q-form φ on a principal bundle $P(M, G)$ with values in a vector space V

$$\varphi : \underbrace{T(P) \otimes \ldots \otimes T(P)}_{q \text{ times}} \rightarrow V \tag{2.483}$$

(generalizing the previous Lie algebra valued forms) and V is a vector space of dimension k. Then the form can be decomposed by

$$\varphi = \varphi^a \otimes e_a \in \Lambda^q(P) \otimes V, \tag{2.484}$$

which is a sum over $a = 1, 2, \ldots, k$. The set $\{e_a\}$ represents the basis of V and $\varphi^a \in \Lambda^q(P)$.

Let $X_1, \ldots, X_{q+1} \in T_u(P)$ be $(q+1)$ tangent vectors on the bundle P, then the **exterior covariant derivative** of the q-form φ is defined by

$$D\varphi(X_1, \ldots, X_{q+1}) = d_p\varphi(X_1^H, \ldots, X_{q+1}^H), \tag{2.485}$$

where $X_i^H \in H_u(P)$ means the horizontal component of $X_i \in T_u(P)$ determined by the connection (recall equations (2.455)–(2.457)) and $d_p\varphi = d_p\varphi^a \otimes e_a$ is the exterior derivative on P.

Curvature: The **curvature 2-form** Ω on the bundle P is the covariant derivative of the connection 1-form ω

$$\Omega := D\omega \in \Lambda^2(P) \otimes \text{Lie } G. \tag{2.486}$$

Theorem: The curvature 2-form has the property

$$R_g{}^*\Omega = Ad_{g^{-1}}\Omega = g^{-1}\Omega g, \qquad g \in G. \tag{2.487}$$

Proof. Recall that

$$\begin{aligned} (R_{g*}X)^H &= R_{g*}X^H && R_{g*} \text{ preserves the horizontal subspace} \\ d_pR_g{}^* &= R_g{}^*d_p && \text{commutation} \\ d_pg &= 0 && \text{since } g \in G \text{ is a constant element} \\ R_g{}^*\omega &= Ad_{g^{-1}}\omega = g^{-1}\omega g && \text{connection property.} \end{aligned} \tag{2.488}$$

Let be $X, Y \in T_u(P)$ then we have

$$\begin{aligned} R_g{}^*\Omega(X,Y) &= \Omega(R_{g*}X, R_{g*}Y) \\ &= d_p\omega((R_{g*}X)^H, (R_{g*}Y)^H) = R_g{}^*d_p\omega(X^H, Y^H) \\ &= d_pR_g{}^*\omega(X^H, Y^H) = d_p(g^{-1}\omega g)(X^H, Y^H) \\ &= g^{-1}d_p\omega(X^H, Y^H)g = g^{-1}\Omega(X,Y)g. \qquad \text{Q.E.D.} \end{aligned} \tag{2.489}$$

Theorem: Ω and ω satisfy the **Cartan structure equation**

$$\Omega(X,Y) = d_p\omega(X,Y) + [\omega(X), \omega(Y)] \tag{2.490}$$

for any $X, Y \in T_u(P)$ and $[\ ,\ ]$ denotes the Lie bracket between the Lie algebra elements of $\omega(X)$ and $\omega(Y)$; or in terms of 2-forms on P

$$\Omega = d_p\omega + \omega^2, \tag{2.491}$$

where

$$\omega^2 = \omega\omega = \frac{1}{2}[\omega, \omega] \tag{2.492}$$

means the wedge product $\omega\omega$ of the 1-form ω or half the commutator.

Expression (2.491) follows from equation (2.490) since

$$\begin{aligned}[\omega,\omega](X,Y) &= [T_a,T_b]\omega^a\omega^b(X,Y)\\ &= [T_a,T_b](\omega^a(X)\omega^b(Y)-\omega^a(Y)\omega^b(X))\\ &= [\omega(X),\omega(Y)]-[\omega(Y),\omega(X)]\\ &= 2[\omega(X),\omega(Y)].\end{aligned} \tag{2.493}$$

The Cartan structure equation (2.490), (2.491) between the curvature Ω and the connection ω is of fundamental importance and we shall use it throughout the book.

Proof. To prove this theorem we consider the following three cases.

i) Let $X,Y \in H_u(P) \Longrightarrow \omega(X)=\omega(Y)=0$ by definition of the connection 1-form (recall equation (2.464)). From the definitions (2.485), (2.486) we have for the curvature

$$\Omega(X,Y)=D\omega(X,Y)=d_p(X^H,Y^H)=d_p(X,Y). \tag{2.494}$$

ii) Let $X \in H_u(P)$, $Y \in V_u(P) \Longrightarrow \Omega(X,Y)=0$ since $Y^H=0$; and $\omega(X)=0$ by definition (2.464).
So we have to prove that $d_p\omega(X,Y)=0$. We remember relation (2.108) for the bundle

$$\begin{aligned}d_p\omega(X,Y) &= X\omega(Y)-Y\omega(X)-\omega([X,Y])\\ &= X\omega(Y)-\omega([X,Y]).\end{aligned} \tag{2.495}$$

Since $Y \in V_u(P)$ $\exists V \in \text{Lie } G$ such that $Y=X_V$, where X_V is the fundamental vector field. Then we have $\omega(Y)=\omega(X_V)=V=$ constant, and therefore $X\omega(Y)=XV=0$. We also note that $[X,Y]\in H_u(P)$ since

$$[Y,X]=\mathcal{L}_Y X=\lim_{t\to 0}\frac{1}{t}[R_{g^{-1}(t)*}X-X]. \tag{2.496}$$

and $R_{g^{-1}(t)*}X \in H_u(P)$. This implies $\omega([X,Y])=0$ so that we find

$$d_p\omega(X,Y)=0. \tag{2.497}$$

iii) Let $X,Y \in V_u(P) \Longrightarrow \Omega(X,Y)=0$; and we also have

$$d_p\omega(X,Y)=X\omega(Y)-Y\omega(X)-\omega([X,Y])=-\omega([X,Y]). \tag{2.498}$$

Since $[X,Y]\in V_u(P)$ $\exists\, A \in \text{Lie } G$ such that $[X,Y]=X_A$ and we get for the connection $\omega([X,Y])=\omega(X_A)=A$. On the other hand, let $X=X_B$, $Y=X_C$, then the commutator is

$$[\omega(X), \omega(Y)] = [\omega(X_B), \omega(X_C)] = [B, C] = A \tag{2.499}$$

since $[X_B, X_C] = X_{[B,C]} \equiv X_A$ (recall equation (2.453)).
So we find

$$[\omega(X), \omega(Y)] = \omega([X, Y]), \tag{2.500}$$

giving

$$d_p\omega(X, Y) + [\omega(X), \omega(Y)] = d_p\omega(X, Y) + \omega([X, Y]) = 0. \qquad \text{Q.E.D.} \tag{2.501}$$

Local curvature and field strength: Finally we make contact with physics. In a local region we can identify the curvature with the Yang–Mills field strength.

Proposition:

- *The curvature 2-form Ω represents locally the Yang–Mills field strength $\mathcal{F}$*

$$\mathcal{F} \equiv s^*\Omega \in \Lambda^2(U) \otimes \text{Lie } G, \tag{2.502}$$

where s^* is the pullback of a local section $s : U \to \Pi^{-1}(U)$ on a chart $U \subset M$.

In terms of the YM gauge potential $\mathcal{A}$ we then find

$$\mathcal{F} = d\mathcal{A} + \mathcal{A}^2 \tag{2.503}$$

since

$$\begin{aligned} \mathcal{F} &= s^*\Omega = s^*d_p\omega + s^*(\omega\omega) \\ &= ds^*\omega + s^*\omega s^*\omega = d\mathcal{A} + \mathcal{A}\mathcal{A}, \end{aligned} \tag{2.504}$$

where we used $s^*d_p\omega = ds^*\omega$ and $s^*\omega = \mathcal{A}$.

Evaluating the 2-form $\mathcal{F}$ on the vector fields $X, Y \in T_p(M)$ we get

$$\begin{aligned} \mathcal{F}(X,Y) &= d\mathcal{A}(X,Y) + \mathcal{A}\mathcal{A}(X,Y) \\ &= d\mathcal{A}(X,Y) + [\mathcal{A}(X), \mathcal{A}(Y)]. \end{aligned} \tag{2.505}$$

In terms of components $\mathcal{A} = \mathcal{A}_\mu(x)dx^\mu$, $\mathcal{F} = \frac{1}{2}\mathcal{F}_{\mu\nu}(x)dx^\mu dx^\nu$, expression (2.503) gives

$$\mathcal{F}_{\mu\nu}(x) = \partial_\mu\mathcal{A}_\nu(x) - \partial_\nu\mathcal{A}_\mu(x) + [\mathcal{A}_\mu(x), \mathcal{A}_\nu(x)]. \tag{2.506}$$

This is the well-known **YM field strength.**

Transformation property: Considering finally the YM field strength on two charts U_α and U_β

$$\mathcal{F}_\alpha = s_\alpha{}^*\Omega \qquad \text{on } U_\alpha \tag{2.507}$$

$$\mathcal{F}_\beta = s_\beta{}^*\Omega \qquad \text{on } U_\beta, \tag{2.508}$$

then they satisfy on the overlap of the charts the **compatibility condition**

$$\mathcal{F}_\beta = h_{\alpha\beta}^{-1}\mathcal{F}_\alpha h_{\alpha\beta} \qquad \text{on } U_\alpha \cap U_\beta, \tag{2.509}$$

where $h_{\alpha\beta}$ is the **transition function** in the overlap as defined before (equation (2.475)). The proof is straightforward by using formula (2.503) together with the transformation law (2.477) for the YM potentials.

Remark: For a **pure gauge** we have

$$\mathcal{A} = g^{-1}dg \Longleftrightarrow \mathcal{F} = 0. \tag{2.510}$$

3

Path integrals, FP method and BRS transformation

Path integral formalism provides a unified view of nonrelativistic quantum mechanics and relativistic quantum field theory (QFT). It is particularly suited to investigating gauge theories and to formulating their anomalies. In fact, we will use path integrals (PI) throughout the book. For this reason we shall develop the formalism of PI to the extent we need it in this book; for a further study we recommend textbooks like [Itzykson, Zuber 1980], [Ryder 1988], [Cheng, Li 1988], [Lucha, Schöberl 1996] and [Pokorsky 1987].

In Section 3.1 we present Feynman's idea of formulating quantum mechanics as a sum of particle trajectories (paths). We generalize this approach in Section 3.2 to scalar fields with selfinteraction and we calculate the Green functions in a perturbative way. In Section 3.3 we introduce the Grassmann algebra in order to define PI for fermionic systems. In Section 3.4 we explain the Abelian-, in Section 3.5 the non-Abelian field case where we focus on the occurrence of the Faddeev–Popov ghosts which play—together with the BRS transformation of Section 3.6—an important role in our treatment of anomalies.

3.1 Quantum mechanics

3.1.1 Propagator

In quantum mechanics a state vector in the Schrödinger picture is related to that in the Heisenberg picture by

$$|at\rangle_S = \exp\left[-\frac{i}{\hbar}Ht\right]|a\rangle_H, \tag{3.1}$$

where H is the total Hamiltonian

$$H = \frac{p^2}{2\mu} + V \tag{3.2}$$

and a denotes some quantum number. The state vector $|at\rangle_S$ obeys the Schrödinger equation

$$i\hbar\frac{\partial}{\partial t}|at\rangle_S = H|at\rangle_S. \tag{3.3}$$

The moving frame is defined by

$$|xt\rangle = \exp\left[\frac{i}{\hbar}Ht\right]|x\rangle \tag{3.4}$$

so that we obtain

$$\langle x|at\rangle_S = \langle x|\exp\left[-\frac{i}{\hbar}Ht\right]|a\rangle_H = \langle xt|a\rangle_H \tag{3.5}$$

or in terms of the wave functions

$$\psi_a(x,t) = \exp\left[-\frac{i}{\hbar}Ht\right]\psi_a(x) \tag{3.6}$$

(from now on we suppress the symbols S, H).

The basic quantity we deal with in the PI approach is the propagator. Quantum mechanically it is the transition amplitude for a state localized at (x_1, t_1) to a state at (x_N, t_N)

$$K(x_N, t_N; x_1, t_1) := \langle x_N t_N|x_1 t_1\rangle = \langle x_N|\exp\left[-\frac{i}{\hbar}H(t_N - t_1)\right]|x_1\rangle. \tag{3.7}$$

Inserting a complete system of energy eigenstates

$$\sum_n |n\rangle\langle n| = \mathbf{1} \tag{3.8}$$

gives

$$K(x_N, t_N; x_1, t_1) = \sum_n \psi_n(x_N)\psi_n^*(x_1)\exp\left[-\frac{i}{\hbar}E_n(t_N - t_1)\right]. \tag{3.9}$$

This propagator, when acting as an integral kernel, connects the final with the initial wave function

$$\psi_a(x_N, t_N) = \int dx K(x_N, t_N; x, t_1)\psi_a(x, t_1). \tag{3.10}$$

The time evolution of the wave function is completely determined by the propagator. The propagator containing the Hamiltonian (3.2) supplies the whole information for the quantum system.

For $t > t_1$ the propagator satisfies the time-dependent Schrödinger equation

$$i\hbar\frac{\partial}{\partial t}K(x,t;x_1,t_1) = HK(x,t;x_1,t_1), \tag{3.11}$$

which follows immediately by direct differentiation of equation (3.7). In the limit $t \to t_1$ the propagator approaches a delta function

$$\lim_{t\to t_1} K(x,t;x_1;t_1) = \delta(x - x_1) \tag{3.12}$$

due to the normalization

$$\langle x|x_1\rangle = \delta(x - x_1) = \frac{1}{(2\pi\hbar)^d}\int dp \exp\left[\frac{i}{\hbar}p(x - x_1)\right] \tag{3.13}$$

(d denotes the dimension). Therefore the propagator represents the Green function for the Schrödinger equation

$$\left(i\hbar\frac{\partial}{\partial t} - H\right)K(x,t;x_1,t_1) = i\hbar\delta(x - x_1)\delta(t - t_1), \tag{3.14}$$

with initial condition

$$K(x,t;x_1,t_1) = 0 \qquad \text{for } t < t_1. \tag{3.15}$$

3.1.2 Typical examples for the propagator

i) Free propagator

In the case of free motion ($V = 0$) the propagator is

$$K_0(x,t;x_1,t_1) = \langle x|\exp\left[-\frac{i}{\hbar}\frac{p^2}{2\mu}(t - t_1)\right]|x_1\rangle. \tag{3.16}$$

Inserting a complete system of momentum states

$$\int dp|p\rangle\langle p| = \mathbf{1} \tag{3.17}$$

and using the plane waves

$$\langle p|x\rangle = \frac{1}{(2\pi\hbar)^{d/2}}\exp\left[-\frac{i}{\hbar}px\right] \tag{3.18}$$

gives

$$K_0(x,t;x_1,t_1) = \frac{1}{(2\pi\hbar)^d}\int_{-\infty}^{\infty} dp \exp\left[\frac{i}{\hbar}[p(x - x_1) - \frac{p^2}{2\mu}(t - t_1)]\right] \tag{3.19}$$

(again d is the dimension). Evaluating this Gaussian integral with the help of the formula (for 1 dimension)

$$\int_{-\infty}^{\infty} \frac{dx}{2\pi} \exp[-ax^2 + bx + c] = \frac{1}{\sqrt{4\pi a}} \exp\left[\frac{b^2}{4a} + c\right] \tag{3.20}$$

we obtain the well-known result for the **free propagator**

$$K_0(x, t; x_1, t_1) = \left(\frac{\mu}{2\pi\hbar i(t - t_1)}\right)^{d/2} \exp\left[-\frac{\mu(x - x_1)^2}{2\hbar i(t - t_1)}\right]. \tag{3.21}$$

ii) Harmonic oscillator propagator

In the case of the harmonic oscillator

$$H = \frac{p^2}{2\mu} + \frac{\mu\omega^2}{2} x^2 \tag{3.22}$$

or

$$H = \omega\left(a^\dagger a + \frac{d}{2}\right), \tag{3.23}$$

with a and $a^\dagger$ being the annihilation and the creation operator we can sum up the oscillator wave functions and energy levels directly in representation (3.9). The procedure is straightforward leading to the following result for the **oscillator propagator** (for imaginary times $it = \tau$) [Bertlmann 1986]

$$K_{\text{osc}}(x, \tau; y) = \langle x| \exp\left[-(H - \frac{d}{2}\omega)\tau\right] |y\rangle \tag{3.24}$$

$$= \quad N^{-1}(\tau) \exp\left[-\frac{\mu\omega}{4}[a(\tau)(x - y)^2 + a^{-1}(\tau)(x + y)^2]\right], \tag{3.25}$$

where

$$N(\tau) \quad = \quad \left[\frac{\pi}{\mu\omega}(1 - e^{-2\omega\tau})\right]^{d/2} \tag{3.26}$$

$$a(\tau) \quad = \quad \frac{1 + e^{-\omega\tau}}{1 - e^{-\omega\tau}}. \tag{3.27}$$

Clearly, for short times $\tau \to 0$ or weak potential $\omega \to 0$ the propagator (3.25) approaches the free case (3.21).

iii) Moments

We call the propagation of a state localized at x (say at $x = 0$) back to that point a **moment** (again for imaginary times τ; and $\hbar = 1$)

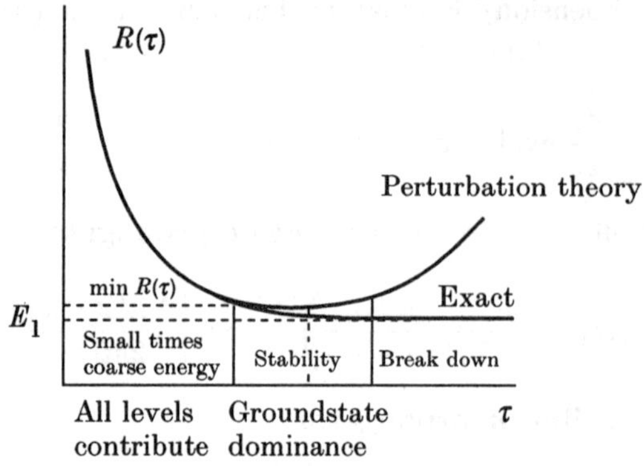

Fig. 3.1. A typical ratio of moments

$$
\begin{aligned}
M(\tau) &:= K(x=0,\tau;y=0) = \langle x=0|e^{-H\tau}|y=0\rangle \\
&= \sum_n |\psi_n(0)|^2 \exp[-E_n\tau].
\end{aligned} \tag{3.28}
$$

Its logarithmic derivative—a **ratio of moments**—serves as a practicable quantity to project just the groundstate energy E_1

$$
R(\tau) := -\frac{d}{d\tau}\ln M(\tau) = -\frac{M'(\tau)}{M(\tau)} \tag{3.29}
$$

$$
= \frac{\sum_n E_n|\psi_n(0)|^2 e^{-E_n\tau}}{\sum_n |\psi_n(0)|^2 e^{-E_n\tau}} \stackrel{\tau\to\infty}{\longrightarrow} E_1. \tag{3.30}
$$

For example, in the free case (3.21) we have (for $d=3$)

$$
M_0(\tau) = \left(\frac{\mu}{2\pi\tau}\right)^{3/2} \tag{3.31}
$$

$$
R_0(\tau) = \frac{3}{2\tau} \stackrel{\tau\to\infty}{\longrightarrow} 0 \tag{3.32}
$$

and for the harmonic oscillator we obtain from formulae (3.24)–(3.27)

$$
M_{\text{osc}}(\tau) = \left[\frac{\mu\omega}{2\pi\sinh\omega\tau}\right]^{3/2} \tag{3.33}
$$

$$
R_{\text{osc}}(\tau) = \frac{3}{2}\omega\coth\omega\tau \stackrel{\tau\to\infty}{\longrightarrow} \frac{3}{2}\omega = E_1. \tag{3.34}
$$

Such moments have been investigated by Bell and Bertlmann [Bell, Bertlmann 1981a,b], [Bertlmann 1982] by considering the minimum of the perturbative result as an approximation to the exact ground state

$$\min_{\tau} R(\tau) = E_1. \tag{3.35}$$

We have plotted the typical feature—the approximation works much better than expected ('magic moments')—in Figure 3.1.

iv) Partition function

We find another important application of the above discussed propagator, which can also be generalized to QFT, in thermodynamics. There the basic quantity is the **partition function** defined by summing the above moments over space

$$Z(\beta) := \int dx\; K(x, \beta; x) = \int dx\; \langle x|e^{-H\beta}|x\rangle. \tag{3.36}$$

Instead of a time τ we work here with the parameter

$$\beta = \frac{1}{kT} \tag{3.37}$$

where T describes the temperature of the system and k is the Boltzmann constant.

Inserting a complete system of energy eigenstates (3.8) and using the normalization of the wave function we can rewrite equation (3.36)

$$\begin{aligned} Z(\beta) &= \sum_n \int dx\; |\psi_n(x)|^2 \exp[-E_n\beta] \\ &= \sum_n \exp[-E_n\beta] = \sum_n \langle n|e^{-H\beta}|n\rangle \end{aligned} \tag{3.38}$$

so that the partition function can generally be defined as the trace of the exponentiated Hamiltonian

$$Z(\beta) = \mathrm{tr}\; e^{-H\beta}. \tag{3.39}$$

The trace is clearly independent of a special representation, say, of equation (3.36) or (3.38). For example, we can calculate the partition function of the harmonic oscillator (3.23) explicitly

$$\begin{aligned} Z_{\mathrm{osc}}(\beta) &= \mathrm{tr}\; e^{-H\beta} = \sum_n \langle n| \exp\left[-\omega(a^\dagger a + \frac{d}{2})\beta\right] |n\rangle \\ &= \exp\left[-\frac{d}{2}\omega\beta\right] \sum_n \left(e^{-\omega\beta}\right)^n. \end{aligned} \tag{3.40}$$

Summing up the geometric series gives

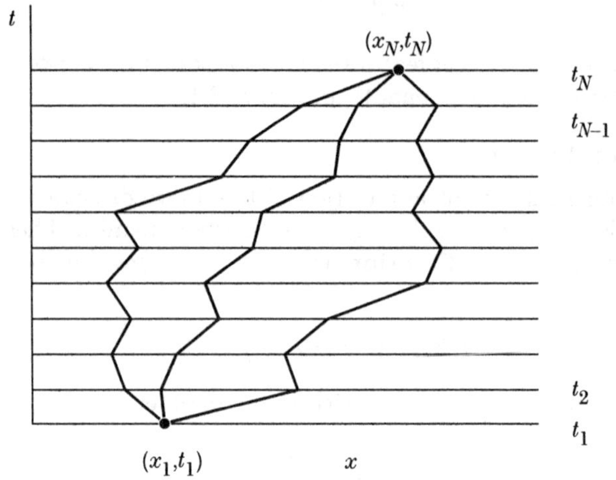

Fig. 3.2. Possible paths from (x_1, t_1) to (x_N, t_N)

$$Z_{\text{osc}}(\beta) = \frac{\exp\left[-\frac{d}{2}\omega\beta\right]}{1 - \exp[-\omega\beta]}. \tag{3.41}$$

In the limit of weak coupling $\omega \to 0$ or, alternatively, in the high temperature limit $T \to \infty$ $(\beta \to 0)$ we approach the free case

$$Z_{\text{osc}}(\beta) \stackrel{\beta\to 0}{\longrightarrow} \frac{1}{\omega\beta} = \frac{kT}{\omega} = Z_0(\beta). \tag{3.42}$$

3.1.3 Feynman's path integral

An alternative formulation of quantum mechanics is based on the propagator which is then represented as a sum—integral—over all possible paths of a particle. This concept was developed by Feynman as a graduate student and the standard reference for this approach is certainly the textbook of Feynman and Hibbs [Feynman, Hibbs 1965].

The idea is to insert complete states into the propagator or transition amplitude (3.7). We discretize the time between the fixed t_1 and t_N

$$t_N > t_{N-1} > \ldots > t_2 > t_1 \tag{3.43}$$

into $N-1$ equal pieces

$$\Delta t = t_j - t_{j-1} = \frac{t_N - t_1}{N-1}. \tag{3.44}$$

Then the transition amplitude can be decomposed into the following integral—for simplicity we consider just one space dimension—

$$\begin{aligned} K(x_N,t_N;x_1,t_1) &= \langle x_N t_N | x_1 t_1 \rangle \\ &= \int dx_{N-1} dx_{N-2} \dots dx_2 \, \langle x_N t_N | x_{N-1} t_{N-1} \rangle \, \langle x_{N-1} t_{N-1} | x_{N-2} t_{N-2} \rangle \\ &\quad \dots \langle x_2 t_2 | x_1 t_1 \rangle . \end{aligned} \tag{3.45}$$

We integrate over all intermediate states, which means we sum over all possible paths (see Figure 3.2). So in Feynman's quantum mechanics *all possible paths* of a particle play a role.

Now, how can we describe an individual path within a small segment? There, according to Dirac [Dirac 1933, 1958] and Feynman [Feynman 1948], the *classical action* S determines the path. So we may try the following equation

$$\langle x_j t_j | x_{j-1} t_{j-1} \rangle = A \exp\left[\frac{i}{\hbar} S(j, j-1)\right] , \tag{3.46}$$

with A being some normalization constant.

In order to calculate, finally, the classical action S in the limit $\Delta t \to 0$ it is certainly legitimate to approximate each path segment by a straight line

$$\begin{aligned} S(j,j-1) &= \int_{t_{j-1}}^{t_j} dt \, L_{\text{class}}(x,\dot{x}) = \int_{t_{j-1}}^{t_j} dt \left[\frac{\mu \dot{x}^2}{2} - V(x)\right] \\ &= \Delta t \left[\frac{\mu}{2}\left(\frac{x_j - x_{j-1}}{\Delta t}\right)^2 - V\left(\frac{x_j + x_{j+1}}{2}\right)\right] . \end{aligned} \tag{3.47}$$

The normalization A is independent of the potential V. Therefore we may evaluate it from the free particle case ($V = 0$), equation (3.21),

$$A = \left(\frac{\mu}{2\pi\hbar i \Delta t}\right)^{1/2} . \tag{3.48}$$

Now we insert equation (3.46) into the propagator decomposition (3.45) and we refine the time discretization. We consider the limit $N \to \infty$, $\Delta t \to 0$ by keeping the endpoints fixed

$$\begin{aligned} K(x_N,t_N;x_1,t_1) &= \langle x_N t_N | x_1 t_1 \rangle \\ &= \lim_{N\to\infty,\Delta t \to 0} \left(\frac{\mu}{2\pi\hbar i \Delta t}\right)^{(N-1)/2} \\ &\quad \cdot \int dx_{N-1} dx_{N-2} \dots dx_2 \prod_{j=2}^{N} \exp\left[\frac{i}{\hbar} S(j,j-1)\right] . \end{aligned} \tag{3.49}$$

Defining the multidimensional integral operator (assuming its existence) as the **path integral**

$$\int_{x_1}^{x_N} d[x(t)] := \lim_{N\to\infty,\Delta t\to 0} \left(\frac{\mu}{2\pi\hbar i\Delta t}\right)^{(N-1)/2} \int \prod_{j=2}^{N-1} dx_j \tag{3.50}$$

we obtain **Feynman's path integral formula** for the propagator

$$\begin{aligned} K(x_N, t_N; x_1, t_1) &= \langle x_N t_N | x_1 t_1 \rangle \\ &= \int_{x_1}^{x_N} d[x(t)] \exp\left[\frac{i}{\hbar}\int_{t_1}^{t_N} dt\, L_{\text{class}}(x, \dot{x})\right] . \end{aligned} \tag{3.51}$$

Proposition:

- *Feynman's path integral is equivalent to Schrödinger's theory!*

Features: Feynman's path integral relies on:

i) the decomposition property (3.45) of the transition amplitude,
ii) the superposition principle—summing of all possible paths,
iii) the classical correspondence for $\hbar \to 0$.

3.1.4 Connection between Lagrange- and Hamilton formalism

What we finally have to show is that the Dirac–Feynman equation (3.46) for the transition amplitude which involves the classical Lagrangian L_{class} is indeed equivalent to the propagator (3.7) discussed at the beginning which contains the Hamiltonian H. For that purpose we derive the path integral in the Hamilton formalism.

Hamilton formalism: We discretize the time as before and we decompose the transition amplitude as we did in equation (3.45). But now we calculate the transition in a small segment in terms of the Hamiltonian (again we just consider one space dimension)

$$\langle x_{j+1} t_{j+1} | x_j t_j \rangle = \langle x_{j+1} | \exp\left[-\frac{i}{\hbar} H (t_{j+1} - t_j)\right] | x_j \rangle . \tag{3.52}$$

We expand for small $\Delta t = t_{j+1} - t_j$

$$\begin{aligned} \langle x_{j+1} t_{j+1} | x_j t_j \rangle &= \langle x_{j+1} | 1 - \frac{i}{\hbar} H \Delta t | x_j \rangle \\ &= \frac{1}{2\pi\hbar} \int dp \exp\left[\frac{i}{\hbar} p (x_{j+1} - x_j)\right] - \frac{i}{\hbar}\Delta t\, \langle x_{j+1} | H | x_j \rangle . \end{aligned} \tag{3.53}$$

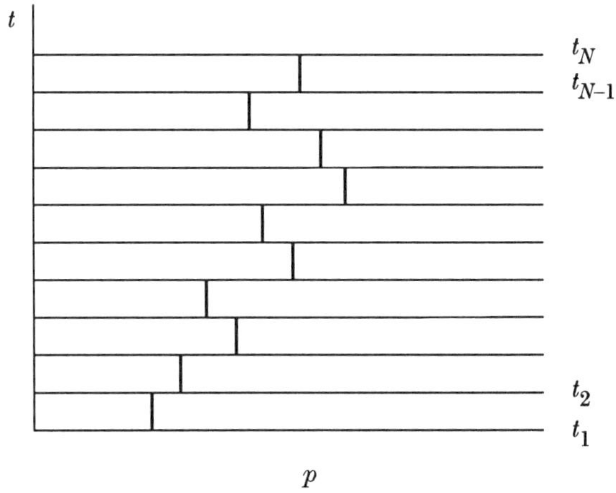

Fig. 3.3. Segments in momentum space

Evaluating the matrix element we begin with the kinetic term

$$\langle x_{j+1} \left| \frac{p^2}{2\mu} \right| x_j \rangle = \frac{1}{2\pi\hbar} \int dp \, \exp\left[\frac{i}{\hbar} p(x_{j+1} - x_j)\right] \frac{p^2}{2\mu} \tag{3.54}$$

and next treat the potential term

$$\begin{aligned} \langle x_{j+1} | V(x) | x_j \rangle &= V(\bar{x}_j) \, \langle x_{j+1} | x_j \rangle = V(\bar{x}_j) \delta(x_{j+1} - x_j) \\ &= \frac{1}{2\pi\hbar} \int dp \exp\left[\frac{i}{\hbar} p(x_{j+1} - x_j)\right] V(\bar{x}_j), \end{aligned} \tag{3.55}$$

where $\bar{x}_j$ is the mean value

$$\bar{x}_j = \frac{x_{j+1} + x_j}{2}. \tag{3.56}$$

Altogether we have

$$\begin{aligned} \langle x_{j+1} | H | x_j \rangle &= \langle x_{j+1} \left| \frac{p^2}{2\mu} + V(x) \right| x_j \rangle \\ &= \frac{1}{2\pi\hbar} \int dp \exp\left[\frac{i}{\hbar} p(x_{j+1} - x_j)\right] H(p, \bar{x}_j). \end{aligned} \tag{3.57}$$

Therefore we get for the transition (3.53) in a small segment of one possible path

$$\langle x_{j+1}t_{j+1}|x_jt_j\rangle$$
$$= \frac{1}{2\pi\hbar}\int dp_j \exp\left[\frac{i}{\hbar}p_j(x_{j+1}-x_j)\right]\left[1-\frac{i}{\hbar}\Delta t\, H(p_j,\bar{x}_j)\right], \quad (3.58)$$

where we have denoted by p_j the momentum between t_{j+1} and t_j (see Figure 3.3). Including also the higher orders in Δt implies

$$\langle x_{j+1}t_{j+1}|x_jt_j\rangle = \frac{1}{2\pi\hbar}\int dp_j \exp\left[\frac{i}{\hbar}[p_j(x_{j+1}-x_j) - \Delta t\, H(p_j,\bar{x}_j)]\right]. \quad (3.59)$$

We insert expression (3.59) into the decomposition of the propagator (3.45) and we refine the segments and finally we perform the limit $N\to\infty$, $\Delta t\to 0$ with fixed endpoints. Defining again the multidimensional integral operators as **path integrals** (assuming their existence)

$$\int_{x_1}^{x_N} d[x(t)] = \lim_{N\to\infty,\Delta t\to 0}\int\prod_{i=2}^{N-1} dx_i \quad (3.60)$$

$$\int d\left[\frac{p(t)}{2\pi\hbar}\right] = \lim_{N\to\infty,\Delta t\to 0}\int\prod_{i=1}^{N-1}\frac{dp_j}{2\pi\hbar} \quad (3.61)$$

we obtain the **path integral for the propagator in the Hamilton formalism**

$$K(x_N,t_N;x_1,t_1) = \langle x_Nt_N|x_1t_1\rangle$$
$$= \int_{x_1}^{x_N} d[x(t)]\int d\left[\frac{p(t)}{2\pi\hbar}\right]\exp\left\{\frac{i}{\hbar}\int_{t_1}^{t_N} dt[p\dot{x} - H(p,x)]\right\}. \quad (3.62)$$

Proposition:

- *The Hamilton representation is equivalent to Feynman's Lagrange formalism !*

Proof.

$$K(x_N,t_N;x_1,t_1) = \langle x_Nt_N|x_1t_1\rangle$$
$$= \lim_{N\to\infty,\Delta t\to 0}\int\prod_{i=2}^{N-1} dx_i\int\prod_{j=1}^{N-1}\frac{dp_j}{2\pi\hbar}$$
$$\cdot\exp\left\{\frac{i}{\hbar}\sum_{j=1}^{N-1}\Delta t\left[p_j\frac{x_{j+1}-x_j}{\Delta t} - \frac{p_j^2}{2\mu} - V(\bar{x}_j)\right]\right\}, \quad (3.63)$$

the Gaussian momentum integration, gives (formula (3.20))

$$\begin{aligned} K(x_N,t_N;x_1,t_1) &= \langle x_N t_N | x_1 t_1 \rangle \\ &= \lim_{N\to\infty,\Delta t\to 0} \left(\frac{\mu}{2\pi\hbar i\Delta t}\right)^{(N-1)/2} \int \prod_{i=2}^{N-1} dx_i \\ &\quad \cdot \exp\left\{ \frac{i}{\hbar} \sum_{j=1}^{N-1} \Delta t \left[\frac{\mu}{2} \left(\frac{x_{j+1}-x_j}{\Delta t}\right)^2 - V(\bar{x}_j) \right] \right\} \\ &= \int_{x_1}^{x_N} d[x(t)] \exp\left[\frac{i}{\hbar} \int_{t_1}^{t_N} dt L_{\text{class}}(x,\dot{x}) \right]. \quad \text{Q.E.D.} \end{aligned} \tag{3.64}$$

Remark: However, this strict equivalence is only valid for Hamiltonians of the above type. For general Hamiltonians (e.g. with velocity dependent potentials) the above momentum integration creates an additional term to the classical Lagrangian which can be subsummed to an effective action [Lee, Yang 1962] (see e.g. [Ryder 1988] and [Lucha, Schöberl 1996]).

3.1.5 Field theory analogies

Our aim is to generalize the quantum mechanical PI formalism to field theory. But there we have different boundary conditions. In QFT particles are created and destroyed. A formalism which describes such features is that of Schwinger [Schwinger 1969].

$Z[J]$ functional or vacuum-to-vacuum transition: In the spirit of Schwinger we study a vacuum state at a time $t \to -\infty$ which evolves into a vacuum state at $t \to +\infty$ while creating and annihilating a particle through the action of a source. The source is introduced into the Lagrangian by replacing

$$L \to L + \hbar J(t)x(t). \tag{3.65}$$

Denoting the vacuum state—the groundstate with zero quantum number—in the presence of the source by $|0,t\rangle^J$ we then consider the vacuum-to-vacuum transition amplitude

$$\langle 0,\infty|0,-\infty\rangle^J \sim Z[J]. \tag{3.66}$$

This represents a functional $Z[J]$ depending on the source J.

What is an explicit expression for that functional $Z[J]$?

We imagine the following situation. We split the time into intervals

$$T < t < t' < T', \tag{3.67}$$

where the source is nonzero in the inner interval

$$J(\bar{t}) \neq 0 \qquad \text{for } \bar{t} \in [t, t'] \tag{3.68}$$

and vanishes outside

$$J(\bar{t}) = 0 \qquad \text{for } \bar{t} \in [T, t) \text{ and } \bar{t} \in (t', T']. \tag{3.69}$$

We represent the transition amplitude of a particle localized in space and time in the presence of a source by Feynman's path integral formula (3.51)

$$\langle X'T'|XT\rangle^J = \int d[x(t)] \exp\left[\frac{i}{\hbar}\int_T^{T'} dt(L + \hbar Jx)\right]. \tag{3.70}$$

On the other hand, we can decompose this transition amplitude by the following intermediate states

$$\langle X'T'|XT\rangle^J = \int dx'dx \,\langle X'T'|x't'\rangle \,\langle x't'|xt\rangle^J \,\langle xt|XT\rangle. \tag{3.71}$$

We express the matrix elements with vanishing current by wave functions and energy levels in the standard way (recall equation (3.9))

$$\langle X'T'|x't'\rangle = \sum_n \psi_n(X')\psi_n^*(x') \exp\left[-\frac{i}{\hbar}E_n(T' - t')\right], \tag{3.72}$$

$$\langle xt|XT\rangle = \sum_m \psi_m(x)\psi_m^*(X) \exp\left[-\frac{i}{\hbar}E_m(t - T)\right]. \tag{3.73}$$

Now we rotate the time axis by an angle of $\pi/2$—we consider imaginary times (analytic continuation)—and we take the limits $T \to i\infty$, $T' \to -i\infty$. Then we just project the groundstate contribution

$$\lim_{T' \to -i\infty} \exp\left[\frac{i}{\hbar}E_0 T'\right] \langle X'T'|x't'\rangle = \psi_0(X')\psi_0^*(x', t'), \tag{3.74}$$

$$\lim_{T \to i\infty} \exp\left[-\frac{i}{\hbar}E_0 T\right] \langle xt|XT\rangle = \psi_0(x, t)\psi_0^*(X). \tag{3.75}$$

For the transition amplitude (3.71) hence it follows that

$$\begin{aligned} &\lim_{T' \to -i\infty, T \to i\infty} \langle X'T'|XT\rangle^J \\ &= \psi_0(X')\psi_0^*(X) \exp\left[-\frac{i}{\hbar}E_0(T' - T)\right]_{T' \to -i\infty, T \to i\infty} \\ &\quad \cdot \int dx'dx \,\psi_0^*(x', t')\,\langle x't'|xt\rangle^J \psi_0(x, t). \end{aligned} \tag{3.76}$$

The integral now represents the quantity we are finally interested in

$$\begin{aligned}\int dx'dx\psi_0^*(x',t')\langle x't'|xt\rangle^J\psi_0(x,t) \\ &= \int dx'dx\ \langle 0,t'|x'\rangle\ \langle x't'|xt\rangle^J\ \langle x|0,t\rangle \\ &= \langle 0,t'|0,t\rangle^J \stackrel{t'\to\infty,t\to-\infty}{\longrightarrow} \langle 0,\infty|0,-\infty\rangle^J. \qquad (3.77)\end{aligned}$$

Since the wavefunction–energy factor in equation (3.76) is just a number which does not matter we get

$$\langle 0,\infty|0,-\infty\rangle^J \sim \lim_{T'\to -i\infty, T\to i\infty}\langle X'T'|XT\rangle^J \qquad (3.78)$$

and inserting the path integral (3.70) we find the $Z[J]$ functional explicitly

$$Z[J] = \lim_{T'\to -i\infty, T\to i\infty}\int d[x(t)]\exp\left[\frac{i}{\hbar}\int_T^{T'} dt(L+\hbar Jx)\right]. \qquad (3.79)$$

Alternatively, instead of rotating the time axis we could also add a term $-\frac{1}{2}i\varepsilon x^2$ to the Hamiltonian, or $\frac{1}{2}i\varepsilon x^2$ to the Lagrangian, in order to isolate the groundstate contribution. Then the $Z[J]$ functional denotes

$$Z[J] = \int d[x(t)]\exp\left[\frac{i}{\hbar}\int_{-\infty}^{\infty} dt(L+\hbar Jx+\frac{1}{2}i\varepsilon x^2)\right]. \qquad (3.80)$$

We will use this last form for a generalization to QFT.

Derivatives of $Z[J]$ or time ordered products of operators: What we finally need as analogues for QFT are time ordered products of operators. Let us begin with the matrix element of the operator $x(t_K)$. Discretizing the time analogously to before (see equations (3.43) and (3.44)) into $N+1$ equal pieces between a fixed initial point (x_i,t_i) and a fixed final point (x_f,t_f), we decompose the matrix element

$$\begin{aligned}\langle x_f t_f|x(t_K)|x_i t_i\rangle &= \int dx_N\dots dx_1\ \langle x_f t_f|x_N t_N\rangle\ \langle x_N t_N|x_{N-1}t_{N-1}\rangle\dots \\ &\quad \dots\langle x_K t_K|x(t_K)|x_{K-1}t_{K-1}\rangle\dots\langle x_1 t_1|x_i t_i\rangle. \qquad (3.81)\end{aligned}$$

We now replace the operator by its scalar eigenvalue which we denote by the same symbol. We use for each segment transition the Dirac–Feynman equation (3.46) and we perform the continuum limit $N\to\infty$, $\Delta t\to 0$ which implies the path integral

$$\langle x_f t_f | x(t_K) | x_i t_i \rangle = \int_{x_i}^{x_f} d[x(t)] x(t_K) \exp \left[\frac{i}{\hbar} \int_{t_i}^{t_f} dt \, L_{\text{class}}(x, \dot{x}) \right] . \quad (3.82)$$

Note, on the l.h.s. $x(t_K)$ represents an operator whereas on the r.h.s. it is a scalar function.

Next we consider the matrix element of two operators $x(t_K)x(t_L)$. For $t_f > t_K > t_L > t_i$ we get

$$\begin{aligned} & \langle x_f t_f | x(t_K) x(t_L) | x_i t_i \rangle \\ = & \int dx_N \dots dx_1 \langle x_f t_f | x_N t_N \rangle \langle x_N t_N | x_{N-1} t_{N-1} \rangle \dots \\ & \dots \langle x_K t_K | x(t_K) | x_{K-1} t_{K-1} \rangle \dots \langle x_L t_L | x(t_L) | x_{L-1} t_{L-1} \rangle \dots \\ & \dots \langle x_1 t_1 | x_i t_i \rangle \end{aligned} \quad (3.83)$$

which leads by analogous argument to the path integral

$$\begin{aligned} & \langle x_f t_f | x(t_K) x(t_L) | x_i t_i \rangle \\ = & \int_{x_i}^{x_f} d[x(t)] x(t_K) x(t_L) \exp \left[\frac{i}{\hbar} \int_{t_i}^{t_f} dt L_{\text{class}}(x, \dot{x}) \right] . \end{aligned} \quad (3.84)$$

On the other hand, for $t_f > t_L > t_K > t_i$ the operators on the l.h.s. are reversed whereas the PI on the r.h.s. remains the same (there $x(t_L)x(_K)$ are interchangeable scalars). Consequently, the time discretization provides a time ordered product automatically

$$\begin{aligned} & \langle x_f t_f | T[x(t_1) x(t_2)] | x_i t_i \rangle \\ = & \int_{x_i}^{x_f} d[x(t)] x(t_1) x(t_2) \exp \left[\frac{i}{\hbar} \int_{t_i}^{t_f} dt L_{\text{class}}(x, \dot{x}) \right] , \end{aligned} \quad (3.85)$$

where T represents the familiar **time ordering for the operators**, say A and B,

$$T[A(t_1)B(t_2)] = \Theta(t_1 - t_2)A(t_1)B(t_2) + \Theta(t_2 - t_1)B(t_2)A(t_1). \quad (3.86)$$

The generalization to n operators is straightforward

$$\begin{aligned} & \langle x_f t_f | T[x(t_1) x(t_2) \dots x(t_n)] | x_i t_i \rangle \\ = & \int_{x_i}^{x_f} d[x(t)] x(t_1) x(t_2) \dots x(t_n) \exp \left[\frac{i}{\hbar} \int_{t_i}^{t_f} dt L_{\text{class}}(x, \dot{x}) \right] . \end{aligned} \quad (3.87)$$

Functional derivatives: Next we want to explain the derivatives of a functional and, respectively, its Taylor series expansion.

Definition: Let f be a well-behaved test function (infinitely differentiable and decreasing faster than any power of $|x|^{-1}$) then a functional F is defined by

$$F : f \in C^{\infty}(M) \to F[f] \in \mathbf{R}, \mathbf{C}. \tag{3.88}$$

By analogy with ordinary functions the expansion of a functional in the vicinity of an element f

$$F[f + \varepsilon h] = F[f] + \varepsilon \frac{\delta F[f]}{\delta f} h + O(\varepsilon^2) \tag{3.89}$$

defines the **derivative**

$$\int dx \frac{\delta F[f]}{\delta f(x)} h(x) = \frac{\delta F[f]}{\delta f} h = \lim_{\varepsilon \to 0} \frac{1}{\varepsilon} \{F[f + \varepsilon h] - F[f]\}. \tag{3.90}$$

For the special choice of

$$h(x) = \delta(x - y) \tag{3.91}$$

the **functional derivative** can be written as

$$\frac{\delta F[f](x)}{\delta f(y)} = \lim_{\varepsilon \to 0} \frac{1}{\varepsilon} \{F[f(x) + \varepsilon \delta(x - y)] - F[f](x)\}. \tag{3.92}$$

Examples:

i)

$$F[f] = \int dx f(x) \tag{3.93}$$

$$\begin{aligned} \frac{\delta F[f]}{\delta f(y)} &= \lim_{\varepsilon \to 0} \frac{1}{\varepsilon} \{ \int dx [f(x) + \varepsilon \delta(x - y)] - \int dx f(x)\} \\ &= \int dx\, \delta(x - y) = 1. \end{aligned} \tag{3.94}$$

ii)

$$F[f](x) = \int dz\, G(x, z) f(z) \tag{3.95}$$

$$\begin{aligned} \frac{\delta F[f](x)}{\delta f(y)} &= \lim_{\varepsilon \to 0} \frac{1}{\varepsilon} \{ \int dz\, G(x, z)[f(z) + \varepsilon \delta(z - y)] \\ &\quad - \int dz\, G(x, z) f(z)\} \\ &= \int dz\, G(x, z) \delta(z - y) = G(x, y). \end{aligned} \tag{3.96}$$

iii)

$$F = \mathbf{1}, \qquad F[f](x) = \int dy\, \delta(y-x) f(y) = f(x) \tag{3.97}$$

$$\frac{\delta f(x)}{\delta f(y)} = \lim_{\varepsilon \to 0} \frac{1}{\varepsilon}\{f(x) + \varepsilon \delta(x-y) - f(x)\} = \delta(x-y). \tag{3.98}$$

We again consider the derivative of a functional as a functional; we differentiate once more and we obtain the second derivative and successively all higher derivatives. Then the **Taylor series expansion** of a functional is given by

$$\begin{aligned} F[f+h] &= \sum_{n=0}^{\infty} \frac{1}{n!} \frac{\delta^n F[f]}{\delta f^n} h^n \\ &= \sum_{n=0}^{\infty} \frac{1}{n!} \int dx_1 \dots dx_n \frac{\delta^n F[f]}{\delta f(x_1) \dots \delta f(x_n)} h(x_1) \dots h(x_n). \end{aligned} \tag{3.99}$$

Now we return to our $Z[J]$ functional expressed as a PI (3.80)

$$Z[J] = \int d[x(t)] \exp\left\{ \frac{i}{\hbar} \int_{-\infty}^{\infty} dt [L(x,\dot{x}) + \hbar J(t) x(t) + \frac{1}{2} i\varepsilon x^2(t)] \right\} \tag{3.100}$$

and we differentiate according to the above rules

$$\begin{aligned} \frac{\delta Z[J]}{\delta J(t_1)} &= \lim_{\varepsilon \to 0} \frac{1}{\varepsilon} \int d[x(t)] \\ &\quad \cdot \left\{ \exp\left[\frac{i}{\hbar} \int_{-\infty}^{\infty} dt [\hbar (J(t) + \varepsilon \delta(t-t_1)) x(t) + \dots] \right] \right. \\ &\quad \left. - \exp\left[\frac{i}{\hbar} \int_{-\infty}^{\infty} dt [\hbar J(t) x(t) + \dots] \right] \right\}. \end{aligned} \tag{3.101}$$

Expanding the ε exponential provides

$$\frac{\delta Z[J]}{\delta J(t_1)} = i \int d[x(t)] x(t_1) \exp\left\{ \frac{i}{\hbar} \int_{-\infty}^{\infty} dt \left[L(x,\dot{x}) + \hbar J x + \frac{1}{2} i \varepsilon x^2 \right] \right\}. \tag{3.102}$$

The higher derivatives then supply the result

$$\begin{aligned} \frac{\delta^n Z[J]}{\delta J(t_1) \dots \delta J(t_n)} &= i^n \int d[x(t)] x(t_1) \dots x(t_n) \\ &\quad \cdot \exp\left\{ \frac{i}{\hbar} \int_{-\infty}^{\infty} dt \left[L(x,\dot{x}) + \hbar J x + \frac{1}{2} i \varepsilon x^2 \right] \right\}. \end{aligned} \tag{3.103}$$

This PI represents the T product of operators (recall equation (3.87)) and the additional term $\frac{1}{2}i\varepsilon x^2$ just projects the groundstate (recall equation (3.80)) so that we end up with the vacuum expectation value of the time ordered operators

$$\left.\frac{\delta^n Z[J]}{\delta J(t_1)\dots\delta J(t_n)}\right|_{J=0} \sim i^n\langle 0,\infty|T[x(t_1)\dots x(t_n)]|0,-\infty\rangle. \tag{3.104}$$

The symbol $\sim$ means equality up to a normalization factor. The relation (3.104) represents the important analogue for QFT which we are going to discuss in the next sections.

3.2 Scalar field theory

3.2.1 Free scalar fields

Instead of trajectories $x(t)$ of a particle we now consider real scalar fields $\phi(x)$ which depend on the space–time point $x = (t, x, y, z)$. The action is given by

$$S = \int d^4x \mathcal{L}(\phi, \partial_\mu\phi), \tag{3.105}$$

with $\mathcal{L}$ being the Lagrangian density. We also suppose that the fields $\phi(x)$ have a source $J(x)$ in the sense of Schwinger. Let us now find a path integral over these scalar fields. We search the $Z[J]$ functional, an expression analogous to equation (3.80), by proceeding as in the quantum mechanical case. We discretize the 4-dimensional space–time and we refine the discretization considering the limiting case as the PI. So we can essentially take over formula (3.80) by replacing $d[x(t)] \to d[\phi(x)] =: d\phi$ (we also choose $\hbar = 1$) and we obtain the **path integral over scalar fields**

$$Z[J] = \int d\phi \exp\left\{i\int d^4x[\mathcal{L}(\phi, \partial_\mu\phi) + J\phi + \frac{1}{2}i\varepsilon\phi^2]\right\}. \tag{3.106}$$

Again, it represents the vacuum-to-vacuum amplitude in the presence of the source J

$$Z[J] \sim \langle 0,\infty|0,-\infty\rangle^J. \tag{3.107}$$

Now we study the free case

$$Z_0[J] = \int d\phi \exp\left\{i\int dx[\mathcal{L}_0 + \phi J + \frac{1}{2}i\varepsilon\phi^2]\right\}, \tag{3.108}$$

with $\mathcal{L}_0$ being the free Lagrangian

$$\mathcal{L}_0 = \frac{1}{2}(\partial_\mu\phi\partial^\mu\phi - m^2\phi^2) \tag{3.109}$$

and with the equation of motion, the Klein–Gordon equation

$$(\Box + m^2)\phi = 0. \tag{3.110}$$

For simplicity we write from now on just dx for the 4-dimensional (n-dimensional in general) volume element. We integrate the kinetic term in $\mathcal{L}_0$ by parts and we obtain the $Z_0[J]$ **functional for free scalar fields**

$$Z_0[J] = \int d\phi \exp\left\{-i\int dx[\frac{1}{2}\phi(\Box + m^2 - i\varepsilon)\phi - J\phi]\right\}. \tag{3.111}$$

Note: The ε-regulator term which guarantees the vacuum-to-vacuum transition also provides the correct Feynman propagator for the field ϕ where the mass acquires an imaginary part $m^2 \to m^2 - i\varepsilon$.

Euclidean space: On the other hand, we also achieve the vacuum-to-vacuum transition by rotating the time axis by an angle of $\pi/2$ (see Section 3.1.5). Defining

$$ix_0 = x_4 \qquad \text{and} \qquad \begin{array}{rcl} -i\partial_0 &=& \partial_4 \\ -ik_0 &=& k_4 \end{array} \tag{3.112}$$

we switch to Euclidean space so that we get

$$\begin{array}{rcl} x^2 &=& x_0^2 - \vec{x}^2 = -(x_1^2 + x_2^2 + x_3^2 + x_4^2) = -x_E^2 \\ k^2 &=& k_0^2 - \vec{k}^2 = -(k_1^2 + k_2^2 + k_3^2 + k_4^2) = -k_E^2 \\ \Box &=& \partial_0^2 - \vec{\nabla}^2 = -(\partial_1^2 + \partial_2^2 + \partial_3^2 + \partial_4^2) = -\Box_E \end{array} \tag{3.113}$$

and for the volume elements

$$\begin{array}{rcl} id^4x &=& d^4x_E =: dx_E \\ -id^4k &=& d^4k_E =: dk_E. \end{array} \tag{3.114}$$

Then the $Z_0[J]$ **functional in Euclidean space** is

$$Z_0[J] = \int d\phi \exp\left\{-\int dx_E \left[\frac{1}{2}(\partial_\mu^E\phi)^2 + \frac{1}{2}m^2\phi^2 - J\phi\right]\right\}. \tag{3.115}$$

The exponent is negative definite so that the path integral can converge.

Alternative representation: When working with the $Z_0[J]$ functional it is more convenient to modify it. In the PI the ϕ-integration extends over all possible fields, they do not obey the Klein–Gordon equation. Now we shift the field

$$\phi \to \phi + \phi_0 \tag{3.116}$$

by an additional field ϕ_0 which satisfies the Klein–Gordon equation with source

$$\mathcal{D}\phi_0 = J, \tag{3.117}$$

where

$$\mathcal{D} := \Box + m^2 - i\varepsilon. \tag{3.118}$$

The solution is

$$\phi_0(x) = -\int dy \Delta_F(x-y)J(y), \tag{3.119}$$

where the kernel $\Delta_F(x-y)$ represents the scalar Feynman propagator, the causal Green function defined by

$$\mathcal{D}\Delta_F(x-y) = -\delta(x-y) \quad \text{or} \quad \mathcal{D}^{-1}(x) = -\Delta_F(x) \tag{3.120}$$

and has the integral representation

$$\Delta_F(x) = \frac{1}{(2\pi)^4}\int \frac{d^4k e^{-ikx}}{k^2 - m^2 + i\varepsilon}. \tag{3.121}$$

Then the exponential in $Z_0[J]$ (3.111) changes to

$$\frac{1}{2}\phi\mathcal{D}\phi - J\phi \overset{\phi\to\phi+\phi_0}{\longrightarrow} \frac{1}{2}\phi\mathcal{D}\phi + \frac{1}{2}J\int dy\Delta_F(x-y)J(y). \tag{3.122}$$

so that the $Z_0[J]$ **functional for the free case** can be rewritten

$$Z_0[J] = \mathcal{N}\exp\left[-\frac{i}{2}\int dxdyJ(x)\Delta_F(x-y)J(y)\right], \tag{3.123}$$

with the **normalization**

$$\mathcal{N} = \int d\phi \exp\left[-\frac{i}{2}\int dx\phi\mathcal{D}\phi\right]. \tag{3.124}$$

The ϕ-field dependence has factorized and can be absorbed by the normalization. What remains is just an exponential of the sources connected by the scalar propagator. This expression (3.123) is especially suitable for series expansions and will be needed later on.

Gaussian integration: Another notion which proves important, especially in connection with gauge theories and anomalies, is that of a determinant of a differential operator. We will introduce it here. Path integrals of the above type, equation (3.124), are called Gaussian and can be expressed

in the following way:
A Gaussian formula can be written as the determinant of the matrix A

$$\int \frac{d^n x}{(2\pi)^{n/2}} \exp\left[-\frac{1}{2}(x, Ax)\right] = (\det A)^{-1/2}. \tag{3.125}$$

Actually formula (3.125) is valid for any real symmetric matrix A.

Extending the above formula to the complex space $\mathbf{C}^n$ we begin with $n = 2$ in equation (3.125) and transform the variables to complex ones $z = x + iy, z^* = x - iy$, then we get

$$\int \frac{dz^*}{(2\pi i)^{1/2}} \frac{dz}{(2\pi i)^{1/2}} e^{-az^* z} = a^{-1}. \tag{3.126}$$

The **generalization to n dimensions** is straightforward and gives

$$\int \frac{d^n z^*}{(2\pi i)^{n/2}} \frac{d^n z}{(2\pi i)^{n/2}} e^{-(z^*, Az)} = (\det A)^{-1}, \tag{3.127}$$

where A is a positive definite Hermitian matrix operator.

Shifting the integration by

$$\begin{aligned} x &\rightarrow x + A^{-1}b \\ z &\rightarrow z - A^{-1}u \end{aligned} \tag{3.128}$$

we obtain the **Gauss-like formulae**

$$\int \frac{d^n x}{(2\pi)^{n/2}} e^{-\frac{1}{2}(x,Ax)-(b,x)} = (\det A)^{-1/2} e^{\frac{1}{2}(b,A^{-1}b)} \tag{3.129}$$

$$\int \frac{d^n z^*}{(2\pi i)^{n/2}} \frac{d^n z}{(2\pi i)^{n/2}} e^{-(z^*,Az)+(u^*,z)+(z^*,u)} = (\det A)^{-1} e^{(u^*,A^{-1}u)}. \tag{3.130}$$

These formulae, valid in a finite, n-dimensional vector space, can be generalized to an infinite-dimensional function space $\{\phi\}$ with the measure $d\phi$ or $d\phi^*$. There the inner product with some operator A is defined by

$$(\phi, A\phi) = \int dx dy\, \phi^*(x) A(x-y) \phi(y) \tag{3.131}$$

or

$$(\phi, A\phi) = \int dx\, \phi^*(x) A(x) \phi(x) \tag{3.132}$$

if the operator A is diagonal

$$A(x - y) = A(x)\delta(x - y). \tag{3.133}$$

Returning to our $Z_0[J]$ functional (3.123) we consider the normalization $\mathcal{N}$ (3.124) and notice that it is a Gauss integral

$$\mathcal{N} = \int d\phi \exp\left[-\frac{1}{2}\int dx\ \phi i\mathcal{D}\phi\right] = (\det i\mathcal{D})^{-1/2}, \tag{3.134}$$

which can be expressed by the determinant of the Klein–Gordon differential operator $i\mathcal{D}$ (3.118).

If we assumed complex scalar fields the normalization would be

$$\mathcal{N} = \int d\phi^* d\phi \exp\left[-\int dx\ \phi^* i\mathcal{D}\phi\right] = (\det i\mathcal{D})^{-1}. \tag{3.135}$$

Note: The Gaussian formula (3.129), generalized to infinite dimensions, serves as a connection between the two expressions (3.111) and (3.123) for the $Z_0[J]$ functional.

3.2.2 Free Green functions

Now we show that the $Z_0[J]$ functional—the vacuum-to-vacuum transition amplitude—just represents the generating functional for all free Green functions.

First we normalize expression (3.107) according to our boundary conditions. We choose the strict identity

$$Z_0[J] \equiv \langle 0, \infty | 0, -\infty\rangle^J \tag{3.136}$$

then $Z_0[J]$ is automatically normalized to unity for $J = 0$

$$Z_0[0] = 1. \tag{3.137}$$

Therefore the PI (3.111) must be normalized by the constant factor $\mathcal{N}$ (3.124)

$$Z_0[J] = \frac{1}{\mathcal{N}}\int d\phi \exp\left\{-i\int dx[\frac{1}{2}\phi\mathcal{D}\phi - J\phi]\right\} \tag{3.138}$$

so that

$$Z_0[J] = \exp\left\{-\frac{i}{2}\int dxdy\ J(x)\Delta_F(x-y)J(y)\right\}. \tag{3.139}$$

How can we interpret this exponential (3.139)? Expanding the exponential we obtain the following series

$$Z_0[J] = 1 - \frac{i}{2}\int J\Delta_F J + \frac{1}{2!}\left(\frac{i}{2}\right)^2\left[\int J\Delta_F J\right]^2 - \frac{1}{3!}\left(\frac{i}{2}\right)^3\left[\int J\Delta_F J\right]^3 + \dots \tag{3.140}$$

(the integrals are written in an obvious short-hand notation) which we describe diagrammatically as

$$Z_0[J] = 1 - \frac{i}{2} \times\!\!\longrightarrow\!\!\times + \frac{1}{2!}\left(\frac{i}{2}\right)^2 \begin{array}{c}\times\!\!\longrightarrow\!\!\times\\ \times\!\!\longrightarrow\!\!\times\end{array} - \frac{1}{3!}\left(\frac{i}{2}\right)^3 \begin{array}{c}\times\!\!\longrightarrow\!\!\times\\ \times\!\!\longrightarrow\!\!\times\\ \times\!\!\longrightarrow\!\!\times\end{array} + \ldots \tag{3.141}$$

and

$$\begin{aligned} \Delta_F(x-y) &= \underset{x}{}\overline{\qquad\quad}\underset{y}{} \\ J(y) &= \underset{y}{\times} \end{aligned} \tag{3.142}$$

So the vacuum-to-vacuum amplitude contains the propagation of 0, 1, 2, 3, ... particles between sources.

Now we turn to the **Green functions** or **n-point functions** of the theory. They are defined by the vacuum expectation value of the time ordered field operators. Generalizing the relation (3.104) to field theory we obtain

$$\begin{aligned} \langle 0|T\phi(x_1)\ldots\phi(x_n)|0\rangle &= \frac{1}{i^n} \left.\frac{\delta^n Z_0[J]}{\delta J(x_1)\ldots\delta J(x_n)}\right|_{J=0} \\ &=: \tau_0(x_1,\ldots,x_n). \end{aligned} \tag{3.143}$$

So the quantization of the fields is given by a path integral.

On the other hand, the Taylor series of the functional $Z_0[J]$ is determined by its derivatives (recall equation (3.99))

$$\begin{aligned} Z_0[J] &= \sum_{n=0}^{\infty} \frac{1}{n!} \left.\frac{\delta^n Z_0[J]}{\delta J^n}\right|_{J=0} J^n \\ &= \sum_{n=0}^{\infty} \frac{1}{n!} \int dx_1 \ldots dx_n \left.\frac{\delta^n Z_0[J]}{\delta J(x_1)\ldots\delta J(x_n)}\right|_{J=0} \\ &\quad \cdot J(x_1)\ldots J(x_n) \end{aligned} \tag{3.144}$$

and with the above definition (3.143) for the Green functions $Z_0[J]$ becomes the **generating functional for all** (free) **Green functions**

$$Z_0[J] = \sum_{n=0}^{\infty} \frac{i^n}{n!} \int dx_1 \ldots dx_n \; J(x_1)\ldots J(x_n)\tau_0(x_1,\ldots,x_n). \tag{3.145}$$

This series has its diagrammatic correspondence ($n = 2$ for 1. diagram, $n = 4$ for 2. diagram, etc.) in equation (3.141).

It is also convenient to rewrite the series (3.145) in the following way

$$
\begin{aligned}
Z_0[J] &= \sum_{n=0}^{\infty} \frac{i^n}{n!} \int dx_1 \dots dx_n \, J(x_1) \dots J(x_n) \langle 0|T\phi(x_1)\dots\phi(x_n)|0\rangle \\
&= \langle 0|T \exp\left[i \int dx J(x)\phi(x) \right] |0\rangle. \qquad (3.146)
\end{aligned}
$$

Examples:

i) *2-point function* (for functional differentiation see Section 3.1.5)

$$
\begin{aligned}
\tau_{12}^0 &= \left. \frac{\delta^2 Z_0[J]}{i\delta J_1 i\delta J_2} \right|_{J=0} \\
&= \left. \frac{\delta}{i\delta J_1}(-) \int J\Delta_F(x - x_2) \exp\left[-\frac{i}{2} \int J\Delta_F J \right] \right|_{J=0} \\
&= i\Delta_F(x_1 - x_2) \\
&= i \; \overline{x_1 \quad x_2} \; , \qquad (3.147)
\end{aligned}
$$

where we use the short-hand notation $\tau_{12}^0 = \tau_0(x_1, x_2)$ and $J_i = J(x_i)$, $i = 1, 2, \dots$.

ii) *4-point function*

$$
\begin{aligned}
\tau_{1234}^0 &= \left. \frac{\delta^4 Z_0[J]}{i\delta J_1 i\delta J_2 i\delta J_3 i\delta J_4} \right|_{J=0} \\
&= -[\Delta_F(x_1 - x_2)\Delta_F(x_3 - x_4) + \Delta_F(x_1 - x_3)\Delta_F(x_2 - x_4) \\
&\quad + \Delta_F(x_1 - x_4)\Delta_F(x_2 - x_3)] \qquad (3.148)
\end{aligned}
$$

or diagrammatically

$$
\tau_{1234}^0 = -\left[\; = \; + \; \| \; + \; \times \; \right] = -3 \; = \; . \qquad (3.149)
$$

Remark: We notice that any odd n-point function vanishes

$$
\tau_0(x_1, \dots, x_{2n-1}) = 0 \qquad (3.150)
$$

since there exists a factor proportional to the source $J \to 0$. So we are left only with even n-point functions which are products of 2-point functions

$$
\tau_0(x_1, \dots, x_{2n}) = \sum_{\text{perm}} \tau_0(x_{p_1}, x_{p_2}) \dots \tau_0(x_{p_{2n-1}}, x_{p_{2n}}). \qquad (3.151)
$$

This result of the PI formalism relying on the rules of functional differentiation corresponds precisely to the Wick theorem in the canonical quantization formalism.

3.2.3 Interacting fields

Next we assume that the scalar fields also interact and for simplicity we choose the ϕ^4 selfinteraction

$$\mathcal{L} = \mathcal{L}_0 + \mathcal{L}_I = \frac{1}{2}\partial_\mu\phi\partial^\mu\phi - \frac{1}{2}m^2\phi^2 - \frac{g}{4!}\phi^4. \tag{3.152}$$

Then analogous to the free case (3.136), (3.138), the **generating functional of the interacting fields** is

$$\begin{aligned} Z[J] &= \langle 0,\infty|0,-\infty\rangle^J & (3.153)\\ &= \frac{1}{\mathcal{N}}\int d\phi \exp\left\{i\int dx[\mathcal{L}_0 + \mathcal{L}_I + J\phi]\right\}, & (3.154)\end{aligned}$$

with the normalization (numerator for $J = 0$)

$$\mathcal{N} = \int d\phi \exp\left\{i\int dx[\mathcal{L}_0 + \mathcal{L}_I]\right\} \tag{3.155}$$

(or $Z[0] = 1$). For the vanishing interaction we come back to the free case which we discussed before

$$Z[J] \overset{g\to 0}{\longrightarrow} Z_0[J]. \tag{3.156}$$

Perturbation theory: In order to calculate the Green functions perturbatively to a certain order in the coupling constant g we have to reformulate the path integral (3.154) in a suitable way.

Proposition:

- *The perturbative series of $Z[J]$ is supplied by*

$$Z[J] = \frac{1}{\mathcal{N}}\exp\left\{i\int dx\mathcal{L}_I\left(\frac{\delta}{i\delta J}\right)\right\} Z_0[J], \tag{3.157}$$

  where $Z_0[J]$ is given by expression (3.139) and the normalization $\mathcal{N}$ represents the numerator for $J = 0$.

Proof. For quick reasoning we separate the interaction

$$Z[J] = \frac{1}{\mathcal{N}}\int d\phi \exp\left\{i\int dx\mathcal{L}_I(\phi)\right\}\exp\left\{i\int dx[\mathcal{L}_0 + J\phi]\right\}. \tag{3.158}$$

Considering now the first exponential as a power series in the variable ϕ we may replace ϕ by the derivative with respect to J

$$\phi \rightarrow \frac{\delta}{i\delta J}, \tag{3.159}$$

which will act on the second exponential

$$Z[J] = \frac{1}{\mathcal{N}} \int d\phi \exp\left\{ i \int dx \mathcal{L}_I \left(\frac{\delta}{i\delta J} \right) \right\} \exp\left\{ i \int dx [\mathcal{L}_0 + J\phi] \right\}. \text{ Q.E.D.} \tag{3.160}$$

Recipe: Perturbation theory is established by treating the exponential as a power series in the interaction, in the coupling constant g. We apply the functional derivatives to $Z_0[J]$ to each order in g. The Green functions finally are calculated from the derivatives of the total functional $Z[J]$.

Example: To order g we have

$$Z[J] = \frac{1}{\mathcal{N}} \left[1 - \frac{ig}{4!} \int dz \left(\frac{\delta}{i\delta J(z)} \right)^4 \right] Z_0[J]. \tag{3.161}$$

We must evaluate the fourth derivative of $Z_0[J]$. This we do successively:

$$\begin{aligned}
Z_0[J] &= \exp\left[-\frac{i}{2} \int J\Delta_F J \right] \\
\frac{\delta}{i\delta J(z)} Z_0[J] &= -\int J\Delta_F(x-z) \exp\left[-\frac{i}{2} \int J\Delta_F J \right] \\
\left(\frac{\delta}{i\delta J(z)} \right)^2 Z_0[J] &= \left\{ -\frac{1}{i}\Delta_F(0) + \left[\int J\Delta_F(x-z) \right]^2 \right\} \\
&\quad \cdot \exp\left[-\frac{i}{2} \int J\Delta_F J \right] \\
\left(\frac{\delta}{i\delta J(z)} \right)^3 Z_0[J] &= \left\{ 3\frac{1}{i}\Delta_F(0) \int J\Delta_F(x-z) \right. \\
&\quad \left. - \left[\int J\Delta_F(x-z) \right]^3 \right\} \exp\left[-\frac{i}{2} \int J\Delta_F J \right] \\
\left(\frac{\delta}{i\delta J(z)} \right)^4 Z_0[J] &= \left\{ -3\Delta_F^2(0) + 6i\Delta_F(0) \left[\int J\Delta_F(x-z) \right]^2 \right. \\
&\quad \left. + \left[\int J\Delta_F(x-z) \right]^4 \right\} \exp\left[-\frac{i}{2} \int J\Delta_F J \right].
\end{aligned} \tag{3.162}$$

In diagrams the fourth derivative is

$$\left(\frac{\delta}{i\delta J(z)}\right)^4 Z_0[J] \tag{3.163}$$

$= \Big\{-3$  $+\ 6i$  $+$  $\Big\}\exp\left[-\frac{i}{2}\int J\Delta_F J\right].$

The coefficients 3 and 6 are symmetry factors and arise from joining the lines of the third diagram in all possible ways.

The normalization $\mathcal{N}$ equals the numerator of equation (3.161) for $J = 0$ so that all diagrams including the sources vanish and we are left with

$\mathcal{N} = 1 - \frac{ig}{4!}\int dz\,[-3$  $]\,.$ (3.164)

Finally, expanding $\mathcal{N}^{-1}$ to order g we obtain the final result for the **generating functional $Z[J]$ to order g**

$Z[J] = \Big\{1 - \frac{ig}{4!}\int dz\,\Big[6i$  $+$  $\Big]\Big\}\exp\left[-\frac{i}{2}\int J\Delta_F J\right].$ (3.165)

Remark: All vacuum diagrams have *cancelled* and do not occur in the properly normalized functional! This is valid to all orders in perturbation theory.

3.2.4 Green functions for interacting fields

Analogous to the free case we gain the **Green functions** for interacting fields by calculating the derivatives of the functional $Z[J]$ (3.154)

$$\begin{aligned}\langle 0|T\phi(x_1)\ldots\phi(x_n)|0\rangle &= \frac{1}{i^n}\frac{\delta^n Z[J]}{\delta J(x_1)\ldots\delta J(x_n)}\bigg|_{J=0}\\ &=: \tau(x_1,\ldots,x_n)\end{aligned} \tag{3.166}$$

and $Z[J]$ represents the **generating functional for all Green functions**

$$\begin{aligned}Z[J] &= \sum_{n=0}^{\infty}\frac{i^n}{n!}\int dx_1\ldots dx_n\; J(x_1)\ldots J(x_n)\tau(x_1,\ldots,x_n)\\ &= \langle 0|T\exp\left[i\int dx\; J(x)\phi(x)\right]|0\rangle,\end{aligned} \tag{3.167}$$

with $\phi(x)$ an interacting field.

2-point function: Calculating, for example, the 2-point function, the propagator, to order g we start from the expanded generating functional (3.165) and we evaluate the second derivative with respect to the source. The first term corresponds to the free propagator and the second term containing two sources provides the correction to order g. The third term with 4 sources does not contribute since after differentiation there always remains a source factor which vanishes in the limit $J \to 0$. Then the correction to order g is

$$\tau_{12} = \left.\frac{\delta^2 Z[J]}{i\delta J_1 i\delta J_2}\right|_{J=0}$$

$$\widehat{=} \left.\frac{\delta^2}{i\delta J_1 i\delta J_2}\right|_{J=0} (-)\frac{ig}{4!}\int dz\, 6i$$ 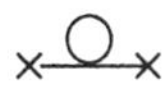 $$\exp\left[-\frac{i}{2}\int J\Delta_F J\right]$$

$$\widehat{=} -2!\frac{g}{4}\Delta_F(0)\int dz\, \Delta_F(x_1 - z)\Delta_F(x_2 - z) \tag{3.168}$$

or diagrammatically

$$\widehat{=} -\frac{g}{2}$$  . (3.169)

Altogether we get for the propagator to order g

$$\tau_{12} = i$$ $$-\frac{g}{2}$$  . (3.170)

Mass renormalization: The second term, however, is a quadratically divergent quantity. It shifts the mass of the scalar particle. Here we meet the simple example of **mass renormalization**.

In momentum space the propagator (3.170) gives

$$\tau_{12}(k) = i\left[\frac{1}{k^2 - m^2} + \frac{1}{k^2 - m^2}\,\frac{i}{2}g\Delta_F(0)\,\frac{1}{k^2 - m^2}\right] \tag{3.171}$$

using the operator expansion

$$\frac{1}{A - B} = \frac{1}{A} + \frac{1}{A}B\frac{1}{A} + \dots \tag{3.172}$$

we get

$$\tau_{12}(k) = \frac{i}{k^2 - m^2 - \delta m^2} = \frac{i}{k^2 - m_R^2}. \tag{3.173}$$

So we have **renormalized** the mass m^2 by the amount

$$\delta m^2 = \frac{i}{2}g\Delta_F(0) \tag{3.174}$$

to achieve the physical mass

$$m_R^2 = m^2 + \delta m^2. \tag{3.175}$$

Similarly other divergent terms which occur in perturbation theory can be removed by renormalization. We consider the general procedure of renormalization, however, beyond the scope of this book and we refer the reader to the standard textbooks on QFT.

4-point function: The other instructive example is the 4-point function. Let us calculate it to order g. Starting again from the generating functional (3.165) we calculate each term separately:

$$\tau_{1234}(1.\ \text{term}) = \left.\frac{\delta^4}{i\delta J_1 i\delta J_2 i\delta J_3 i\delta J_4}\right|_{J=0} \exp\left[-\frac{i}{2}\int J\Delta_F J\right]$$

$$= \tau^0_{1234} = -3$$  (3.176)

$\tau_{1234}(2.\ \text{term})$

$$= \left.\frac{\delta^4}{i\delta J_1 i\delta J_2 i\delta J_3 i\delta J_4}\right|_{J=0} (-)\frac{ig}{4!}\int dz\, 6i$$  $$\exp\left[-\frac{i}{2}\int J\Delta_F J\right]$$

$$= -\frac{i}{2}4!\frac{g}{4}\Delta_F(0)\int dz\, \Delta_F(x_1 - z)\Delta_F(x_2 - z)\Delta_F(x_3 - x_4) \tag{3.177}$$

or diagrammatically

$$= -3ig$$  . (3.178)

$\tau_{1234}(3.\ \text{term})$

$$= \left.\frac{\delta^4}{i\delta J_1 i\delta J_2 i\delta J_3 i\delta J_4}\right|_{J=0} (-)\frac{ig}{4!}\int dz$$  $$\exp\left[-\frac{i}{2}\int J\Delta_F J\right]$$

$$= -4!\frac{ig}{4!}\int dz\, \Delta_F(x_1 - z)\Delta_F(x_2 - z)\Delta_F(x_3 - z)\Delta_F(x_4 - z) \tag{3.179}$$

or diagrammatically

$$= -ig$$  . (3.180)

The factorial 4! results from the permutations of equal terms.

Altogether we obtain for the 4-point function to order g

$$\tau_{1234} = \left.\frac{\delta^4}{i\delta J_1 i\delta J_2 i\delta J_3 i\delta J_4}\right|_{J=0}$$

$$= -3$$  $$- 3ig$$ 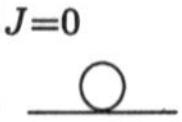 $$- ig$$  . (3.181)

We observe that the first and second diagrams are disconnected. The second diagram just modifies the propagator. The third diagram is the connected one and yields the nontrivial part for the transition amplitude. (If we had not normalized the generating functional (3.161) by $\mathcal{N}$ we would have to add an additional diagram

$$-\frac{ig}{4!}\,3\cdot 3$$

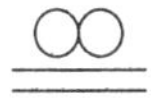

containing a vacuum diagram.)

n-point function: Generally, we find the **n-point function** within perturbation theory in the following systematic way:

$$\begin{aligned}
\tau(x_1,\ldots,x_n) &= \frac{1}{i^n}\left.\frac{\delta^n Z[J]}{\delta J(x_1)\ldots\delta J(x_n)}\right|_{J=0} \\
&= \frac{\int d\phi\;\phi(x_1)\ldots\phi(x_n)e^{i\int dx\;\mathcal{L}_I(\phi)}\;e^{i\int dx\;\mathcal{L}_0}}{\int d\phi\;e^{i\int dx\;\mathcal{L}_I(\phi)}\;e^{i\int dx\;\mathcal{L}_0}} \\
&= \frac{\left.\int d\phi\phi(x_1)\ldots\phi(x_n)\sum_{k=0}^{\infty}\frac{i^k}{k!}\left[\int dz\;\mathcal{L}_I(\phi)\right]^k e^{i\int dx[\mathcal{L}_0+J\phi]}\right|_{J=0}}{\left.\int d\phi\sum_{k=0}^{\infty}\frac{i^k}{k!}\left[\int dz\;\mathcal{L}_I(\phi)\right]^k e^{i\int dx[\mathcal{L}_0+J\phi]}\right|_{J=0}} \\
&= \frac{\frac{\delta}{i\delta J(x_1)}\cdots\frac{\delta}{i\delta J(x_n)}\sum_{k=0}^{\infty}\frac{i^k}{k!}\left.\left[\int dz\;\mathcal{L}_I\left(\frac{\delta}{i\delta J(z)}\right)\right]^k\right|_{J=0} Z_0[J]}{\sum_{k=0}^{\infty}\frac{i^k}{k!}\left.\left[\int dz\;\mathcal{L}_I\left(\frac{\delta}{i\delta J(z)}\right)\right]^k\right|_{J=0} Z_0[J]}.
\end{aligned} \tag{3.182}$$

The functional derivatives are applied to $Z_0[J]$ which is given by expression (3.139). The series we evaluate to the order k finding the Green functions to k-th order.

3.2.5 Connected Green functions

We are only interested in nontrivial transitions—in connected Green functions.

Definition: The '**quantum action**' $W[J]$ is defined by

$$W[J] := -i\ln Z[J] \quad \text{or} \quad Z[J] =: e^{iW[J]}. \tag{3.183}$$

Proposition:

- *The 'quantum action' generates only connected Green functions!*

This remarkable feature is true quite generally in quantum field theories. However, we want to demonstrate it just for our previous examples.

Examples:

i) *2-point function*

$$\frac{\delta^2 W[J]}{\delta J_1 \delta J_2} = \frac{\delta}{\delta J_1} \frac{-i}{Z} \frac{\delta Z[J]}{\delta J_2}$$
$$= \frac{i}{Z^2[J]} \frac{\delta Z[J]}{\delta J_1} \frac{\delta Z[J]}{\delta J_2} - \frac{i}{Z[J]} \frac{\delta^2 Z[J]}{\delta J_1 \delta J_2}. \quad (3.184)$$

In the limit $J \to 0$ the first term vanishes (odd derivatives vanish) and recalling the normalization $Z[0] = 1$ we have

$$\left.\frac{\delta^2 W[J]}{\delta J_1 \delta J_2}\right|_{J=0} = -i \left.\frac{\delta^2 Z[J]}{\delta J_1 \delta J_2}\right|_{J=0} = i\tau_{12} \quad (3.185)$$

$$= -$$  $$- \frac{ig}{2}$$  . (3.186)

Thus we get from $W[J]$ the propagator (to all orders in perturbation theory) which is clearly a connected diagram.

ii) *4-point function*

$$\left.\frac{\delta^4 W[J]}{\delta J_1 \delta J_2 \delta J_3 \delta J_4}\right|_{J=0} = i\left[\frac{1}{Z^2[J]} \frac{\delta^2 Z[J]}{\delta J_1 \delta J_3} \frac{\delta^2 Z[J]}{\delta J_2 \delta J_4}\right.$$
$$\left.\left. + 2 \text{ permutations} - \frac{1}{Z[J]} \frac{\delta^4 Z[J]}{\delta J_1 \delta J_2 \delta J_3 \delta J_4}\right]\right|_{J=0}$$
$$= i[\tau_{13}\tau_{24} + \tau_{12}\tau_{34} + \tau_{14}\tau_{23} - \tau_{1234}]. \quad (3.187)$$

Considering diagrams up to order g we find from the propagator (3.170) the product

$$\tau_{12}\tau_{34} = -$$ 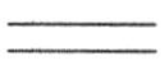 $$- ig$$ 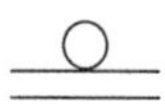 . (3.188)

Using also result (3.181) for τ_{1234} we obtain

$$\left.\frac{\delta^4 W[J]}{\delta J_1 \delta J_2 \delta J_3 \delta J_4}\right|_{J=0}$$

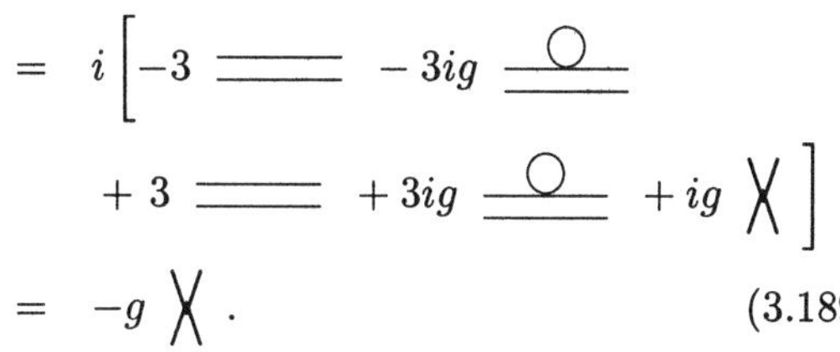

(3.189)

Only the connected (or irreducible) diagram remains! This continues to all orders in perturbation theory.

Connected n-point function: Generally the **connected n-point function** is

$$\tau_{\text{conn}}(x_1, \ldots, x_n) = \frac{1}{i^n} \left.\frac{\delta^n i W[J]}{\delta J(x_1) \ldots \delta J(x_n)}\right|_{J=0} \tag{3.190}$$

and, conversely, the 'quantum action' ($W[0] = 0$ by normalization)

$$i W[J] = \sum_{n=1}^{\infty} \frac{i^n}{n!} \int dx_1 \ldots dx_n J(x_1) \ldots J(x_n) \tau_{\text{conn}}(x_1, \ldots, x_n). \tag{3.191}$$

Green functions which are connected proper vertex functions—Feynman diagrams that remain connected when cutting an arbitrary internal line—are called **'one-particle irreducible Green functions'**.

3.3 Fermion field theory

In QFT the Fermi field operators obey the familiar anticommutation relations. In the PI approach, however, we deal with classical fields and classical sources which are just c-number functions. Nevertheless, in the path integral over Fermi fields these fields (and the sources) must be regarded as anticommuting. Such anticommuting functions are elements of the so-called Grassmann algebra which we will introduce in the following section.

3.3.1 Grassmann algebra

Definition: The generators θ_i $(i = 1, \ldots, n)$ of an n-dimensional Grassmann algebra are elements which satisfy the anticommutation relation

$$\{\theta_i, \theta_j\} = \theta_i \theta_j + \theta_j \theta_i = 0. \tag{3.192}$$

This implies the important identity

$$\theta_i^2 = 0. \tag{3.193}$$

For this reason a function depending on Grassmann variables can be expanded into a series which contains only a *finite* number of terms; for example in one dimension

$$f(\theta) = a + b\theta. \tag{3.194}$$

Differentiation: For the operation of differentiation we can introduce two types:
the **left derivative**

$$\frac{\partial^L}{\partial\theta_i}\theta_k\theta_\ell = \delta_{ik}\theta_\ell - \theta_k\delta_{i\ell} \tag{3.195}$$

the **right derivative**

$$\frac{\partial^R}{\partial\theta_i}\theta_k\theta_\ell = \theta_k\delta_{i\ell} - \delta_{ik}\theta_\ell. \tag{3.196}$$

For both left and right differentiation we have

$$\left\{\frac{\partial}{\partial\theta_i}, \theta_k\right\} = \delta_{ik} \tag{3.197}$$

$$\left\{\frac{\partial}{\partial\theta_i}, \frac{\partial}{\partial\theta_k}\right\} = 0. \tag{3.198}$$

Integration: For the integration over Grassmann variables there occurs a problem, however. Since

$$\left(\frac{\partial}{\partial\theta_i}\right)^2 = 0 \tag{3.199}$$

for both L and R there exists no inverse operation to the differentiation. But we can overcome this problem by imposing the following **Grassmann integration rules**:

$$\int d\theta_i\theta_i = 1 \qquad \text{and} \qquad \int d\theta_i = 0. \tag{3.200}$$

The so-defined Grassmann integration acts like the differentiation. It is defined as a linear algebraic operation.

We shall find finally that a Grassmann PI formalism mimics the boson PI approach described before and that we achieve a Grassmann integral representation for the determinant of a differential operator which we are interested in.

For example, we can define a Grassmann δ-function by

$$\int d\theta\; \delta(\theta - \theta')f(\theta) = f(\theta') \tag{3.201}$$

implying

$$\delta(\theta - \theta') = \theta - \theta' \tag{3.202}$$

and a Grassmann momentum π by

$$\int d\pi\, e^{i\pi\theta} = \int d\pi(1 + i\pi\theta) = i\theta = i\delta(\theta). \tag{3.203}$$

Both equations (3.201) and (3.203) simulate the usual result for **R**- or **C**-functions (in equation (3.203) up to a normalization constant).

Gaussian integration: Next we evaluate Gaussian integrals which we will need in our PI formalism.

Lemma: The value of a Gauss integral is

$$\int d\bar{\theta}d\theta\, e^{-\bar{\theta}\theta} = \int d\theta d\bar{\theta}\, e^{\bar{\theta}\theta} = 1. \tag{3.204}$$

Proof. Let us work in two dimensions; there we have

$$\theta = \begin{pmatrix} \theta_1 \\ \theta_2 \end{pmatrix}, \qquad \bar{\theta} = (\bar{\theta}_1, \bar{\theta}_2) \tag{3.205}$$

yielding the products

$$\bar{\theta}\theta = (\bar{\theta}_1, \bar{\theta}_2)\begin{pmatrix} \theta_1 \\ \theta_2 \end{pmatrix} = \bar{\theta}_1\theta_1 + \bar{\theta}_2\theta_2 \tag{3.206}$$

$$(\bar{\theta}\theta)^2 = (\bar{\theta}_1\theta_1 + \bar{\theta}_2\theta_2)(\bar{\theta}_1\theta_1 + \bar{\theta}_2\theta_2) = 2\bar{\theta}_1\theta_1\bar{\theta}_2\theta_2. \tag{3.207}$$

Higher powers of the product $\bar{\theta}\theta$ vanish because of identity (3.193). Therefore the Gauss function terminates with the quadratic term

$$e^{-\bar{\theta}\theta} = 1 - \bar{\theta}\theta + \frac{1}{2!}(\bar{\theta}\theta)^2. \tag{3.208}$$

We then evaluate the Gauss integral by the expansion

$$\begin{aligned} \int d\bar{\theta}d\theta\, e^{-\bar{\theta}\theta} &= \int d\bar{\theta}_1 d\theta_1 d\bar{\theta}_2 d\theta_2 (1 - \bar{\theta}_1\theta_1 - \bar{\theta}_2\theta_2 + \bar{\theta}_1\theta_1\bar{\theta}_2\theta_2) \\ &= \int d\bar{\theta}_2 d\bar{\theta}_1 d\theta_1 d\theta_2\, \theta_2\theta_1\bar{\theta}_1\bar{\theta}_2 = 1. \qquad \text{Q.E.D.} \end{aligned} \tag{3.209}$$

Lemma: For a change of the integration variables like $\theta = X\alpha$ the Grassmann measure transforms with the *inverse* Jacobian (e.g. in 2 dimensions)

$$d\theta_1 d\theta_2 = \frac{1}{\det X} d\alpha_1 d\alpha_2. \tag{3.210}$$

Proof. Let $\theta = X\alpha$; for the product $\theta_1\theta_2$ we find

$$\begin{aligned}\theta_1\theta_2 &= (X_{11}\alpha_1 + X_{12}\alpha_2)(X_{21}\alpha_1 + X_{22}\alpha_2) \\ &= (X_{11}X_{22} - X_{12}X_{21})\alpha_1\alpha_2 \\ &= \det X \cdot \alpha_1\alpha_2. \end{aligned} \tag{3.211}$$

Preserving the identity

$$\int d\theta_2 d\theta_1\, \theta_1\theta_2 = \int d\alpha_2 d\alpha_1\, \alpha_1\alpha_2 \tag{3.212}$$

implies for the measure

$$d\theta_2 d\theta_1 = \frac{1}{\det X} d\alpha_2 d\alpha_1. \qquad \text{Q.E.D.} \tag{3.213}$$

Formulae: Now we transform the Gauss integral (3.204) according to the changes $\theta = X\alpha$ and $\bar\theta = \bar\alpha Y$

$$\int d\bar\theta d\theta\, e^{-\bar\theta\theta} = \frac{1}{\det X \det Y} \int d\bar\alpha d\alpha\, e^{-\bar\alpha Y X \alpha} = 1, \tag{3.214}$$

we define

$$YX =: D \tag{3.215}$$

and recall the properties of determinants

$$\det X \det Y = \det XY = \det D \tag{3.216}$$

and we end up with the **Gauss integration formula** for Grassmann variables

$$\int d\alpha d\bar\alpha\, e^{\bar\alpha D\alpha} = \int d\bar\alpha d\alpha\, e^{-\bar\alpha D\alpha} = \det D. \tag{3.217}$$

So here the Gaussian integral is directly proportional to the determinant of the operator D in contrast to the usual case, equation (3.127). Clearly, formula (3.217) can be generalized to n-dimensional Grassmann vectors. For $D = \mathbf{1}$ we reproduce the normalization (3.204).

Shifting the integration variables next by

$$\alpha \;\rightarrow\; \alpha - D^{-1}\xi = \alpha' \tag{3.218}$$

$$\bar\alpha \;\rightarrow\; \bar\alpha - \bar\xi D^{-1} = \bar\alpha' \tag{3.219}$$

the new exponent contains

$$\begin{aligned}\bar\alpha' D\alpha' &= (\bar\alpha - \bar\xi D^{-1}) D (\alpha - D^{-1}\xi) \\ &= \bar\alpha D\alpha - \bar\xi\alpha - \bar\alpha\xi + \bar\xi D^{-1}\xi \end{aligned} \tag{3.220}$$

and we obtain the **Gauss-like formula**

$$\int d\bar{\alpha}d\alpha\ \exp[-\bar{\alpha}D\alpha+\bar{\xi}\alpha+\bar{\alpha}\xi]=\det D\cdot\exp[\bar{\xi}D^{-1}\xi], \tag{3.221}$$

which should be contrasted with the result (3.130) for **C**-functions.

Generalization to infinite dimensions: The above formalism can be generalized to infinite dimensions. Let $\theta(x)$ be the generator of such an infinite dimensional Grassmann algebra obeying the anticommutation relation

$$\{\theta(x),\theta(y)\}=0, \tag{3.222}$$

with

$$\theta^2(x)=0, \tag{3.223}$$

then we have the following **rules for differentiation and integration**

$$\frac{\delta^{L,R}\theta(x)}{\delta\theta(y)}=\delta(x-y), \tag{3.224}$$

$$\frac{\delta^L}{\delta\theta(z)}\theta(x)\theta(y) = \delta(x-z)\theta(y)-\theta(x)\delta(y-z), \tag{3.225}$$

$$\frac{\delta^R}{\delta\theta(z)}\theta(x)\theta(y) = \theta(x)\delta(y-z)-\delta(x-z)\theta(y), \tag{3.226}$$

and

$$\int d\theta(x)\ \theta(x)=1, \qquad \int d\theta(x)=0. \tag{3.227}$$

The **Gaussian integrals** give

$$\int d\bar{\theta}(x)d\theta(x)\ e^{-(\bar{\theta},D\theta)} = \det D, \tag{3.228}$$

$$\int d\bar{\theta}(x)d\theta(x)\ e^{-(\bar{\theta},D\theta)+(\bar{\xi},\theta)+(\bar{\theta},\xi)} = \det D\cdot e^{(\bar{\xi},D^{-1}\xi)}, \tag{3.229}$$

with

$$\begin{aligned}(\bar{\theta},D\theta) &= \int dxdy\ \bar{\theta}(x)D(x,y)\theta(y)\\ &= \int dx\ \bar{\theta}(x)D(x)\theta(x) \quad \text{if } D(x,y)=D(x)\delta(x-y),\\ (\bar{\xi},\theta) &= \int dx\ \bar{\xi}(x)\theta(x),\end{aligned}$$

$$(\bar{\theta}, \xi) = \int dx\, \bar{\theta}(x)\xi(x),$$
$$(\bar{\xi}, D^{-1}\xi) = \int dxdy\, \bar{\xi}(x)D^{-1}(x,y)\xi(y), \tag{3.230}$$

and D some invertible operator.

3.3.2 Dirac fields

We begin with the free Lagrangian for fermions

$$\mathcal{L}_0 = \bar{\psi} i \not\partial \psi - m\bar{\psi}\psi \tag{3.231}$$

corresponding to the equation of motion, the Dirac equation

$$(i \not\partial - m)\psi = 0. \tag{3.232}$$

We develop the path integral formalism analogously to the scalar case (3.108), (3.124), (3.138), (3.139), and construct the **generating functional for free Dirac fields**

$$Z_0[\eta, \bar{\eta}] = \frac{1}{\mathcal{N}} \int d\bar{\psi} d\psi \exp\{i \int dx [\bar{\psi}(i \not\partial - m)\psi + \bar{\eta}\psi + \bar{\psi}\eta]\}, \tag{3.233}$$

with the **normalization**

$$\mathcal{N} = \int d\bar{\psi} d\psi \exp[i \int dx\, \bar{\psi}(i \not\partial - m)\psi]. \tag{3.234}$$

The Grassmann functions $\eta(x)$, $\bar{\eta}(x)$ represent the Schwinger sources for the Dirac fields $\bar{\psi}(x)$, $\psi(x)$.

We search an appropriate form to evaluate a perturbation expansion. Defining the Dirac operator by

$$i \not\partial - m =: D, \tag{3.235}$$

its inverse expresses the Feynman propagator

$$DS_F(x-y) = \delta(x-y) \quad \text{or} \quad D^{-1}(x) = S_F(x). \tag{3.236}$$

The fermion propagator S_F is connected with the scalar propagator Δ_F, equations (3.120), (3.121), by

$$S_F(x) = (i \not\partial + m)\Delta_F(x) \tag{3.237}$$

and has the following integral representation

$$S_F(x) = \frac{1}{(2\pi)^4}\int \frac{d^4p\, e^{-ipx}}{\not p - m} = \frac{1}{(2\pi)^4}\int \frac{d^4p(\not p + m)e^{-ipx}}{p^2 - m^2} \tag{3.238}$$

(we suppress the ε-term in the Feynman propagator). Applying the Gaussian formulae (3.228)–(3.230) we can rewrite the normalization (3.234)

$$\mathcal{N} = \int d\bar\psi d\psi \exp\left[-(\bar\psi, \frac{1}{i}D\psi)\right] = \det\frac{1}{i}D \tag{3.239}$$

and we obtain the convenient **formula for perturbation expansions**

$$Z_0[\eta,\bar\eta] = \exp\left[-i\int dxdy\ \bar\eta(x)S_F(x-y)\eta(y)\right]. \tag{3.240}$$

Expression (3.240) is the fermionic analogue of the scalar case (3.139).

Propagator: For example, the 2-point function—the free propagator—is determined by the second derivative of Z_0

$$\begin{aligned}\tau_0(x_1,x_2) &= \left.\frac{\delta^2 Z_0[\eta,\bar\eta]}{i\delta\bar\eta(x_1)i\delta\eta(x_2)}\right|_{\eta=\bar\eta=0} \\ &= \left.\frac{\delta}{i\delta\bar\eta(x_1)}(-)\int dx\ \bar\eta(x)S_F(x-x_2)\exp[-i\int \bar\eta S_F\eta]\right|_{\eta=\bar\eta=0} \\ &= iS_F(x_1-x_2).\end{aligned} \tag{3.241}$$

Interacting fields: Now we incorporate the interaction described by the Lagrangian

$$\mathcal{L} = \mathcal{L}_0 + \mathcal{L}_I \tag{3.242}$$

and

$$\mathcal{L}_I = -e\bar\psi\gamma^\mu\psi A_\mu \qquad \text{for QED,} \tag{3.243}$$

$$\mathcal{L}_I = ig_s\bar\psi\gamma^\mu T^a\psi A^a_\mu \qquad \text{for YM.} \tag{3.244}$$

We consider the photon field A_μ or the Yang–Mills field (YM) A^a_μ for the moment as external fields. T^a denote the generators of some group.

Then the **generating functional for interacting Dirac fields** is

$$Z[\eta,\bar\eta] = \frac{1}{\mathcal{N}}\int d\bar\psi d\psi \exp\left\{i\int dx[\mathcal{L}_0 + \mathcal{L}_I + \bar\eta\psi + \bar\psi\eta]\right\} \tag{3.245}$$

with

$$\mathcal{N} = \int d\bar\psi d\psi \exp\left\{i\int dx[\mathcal{L}_0 + \mathcal{L}_I]\right\}. \tag{3.246}$$

We derive the suitable form for perturbation expansions analogously to the scalar case (see Section 3.2.3). We separate the interaction term $\mathcal{L}_I(\bar{\psi}, \psi)$ and replace the fields there by the derivatives with respect to the sources

$$\bar{\psi} \to \frac{\delta}{i\delta\eta}, \qquad \psi \to \frac{\delta}{i\delta\bar{\eta}}, \tag{3.247}$$

and we obtain the **formula for perturbation expansions**

$$Z[\eta, \bar{\eta}] = \frac{1}{\mathcal{N}} \exp\left[i \int dx\, \mathcal{L}_I\left(\frac{\delta}{i\delta\eta}, \frac{\delta}{i\delta\bar{\eta}}\right)\right] Z_0[\eta, \bar{\eta}], \tag{3.248}$$

with $Z_0[\eta, \bar{\eta}]$ being given by expression (3.240) and the normalization $\mathcal{N}$ represents the numerator for $\eta = \bar{\eta} = 0$.

The Green functions to a given order in the interaction are evaluated from the functional derivatives of $Z[\eta, \bar{\eta}]$. The connected Green functions follow from the W-functional

$$Z[\eta, \bar{\eta}] = \begin{cases} e^{iW[\eta,\bar{\eta}]} & \text{in Minkowski space,} \\ e^{-W[\eta,\bar{\eta}]} & \text{in Euclidean space.} \end{cases} \tag{3.249}$$

3.4 Abelian gauge fields

Next we turn to the gauge fields which cause a new complication. Let us begin with the simple case of Abelian fields, with QED. When constructing a PI for the generating functional Z we finally want to evaluate the propagator of the fields. The propagator is the inverse of the operator in the quadratic field part of the Lagrangian. Now, for gauge fields we also have the liberty of gauge transformations

$$A_\mu(x) \to A_\mu(x) + \partial_\mu \Lambda(x). \tag{3.250}$$

This, however, creates difficulties such that the inverse of the operator might not exist.

From the previous discussion we would construct the following **gauge field path integral** for the generating functional of the Green functions

$$Z[J_\mu] = \int dA_\mu \exp\left\{i \int dx[\mathcal{L}_0 + J^\mu A_\mu]\right\}, \tag{3.251}$$

with the free Lagrangian

$$\mathcal{L}_0 = -\frac{1}{4} F_{\mu\nu} F^{\mu\nu} \tag{3.252}$$

and the field strength tensor

$$F_{\mu\nu} = \partial_\mu A_\nu - \partial_\nu A_\mu. \tag{3.253}$$

Here the external current J^μ represents the source for the gauge field introduced in the spirit of Schwinger. We consider for our purpose a pure gauge field theory and disregard the interaction with the fermion fields. The Lagrangian (3.252) can be rewritten into a quadratic field term by partial integration

$$\mathcal{L}_0 = \frac{1}{2} A^\mu (g_{\mu\nu}\Box - \partial_\mu \partial_\nu) A^\nu. \tag{3.254}$$

Proposition:

- *The operator* $g_{\mu\nu}\Box - \partial_\mu\partial_\nu$ *has no inverse!*

Proof. Choose, for instance, a pure gauge

$$A^\nu = \partial^\nu \Lambda \tag{3.255}$$

then the operator (3.254) has a zero eigenvalue

$$(g_{\mu\nu}\Box - \partial_\mu\partial_\nu)\partial^\nu\Lambda = (\Box\partial_\mu - \partial_\mu\Box)\Lambda = 0 \tag{3.256}$$

and cannot be inverted for this reason. Or consider the momentum space

$$g_{\mu\nu}\Box - \partial_\mu\partial_\nu \overset{i\partial_\mu \to k_\mu}{\longrightarrow} -g_{\mu\nu}k^2 + k_\mu k_\nu \tag{3.257}$$

then the inverse must be of the form

$$a g^{\nu\sigma} + b k^\nu k^\sigma \tag{3.258}$$

and we have by definition

$$(-g_{\mu\nu}k^2 + k_\mu k_\nu)(a g^{\nu\sigma} + b k^\nu k^\sigma) = \delta_\mu{}^\sigma. \tag{3.259}$$

This gives

$$-ak^2\delta_\mu{}^\sigma + a k_\mu k^\sigma = \delta_\mu{}^\sigma \tag{3.260}$$

which has no solution for a. Q.E.D.

Thus we cannot find the photon propagator in a straightforward way. The reason is the following. In the PI (3.251) the field integration covers *all* possible field configurations, also those which are just gauge transformed fields. Since the action remains invariant under gauge transformations—constant along a gauge orbit—the path integral over the fields produces an infinity (the volume of the gauge group) for the functional Z and consequently for the resulting Green functions. So it is the integration over these redundant gauge equivalent fields which causes the PI to diverge. How can we solve this problem? Obviously by fixing the gauge.

Gauge fixing: We select only one representative from each gauge orbit which contributes to the PI.

It is convenient to impose the Lorentz condition

$$\partial_\mu A^\mu = 0 \tag{3.261}$$

then we have the Lagrangian

$$\mathcal{L} = \frac{1}{2} A^\mu g_{\mu\nu} \Box A^\nu . \tag{3.262}$$

Proposition:

- *The operator $g_{\mu\nu}\Box$ has an inverse which is the Feynman propagator of the photon*
$$D^F{}_{\mu\nu}(x) = -g_{\mu\nu}\Delta_F(x, m=0), \tag{3.263}$$
with Δ_F defined by equations (3.120), (3.121).

The Lagrangian (3.262) can be rewritten

$$\begin{aligned} \mathcal{L} &= -\frac{1}{4} F_{\mu\nu}F^{\mu\nu} - \frac{1}{2}(\partial_\mu A^\mu)^2 & (3.264)\\ &= \mathcal{L}_0 + \mathcal{L}_{G\,\text{fix}}, & (3.265) \end{aligned}$$

where $\mathcal{L}_{G\,\text{fix}}$ represents the **gauge fixing term.** Generally we may add

$$\mathcal{L}_{G\,\text{fix}} = -\frac{1}{2\alpha}(\partial_\mu A^\mu)^2 \tag{3.266}$$

with α some parameter free to our disposal. Then the Lagrangian is

$$\begin{aligned} \mathcal{L} &= -\frac{1}{4} F_{\mu\nu}F^{\mu\nu} - \frac{1}{2\alpha}(\partial_\mu A^\mu)^2 \\ &= \frac{1}{2} A^\mu \left[g_{\mu\nu}\Box + \left(\frac{1}{\alpha} - 1\right)\partial_\mu\partial_\nu \right] A^\nu . & (3.267) \end{aligned}$$

Proposition:

- *The operator*

$$g_{\mu\nu}\Box + \left(\frac{1}{\alpha} - 1\right)\partial_\mu\partial_\nu \overset{i\partial_\mu \to k_\mu}{\longrightarrow} -g_{\mu\nu}k^2 - \left(\frac{1}{\alpha} - 1\right)k_\mu k_\nu =: K_{\mu\nu}(k)$$

has an inverse which is the photon propagator

$$D_{\mu\nu}(k) = -\frac{1}{k^2}\left[g_{\mu\nu} + (\alpha - 1)\frac{k_\mu k_\nu}{k^2}\right]. \tag{3.268}$$

We can check explicitly that

$$K_{\mu\nu}(k)D^{\nu\sigma}(k) = \delta_\mu{}^\sigma. \tag{3.269}$$

In the literature we frequently meet the following parameter selection

$$\begin{aligned}
\alpha \to 1: &\quad D^F{}_{\mu\nu}(k) = -\frac{g_{\mu\nu}}{k^2 + i\varepsilon} &&\text{Feynman gauge}\\
\alpha \to 0: &\quad D^L{}_{\mu\nu}(k) = -\frac{g_{\mu\nu} - \frac{k_\mu k_\nu}{k^2}}{k^2 + i\varepsilon} &&\text{Landau gauge.}
\end{aligned} \tag{3.270}$$

The physical quantities do not depend on the parameter α.

3.5 Faddeev–Popov method for non-Abelian gauge fields

3.5.1 Yang–Mills theory

When 1954 Yang and Mills (YM) [Yang, Mills 1954] developed a gauge theory for non-Abelian fields it was considered rather a curiosity. But nowadays non-Abelian gauge theories turn out as *the* fundamental field theories due to their rich underlying structure.

Let us start with a free fermion theory

$$\mathcal{L}^0_\psi = \bar\psi(i\,\partial\!\!\!/ - m)\psi \tag{3.271}$$

but in addition we assume that the fermions still have an internal degree of freedom represented by a vector in some group space G. The famous example describing the strong interactions is colour $SU(3)$ where

$$\psi = \begin{pmatrix} q_p \\ q_y \\ q_b \end{pmatrix} \quad \begin{matrix}\text{pink}\\ \text{yellow}\\ \text{blue}\end{matrix} \tag{3.272}$$

and q denotes the spinor field for a certain flavour kind of quark ($q = u, d, s, c, b, t$).

Gauge group: A group element $g(x) \in G$ transforms the spinor fields as

$$\begin{aligned}
\psi^g(x) &= g^{-1}(x)\psi(x)\\
\bar\psi^g(x) &= \bar\psi(x)g(x),
\end{aligned} \tag{3.273}$$

with

$$g(x) = e^{\Lambda(x)}, \qquad \Lambda(x) = \Lambda^a(x)T^a. \tag{3.274}$$

The matrices T^a are the **generators of the group** and satisfy the **algebra**

$$[T^a, T^b] = f^{abc}T^c, \tag{3.275}$$

where the f^{abc} are the totally **antisymmetric structure constants.** If $f^{abc} = 0$ we return to the Abelian case.

Here the matrices are chosen anti-Hermitian

$$T^{a\dagger} = -T^a \tag{3.276}$$

and they are normalized by

$$\text{tr}\ T^aT^b = -\frac{1}{2}\delta^{ab}. \tag{3.277}$$

Examples: The **special unitary groups** in N dimensions $SU(N)$, where

$$gg^\dagger = \mathbf{1}, \qquad \det g = 1 \tag{3.278}$$

play an important role in physics. There we have

$$T^a = \frac{\sigma^a}{2i} \quad \text{for } SU(2) \tag{3.279}$$

$$T^a = \frac{\lambda^a}{2i} \quad \text{for } SU(3), \tag{3.280}$$

σ^a denote the Pauli matrices $a = 1, 2, 3$

$$\sigma^1 = \begin{pmatrix} 0 & 1 \\ 1 & 0 \end{pmatrix} \quad \sigma^2 = \begin{pmatrix} 0 & -i \\ i & 0 \end{pmatrix} \quad \sigma^3 = \begin{pmatrix} 1 & 0 \\ 0 & -1 \end{pmatrix} \tag{3.281}$$

and λ^a express the Gell-Mann matrices $a = 1, \ldots, 8$

$$\lambda^1 = \begin{pmatrix} 0 & 1 & 0 \\ 1 & 0 & 0 \\ 0 & 0 & 0 \end{pmatrix} \quad \lambda^2 = \begin{pmatrix} 0 & -i & 0 \\ i & 0 & 0 \\ 0 & 0 & 0 \end{pmatrix} \quad \lambda^3 = \begin{pmatrix} 1 & 0 & 0 \\ 0 & -1 & 0 \\ 0 & 0 & 0 \end{pmatrix}$$
$$\lambda^4 = \begin{pmatrix} 0 & 0 & 1 \\ 0 & 0 & 0 \\ 1 & 0 & 0 \end{pmatrix} \quad \lambda^5 = \begin{pmatrix} 0 & 0 & -i \\ 0 & 0 & 0 \\ i & 0 & 0 \end{pmatrix} \tag{3.282}$$
$$\lambda^6 = \begin{pmatrix} 0 & 0 & 0 \\ 0 & 0 & 1 \\ 0 & 1 & 0 \end{pmatrix} \quad \lambda^7 = \begin{pmatrix} 0 & 0 & 0 \\ 0 & 0 & -i \\ 0 & i & 0 \end{pmatrix} \quad \lambda^8 = \frac{1}{\sqrt{3}}\begin{pmatrix} 1 & 0 & 0 \\ 0 & 1 & 0 \\ 0 & 0 & -2 \end{pmatrix}.$$

For $SU(2)$ the structure constants represent the antisymmetric ε-tensor in 3 dimensions ($a = 1, 2, 3$)

$$f^{abc} \to \varepsilon^{abc} = \begin{cases} 1 & \text{even permutation of } 123 \\ 0 & \text{indices equal} \\ -1 & \text{odd permutation of } 123 \end{cases} \tag{3.283}$$

and in case of $SU(3)$ the nonvanishing structure constants have the following values

$$\begin{aligned} f_{123} &= 1 \\ f_{147} &= -f_{156} = f_{246} = f_{257} = f_{345} = -f_{367} = \frac{1}{2} \\ f_{458} &= f_{678} = \frac{\sqrt{3}}{2}. \end{aligned} \tag{3.284}$$

The anticommutator, on the other hand, satisfies the relation

$$\{\lambda^a, \lambda^b\} = \frac{4}{3}\delta^{ab}\mathbf{1} + 2d^{abc}\lambda^c, \tag{3.285}$$

where the d^{abc} denote the **symmetric structure constants**

$$d^{abc} = \frac{1}{4} \text{ tr } \{\lambda^a, \lambda^b\}\lambda^c. \tag{3.286}$$

They follow from relation (3.285) by multiplication by λ^c and taking the trace. Analogously we calculate from the commutator (3.275)

$$f^{abc} = \frac{1}{4i} \text{ tr } [\lambda^a, \lambda^b]\lambda^c. \tag{3.287}$$

The **Jacobi identity** of the generators

$$[T^a, [T^b, T^c]] + [T^b, [T^c, T^a]] + [T^c, [T^a, T^b]] = 0 \tag{3.288}$$

implies for the structure constants the relation

$$f^{abd} f^{dce} + f^{bcd} f^{dae} + f^{cad} f^{dbe} = 0. \tag{3.289}$$

One also can construct matrices T^a from the structure constants f^{abc} according to

$$(T^a)_{bc} = -f^{abc}. \tag{3.290}$$

Such a representation of the Lie algebra, which is of the group dimension, is called **adjoint representation**.

Gauge transformations: Now let us return to the Lagrangian (3.271). We want to keep this Lagrangian invariant under a local gauge group transformation

$$\begin{aligned}\mathcal{L}^0_\psi \to \mathcal{L}^0_{\psi^g} &= \bar{\psi}^g(i\,\partial\!\!\!/ - m)\psi^g \\ &= \bar{\psi}(i\,\partial\!\!\!/ - m)\psi + i\bar{\psi}g(\partial\!\!\!/ g^{-1})\psi. \end{aligned} \tag{3.291}$$

Analogous to the Abelian case of QED we introduce a **gauge potential field**

$$A_\mu(x) = A^a_\mu(x)T^a, \tag{3.292}$$

which **gauge transforms** as

$$A^g_\mu = g^{-1}A_\mu g + g^{-1}\partial_\mu g. \tag{3.293}$$

The field is brought in via **minimal substitution**

$$\partial_\mu \to D_\mu = \partial_\mu + A_\mu, \tag{3.294}$$

where the **covariant derivative** D_μ transforms as

$$D^g_\mu = g^{-1}D_\mu g. \tag{3.295}$$

Then the Lagrangian

$$\mathcal{L}_\psi = \bar{\psi}(i\,D\!\!\!\!/ - m)\psi \tag{3.296}$$

remains invariant under the gauge group transformations (3.273), (3.295).

In particular, for colour $SU(3)$, QCD, the gauge potential is a matrix in colour–group space which contains all 8 gluon fields

$$A_\mu = A^a_\mu \frac{\lambda^a}{2i} = \frac{1}{2i}\begin{pmatrix} A^3_\mu + \frac{1}{\sqrt{3}}A^8_\mu & A^1_\mu - iA^2_\mu & A^4_\mu - iA^5_\mu \\ A^1_\mu + iA^2_\mu & -A^3_\mu + \frac{1}{\sqrt{3}}A^8_\mu & A^6_\mu - iA^7_\mu \\ A^4_\mu + iA^5_\mu & A^6_\mu + iA^7_\mu & -\frac{2}{\sqrt{3}}A^8_\mu \end{pmatrix}. \tag{3.297}$$

In addition, we have to consider the motion of the YM fields given by the Lagrangian

$$\mathcal{L}_{YM} = -\frac{1}{4}F^a_{\mu\nu}F^{a\mu\nu} = \frac{1}{2}\,\mathrm{tr}\,F_{\mu\nu}F^{\mu\nu}, \tag{3.298}$$

where

$$F_{\mu\nu} = F^a_{\mu\nu}T^a \tag{3.299}$$

$$F_{\mu\nu} = \partial_\mu A_\nu - \partial_\nu A_\mu + [A_\mu, A_\nu] \tag{3.300}$$

describes the **YM field strength tensor**. Since the $F_{\mu\nu}$ **gauge transforms** as

$$F^g_{\mu\nu} = g^{-1}F_{\mu\nu}g \tag{3.301}$$

the YM Lagrangian (3.298) remains gauge invariant.

Note: A quantity transforming like equation (3.293)—the gauge potential—represents, geometrically, a connection and the corresponding quantity (3.300)—the field strength tensor—describes the curvature (recall Section 2.7).

Theorem:

$$[D_\mu, D_\nu] = F_{\mu\nu}. \tag{3.302}$$

If, however, the covariant derivative D_μ acts on the **adjoint group space**—on elements containing the matrix generators T^a—then there occurs a commutator in the definition of D_μ.

Definition:

$$\begin{aligned} D_\mu &= \partial_\mu + [A_\mu, \;] & (3.303) \\ D_\mu \Lambda &= \partial_\mu \Lambda + [A_\mu, \Lambda], & (3.304) \end{aligned}$$

with the Lie algebra valued element

$$\Lambda(x) = \Lambda^a(x) T^a. \tag{3.305}$$

In components we have

$$D_\mu^{ba} = \partial_\mu \delta^{ba} + f^{bac} A_\mu^c \tag{3.306}$$

$$(D_\mu \Lambda)^a = D_\mu^{ba} \Lambda^b = \partial_\mu \Lambda^a + f^{acb} A_\mu^c \Lambda^b. \tag{3.307}$$

Theorem:

$$[D_\mu, D_\nu]\Lambda = [F_{\mu\nu}, \Lambda]. \tag{3.308}$$

Proof.

$$\begin{aligned} [D_\mu, D_\nu]\Lambda &= D_\mu D_\nu \Lambda - \mu \leftrightarrow \nu \\ &= D_\mu(\partial_\nu \Lambda + [A_\nu, \Lambda]) - \mu \leftrightarrow \nu \\ &= \partial_\mu(\partial_\nu \Lambda + [A_\nu, \Lambda]) + [A_\mu, (\partial_\nu \Lambda + [A_\nu, \Lambda])] - \mu \leftrightarrow \nu \\ &= [(\partial_\mu A_\nu - \partial_\nu A_\mu), \Lambda] + [A_\mu, [A_\nu, \Lambda]] - [A_\nu, [A_\mu, \Lambda]]. \end{aligned}$$

Using the Jacobi identity (3.288) for the last two terms gives

$$\begin{aligned} [D_\mu, D_\nu]\Lambda &= [(\partial_\mu A_\nu - \partial_\nu A_\mu), \Lambda] + [[A_\mu, A_\nu], \Lambda] \\ &= [F_{\mu\nu}, \Lambda]. \qquad \text{Q.E.D.} & (3.309) \end{aligned}$$

Lemma:

i)

$$D_\mu(\Lambda_1\Lambda_2) = (D_\mu\Lambda_1)\Lambda_2 + \Lambda_1 D_\mu\Lambda_2 \tag{3.310}$$

ii)

$$\partial_\mu \text{ tr } \Lambda_1\Lambda_2 = \text{tr } (D_\mu\Lambda_1)\Lambda_2 + \text{tr } \Lambda_1 D_\mu\Lambda_2. \tag{3.311}$$

Infinitesimal gauge transformations: If we just consider infinitesimal gauge elements

$$g(x) = \mathbf{1} + \Lambda(x) \tag{3.312}$$

then the above transformations lead to the **infinitesimal gauge transformations**

$$\begin{aligned} \delta\psi &= -\Lambda\psi \\ \delta\bar{\psi} &= \bar{\psi}\Lambda \\ \delta A_\mu &= D_\mu\Lambda \\ \delta F_{\mu\nu} &= [F_{\mu\nu}, \Lambda]. \end{aligned} \tag{3.313}$$

In contrast to the Abelian case the non-Abelian field strength is *not* gauge invariant. It transforms according to the adjoint representation (3.301), (3.313) and is therefore termed gauge covariant.

Equations of motion: Next we determine the equations of motion for the YM field. We consider the total Lagrangian

$$\mathcal{L} = \mathcal{L}_\psi + \mathcal{L}_{YM} \tag{3.314}$$

then the Euler–Lagrange equations

$$\frac{\delta\mathcal{L}}{\delta A^a_\nu} - \partial_\mu \frac{\delta\mathcal{L}}{\delta\partial_\mu A^a_\nu} = 0 \tag{3.315}$$

provide the following **YM field equations**

$$D_\mu F^{\mu\nu} = j^\nu \tag{3.316}$$

with the non-Abelian vector current

$$j_\nu = j^a_\nu T^a \tag{3.317}$$

$$j^a_\nu = -i\bar{\psi}\gamma_\nu T^a\psi. \tag{3.318}$$

(Note that we have chosen the coupling strength of the YM fields $g_s = 1$.)

Theorem: The non-Abelian current is covariantly conserved

$$D_\nu j^\nu = 0. \tag{3.319}$$

Proof.

$$\begin{aligned} D_\nu j^\nu &= D_\nu D_\mu F^{\mu\nu} = \frac{1}{2}[D_\nu, D_\mu]F^{\mu\nu} \\ &= -\frac{1}{2}[F_{\mu\nu}, F^{\mu\nu}] = 0. \qquad \text{Q.E.D.} \end{aligned} \tag{3.320}$$

In the absence of matter fields ($j_\nu = 0$), in the case of a pure YM theory we have

$$D_\mu F^{\mu\nu} = 0, \tag{3.321}$$

which is still a nontrivial theory with selfinteracting fields.

Finally we mention the **Bianchi identity** for the YM field strength tensor

$$D_\lambda F_{\mu\nu} + D_\mu F_{\nu\lambda} + D_\nu F_{\lambda\mu} = 0. \tag{3.322}$$

It is a consequence of the Jacobi identity for the covariant derivative and the commutator relation (3.302), (3.308).

The Bianchi identity can be rewritten in terms of the **dual field strength tensor**

$$*F^{\mu\nu} = \frac{1}{2}\varepsilon^{\mu\nu\alpha\beta}F_{\alpha\beta} \tag{3.323}$$

($\varepsilon^{0123} = 1$) then equation (3.322) simplifies to

$$D_\mu * F^{\mu\nu} = 0. \tag{3.324}$$

3.5.2 Faddeev–Popov determinant and ghosts

Now we are prepared to construct a path integral for non-Abelian gauge fields. However, in the non-Abelian case we are confronted with additional problems which have to be mastered. The popular procedure with far-reaching consequences for physics and geometry has been invented by Faddeev and Popov (FP) in 1967 in order to quantize YM fields [Faddeev, Popov 1967], [Faddeev, Slavnov 1980]. We shall describe it in the following.

As in the Abelian case the **PI over non-Abelian fields**

$$Z = \int dA_\mu e^{iS[A_\mu]} \tag{3.325}$$

is not well defined for the gauge invariant YM action

$$S[A_\mu] = \int dx \mathcal{L}_{YM}(A_\mu(x)) \tag{3.326}$$

(we neglect the Schwinger sources for the moment) due to the gauge freedom of the fields. In order to cure this defect we impose a **gauge condition**

$$f[A_\mu^g] = 0. \tag{3.327}$$

We also assume that the functional equation (3.327) has a *unique* solution $g(x)$ for any given gauge potential $A_\mu(x)$.

We incorporate this gauge condition via the following trick. Since

$$\int dg \Delta_{FP}[A_\mu^g]\delta(f[A_\mu^g]) = 1, \tag{3.328}$$

with

$$\Delta_{FP}[A_\mu^g] = \det \frac{\delta f[A_\mu^g]}{\delta g}, \tag{3.329}$$

we may write

$$Z = \int dA_\mu \int dg \Delta_{FP}[A_\mu^g]\delta(f[A_\mu^g]) e^{iS[A_\mu]}. \tag{3.330}$$

Now we perform a gauge transformation $A_\mu^g \to A_\mu$ (or $g \to g^{-1}$); we use the gauge invariance of

i) $dA_\mu^g = dA_\mu$, gauge potential measure,
ii) $S[A_\mu^g] = S[A_\mu]$, classical action,
iii) $dg = d(gg'), g' \in G$, group measure,

and we obtain

$$Z = \int dg \cdot \int dA_\mu \Delta_{FP}[A_\mu]\delta(f[A_\mu])\, e^{iS[A_\mu]}. \tag{3.331}$$

We observe that this expression is independent of the group element g so that the group integration factorizes and produces an infinite constant (the infinite volume of the gauge group). We remove this 'constant' by redefining Z which we have to normalize anyhow.

Then the correct **PI for the generating functional of pure YM fields** is given by

$$Z = \int dA_\mu \Delta_{FP}[A_\mu]\delta(f[A_\mu])\, e^{iS[A_\mu]}. \tag{3.332}$$

This is Faddeev–Popov's equation. We still have to evaluate the FP determinant.

Faddeev–Popov determinant:

$$\Delta_{FP}[A_\mu] = \det \left. \frac{\delta f[A_\mu^g]}{\delta g} \right|_{g=1}. \tag{3.333}$$

It is enough to consider just infinitesimal gauge transformations $g = \mathbf{1} + \Lambda$, equations (3.312), (3.313), then we have

$$\Delta_{FP}[A_\mu] = \det \left. \frac{\delta f[A_\mu^\Lambda]}{\delta \Lambda} \right|_{\Lambda=0} \tag{3.334}$$

and

$$\frac{\delta f}{\delta \Lambda} = \frac{\delta}{\delta \Lambda} \frac{\delta f}{\delta A_\mu} \delta A_\mu = \frac{\delta f}{\delta A_\mu} D_\mu. \tag{3.335}$$

Choosing for definiteness the Lorentz condition

$$f[A_\mu] = \partial^\mu A_\mu = 0 \tag{3.336}$$

we find

$$\frac{\delta f}{\delta \Lambda} = \partial^\mu D_\mu \tag{3.337}$$

and in components

$$\frac{\delta f^a[A_\mu^\Lambda](x)}{\delta \Lambda^b(y)} = \partial^\mu D_\mu^{ba}(x) \delta^4(x-y). \tag{3.338}$$

Gauge fixing: The previous treatment of Z can be immediately extended to a new **gauge fixing function**

$$f[A_\mu^g](x) - c(x) = 0, \tag{3.339}$$

with

$$c(x) = c^a(x) T^a \tag{3.340}$$

some function *not* depending on A_μ.

Is there any change?

Lemma: The dependence of $\Delta_{FP}(A_\mu^g)$ on $c(x)$ can be removed.

Proof: The dependence on c comes from a gauge element $g_0 : \exists g_0$ such that $f[A_\mu^{g_0}] = c$, then we have $\Delta_{FP}(A_\mu^{g_0})$. But we may write

$$1 = \int dg \Delta_{FP}[A_\mu^{g_0}] \delta(f[A_\mu^g] - c) = \int dg \Delta_{FP}[A_\mu^g] \delta(f[A_\mu^g] - c),$$

where the second equality is true because the δ-function implies $g = g_0$, and the dependence of Δ_{FP} on c has been removed. Q.E.D.

So we can prescribe the following generating functional

$$Z = \int dA_\mu \Delta_{FP}[A_\mu]\delta(f[A_\mu] - c)\, e^{iS[A_\mu]}. \tag{3.341}$$

Next we integrate over the functions $c(x)$ with some, say, Gaussian weight which changes just the overall normalization constant

$$\begin{aligned} Z &= \int dc \exp\left[-\frac{i}{2\alpha}\int dx\, c^a(x)c^a(x)\right]\int dA_\mu \det\frac{\delta f}{\delta\Lambda}\,\delta(f[A_\mu]-c)e^{iS[A_\mu]} \\ &= \int dA_\mu \det\frac{\delta f}{\delta\Lambda}\exp\left(iS[A_\mu] - \frac{i}{2\alpha}\int dx\, f^a[A_\mu](x)f^a[A_\mu](x)\right). \end{aligned} \tag{3.342}$$

Faddeev–Popov ghosts: Now we use another Faddeev–Popov trick. We re-express the FP determinant by an integral over Grassmann fields. Recalling formula (3.228) of Section 3.3 we introduce the Grassmann fields

$$\textbf{FP ghost}: \quad v(x) = v^a(x)T^a \tag{3.343}$$

$$\textbf{FP antighost}: \quad \bar{v}(x) = \bar{v}^a(x)T^a \tag{3.344}$$

so that we get a Gaussian integral

$$\det\frac{\delta f}{\delta\Lambda} = \int d\bar{v}dv \exp\left(-i\int dx\, \bar{v}\frac{\delta f}{\delta\Lambda}v\right) \tag{3.345}$$

(the factor i is absorbed by Euclidean space calculations). The fields $v^a(x)$, $\bar{v}^a(x)$ are scalar fields obeying Fermi–Dirac statistics. Note that $v^a(x)$ and $\bar{v}^a(x)$ are *not* related by Hermitian conjugation but represent two independent Grassmann fields.

In this way we arrive at the following **generating functional for pure YM fields**

$$\begin{aligned} Z &= \int dA_\mu d\bar{v}dv \\ &\quad\cdot \exp\left(i\int dx\left[-\frac{1}{4}F^a_{\mu\nu}F^{a\mu\nu} - \frac{1}{2\alpha}(\partial^\mu A^a_\mu)^2 - \bar{v}^a\partial^\mu D^{ba}_\mu v^b\right]\right). \end{aligned} \tag{3.346}$$

If we finally insert the external sources à la Schwinger and include the fermions we obtain the **generating functional for an interacting non-Abelian gauge theory**

$$Z[J,\omega,\bar{\omega},\eta,\bar{\eta}] = \frac{1}{\mathcal{N}}\int dA_\mu d\bar{v}dv d\bar{\psi}d\psi$$
$$\cdot \exp\left(i\int dx[\mathcal{L}_{\text{tot}} + J^{a\mu}A^a_\mu + \bar{\omega}^a v^a + \bar{v}^a\omega^a + \bar{\eta}\psi + \bar{\psi}\eta]\right). \tag{3.347}$$

The **normalization** $\mathcal{N}$ denotes the numerator for vanishing sources $J^a_\mu = \omega^a = \bar{\omega}^a = \eta = \bar{\eta} = 0$.

The **total Lagrangian** consists of

$$\mathcal{L}_{\text{tot}} = \mathcal{L}_\psi + \mathcal{L}_{YM} + \mathcal{L}_{G\,\text{fix}} + \mathcal{L}_{FP}, \tag{3.348}$$

where the several parts represent:

$\mathcal{L}_\psi$	**fermion term**	(3.296)
$\mathcal{L}_{YM}$	**pure Yang–Mills term**	(3.298)
$\mathcal{L}_{G\,\text{fix}}$	**gauge fixing term**	(3.349)
$\mathcal{L}_{FP}$	**FP ghost term**	(3.350).

Résumé: We have reached our aim! By restricting the functional measure with a gauge condition we obtain a **gauge fixing term**

$$\mathcal{L}_{G\,\text{fix}} = -\frac{1}{2\alpha}(\partial^\mu A^a_\mu)^2 = \frac{1}{\alpha}\ \text{tr}\ (\partial^\mu A_\mu)^2 \tag{3.349}$$

and a nonsingular gauge boson propagator. We get rid of the unwanted redundant gauge equivalent fields and the dependence on the gauge group just drops. But there occurs the FP determinant which creates an additional term in the Lagrangian, the **FP ghost term**

$$\mathcal{L}_{FP} = -\bar{v}^a\partial^\mu D^{ba}_\mu v^b = -\bar{v}^a\partial^\mu(D_\mu v)^a = 2\ \text{tr}\ \bar{v}\partial^\mu D_\mu v. \tag{3.350}$$

The ghost and antighost are massless scalar fields which propagate because of the kinetic part

$$-\bar{v}^a\Box v^a \tag{3.351}$$

and they interact with the gauge potential according to

$$-\bar{v}^a\partial^\mu[A_\mu, v]^a. \tag{3.352}$$

Remark: There exists, however, a gauge where the FP ghosts disappear. This is the **axial gauge** defined by

$$f[A_\mu] = n^\mu A_\mu = 0, \qquad n^\mu n_\mu = -1. \tag{3.353}$$

Then the FP determinant does not depend on the gauge potential since the functional derivative is

$$\frac{\delta f}{\delta \Lambda} = \frac{\delta f}{\delta A_\mu} \cdot \frac{\delta A_\mu}{\delta \Lambda} = n^\mu D_\mu = n^\mu \partial_\mu \tag{3.354}$$

due to condition (3.353).

In this case the FP ghosts decouple from the gauge fields and may be integrated out and put into the normalization of the generating functional Z. But the prize one has to pay is the rather complicated gauge boson propagator in this gauge. For a field theory in the axial gauge we refer to Kummer [Kummer 1976] and Leibbrandt [Leibbrandt 1990].

In QED the gauge boson—the photon—is group 'colour' blind and does not couple to the FP ghosts. Here the FP ghosts disappear without reference to a special gauge. Therefore the introduction of FP ghosts in QED is not necessary—and not meaningful.

Physics and ghosts: The importance of the FP (anti-) ghosts for physics and geometry is the following:

- FP ghosts are needed to preserve unitarity of the S-matrix elements. They precisely cancel the unphysical degrees of freedom of the gauge bosons. For a discussion we refer to Aitchison and Hey [Aitchison, Hey 1989].
- FP ghosts re-establish an invariance of the total Lagrangian. The gauge invariance (symmetry) is replaced by another adequate invariance (symmetry)—the BRS transformation. This is essential to establish the Ward identities (Slavnov–Taylor identities) in order to prove the renormalizability of a non-Abelian QFT. We shall discuss the BRS transformation next in Section 3.6.
- FP ghosts play an important role in the differential geometry of gauge potentials. They are demanded in the study of anomalies in QFT. This will be elaborated in this book. Together with the BRS operator the FP ghosts have a geometric interpretation: the BRS operator represents a derivative, the FP ghost a Maurer–Cartan form on the group (see Section 8.1).
- Also the FP determinant has geometric roots. Babelon and Viallet found that it corresponds to a 'natural' Riemannian metric on the moduli space of gauge potentials [Babelon, Viallet 1979, 1981].

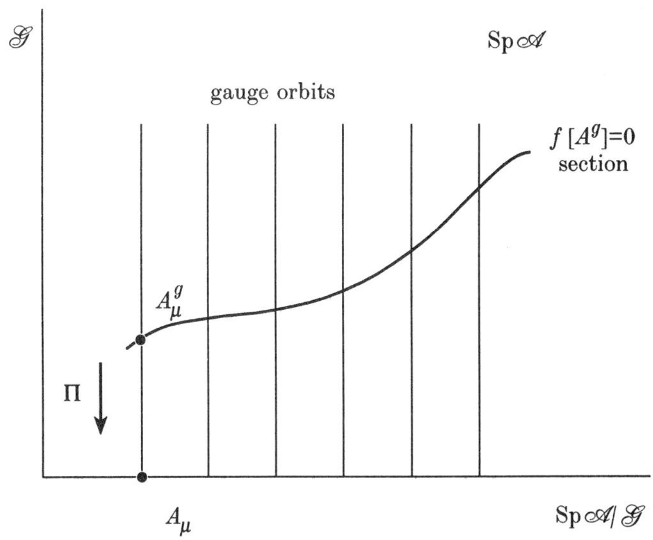

Fig. 3.4. The principal bundle in Sp $\mathcal{A}$, gauge orbits and gauge section

Geometry: How can we describe the gauge fixing $f[A^g_\mu] = 0$, equation (3.327), in a geometric way? Let us consider the **affine space of all gauge connections** Sp $\mathcal{A}$. We denote the space of all gauge elements by $\mathcal{G} = \{g(x)\}$—the **gauge group space**. Then we can construct a (infinite dimensional) principal bundle $\mathcal{P}(\mathcal{A}, \Pi, M, \mathcal{G})$ with fibre $\mathcal{G}$ and the projection Π : Sp $\mathcal{A} \to$ Sp $\mathcal{A}/\mathcal{G}$. The base space Sp $\mathcal{A}/\mathcal{G}$—the **moduli space**—represents the physical space of the YM theory. The field configurations A^g_μ determined by the gauge transformations (3.293) for all possible elements g to a given A_μ are called an **orbit** of the gauge group $\mathcal{G}$. The gauge condition (3.327) then defines a surface in the space of all gauge fields—a **gauge section**—which crosses each gauge orbit precisely once (see Figure 3.4).

Gribov ambiguity and geometry: In the FP derivation of the gauge field PI it was crucial to assume that the gauge condition (3.327) has precisely one solution. Gribov [Gribov 1978], however, discovered that non-Abelian fields—in contrast to Abelian fields—can have values such that the gauge condition (3.327) does *not* provide a unique solution for the gauge element $g(x)$—the Gribov ambiguity. This happens beyond perturbation theory and the PI representation (3.346), (3.347) is not meaningful there.

The geometric understanding of this phenomenon has been found by Singer [Singer 1978] and is due to the *nontrivial* topology of the principal bundle $\mathcal{P}(\mathcal{A}, \Pi, M, \mathcal{G})$ (for $M = S^4$ and $SU(N)$ gauge groups). Gauge fixing corresponds to choosing only one representative connection in the bundle for each element of the base Sp $\mathcal{A}/\mathcal{G}$, see Figure 3.4. Locally this is always possible—the delta function in the FP procedure selects from the PI over the gauge fields only those connections which lie on a section—but globally *no* such section exists due to the nontriviality of the principal bundle. So there exist gauge configurations where a gauge fixing section is not well defined—ambiguous á la Gribov.

3.6 BRS transformation

We consider again the total Lagrangian for a non-Abelian gauge theory

$$\mathcal{L}_{\text{tot}} = \mathcal{L}_{\psi} + \mathcal{L}_{YM} + \mathcal{L}_{G\,\text{fix}} + \mathcal{L}_{FP}. \tag{3.355}$$

The fermion term $\mathcal{L}_{\psi}$ and the YM part $\mathcal{L}_{YM}$ are gauge invariant by construction. However, in the total Lagrangian gauge invariance is obviously lost because of the gauge fixing term $\mathcal{L}_{G\,\text{fix}}$. But there is still the freedom to work with the FP ghosts. It was Becchi, Rouet and Stora (BRS) [Becchi, Rouet, Stora 1974, 1975] who discovered an 'enlarged' gauge transformation—the BRS transformation—which also involves the FP (anti-) ghosts and leaves the total Lagrangian invariant.

Nowadays this transformation is also called BRST transformation to refer to a work of Tyutin [Tyutin 1975].

How is the BRS transformation defined?

BRS transformation: Let us first assign to all fields a **ghost number**:

ghost number	
$+1$	$v^a(x)$
-1	$\bar{v}^a(x)$
0	$A_\mu(x), \psi(x), \bar{\psi}(x)$.

We define an operator—the **BRS operator** s—which acts on the usual fields like a usual infinitesimal gauge transformation with parameter v, but on the (anti-) ghost fields such that s is nilpotent

$$s^2 = 0. \tag{3.356}$$

This is achieved by the **BRS transformations**

$$\begin{aligned} sA_\mu &= D_\mu v \\ sF_{\mu\nu} &= [F_{\mu\nu}, v] \\ sv &= -\frac{1}{2}[v, v] = -v^2 \\ s\psi &= -v\psi \\ s\bar{\psi} &= -\bar{\psi}v. \end{aligned} \tag{3.357}$$

Examples:

$$\begin{aligned} s^2 A_\mu &= sD_\mu v = \partial_\mu sv + [sA_\mu, v] + [A_\mu, sv] \\ &= D_\mu(sv + \frac{1}{2}[v,v]) = 0 \end{aligned} \quad (3.358)$$

$$s^2 v = -svv = v^3 - v^3 = 0. \quad (3.359)$$

Graded algebra: In fact, the BRS operator s increases the ghost number by one unit. This happens analogously to the exterior differential d which raises the form degree by one unit. We can combine the exterior algebra structure with the BRS structure. We add the form degree and the ghost degree to get a **total degree** and we consider the algebra (generated by all fields) graded by this total degree. Both operators s and d act as **antiderivations** on this algebra and they are required to satisfy

$$sd + ds = 0. \quad (3.360)$$

Generally, a commutator means

$$[P, Q] = PQ - (-)^{\deg P \cdot \deg Q} QP \quad (3.361)$$

and specifically for the 1-form $A = A_\mu dx^\mu$, 2-form $F = \frac{1}{2} F_{\mu\nu} dx^\mu dx^\nu$ and FP ghost v

$$[A, v] = Av + vA \quad (3.362)$$

$$[F, v] = Fv - vF \quad (3.363)$$

$$[v, v] = vv + vv = 2v^2. \quad (3.364)$$

Within differential forms we have the BRS transformations

$$\begin{aligned} sA &= -Dv \\ sF &= [F, v]. \end{aligned} \quad (3.365)$$

Note: The third BRS equation (3.357) reminds us of the Maurer–Cartan equation on the Lie group (Section 2.6.6). We discuss Stora's [Stora 1986] algebraic interpretation of the FP ghost separately in Section 8.3.2.

Now we are still left with the treatment of the antighost $\bar{v}$. We want to define its transformation such that the total Lagrangian remains invariant and that $s^2 = 0$ is respected. We can achieve this with the help of an auxiliary field.

Auxiliary field: We reformulate the Lagrangian by introducing a scalar and commuting **auxiliary field** (for a review see e.g. [Gieres 1988], [Baulieu 1985])

$$b(x) = b^a(x)T^a \tag{3.366}$$

and we incorporate the gauge fixing condition

$$f[A_\mu] = \partial^\mu A_\mu = 0 \tag{3.367}$$

in the following way. We choose as the gauge fixing term

$$\mathcal{L}_{G\,\mathrm{fix}} = b^a f^a[A_\mu] + \frac{\alpha}{2} b^a b^a. \tag{3.368}$$

Taking into account the equations of motion for the field b

$$\frac{\delta \mathcal{L}}{\delta b^a} = 0, \tag{3.369}$$

with its solution

$$b^a = -\frac{1}{\alpha} f^a[A_\mu] = -\frac{1}{\alpha}\partial^\mu A^a_\mu \tag{3.370}$$

we recover the original gauge fixing

$$\mathcal{L}_{G\,\mathrm{fix}} = \left(-\frac{1}{\alpha} + \frac{\alpha}{2\alpha^2}\right)(f^a[A_\mu])^2 = -\frac{1}{2\alpha}(\partial^\mu A^a_\mu)^2. \tag{3.371}$$

So the auxiliary field has the interpretation of a **Lagrange multiplier** for the gauge fixing condition. Due to its canonical dimension 2 it cannot acquire a kinetic term in a (renormalizable) Lagrangian.

Rewriting the FP part next

$$\mathcal{L}_{FP} = -\bar{v}^a \partial^\mu (D_\mu v)^a = -\bar{v}^a \partial^\mu s A^a_\mu = -\bar{v}^a s f^a[A_\mu] \tag{3.372}$$

we have for the sum

$$\mathcal{L}_{G\,\mathrm{fix}} + \mathcal{L}_{FP} = b^a f^a[A_\mu] + \frac{\alpha}{2} b^a b^a - \bar{v}^a s f^a[A_\mu]. \tag{3.373}$$

In order to get it invariant under the BRS operator s we choose the transformations

$$\begin{aligned} s\bar{v} &= b \\ sb &= 0, \end{aligned} \tag{3.374}$$

which imply

BRS transformations	anti-BRS transformations
$sA = -Dv$ $sF = [F, v]$ $sv = -\frac{1}{2}[v, v] = -v^2$ $s\bar{v} = b$ $sb = 0$ $s\psi = -v\psi$ $s\bar{\psi} = -\bar{\psi}v$	$\bar{s}A = -D\bar{v}$ $\bar{s}F = [F, \bar{v}]$ $\bar{s}\bar{v} = -\frac{1}{2}[\bar{v}, \bar{v}] = -\bar{v}^2$ $\bar{s}v = -[\bar{v}, v] - b$ $\bar{s}b = -[\bar{v}, b]$ $\bar{s}\psi = -\bar{v}\psi$ $\bar{s}\bar{\psi} = -\bar{\psi}\bar{v}$
$s^2 = sd + ds = 0$	$\bar{s}^2 = \bar{s}d + d\bar{s} = 0$
$s\bar{s} + \bar{s}s = 0$ $s\bar{v} + \bar{s}v = -[\bar{v}, v]$	

Table 3.1.

$$\begin{aligned} \mathcal{L}_{G\,\text{fix}} + \mathcal{L}_{FP} &= s\bar{v}^a f^a[A_\mu] - \bar{v}^a s f^a[A_\mu] + \frac{\alpha}{2} b^a b^a \\ &= s(\bar{v}^a f^a[A_\mu]) + \frac{\alpha}{2} b^a b^a \end{aligned} \tag{3.375}$$

and consequently

$$s(\mathcal{L}_{G\,\text{fix}} + \mathcal{L}_{FP}) = 0. \tag{3.376}$$

Thus we derived the following proposition.

Proposition:

- *The total Lagrangian* (3.355) *is BRS transformation invariant!*

Anti-BRS transformation: In analogy with the BRS operator s we can construct an **anti-BRS operator** $\bar{s}$ which decreases the ghost number by one unit and has similar mathematical properties (for a review see [Baulieu 1985]). We collect the BRS and anti-BRS transformations in Table 3.1.

Remark: The algebra of Table 3.1 is called **(anti-) BRS algebra**. Its structure is highly symmetric and holds independent of a Lagrangian. Thus we can regard the BRS symmetry as a fundamental symmetry that a gauge theory has to obey. In fact, the **principle of BRS symmetry** is used as an alternative approach to construct generalized YM field theories (see e.g. [Baulieu 1985], [Baulieu, Thierry-Mieg 1982], [Alvarez-Gaumé, Baulieu 1983]).

Outlook: As we have demonstrated the importance of the FP ghosts and the BRS transformation lies in the restoration of a symmetry of the total Lagrangian or of the classical action.

What about quantum field theory? Does the 'quantum action' remain BRS invariant too?

The invariance of $W = -i \ln Z$

$$sW = 0, \tag{3.377}$$

called the **Ward identity** (Slavnov–Taylor identity) is necessary to get a non-Abelian gauge theory renormalized [Becchi, Rouet, Stora 1976 a,b,c], [Stora 1977].

However, it might happen that this invariance is broken. Then an anomaly appears and renormalizability is lost

$$sW = G \neq 0. \tag{3.378}$$

Thus the BRS transformations are on the one hand a key to showing the renormalizability of gauge theories and on the other they play an important role in the determination of the anomalies. We want to discuss this role in detail in the following chapters.

4

Anomalies in QFT

This chapter is a principal chapter in which we investigate the anomaly within standard mathematical techniques. That means we work predominantly in perturbation theory calculating Feynman diagrams with which we assume the reader has a working knowledge. We show in great detail the occurrence of the anomaly in QFT and discuss its meaning.

In Section 4.1 we start with the classical conservation laws and symmetries for Abelian and non-Abelian fields. Then we describe the quantum partners—the Ward identities—in Section 4.2 and we discuss their violation due to the anomaly. We devote Section 4.3 to the calculation of the Adler–Bell–Jackiw anomaly and we demonstrate in detail how the anomaly arises in several renormalization schemes. Although this part is somewhat technical, such a treatment is necessary for an understanding of the anomaly and will repay careful study. Next (Section 4.4) we describe the anomaly by means of dispersion relations and, furthermore, as a simple, yet illuminating, example, we study the anomaly in the Dirac sea (Section 4.5). In Section 4.6, making contact with physics, we calculate explicitly the famous $\pi^0 \to \gamma\gamma$ decay, the milestone of the anomaly. In the next sections we extend our discussion to non-Abelian fields, we determine the singlet anomaly (Section 4.7) and present the non-Abelian anomaly—Bardeen's result (Section 4.8). A final summary (Section 4.9) on the importance of the anomalies for physics closes this chapter.

4.1 Classical conservation laws and symmetries

Let us first discuss the classical conservation laws and symmetries we actually need in connection with the anomaly.

4.1.1 Abelian fields

We consider QED with the Lagrangian

$$\mathcal{L}_{QED} = \bar{\psi}(i\, \partial\!\!\!/ - m + e\, A\!\!\!/)\psi - \frac{1}{4}F_{\mu\nu}F^{\mu\nu} \tag{4.1}$$

and the field strength

$$F_{\mu\nu} = \partial_\mu A_\nu - \partial_\nu A_\mu. \tag{4.2}$$

We construct the currents:

$$\begin{array}{lll} \text{vector} & j_\mu = \bar{\psi}\gamma_\mu\psi & (4.3) \\ \text{axial} & j^5_\mu = \bar{\psi}\gamma_\mu\gamma_5\psi & (4.4) \\ \text{pseudoscalar} & P = \bar{\psi}\gamma_5\psi. & (4.5) \end{array}$$

Conservation laws: With the help of the equations of motion—the Dirac equations—

$$\begin{aligned} (i \not\partial - m + e \not A)\psi &= 0 \\ \bar{\psi}(i \overleftarrow{\not\partial} + m - e \not A) &= 0 \end{aligned} \tag{4.6}$$

we can establish the following **conservation laws**

$$\begin{aligned} \partial^\mu j_\mu &= \bar{\psi} \overleftarrow{\not\partial} \psi + \bar{\psi} \not\partial\psi \\ &= i\bar{\psi}(m - e \not A)\psi + i\bar{\psi}(-m + e \not A)\psi \\ &= 0 \end{aligned} \tag{4.7}$$

$$\begin{aligned} \partial^\mu j^5_\mu &= i\bar{\psi}(m - e \not A)\gamma_5\psi - i\bar{\psi}\gamma_5(-m + e \not A)\psi \\ &= 2imP. \end{aligned} \tag{4.8}$$

In equation (4.8) we used the anticommutativity of the Dirac matrices

$$\{\gamma_5, \gamma_\mu\} = 0. \tag{4.9}$$

The vector current is conserved, equation (4.7), the axial current too for massless fermions

$$\partial^\mu j^5_\mu = 0 \qquad \text{for } m = 0. \tag{4.10}$$

Symmetries: Conservation laws are connected with symmetries, which is the basic statement of the Noether theorem. Let us consider now as gauge fields a vector field and an axial field which we denote by $\mathcal{V}_\mu$ and $\mathcal{A}_\mu$. Both fields are assumed to be external fields, and for simplicity we treat just the massless case $m = 0$. So we start from the Lagrangian

$$\mathcal{L}(\mathcal{V}_\mu, \mathcal{A}_\mu) = \bar{\psi}(i \not\partial + \not\mathcal{V} + \not\mathcal{A}\, \gamma_5)\psi \tag{4.11}$$

(the coupling constants are chosen unity). This Lagrangian is invariant under the **local vector gauge transformation** $U_V(1)$

$$\begin{aligned}\psi &\rightarrow e^{i\alpha(x)}\psi \\ \bar{\psi} &\rightarrow \bar{\psi}e^{-i\alpha(x)} \\ \mathcal{V}_\mu &\rightarrow \mathcal{V}_\mu + \partial_\mu \alpha(x).\end{aligned} \tag{4.12}$$

Consequently, the vector current is conserved, equation (4.7).

In addition, the Lagrangian (4.11) remains invariant under a **local axial gauge transformation** $U_A(1)$

$$\begin{aligned}\psi &\rightarrow e^{i\beta(x)\gamma_5}\psi \\ \bar{\psi} &\rightarrow \bar{\psi}e^{i\beta(x)\gamma_5} \\ \mathcal{A}_\mu &\rightarrow \mathcal{A}_\mu + \partial_\mu \beta(x),\end{aligned} \tag{4.13}$$

implying a conserved axial current, equation (4.10). Altogether we have the **symmetry** $U_V(1) \times U_A(1)$.

It is also customary to decompose the fields into left-handed (L-) and right-handed (R-) components. Introducing the **projection operator** onto positive and negative chirality fields

$$P_\pm = \frac{1}{2}(1 \pm \gamma_5), \tag{4.14}$$

with properties

$$P_\pm^2 = P_\pm, \qquad P_+ P_- = 0, \qquad P_+ + P_- = \mathbf{1}, \tag{4.15}$$

we have

$$\psi_{L,R} \equiv \psi_\pm := P_\pm \psi, \tag{4.16}$$

with **chirality** ± 1

$$\gamma_5 \psi_\pm = \pm \psi_\pm. \tag{4.17}$$

We use here the γ_5-convention [Pietschmann 1983]

$$\gamma_5 = i\gamma_0\gamma_1\gamma_2\gamma_3 \tag{4.18}$$

for the identity (4.16).

Then the above Lagrangian (4.11) can be rewritten in terms of L- and R-handed fields

$$\mathcal{L}(\mathcal{V}_\mu, \mathcal{A}_\mu) \equiv \mathcal{L}_{L,R} := \bar{\psi}_L(i\,\partial\!\!\!/ + A\!\!\!/^L)\psi_L + \bar{\psi}_R(i\,\partial\!\!\!/ + A\!\!\!/^R)\psi_R, \tag{4.19}$$

where the L- and R-gauge fields are defined by

$$\begin{aligned}A_\mu^L &= \mathcal{V}_\mu + \mathcal{A}_\mu \qquad & \mathcal{V}_\mu &= \frac{1}{2}(A_\mu^L + A_\mu^R) \\ A_\mu^R &= \mathcal{V}_\mu - \mathcal{A}_\mu \qquad & \mathcal{A}_\mu &= \frac{1}{2}(A_\mu^L - A_\mu^R).\end{aligned} \tag{4.20}$$

Obviously we can transform the L- and R-fields separately in Lagrangian (4.19):

left-handed gauge transformation $U_L(1)$

$$\begin{aligned} \psi_L &\to e^{i\Lambda_L(x)}\psi_L, \qquad \delta\psi_L = i\Lambda_L\psi_L, \\ A_\mu^L &\to A_\mu^L + \partial_\mu\Lambda_L(x), \end{aligned} \tag{4.21}$$

right-handed gauge transformation $U_R(1)$

$$\begin{aligned} \psi_R &\to e^{i\Lambda_R(x)}\psi_R, \qquad \delta\psi_R = i\Lambda_R\psi_R, \\ A_\mu^R &\to A_\mu^R + \partial_\mu\Lambda_R(x). \end{aligned} \tag{4.22}$$

So altogether the Lagrangian $\mathcal{L}_{L,R}$ (4.19) has the **symmetry** $U_L(1)\times U_R(1)$.

Note: For $\Lambda_L = \Lambda_R$ we recover the vector symmetry $U_V(1)$ and for $\Lambda_L = -\Lambda_R$ the axial symmetry $U_A(1)$.

Conservation laws: Finally we consider the L- and R-handed currents

$$\begin{aligned} j_\mu^L &= \bar\psi_L\gamma_\mu\psi_L = \bar\psi\gamma_\mu\frac{1}{2}(1+\gamma_5)\psi \\ j_\mu^R &= \bar\psi_R\gamma_\mu\psi_R = \bar\psi\gamma_\mu\frac{1}{2}(1-\gamma_5)\psi, \end{aligned} \tag{4.23}$$

which can be expressed by

$$\begin{aligned} j_\mu^L &= \frac{1}{2}(j_\mu + j_\mu^5) \qquad & j_\mu &= j_\mu^L + j_\mu^R \\ j_\mu^R &= \frac{1}{2}(j_\mu - j_\mu^5) \qquad & j_\mu^5 &= j_\mu^L - j_\mu^R. \end{aligned} \tag{4.24}$$

Then due to identities (4.7) and (4.10) the **L- and R-currents are conserved** separately

$$\partial^\mu j_\mu^L = \partial^\mu j_\mu^R = 0. \tag{4.25}$$

4.1.2 Non-Abelian fields

Our study of fermions interacting with Abelian fields can be extended to non-Abelian gauge fields, so-called Yang–Mills fields (recall Section 3.5.1). In this case the Lagrangian

$$\mathcal{L}_{\text{na}}(\mathcal{V}_\mu, \mathcal{A}_\mu) = i\bar\psi(\partial\!\!\!/ + \mathcal{V}\!\!\!/ + \mathcal{A}\!\!\!/\,\gamma_5)\psi - m\bar\psi\psi \tag{4.26}$$

contains non-Abelian vector and axial fields

$$\mathcal{V}_\mu = \mathcal{V}_\mu^a T^a, \qquad \mathcal{A}_\mu = \mathcal{A}_\mu^a T^a, \tag{4.27}$$

where the group generators T^a satisfy the algebra (3.275) discussed in Section 3.5.1. Again, we assume external gauge fields.

Symmetries: The above Lagrangian remains invariant under the **local vector symmetry** $SU_V(N)$

$$\begin{aligned}
\psi &\to \psi^g = g^{-1}(x)\psi \\
\bar\psi &\to \bar\psi^g = \bar\psi(x)g(x) \\
\mathcal{V}_\mu &\to \mathcal{V}_\mu^g = g^{-1}(x)(\mathcal{V}_\mu + \partial_\mu)g(x) \\
\mathcal{A}_\mu &\to \mathcal{A}_\mu^g = g^{-1}(x)\mathcal{A}_\mu g(x),
\end{aligned} \tag{4.28}$$

where $g(x)$ denotes the **gauge group element**

$$g(x) = e^{\alpha(x)}, \qquad \alpha(x) = \alpha^a(x)T^a. \tag{4.29}$$

Note that we have chosen the matrices T^a anti-Hermitian, equation (3.276), so that

$$\alpha^\dagger(x) = -\alpha(x). \tag{4.30}$$

For the **infinitesimal vector gauge transformation** we then have

$$\begin{aligned}
\delta\psi &= -\alpha(x)\psi \\
\delta\bar\psi &= \bar\psi\alpha(x) \\
\delta\mathcal{V}_\mu &= \mathcal{D}_\mu\alpha \\
\delta\mathcal{A}_\mu &= [\mathcal{A}, \alpha]
\end{aligned} \tag{4.31}$$

and $\mathcal{D}_\mu$ expresses the familiar **covariant derivative**

$$\mathcal{D}_\mu = \partial_\mu + [\mathcal{V}_\mu, \ ]. \tag{4.32}$$

Considering next **chiral transformations**

$$\begin{aligned}
\psi &\to e^{-\beta(x)\gamma_5}\psi \\
\bar\psi &\to \bar\psi e^{-\beta(x)\gamma_5},
\end{aligned} \tag{4.33}$$

with

$$\beta(x) = \beta^a(x)T^a \tag{4.34}$$

we find a **local axial symmetry** $SU_A(N)$ in the Lagrangian (4.26) if the fermion mass vanishes, $m = 0$. Its infinitesimal form—the **infinitesimal axial gauge transformation**—is

$$\begin{aligned}\delta\psi &= -\beta(x)\gamma_5\psi \\ \delta\bar{\psi} &= -\bar{\psi}\beta(x)\gamma_5 \\ \delta\mathcal{V}_\mu &= [\mathcal{A}_\mu, \beta] \\ \delta\mathcal{A}_\mu &= \mathcal{D}_\mu\beta. \end{aligned} \tag{4.35}$$

So for massless fermions ($m = 0$) the Lagrangian (4.26) remains invariant under the **combined local transformation** $SU_V(N) \times SU_A(N)$

$$\begin{aligned}\delta\psi &= (-\alpha(x) - \beta(x)\gamma_5)\psi \\ \delta\bar{\psi} &= \bar{\psi}(\alpha(x) - \beta(x)\gamma_5) \\ \delta\mathcal{V}_\mu &= \mathcal{D}_\mu\alpha + [\mathcal{A}_\mu, \beta] \\ \delta\mathcal{A}_\mu &= \mathcal{D}_\mu\beta + [\mathcal{A}_\mu, \alpha]. \end{aligned} \tag{4.36}$$

Analogous to the Abelian case we also introduce here L- and R-fields with help of the chiral projection operator (4.14). Then the above Lagrangian (4.26) can be decomposed into separate L- and R-field terms

$$\mathcal{L}_{\text{na}}(\mathcal{V}_\mu, \mathcal{A}_\mu) \equiv \mathcal{L}_{\text{na}}^{L,R} := \bar{\psi}_L i \not{D}_L\psi_L + \bar{\psi}_R i \not{D}_R\psi_R - m(\bar{\psi}_L\psi_R + \bar{\psi}_R\psi_L). \tag{4.37}$$

The operators $D_\mu^{L,R}$ denote the covariant derivatives containing the L- and R-handed gauge fields

$$D_\mu^{L,R} = \partial_\mu + [A_\mu^{L,R},\ ]. \tag{4.38}$$

Clearly, when applied to the spinors we have

$$\not{D}_{L,R} = \not{\partial} + \not{A}_{L,R}. \tag{4.39}$$

The L- and R-fields are defined as in the Abelian case

$$\begin{aligned} A_\mu^L &= \mathcal{V}_\mu + \mathcal{A}_\mu \qquad & \mathcal{V}_\mu &= \frac{1}{2}(A_\mu^L + A_\mu^R) \\ A_\mu^R &= \mathcal{V}_\mu - \mathcal{A}_\mu \qquad & \mathcal{A}_\mu &= \frac{1}{2}(A_\mu^L - A_\mu^R). \end{aligned} \tag{4.40}$$

But remember that these fields now express non-Abelian fields

$$A_\mu^{L,R} = A_\mu^{a\ L,R} T^a. \tag{4.41}$$

In the massless case ($m = 0$) Lagrangian (4.37) is invariant under the **local L- and R-handed gauge transformation** $SU_L(N) \times SU_R(N)$

$$\begin{aligned}\psi_H &\to \psi_H^g = g_H^{-1}(x)\psi_H, \qquad g_H = e^{\Lambda_H(x)} \\ \bar{\psi}_H &\to \bar{\psi}_H^g = \bar{\psi}_H g_H(x)\end{aligned}$$

$$\begin{aligned} A^H_\mu &\rightarrow A^{gH}_\mu = g_H^{-1}(x)(A^H_\mu + \partial_\mu) g_H(x) \\ D^H_\mu &\rightarrow D^{gH}_\mu = g_H^{-1}(x) D^H_\mu g_H(x) \end{aligned} \tag{4.42}$$

$(H = L, R)$ or in its infinitesimal version

$$\begin{aligned} \delta\psi_H &= -\Lambda_H(x)\psi_H \\ \delta\bar{\psi}_H &= \bar{\psi}\Lambda_H(x) \\ \delta A^H_\mu &= D^H_\mu \Lambda_H(x). \end{aligned} \tag{4.43}$$

Again, the L- and R-gauge elements are non-Abelian

$$\Lambda_H(x) = \Lambda^a_H(x) T^a. \tag{4.44}$$

Note: Defining the transformation parameters

$$\alpha := \frac{1}{2}(\Lambda_L + \Lambda_R) \quad , \quad \beta := \frac{1}{2}(\Lambda_L - \Lambda_R) \tag{4.45}$$

we can connect both transformations (4.43) and (4.36).

Finally we construct several **non-Abelian currents** $(H = L, R)$

$$\begin{aligned} \text{vector} \quad & j^a_\mu = \bar{\psi}\gamma_\mu T^a \psi \\ \text{axial} \quad & j^{5a}_\mu = \bar{\psi}\gamma_\mu\gamma_5 T^a \psi \\ \text{pseudoscalar} \quad & P^a = \bar{\psi}\gamma_5 T^a \psi \\ L\text{-}, R\text{-handed} \quad & j^{Ha}_\mu = \bar{\psi}_H \gamma_\mu T^a \psi_H. \end{aligned} \tag{4.46}$$

These currents are related in the following way

$$\begin{aligned} j^a_\mu &= j^{La}_\mu + j^{Ra}_\mu & j^{La}_\mu &= \frac{1}{2}(j^a_\mu + j^{5a}_\mu) \\ j^{5a}_\mu &= j^{La}_\mu - j^{Ra}_\mu & j^{Ra}_\mu &= \frac{1}{2}(j^a_\mu - j^{5a}_\mu). \end{aligned} \tag{4.47}$$

Conservation laws: Now we search for classical conservation laws. What we need are the equations of motion according to Lagrangian (4.26)

$$\begin{aligned} (i \not\partial - m + i \not\mathcal{V} + i \not\mathcal{A} \gamma_5)\psi &= 0 \\ \bar{\psi}(i \overleftarrow{\not\partial} + m - i \not\mathcal{V} - i \not\mathcal{A} \gamma_5) &= 0. \end{aligned} \tag{4.48}$$

Let us first consider the divergence of the vector current

$$\begin{aligned} \partial_\mu j^{\mu a} &= \bar{\psi} \overleftarrow{\not\partial} T^a \psi + \bar{\psi} T^a \not\partial \psi \\ &= \bar{\psi}\gamma^\mu [T^b, T^a]\psi \mathcal{V}^b_\mu + \bar{\psi}\gamma^\mu\gamma_5 [T^b, T^a]\psi \mathcal{A}^b_\mu. \end{aligned} \tag{4.49}$$

Using the algebra (3.275) satisfied by the group generators we obtain

$$\partial_\mu j^{\mu a} + f^{abc}\mathcal{V}^b_\mu j^{\mu c} + f^{abc}\mathcal{A}^b_\mu j^{5\mu c} = 0. \tag{4.50}$$

Multiplying again by T^a we find in terms of the covariant derivative (4.32) the **classical conservation law of the non-Abelian vector current**

$$\mathcal{D}_\mu j^\mu + [\mathcal{A}_\mu, j^{5\mu}] = 0. \tag{4.51}$$

Next we study the axial current

$$\begin{aligned} \partial_\mu j^{5\mu a} &= \bar{\psi} \overleftarrow{\not\partial} \gamma_5 T^a \psi - \bar{\psi}\gamma_5 T^a \not\partial \psi \\ &= \bar{\psi}\gamma^\mu\gamma_5 [T^b, T^a]\psi \mathcal{V}^b_\mu + \bar{\psi}\gamma^\mu [T^b, T^a]\psi \mathcal{A}^b_\mu + 2im\bar{\psi}\gamma_5 T^a \psi, \end{aligned} \tag{4.52}$$

implying

$$\partial_\mu j^{5\mu a} + f^{abc}\mathcal{V}^b_\mu j^{5\mu c} + f^{abc}\mathcal{A}^b_\mu j^{\mu c} = 2imP^a, \tag{4.53}$$

the **classical conservation law for the non-Abelian axial current**

$$\mathcal{D}_\mu j^{5\mu} + [\mathcal{A}_\mu, j^\mu] = 2imP. \tag{4.54}$$

If the Lagrangian (4.26) contains only vector gauge fields (hence $\mathcal{A}_\mu = 0$) we have

$$\mathcal{D}_\mu j^\mu = 0 \tag{4.55}$$
$$\mathcal{D}_\mu j^{5\mu} = 2imP \tag{4.56}$$

and in the massless case ($m = 0$) both currents, **vector** *and* **axial, are covariantly conserved**

$$\mathcal{D}_\mu j^\mu = \mathcal{D}_\mu j^{5\mu} = 0. \tag{4.57}$$

Finally we turn to the laws for the L- and R-handed currents. From the vector current conservation (4.50) follows when inserting relations (4.47) and (4.38), (4.40)

$$D^L_\mu j^{L\mu} + D^R_\mu j^{R\mu} = 0, \tag{4.58}$$

whereas the axial current law (4.53) implies

$$D^L_\mu j^{L\mu} - D^R_\mu j^{R\mu} = 2imP. \tag{4.59}$$

Altogether we obtain as the **classical conservation law for the L-handed (or R-handed) non-Abelian current**

$$-D^R_\mu j^{R\mu} = D^L_\mu j^{L\mu} = imP. \tag{4.60}$$

In the massless case ($m = 0$) both currents, L- **and** R**-handed, are covariantly conserved**

$$D^L_\mu j^{L\mu} = D^R_\mu j^{R\mu} = 0. \tag{4.61}$$

4.2 Ward identities and anomaly

How are the classical conservation laws for the currents transferred to a quantized field theory? In QFT we are dealing with Green functions and with the generating functional of all these Green functions respectively (recall Chapter 3). The validity of the classical conservation laws induces relations among various Green functions, so-called Ward identities. They form the basic equations for a consistently quantized theory.

4.2.1 Green functions

Let us consider the vacuum expectation value of the time ordered product of a current j^μ and some operators O^i

$$\langle 0|Tj^\mu(x)O^1(y_1)\dots O^n(y_n)|0\rangle. \tag{4.62}$$

Differentiating such a quantity we obtain

$$\begin{aligned}\partial^x_\mu\langle 0|Tj^\mu(x)O^1(y_1)\dots O^n(y_n)|0\rangle &= \\ = \quad \langle 0|T\partial^x_\mu j^\mu(x)O^1(y_1)\dots O^n(y_n)|0\rangle & \qquad (4.63)\\ +\sum_{i=1}^{n}\langle 0|T[j^0(x),O^i(y_i)]\delta(x_0-y_{i0})O^1\dots O^{i-1}O^{i+1}\dots O^n|0\rangle,\end{aligned}$$

where the commutator terms arise from the differentiation of the θ-functions in the T-product (recall equation (3.86) of Chapter 3). When inserting the classical conservation law for the current and respecting the canonical algebra for the commutator (neglecting so-called Schwinger terms) we call the above relation (4.63) a **Ward identity** (WI). Such relations between Green functions have to be satisfied in order to ensure the renormalizability of QFT. This is the importance of the Ward identities. If they are broken, renormalizability is lost.

Ward's identity: The simplest example which is naming the above relation is to be found from the old days of QED. It is the important relation between the vertex and the propagator established by Ward [Ward 1950] and Takahashi [Takahashi 1957] which implies the equality of the renormalization constants of the vertex and the fermion wave function.

Let us consider the following 3-point function

$$\tau^\mu(x,y,z) = \langle 0|Tj^\mu(z)\psi(x)\bar{\psi}(y)|0\rangle \tag{4.64}$$

and we shall need the equal-time commutator

$$\begin{aligned}[j^0(z),\psi(x)]\delta(z_0-x_0) &= [\psi^\dagger(z)\psi(z),\psi(x)]\delta(z_0-x_0)\\ &= -\{\psi(x),\psi^\dagger(z)\}\psi(z)\delta(z_0-x_0)\\ &= -\psi(z)\delta^4(z-x), \end{aligned} \tag{4.65}$$

where we have used the familiar anticommutation relations for the fermion fields.

Analogously we have

$$[j^0(z),\bar{\psi}(y)]\delta(z_0-y_0) = \bar{\psi}(z)\delta^4(z-y). \tag{4.66}$$

Now we differentiate the 3-point function (4.64)

$$\begin{aligned}\partial^z_\mu\tau^\mu(x,y,z) &= \partial^z_\mu\langle 0|Tj^\mu(z)\psi(x)\bar{\psi}(y)|0\rangle\\ &= \langle 0|T\partial^z_\mu j^\mu(z)\psi(x)\bar{\psi}(y)|0\rangle\\ &\quad + \langle 0|T[j^0(z),\psi(x)]\delta(z_0-x_0)\bar{\psi}(y)|0\rangle\\ &\quad + \langle 0|T\psi(x)[j^0(z),\bar{\psi}(y)]\delta(z_0-y_0)|0\rangle \end{aligned} \tag{4.67}$$

and we insert the vector current conservation (4.7) and the commutator results (4.65), (4.66), then we obtain the **vector Ward identity** in x-space

$$\begin{aligned}\partial^z_\mu\tau^\mu(x,y,z) &= -\langle 0|T\psi(z)\bar{\psi}(y)|0\rangle\delta^4(z-x) + \langle 0|T\psi(x)\bar{\psi}(z)|0\rangle\delta^4(z-y)\\ &= -iS_F(z-y)\delta^4(z-x) + iS_F(x-z)\delta^4(z-y). \end{aligned} \tag{4.68}$$

For the more familiar version we transform equation (4.68) à la Fourier into momentum space

$$(p_\mu - p'_\mu)\tau^\mu(p,p') = S_F(p) - S_F(p'). \tag{4.69}$$

When amputating the 3-point function, the vertex remains

$$-\frac{\tau^\mu(p,p')}{S_F(p)S_F(p')} =: \Gamma^\mu(p,p') \tag{4.70}$$

and we arrive at **Takahashi's identity**

$$(p_\mu - p'_\mu)\Gamma^\mu(p,p') = S_F^{-1}(p) - S_F^{-1}(p'). \tag{4.71}$$

Considering the limit $p' \to p$

$$\Gamma^\mu(p,p') = \frac{S_F^{-1}(p) - S_F^{-1}(p')}{p_\mu - p'_\mu} \overset{p'\to p}{=} \frac{\Delta S_F^{-1}(p)}{\Delta p_\mu} \tag{4.72}$$

provides the **Ward identity**

$$\Gamma^\mu(p,p) = \frac{\partial}{\partial p_\mu} S_F^{-1}(p). \tag{4.73}$$

Ward identities: In our case the interesting Green functions—which finally violate the classically expected result—are the following 3-point functions

$$\langle 0|Tj_\mu(x)j_\nu(y)j^5_\lambda(z)|0\rangle \tag{4.74}$$

$$\langle 0|Tj_\mu(x)j_\nu(y)P(z)|0\rangle. \tag{4.75}$$

They correspond to the triangle graphs which we investigate in the next section. Here we study their features in momentum space

$$\begin{aligned} T_{\mu\nu\lambda}(k_1,k_2,q) &:= i\int d^4x d^4y d^4z\; e^{ik_1x+ik_2y-iqz} \\ &\quad \cdot \langle 0|Tj_\mu(x)j_\nu(y)j^5_\lambda(z)|0\rangle \end{aligned} \tag{4.76}$$

$$\begin{aligned} T_{\mu\nu}(k_1,k_2,q) &:= i\int d^4x d^4y d^4z\; e^{ik_1x+ik_2y-iqz} \\ &\quad \cdot \langle 0|Tj_\mu(x)j_\nu(y)P(z)|0\rangle. \end{aligned} \tag{4.77}$$

Actually the above quantities depend on one variable less because of energy–momentum conservation. We may shift, for instance, the z-dependence in the axial quantities to the point $z \to 0$. We ignore these shifts which are not essential for the moment.

In order to establish a WI we have to differentiate the Green function (4.74). In momentum space there is the relation

$$\begin{aligned} q^\lambda T_{\mu\nu\lambda} &= \int d^4x d^4y d^4z\; e^{ik_1x+ik_2y-iqz} \\ &\quad \cdot \partial_z^\lambda \langle 0|Tj_\mu(x)j_\nu(y)j^5_\lambda(z)|0\rangle, \end{aligned} \tag{4.78}$$

which can be quickly checked by partial integration. When differentiating the Green function the commutator terms vanish in our case (we ignore possible Schwinger terms) and we obtain

$$\begin{aligned} q^\lambda T_{\mu\nu\lambda} &= \int d^4x d^4y d^4z\; e^{ik_1x+ik_2y-iqz} \\ &\quad \cdot \langle 0|Tj_\mu(x)j_\nu(y)\partial_z^\lambda j^5_\lambda(z)|0\rangle. \end{aligned} \tag{4.79}$$

Now using the classical conservation law (4.8) for the axial current provides

$$\begin{aligned} q^\lambda T_{\mu\nu\lambda} &= 2mi \int d^4x d^4y d^4z \, e^{ik_1x+ik_2y-iqz} \\ &\quad \cdot \langle 0|Tj_\mu(x)j_\nu(y)P(z)|0\rangle \\ &= 2mT_{\mu\nu}, \end{aligned} \tag{4.80}$$

which is called the **axial Ward identity AWI**

$$q^\lambda T_{\mu\nu\lambda} = 2mT_{\mu\nu}. \tag{4.81}$$

Analogously when differentiating the vector currents and respecting the vector current conservation law (4.7) we find the **vector Ward identity VWI**

$$k_1^\mu T_{\mu\nu\lambda} = k_2^\nu T_{\mu\nu\lambda} = 0. \tag{4.82}$$

Anomaly: On the other hand, within QFT we calculate the amplitudes $T_{\mu\nu\lambda}$ (4.76) and $T_{\mu\nu}$ (4.77) directly with help of the Feynman rules for the corresponding triangle diagrams—we work this out in the next section—and we check if the Ward identities (4.81) and (4.82) are satisfied. The answer is no! An additional term—the **anomaly**—occurs. The actual calculation (Section 4.3) leads to the following proposition.

Proposition:

- *If the VWI is fulfilled the AWI is anomalous or vice versa!*

So both, the VWI (4.82) and the AWI (4.81), cannot be satisfied; there *always* exists an anomaly.

Anomalous AWI:

$$q^\lambda T_{\mu\nu\lambda} = 2mT_{\mu\nu} + \mathcal{A}_{\mu\nu}, \tag{4.83}$$

with the **anomaly**

$$\mathcal{A}_{\mu\nu} = -\frac{1}{2\pi^2}\varepsilon_{\mu\nu\alpha\beta}k_1^\alpha k_2^\beta. \tag{4.84}$$

This corresponds to a modification of the classical conservation law for the axial current.

Anomalous divergence of the axial current:

$$\partial^\mu j_\mu^5 = 2imP + \mathcal{A}, \tag{4.85}$$

with the **ABJ anomaly**

$$\mathcal{A} = \frac{e^2}{16\pi^2}\varepsilon^{\mu\nu\alpha\beta}F_{\mu\nu}F_{\alpha\beta}. \tag{4.86}$$

Both expressions (4.84) and (4.86) are equivalent, which one can verify quickly by considering the expectation value with 2 external photons (with momenta k_1, k_2 and polarization vectors $\varepsilon_1, \varepsilon_2$).

So what went wrong in our naïve use of classical conservation laws for the Green functions? Let us return to equation (4.78); there we have been a bit hasty with the integration by parts. The Green functions are singular quantities which have to be regularized. This creates a 'surface' term which has been neglected so far and leads to the anomaly. We present the detailed investigation in Section 4.3.

Ambiguity: Green functions containing operator products at the same space–time point are singular objects. The question is whether after regularization the Ward identities can be maintained. Let us consider an example given by Leutwyler [Leutwyler 1986a]—the 2-point function of the vector currents

$$\begin{aligned} \tau_{\mu\nu}(x-y) &= \langle 0|Tj_\mu(x)j_\nu(y)|0\rangle \\ &= \mathrm{tr}\ \gamma_\mu S_F(x-y)\gamma_\nu S_F(y-x). \end{aligned} \tag{4.87}$$

$S_F(z)$ denotes the free fermion propagator in coordinate space

$$S_F(z) = \frac{1}{2\pi^2}\frac{z_\mu \gamma^\mu}{(z^2 - i\varepsilon)^2} \tag{4.88}$$

so that the Green function (4.87) has a singular behaviour like (suppressing indices)

$$\tau(z) \overset{z\to 0}{\sim} \frac{1}{z^6}. \tag{4.89}$$

Therefore the Green function represents a distribution that is well defined only on test functions $f_{\mu\nu}(z)$ with

$$f_{\mu\nu}(0) = \partial^\mu f_{\mu\nu}(0) = \partial^\mu\partial^\nu f_{\mu\nu}(0) = 0. \tag{4.90}$$

The VWI is satisfied on these test functions

$$\partial_x^\mu \tau_{\mu\nu}(x-y) = \partial_y^\nu \tau_{\mu\nu}(x-y) = 0. \tag{4.91}$$

However, extending the Green function to arbitrary test functions is not unique. Two Lorentz invariant extensions obeying the VWI (4.91) differ by a term [Leutwyler 1986a]

$$\widetilde{\tau}_{\mu\nu}(z) = \tau_{\mu\nu}(z) + ic(g_{\mu\nu}\Box - \partial_\mu\partial_\nu)\delta^4(z), \tag{4.92}$$

with c some constant. So this Green function—the 2-point function—has a renormalization ambiguity.

The higher Green functions are again unique (the 3-point function vanishes because of Furry's theorem) and obey the VWI

$$\partial_{x_1}^{\mu_1}\tau_{\mu_1\ldots\mu_n}(x_1,\ldots,x_n) = \partial_{x_1}^{\mu_1}\langle 0|Tj_{\mu_1}(x_1)\ldots j_{\mu_n}(x_n)|0\rangle = 0. \tag{4.93}$$

Anomaly for L-currents: In the previous section we derived the classical conservation of the left-handed current (massless, positive chirality fields), equation (4.25). What about the Green functions? Are they conserved?

Up to the 4-point function the Green functions contain ambiguities. After renormalization the L-currents (or analogously the R-currents) remain conserved in the 2- and 4-point functions (a possible choice). These Green functions satisfy the Ward identities. But then the **3-point function WI is anomalous**

$$\begin{aligned} &\partial_x^\lambda\langle 0|Tj_\lambda^L(x)j_\mu^L(y)j_\nu^L(z)|0\rangle \\ &= \frac{1}{12\pi^2}\varepsilon_{\mu\nu\alpha\beta}\partial_y^\alpha\partial_z^\beta[\delta^4(x-y)\delta^4(x-z)]. \end{aligned} \tag{4.94}$$

Equation (4.94) corresponds to the triangle anomaly containing 3 L-currents and expresses Bardeen's result (4.387) (see Section 4.8) when replacing the gauge group $SU(3)$ by $U(1)$.

Result (4.94) we can quickly infer from the ABJ result (4.84) when we re-express the L-currents by the vector- and axial currents, equation (4.24). Suppressing the vector indices we have to consider

$$j^Lj^Lj^L = \frac{1}{8}(j+j^5)^3 \hat{=} \frac{1}{8}(j^5jj + jj^5j + jjj^5 + j^5j^5j^5), \tag{4.95}$$

where we keep only the anomalous contributions. The second term cancels with the third and for pure axials we have by Bose symmetry

$$j^5j^5j^5 = \frac{1}{3}j^5jj. \tag{4.96}$$

Hence we get

$$j^Lj^Lj^L \hat{=} \frac{1}{8}\left(1+\frac{1}{3}\right)j^5jj = \frac{1}{6}j^5jj. \tag{4.97}$$

The L-handed anomaly amounts to $1/6$ of the ABJ anomaly whose coefficient is $1/2\pi^2$ in coordinate space. Therefore the coefficient in equation (4.94) has to be

$$\frac{1}{6}\cdot\frac{1}{2\pi^2} = \frac{1}{12\pi^2}. \tag{4.98}$$

For R-currents the anomalous WI (4.94) just changes the sign (since relation (4.97) changes sign).

4.2.2 Generating functional

In QFT all information is contained in the generating functional for all Green functions (recall Chapter 3)

$$\begin{aligned} Z[A_\mu] = e^{iW[A_\mu]} &= \langle 0|T \exp[i \int dx j^\mu(x) A_\mu(x)]|0\rangle \\ &= \sum_{n=0}^{\infty} \frac{i^n}{n!} \int dx_1 \dots dx_n\, A_{\mu_1}(x_1) \dots A_{\mu_n}(x_n) \langle 0|T j^{\mu_1}(x_1) \dots j^{\mu_n}(x_n)|0\rangle, \end{aligned} \tag{4.99}$$

where A_μ represents the external gauge field. The connected Green functions are determined by the 'quantum action'

$$\begin{aligned} iW[A_\mu] &= \sum_{n=1}^{\infty} \frac{i^n}{n!} \int dx_1 \dots dx_n\, A_{\mu_1}(x_1) \dots A_{\mu_n}(x_n) \\ &\quad \cdot \langle 0|T j^{\mu_1}(x_1) \dots j^{\mu_n}(x_n)|0\rangle_{\text{conn}}. \end{aligned} \tag{4.100}$$

Ambiguity: How do the renormalization ambiguities of the Green functions affect the generating functional?

Proposition:

- *The generating functional is unique up to a local polynomial of the external gauge field and its derivatives!*

So different renormalization methods induce different polynomials $f[A_\mu]$ in the generating functional. In terms of the 'quantum action' we have

$$\widetilde{W}[A_\mu] = W[A_\mu] + f[A_\mu]. \tag{4.101}$$

Proposition:

- *The renormalization ambiguity induces a $U(1)$ transformation in the generating functional $Z[A_\mu]$*

$$\begin{aligned} \widetilde{Z}[A_\mu] &= e^{iW[A_\mu]+if[A_\mu]} = e^{if[A_\mu]}\, e^{iW[A_\mu]} \\ &= e^{if[A_\mu]}\, Z[A_\mu]. \end{aligned} \tag{4.102}$$

Example: The renormalization ambiguity of the 2-point function (4.92) is then reflected by the generating functional for the special polynomial

$$f[A_\mu] = \frac{c}{4} \int d^4x (\partial_\mu A_\nu - \partial_\nu A_\mu)(\partial^\mu A^\nu - \partial^\nu A^\mu). \tag{4.103}$$

Proof. The 2-point function follows from the $Z[A_\mu]$ functional via

$$\tau^{\mu\nu}(x-y) = \left.\frac{\delta^2 Z[A_\mu]}{i\delta A_\mu(x) i\delta A_\nu(y)}\right|_{A_\mu=0} = \langle 0|Tj^\mu(x)j^\nu(y)|0\rangle. \tag{4.104}$$

Rewriting the polynomial (4.103)

$$f[A_\mu] = -\frac{c}{2}\int d^4x\; A_\mu(g^{\mu\nu}\Box - \partial^\mu\partial^\nu)A_\nu \tag{4.105}$$

we differentiate the transformed functional (4.102)

$$\begin{aligned}\widetilde{\tau}^{\mu\nu}(x-y) &= \left.\frac{\delta^2 \widetilde{Z}[A_\mu]}{i\delta A_\mu(x) i\delta A_\nu(y)}\right|_{A_\mu=0} \\ &= \left.\frac{\delta^2 Z[A_\mu]}{i\delta A_\mu(x) i\delta A_\nu(y)} e^{if[A_\mu]}\right|_{A_\mu=0} + Z[A_\mu]\left.\frac{\delta^2 e^{if[A_\mu]}}{i\delta A_\mu(x) i\delta A_\nu(y)}\right|_{A_\mu=0} \\ &= \tau^{\mu\nu}(x-y) + ic(g^{\mu\nu}\Box - \partial^\mu\partial^\nu)\delta^4(x-y). \qquad \text{Q.E.D.}\end{aligned} \tag{4.106}$$

Ward identity: How are the Ward identities described by the generating functional?

Proposition:

- *The Ward identities for the Green functions are equivalent to the gauge invariance of the 'quantum action'*

$$\delta W[A_\mu] = W[A_\mu + \partial_\mu\Lambda] - W[A_\mu] = 0. \tag{4.107}$$

Proof.

$$\begin{aligned} i\delta W[A_\mu] &= iW[A_\mu + \partial_\mu\Lambda] - iW[A_\mu] \\ &= \sum_{n=1}^{\infty}\frac{i^n}{n!}n\int dx_1\ldots dx_n\;\partial^{x_1}_{\mu_1}\Lambda(x_1)A_{\mu_2}(x_2)\ldots A_{\mu_n}(x_n) \\ &\quad \cdot \tau^{\mu_1\mu_2\ldots\mu_n}(x_1,x_2,\ldots,x_n)\end{aligned} \tag{4.108}$$

and we integrate by parts

$$\begin{aligned} i\delta W[A_\mu] &= -\sum_n \frac{i^n}{(n-1)!}\int dx_1\ldots dx_n\;\Lambda(x_1)A_{\mu_2}(x_2)\ldots A_{\mu_n}(x_n) \\ &\quad \cdot \partial^{x_1}_{\mu_1}\tau^{\mu_1\mu_2\ldots\mu_n}(x_1,x_2,\ldots,x_n).\end{aligned} \tag{4.109}$$

Each term in the sum contains different powers in A_μ. If $\delta W = 0$ then each term in the sum vanishes—*all* Green functions are conserved, equation (4.93)—and vice versa. Q.E.D.

Anomaly: If, however, a Green function obeys an anomalous WI then the 'quantum action' reflects this too.

Proposition:

- *The anomalous Ward identity is equivalent to the gauge transformation of the 'quantum action'*

$$\delta W[A_\mu] = G(\Lambda, A) =: \int dx \Lambda^a(x) G^a[A_\mu](x). \tag{4.110}$$

$G(\Lambda, A)$ represents precisely the **anomaly** in the Abelian as well as in the non-Abelian case. The index a denotes some group index ($a = 0$ for $U(1)$).

Note: Equation (4.110) is a very compact notation for *all* WI of the Green functions.

Example: Let us return to our example (4.94) of the previous section. Then the **anomaly for L-fields** is given by

$$G(\Lambda^L, A^L) = -\frac{1}{24\pi^2} \int dx\ \Lambda^L(x) \varepsilon^{\mu\nu\alpha\beta} \partial_\mu A^L_\nu(x) \partial_\alpha A^L_\beta(x). \tag{4.111}$$

From equation (4.110) follows that all L-handed Green functions are conserved

$$\partial^{\mu_1}_{x_1} \langle 0|T j^L_{\mu_1}(x_1) \dots j^L_\mu(x_n)|0\rangle = 0 \tag{4.112}$$

except the 3-point function. For $n = 3$ the anomalous result (4.94) is valid.

Proof. We consider the difference

$$\begin{aligned} i\delta W[A^L_\mu] &= iW[A^L_\mu + \partial_\mu \Lambda^L] - iW[A^L_\mu] \\ &= \sum_{n=1}^{\infty} \frac{i^n}{n!} n \int dx_1 \dots dx_n \partial^{x_1}_{\mu_1} \Lambda^L(x_1) A^L_{\mu_2}(x_2) \dots A^L_{\mu_n}(x_n) \\ &\quad \cdot \langle 0|T j^{\mu_1}_L(x_1) \dots j^{\mu_n}_L(x_n)|0\rangle_{\text{conn}} \end{aligned} \tag{4.113}$$

and we integrate by parts and separate the $n = 3$ term

$$\begin{aligned}
i\delta W[A_\mu^L] &= -\sum_{n\neq 3} \frac{i^n}{(n-1)!} \int dx_1 \dots dx_n \, \Lambda^L(x_1) A^L_{\mu_2}(x_2) \dots A^L_{\mu_n}(x_n) \\
&\cdot \partial^{x_1}_{\mu_1} \langle 0|T j_L^{\mu_1}(x_1) \dots j_L^{\mu_n}(x_n)|0\rangle_{\text{conn}} \\
&- \frac{i^3}{2!} \int dxdydz \, \Lambda^L(x) A^L_\mu(y) A^L_\nu(z) \\
&\cdot \partial^x_\lambda \langle 0|T j^\lambda_L(x) j^\mu_L(y) j^\nu_L(z)|0\rangle_{\text{conn}}. \qquad (4.114)
\end{aligned}$$

Inserting the identities (4.112) and the anomalous result (4.94) we obtain after integrating by parts again

$$\begin{aligned}
\delta W[A_\mu^L] &= \frac{1}{24\pi^2} \int dxdydz \, \Lambda^L(x) \varepsilon^{\mu\nu\alpha\beta} \partial^y_\alpha A^L_\mu(y) \partial^z_\beta A^L_\nu(z) \\
&\cdot \delta(x-y)\delta(x-z) \\
&= -\frac{1}{24\pi^2} \int dx \, \Lambda^L(x) \varepsilon^{\mu\nu\alpha\beta} \partial_\mu A^L_\nu(x) \partial_\alpha A^L_\beta(x) \\
&= G(\Lambda^L, A^L) \qquad (4.115)
\end{aligned}$$

and vice versa. Q.E.D.

Phase and anomaly: In Chapter 3 we have discussed the significance of the generating functional. It describes the vacuum-to-vacuum transition and corresponds via a path integral to the determinant of a differential operator, here the Dirac operator

$$\begin{aligned}
Z[A_\mu] &= e^{iW[A_\mu]} \\
&= \langle 0|T \exp[i \int dx \mathcal{L}_I]|0\rangle = \langle 0, \infty|0, -\infty\rangle \\
&= \int d\bar{\psi} d\psi \exp[i \int dx \bar{\psi} i \not{D} \psi] = \det i \not{D}. \qquad (4.116)
\end{aligned}$$

Now performing a gauge transformation, there may occur—as in our previous example of *L*- or *R*-handed fields (we supress the indices)—an anomaly in the 'quantum action'. This anomaly then is a *local* polynomial in the gauge fields and their derivatives. For this reason a gauge transformation only influences the phase

$$\begin{aligned}
Z[A_\mu + \delta A_\mu] &= e^{iW[A_\mu+\delta A_\mu]} = e^{iW[A_\mu]+i\delta W[A_\mu]} \\
&= e^{iW[A_\mu]+iG(\Lambda,A)} \\
&= Z[A_\mu] \, e^{iG(\Lambda,A)} \qquad (4.117)
\end{aligned}$$

$$\langle 0, \infty|0, -\infty\rangle|_{A_\mu+\delta A_\mu} = \langle 0, \infty|0, -\infty\rangle|_{A_\mu} \, e^{iG(\Lambda,A)} \qquad (4.118)$$

$$\det i\,\not{D}|_{A_\mu+\delta A_\mu} \;=\; \det i\,\not{D}|_{A_\mu}\, e^{iG(\Lambda,A)}. \tag{4.119}$$

We obtain the important result:

Proposition:

- *The generating functional, the vacuum-to-vacuum amplitude or the determinant of the fermionic differential operator are not gauge invariant quantities if there is an anomaly. But the anomaly affects only the phase of these quantities. The modulus or the transition probability is gauge invariant again.*

We certainly also have renormalization ambiguities in the generating functional. As discussed before they also induce a local polynomial in the 'quantum action' (see equation (4.102)).

Proposition:

- *An anomaly does exist if no renormalization polynomial can cancel the anomaly effect!*

The Ward identity remains violated.
The extension to non-Abelian fields is straightforward.

Non-Abelian Ward identities: In the case of non-Abelian fields $A_\mu = A_\mu^a T^a$ the compact notation for the **Ward identities** is again the gauge freedom of the 'quantum action'

$$\delta W[A_\mu] = W[A_\mu + \delta A_\mu] - W[A_\mu] = 0 \tag{4.120}$$

by the infinitesimal gauge transformation $\delta A_\mu = D_\mu \Lambda$.

As in the Abelian case the identity (4.120) implies *all* WI for the non-Abelian Green functions. Due to the non-Abelian nature of the fields there is a slight complication in the derivation. Let us consider the W-functional

$$\begin{aligned} iW[A_\mu] \;&=\; \sum_{n=1}^{\infty} \frac{i^n}{n!} \int dx_1 \ldots dx_n\, A_{\mu_1}^{a_1}(x_1) \ldots A_{\mu_n}^{a_n}(x_n) \\ &\quad \cdot \langle 0|T j_{a_1}^{\mu_1}(x_1) \ldots j_{a_n}^{\mu_n}(x_n)|0\rangle_{\text{conn}} \end{aligned} \tag{4.121}$$

and

$$\tau_{a_1\ldots a_n}^{\mu_1\ldots\mu_n}(x_1,\ldots,x_n) = \langle 0|T j_{a_1}^{\mu_1}(x_1)\ldots j_{a_n}^{\mu_n}(x_n)|0\rangle_{\text{conn}}. \tag{4.122}$$

We take a gauge variation

$$\begin{aligned} i\delta W[A_\mu] &= iW[A_\mu + D_\mu\Lambda] - iW[A_\mu] \\ &= \sum_{n=1}^{\infty} \frac{i^n}{n!} n \int dx_1 \dots dx_n (D_{\mu_1}\Lambda)^{a_1}(x_1) A^{a_2}_{\mu_2}(x_2) \dots A^{a_n}_{\mu_n}(x_n) \\ &\quad \cdot \tau^{\mu_1\mu_2\dots\mu_n}_{a_1a_2\dots a_n}(x_1, x_2, \dots, x_n), \end{aligned} \tag{4.123}$$

we integrate by parts and separate the first term

$$\begin{aligned} i\delta W[A_\mu] &= -\int dx \Lambda^a(x) \left\{ i\partial_\mu \tau^\mu_a(x) + i f_{ac}{}^b A^c_\mu(x) \tau^\mu_b(x) \right. \\ &\quad + \sum_{n=2}^{\infty} \frac{i^n}{(n-1)!} \int dx_2 \dots dx_n\, A^{a_2}_{\mu_2}(x_2) \dots A^{a_n}_{\mu_n}(x_n) \\ &\quad \cdot \partial_\mu \tau^{\mu\mu_2\dots\mu_n}_{aa_2\dots a_n}(x, x_2, \dots, x_n) \\ &\quad + f_{ac}{}^b \sum_{n=2}^{\infty} \frac{i^n}{(n-1)!} A^c_\mu(x) \int dx_2 \dots dx_n\, A^{a_2}_{\mu_2}(x_2) \dots A^{a_n}_{\mu_n}(x_n) \\ &\quad \left. \cdot \tau^{\mu\mu_2\dots\mu_n}_{ba_2\dots a_n}(x, x_2, \dots, x_n) \right\}. \end{aligned} \tag{4.124}$$

Gauge invariance $\delta W = 0$ implies that all terms to a given power in A^a_μ must vanish. However, here these terms mix in the 2 sums.

Let us demonstrate the quadratic terms in A^a_μ. Then from $\delta W = 0$

$$\begin{aligned} \int dx_2 dx_3\, A^{a_2}_{\mu_2}(x_2) A^{a_3}_{\mu_3}(x_3) [\partial_\mu \tau^{\mu\mu_2\mu_3}_{aa_2a_3}(x, x_2, x_3) & \\ - i2 f_{aa_3}{}^b \tau^{\mu_3\mu_2}_{ba_2}(x_3, x_2) \delta(x - x_3)] &= 0. \end{aligned} \tag{4.125}$$

After symmetrizing the last term

$$\frac{1}{2} f_{aa_2}{}^b \tau^{\mu_2\mu_3}_{ba_3}(x_2, x_3) \delta(x - x_2) + \frac{1}{2} f_{aa_3}{}^b \tau^{\mu_3\mu_2}_{ba_2}(x_3, x_2) \delta(x - x_3) \tag{4.126}$$

the **WI for the non-Abelian 3-point function** is

$$\begin{aligned} & \partial_\mu \tau^{\mu\mu_2\mu_3}_{aa_2a_3}(x, x_2, x_3) \\ &= i f_{aa_2}{}^b \tau^{\mu_2\mu_3}_{ba_3}(x, x_3) \delta(x - x_2) + i f_{aa_3}{}^b \tau^{\mu_3\mu_2}_{ba_2}(x, x_2) \delta(x - x_3). \end{aligned} \tag{4.127}$$

Calculating, finally, the n-th terms

$$\begin{aligned} \int dx_2 \dots dx_n\, A^{a_2}_{\mu_2}(x_2) \dots A^{a_n}_{\mu_n}(x_n) [\partial_\mu \tau^{\mu\mu_2\dots\mu_n}_{aa_2\dots a_n}(x, x_2, \dots, x_n) & \\ - i(n-1) f_{aa_n}{}^b \tau^{\mu_n\mu_2\dots\mu_{n-1}}_{ba_2\dots a_{n-1}}(x_n, x_2, \dots, x_{n-1}) \delta(x - x_n)] &= 0 \end{aligned} \tag{4.128}$$

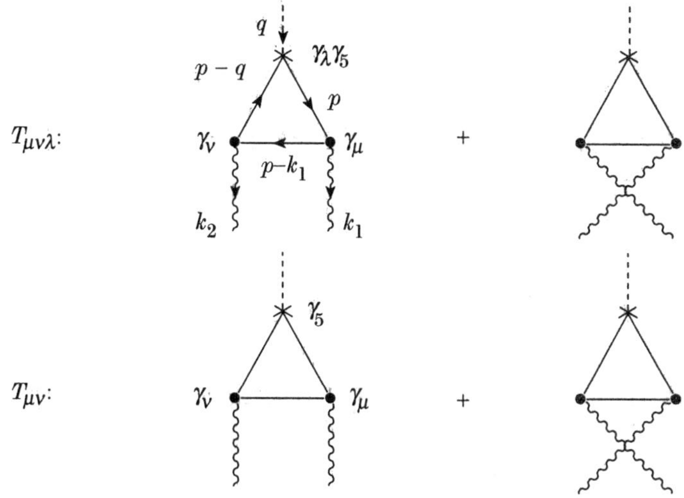

Fig. 4.1. Triangle graphs with vector–vector–axial currents and vector–vector–pseudoscalar currents

implies the **WI for the non-Abelian n-point function**

$$\partial_\mu \tau^{\mu\mu_2\ldots\mu_n}_{aa_2\ldots a_n}(x, x_2, \ldots, x_n) = \sum_{j=2}^{n} i f_{aa_j}{}^{b} \tau^{\mu_2\ldots\mu_j\ldots\mu_n}_{a_2\ldots b\ldots a_n}(x_2, \ldots, x, \ldots x_n)\delta(x - x_j). \quad (4.129)$$

Note: Of course, relation (4.129) must follow directly from differentiating the non-Abelian n-point function (4.122) and using the equal-time commutators of the non-Abelian currents.

4.3 ABJ anomaly calculations

The Adler–Bell–Jackiw anomaly [Adler 1969], [Bell, Jackiw 1969] (ABJ hereafter) is a true phenomenon of quantum physics. Consequently, all renormalization methods must lead to the same result. We will present the customary methods.

4.3.1 Triangle graph regularization

Now let us focus on the famous triangle diagram. In order to understand how the anomaly unavoidably arises, when regularizing this UV-divergent quantity, we go through the calculation step by step. Here we follow the classical lectures on anomalies by Adler [Adler 1970a] and by Jackiw [Jackiw 1985d], [Jackiw 1995]. In addition the reader may find literature like [Cheng, Li 1988], [Hořejší 1992a,b], [Morozov 1986], [Adam 1990], [Shifman 1991] helpful.

With the help of the familiar Feynman rules (for our notation see [Pietschmann 1983]) we find for the amplitudes depicted in Figure 4.1 the following expressions in momentum space

$$\begin{aligned} T_{\mu\nu\lambda} &= i\int \frac{d^4p}{(2\pi)^4}(-)\ \mathrm{tr}\ \frac{i}{\not p - m}\gamma_\lambda\gamma_5\frac{i}{\not p - \not q - m}\gamma_\nu\frac{i}{\not p - \not k_1 - m}\gamma_\mu \\ &\quad + \begin{pmatrix} k_1 \leftrightarrow k_2 \\ \mu \leftrightarrow \nu \end{pmatrix} \end{aligned} \tag{4.130}$$

$$\begin{aligned} T_{\mu\nu} &= i\int \frac{d^4p}{(2\pi)^4}(-)\ \mathrm{tr}\ \frac{i}{\not p - m}\gamma_5\frac{i}{\not p - \not q - m}\gamma_\nu\frac{i}{\not p - \not k_1 - m}\gamma_\mu \\ &\quad + \begin{pmatrix} k_1 \leftrightarrow k_2 \\ \mu \leftrightarrow \nu \end{pmatrix}, \end{aligned} \tag{4.131}$$

with $q = k_1 + k_2$.

Axial Ward identity: In order to establish the axial Ward identity (4.81) of the previous section we use the identity

$$\not q\gamma_5 = \gamma_5(\not p - \not q - m) + (\not p - m)\gamma_5 + 2m\gamma_5. \tag{4.132}$$

Then we obtain the following relation between the amplitudes (4.130) and (4.131)

$$q^\lambda T_{\mu\nu\lambda} = 2mT_{\mu\nu} + R^1_{\mu\nu} + R^2_{\mu\nu}, \tag{4.133}$$

where

$$\begin{aligned} R^1_{\mu\nu} &= \int \frac{d^4p}{(2\pi)^4}\ \mathrm{tr}\ \left[\frac{1}{\not p - \not k_2 - m}\gamma_5\gamma_\nu\frac{1}{\not p - \not q - m}\gamma_\mu \right. \\ &\quad \left. - \frac{1}{\not p - m}\gamma_5\gamma_\nu\frac{1}{\not p - \not k_1 - m}\gamma_\mu\right] \end{aligned} \tag{4.134}$$

$$\begin{aligned} R^2_{\mu\nu} &= \int \frac{d^4p}{(2\pi)^4}\ \mathrm{tr}\ \left[\frac{1}{\not p - \not k_1 - m}\gamma_5\gamma_\mu\frac{1}{\not p - \not q - m}\gamma_\nu \right. \\ &\quad \left. - \frac{1}{\not p - m}\gamma_5\gamma_\mu\frac{1}{\not p - \not k_2 - m}\gamma_\nu\right]. \end{aligned} \tag{4.135}$$

If the rest-terms vanish

$$R^1_{\mu\nu}, R^2_{\mu\nu} \to 0 \tag{4.136}$$

the AWI is satisfied. This happens if we formally shift the integral of the first term in equation (4.134) and in (4.135) by

$$p \to p + k_2 \quad \text{and by} \quad p \to p + k_1. \tag{4.137}$$

However, the integrals are linearly divergent. Note, the seemingly quadratic divergence disappears because of the trace which gives

$$\varepsilon_{\mu\nu\alpha\beta} p^\alpha p^\beta = 0. \tag{4.138}$$

In linear divergent integrals we are *not* allowed to shift the integral variable! So the rest-terms may contribute

$$R^1_{\mu\nu}, R^2_{\mu\nu} \neq 0 \tag{4.139}$$

and the AWI becomes anomalous.

Linear divergent integrals: Let us begin with an example in 1 dimension

$$\begin{aligned} \Delta(a) &= \int_{-\infty}^{\infty} dx [f(x+a) - f(x)] \\ &= \int_{-\infty}^{\infty} dx [a f'(x) + \frac{a^2}{2!} f''(x) + \ldots] \\ &= a[f(\infty) - f(-\infty)] + \frac{a^2}{2!}[f'(\infty) - f'(-\infty)] + \ldots . \end{aligned} \tag{4.140}$$

If the integral

$$\int_{-\infty}^{\infty} dx f(x) \tag{4.141}$$

converges (or is at most logarithmic divergent) then $f(\pm\infty) = f'(\pm\infty) = \ldots = 0$, and the quantity $\Delta(a)$ vanishes, $\Delta(a) = 0$. In this case one can shift the variable $x \to x - a$ in the first integral.

If, however, the integral (4.141) is linear divergent then $f(\pm\infty) \neq 0$, $f'(\pm\infty) = f''(\pm\infty) = \ldots = 0$, and $\Delta(a)$ may create a nonvanishing 'surface term'

$$\Delta(a) = a[f(\infty) - f(-\infty)] \neq 0. \tag{4.142}$$

Generalizing this example to n dimensions (Euclidean) we have

$$\begin{aligned} \Delta(a) &= \int d^n x [f(x+a) - f(x)] \\ &= \int d^n x [a^\mu \partial_\mu f(x) + a^\mu a^\nu \partial_\mu \partial_\nu f(x) + \ldots]. \end{aligned} \tag{4.143}$$

Now we apply the Gauss theorem, we integrate symmetrically (average over all directions) and obtain

$$\Delta(a) = a^{\mu} \lim_{R\to\infty} \frac{R_{\mu}}{R} S^{n-1}(R) f(R), \tag{4.144}$$

where $S^{n-1}(R)$ expresses the surface of an $(n-1)$-dimensional sphere with radius R; for instance for $n=4$

$$S^3(R) = 2\pi^2 R^3. \tag{4.145}$$

In 4-dimensional Minkowski space we get an extra factor i from $x_4 = ix_0$ so that we find the **formula**

$$\Delta(a) = i2\pi^2 a^{\mu} \lim_{R\to\infty} R_{\mu} R^2 f(R). \tag{4.146}$$

We are going to use this formula frequently in this chapter.

Now we return to our rest-terms $R^{1,2}_{\mu\nu}$ (4.134), (4.135) which we have to investigate. They are precisely of above type

$$R^1_{\mu\nu} = \int \frac{d^4p}{(2\pi)^4} [f(p-k_2) - f(p)] \tag{4.147}$$

$$R^2_{\mu\nu} = \int \frac{d^4p}{(2\pi)^4} [f(p-k_1) - f(p)] \tag{4.148}$$

so that we can use formula (4.146). For the term $R^1_{\mu\nu}$ we obtain

$$R^1_{\mu\nu} = i2\pi^2(-k_2^{\lambda}) \lim_{p\to\infty} p_{\lambda} p^2 \frac{\text{tr}\ (\not{p} + m)\gamma_5\gamma_{\nu}(\not{p} - \not{k}_1 + m)\gamma_{\mu}}{(2\pi)^4[p^2 - m^2][(p-k_1)^2 - m^2]}. \tag{4.149}$$

The trace is calculated with help of the familiar formula

$$\text{tr}\ \gamma_5\gamma_{\beta}\gamma_{\nu}\gamma_{\alpha}\gamma_{\mu} = 4i\varepsilon_{\beta\nu\alpha\mu}. \tag{4.150}$$

(Note that the axial trace containing 1, 2, 3 γ_{μ}-matrices vanishes.) Then only the term proportional to $\not{p}\ \not{k}_1$ survives leading to

$$R^1_{\mu\nu} = \frac{1}{2\pi^2} \varepsilon_{\beta\nu\alpha\mu} k_1^{\alpha} k_2^{\lambda} \lim_{p\to\infty} \frac{p_{\lambda} p^{\beta}}{p^2}. \tag{4.151}$$

Performing the symmetric momentum limit

$$\lim_{p\to\infty} \frac{p_{\lambda} p_{\beta}}{p^2} = \frac{g_{\lambda\beta}}{4} \tag{4.152}$$

gives

$$R^1_{\mu\nu} = -\frac{1}{8\pi^2}\varepsilon_{\mu\nu\alpha\beta}k_1^\alpha k_2^\beta. \tag{4.153}$$

The second term $R^2_{\mu\nu}$ arising from the interchange $k_1 \leftrightarrow k_2$, $\mu \leftrightarrow \nu$ provides the same result

$$R^2_{\mu\nu} = -\frac{1}{8\pi^2}\varepsilon_{\mu\nu\alpha\beta}k_1^\alpha k_2^\beta. \tag{4.154}$$

Altogether the AWI receives the following anomalous contribution

$$q^\lambda T_{\mu\nu\lambda} = 2mT_{\mu\nu} - \frac{1}{4\pi^2}\varepsilon_{\mu\nu\alpha\beta}k_1^\alpha k_2^\beta. \tag{4.155}$$

But this is not quite the result we finally want to present. Let us investigate a little bit further.

Ambiguity of the amplitude: For the above calculation we have chosen the momentum route in the triangle graph quite arbitrarily (see Figure 4.1). We can certainly shift this internal momentum by some amount then the linear divergent integral, which is not uniquely defined, also alters its value when regularized like above.

Note: There is no ambiguity in the amplitude $T_{\mu\nu}$ (4.131) since it is a perfect convergent quantity. The seemingly linear and logarithmic divergences disappear because of the vanishing axial trace of 5 γ_μ-matrices or because of identity (4.138).

We shift the internal momentum integration by

$$p \to p + a, \tag{4.156}$$

with

$$a = \alpha k_1 + (\alpha - \beta)k_2 \tag{4.157}$$

and we consider the difference of the amplitudes

$$\Delta_{\mu\nu\lambda}(a) = T_{\mu\nu\lambda}(a) - T_{\mu\nu\lambda}(0) = -\int \frac{d^4p}{(2\pi)^4}\cdot$$

$$\cdot\left[\mathrm{tr}\ \frac{1}{\not p + \not a - m}\gamma_\lambda\gamma_5\frac{1}{\not p + \not a - \not q - m}\gamma_\nu\frac{1}{\not p + \not a - \not k_1 - m}\gamma_\mu\right.$$

$$\left. - \mathrm{tr}\ \frac{1}{\not p - m}\gamma_\lambda\gamma_5\frac{1}{\not p - \not q - m}\gamma_\nu\frac{1}{\not p - \not k_1 - m}\gamma_\mu\right] + \begin{pmatrix} k_1 \leftrightarrow k_2 \\ \mu \leftrightarrow \nu \end{pmatrix}$$

$$=: \Delta^1_{\mu\nu\lambda} + \Delta^2_{\mu\nu\lambda}. \tag{4.158}$$

We regularize the linear divergence of type (4.143) by formula (4.146) which provides

$$\begin{aligned}\Delta^1_{\mu\nu\lambda} &= i2\pi^2 a^\tau \lim_{p\to\infty} p_\tau p^2 \\ &\cdot \frac{(-)\ \mathrm{tr}\ (\not p + m)\gamma_\lambda\gamma_5(\not p - \not q + m)\gamma_\nu(\not p - \not k_1 + m)\gamma_\mu}{(2\pi)^4[p^2 - m^2][(p-q)^2 - m^2][(p-k_1)^2 - m^2]}.\end{aligned} \tag{4.159}$$

In the limit $p \to \infty$ only the leading powers in p survive

$$\Delta^1_{\mu\nu\lambda} = -\frac{i2\pi^2}{(2\pi)^4} a^\tau \lim_{p\to\infty} \frac{p_\tau p^\alpha}{p^2}\ \mathrm{tr}\ \gamma_5\gamma_\nu\gamma_\alpha\gamma_\mu\gamma_\lambda \tag{4.160}$$

and using formulae (4.150) and (4.152) we obtain

$$\Delta^1_{\mu\nu\lambda} = -\frac{1}{8\pi^2}\varepsilon_{\mu\nu\lambda\alpha} a^\alpha. \tag{4.161}$$

The second part $\Delta^2_{\mu\nu\lambda}$ follows from the first by interchanging $k_1 \leftrightarrow k_2$ and $\mu \leftrightarrow \nu$ so that the total difference of the amplitudes (4.158) is

$$\Delta_{\mu\nu\lambda} = \Delta^1_{\mu\nu\lambda} + \Delta^2_{\mu\nu\lambda} = -\frac{\beta}{8\pi^2}\varepsilon_{\mu\nu\lambda\alpha}(k_1^\alpha - k_2^\alpha) \tag{4.162}$$

(recall equation (4.157)).

Considering, finally, the amplitude

$$T_{\mu\nu\lambda}(a) = T_{\mu\nu\lambda}(0) + \Delta_{\mu\nu\lambda}(a) \tag{4.163}$$

we establish the AWI by multiplying equation (4.163) by the momentum q^λ and inserting the results (4.155) and (4.162). This leads to the **anomalous AWI** (for $\beta \neq 1$)

$$q^\lambda T_{\mu\nu\lambda}(\beta) = 2mT_{\mu\nu} - \frac{1-\beta}{4\pi^2}\varepsilon_{\mu\nu\alpha\beta}k_1^\alpha k_2^\beta. \tag{4.164}$$

Vector Ward identity: The next important relation that we have to investigate is the vector Ward identity (4.82). So we calculate

$$\begin{aligned}k_1^\mu T_{\mu\nu\lambda} &= -\int \frac{d^4p}{(2\pi)^4}\ \mathrm{tr}\ \left[\frac{1}{\not p - m}\gamma_\lambda\gamma_5\frac{1}{\not p - \not q - m}\gamma_\nu\frac{1}{\not p - \not k_1 - m}\ \not k_1\right. \\ &\left. + \frac{1}{\not p - m}\gamma_\lambda\gamma_5\frac{1}{\not p - \not q - m}\ \not k_1\frac{1}{\not p - \not k_2 - m}\gamma_\nu\right].\end{aligned} \tag{4.165}$$

Using the identities

$$\begin{aligned} \not{k}_1 \frac{1}{\not{p} - m} &= 1 - (\not{p} - \not{k}_1 - m)\frac{1}{\not{p} - m} \\ \frac{1}{\not{p} - \not{q} - m} \not{k}_1 &= -1 + \frac{1}{\not{p} - \not{q} - m}(\not{p} - \not{k}_2 - m) \end{aligned} \qquad (4.166)$$

only the first terms remain and we obtain

$$\begin{aligned} k_1^\mu T_{\mu\nu\lambda} &= -\int \frac{d^4p}{(2\pi)^4} \operatorname{tr} \left[\gamma_\lambda\gamma_5 \frac{1}{\not{p} - \not{q} - m}\gamma_\nu \frac{1}{\not{p} - \not{k}_1 - m} \right. \\ &\quad \left. - \gamma_\lambda\gamma_5 \frac{1}{\not{p} - \not{k}_2 - m}\gamma_\nu \frac{1}{\not{p} - m}\right] \qquad (4.167) \\ &=: -\int \frac{d^4p}{(2\pi)^4}[f(p - k_1) - f(p)]. \qquad (4.168) \end{aligned}$$

Again, we have a linear divergence of type (4.143) which we regularize by formula (4.146). Recalling further the formulae (4.150) and (4.152) we evaluate

$$\begin{aligned} k_1^\mu T_{\mu\nu\lambda}(0) &= i2\pi^2(-k_1^\tau) \lim_{p\to\infty} p_\tau p^2 \\ &\quad \cdot \frac{(-) \operatorname{tr} \gamma_5(\not{p} - \not{k}_2 + m)\gamma_\nu(\not{p} + m)\gamma_\lambda}{(2\pi)^4[(p - k_2)^2 - m^2][p^2 - m^2]} \\ &= -\frac{i2\pi^2}{(2\pi)^4} k_1^\tau \lim_{p\to\infty} \frac{p_\tau p^\alpha}{p^2} k_2^\beta \operatorname{tr} \gamma_5\gamma_\beta\gamma_\nu\gamma_\alpha\gamma_\lambda \\ &= \frac{1}{8\pi^2}\varepsilon_{\beta\nu\alpha\lambda} k_1^\alpha k_2^\beta. \end{aligned} \qquad (4.169)$$

So the VWI of the unshifted amplitude ($a = 0$) is not satisfied. But also the shifted amplitude (4.163) acquires an **anomalous VWI** (for $\beta \neq -1$)

$$k_1^\mu T_{\mu\nu\lambda}(\beta) = \frac{1+\beta}{8\pi^2}\varepsilon_{\nu\lambda\alpha\beta} k_1^\alpha k_2^\beta. \qquad (4.170)$$

When considering both Ward identities (4.164) and (4.170) we find the very remarkable result:

Proposition:

- *There is no value of β such that both the VWI* (4.170) *and the AWI* (4.164), *can be fulfilled!*

Examples:

i) *VWI and anomalous AWI.*
We can choose, for instance, $\beta = -1$ then the VWI

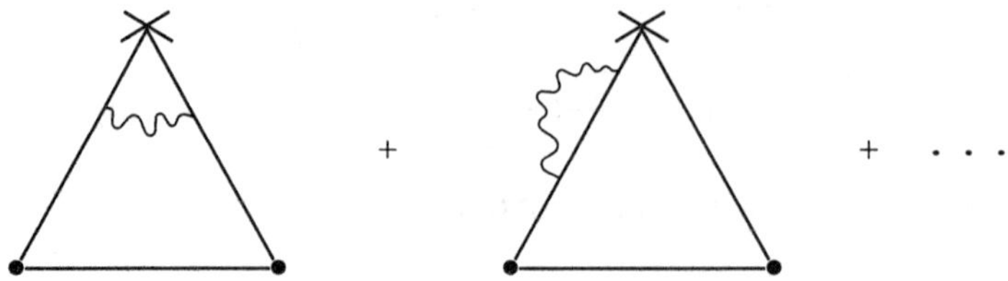

Fig. 4.2. Radiative corrections to the triangle graph

$$k_1^\mu T_{\mu\nu\lambda} = 0 \tag{4.171}$$

is satisfied—the preferable case of gauge invariance—but the AWI turns out to be anomalous

$$q^\lambda T_{\mu\nu\lambda} = 2mT_{\mu\nu} + \mathcal{A}_{\mu\nu}, \tag{4.172}$$

with the **ABJ anomaly** in momentum space

$$\mathcal{A}_{\mu\nu} = -\frac{1}{2\pi^2}\varepsilon_{\mu\nu\alpha\beta}k_1^\alpha k_2^\beta. \tag{4.173}$$

ii) *Anomalous VWI and AWI.*
Or we fix $\beta = 1$ then the AWI is fulfilled

$$q^\lambda T_{\mu\nu\lambda} = 2mT_{\mu\nu} \tag{4.174}$$

but the VWI becomes anomalous

$$k_1^\mu T_{\mu\nu\lambda} = -\frac{1}{2}\mathcal{A}_{\nu\lambda}. \tag{4.175}$$

iii) *Anomalous VWI and anomalous AWI.*
If we select $\beta = 1/3$ the anomaly is equally distributed in the VWI and AWI

$$k_1^\mu T_{\mu\nu\lambda} = -\frac{1}{3}\mathcal{A}_{\nu\lambda} \tag{4.176}$$

$$q^\lambda T_{\mu\nu\lambda} = 2mT_{\mu\nu} + \frac{1}{3}\mathcal{A}_{\mu\nu}. \tag{4.177}$$

This case corresponds to a triangle graph calculation with 3 axial currents where the Bose symmetry provides a factor 1/3 (Section 4.8).

Note: Clearly, the anomaly result is gauge independent. (The gauge independence of anomalies, specifically for string theories, has been discussed by Kummer [Kummer 1990].)

Radiative corrections: The anomaly discussed so far stems from a 1-loop perturbation calculation. Now an important question certainly arises. Does the anomaly receive contributions from radiative corrections (see Figure 4.2) in higher order perturbation theory? The answer is no!

Adler–Bardeen theorem: [Adler, Bardeen 1969] The full structure of the chiral anomaly is given by the triangle graph!

So higher order radiative effects just renormalize the fields and charges. It is enough to calculate only the triangle graph for the anomaly result. This is an important statement and is certainly true in QED and QCD but in more general theories the Adler–Bardeen theorem needs to be confirmed. For further literature we refer to [Piguet, Sorella 1992, 1993].

4.3.2 Pauli–Villars regularization

The classical regularization technique for a divergent amplitude is that of Pauli and Villars (PV) [Pauli, Villars 1949]. How does the anomaly arise in this scheme? (See e.g. [Jackiw 1985d], [Hořejší 1992a,b], [Adam 1990].)

Let us consider the amplitude as a function of the fermion mass—the particle which loops around.

Definition: The **PV-regulated amplitude** is the difference between the given amplitude and the amplitude evaluated at a regulator mass M

$$T^{\text{reg}}_{\mu\nu\lambda} = T_{\mu\nu\lambda}(m) - T_{\mu\nu\lambda}(M). \tag{4.178}$$

The **physical amplitude** follows from the limit $M \to \infty$

$$T^{\text{phys}}_{\mu\nu\lambda} = \lim_{M\to\infty} T^{\text{reg}}_{\mu\nu\lambda}. \tag{4.179}$$

In the case of the $T_{\mu\nu}$ amplitude (4.131) we already have

$$T^{\text{phys}}_{\mu\nu} = \lim_{M\to\infty} T^{\text{reg}}_{\mu\nu} = \lim_{M\to\infty} [T_{\mu\nu}(m) - T_{\mu\nu}(M)] = T_{\mu\nu}(m) \tag{4.180}$$

since $T_{\mu\nu}(M) \sim \frac{1}{M}$ as we shall show below. $T_{\mu\nu}$ is convergent and clearly no regularization is needed.

Vector Ward identity: The **VWI** is automatically fulfilled for the regulated amplitude, hence also for the physical amplitude

$$k_1^{\mu} T^{\text{phys}}_{\mu\nu\lambda} = k_2^{\nu} T^{\text{phys}}_{\mu\nu\lambda} = 0. \tag{4.181}$$

The reason is simply that a mass independent term like equation (4.170) just cancels in the PV difference of the amplitudes.

Axial Ward identity: For the **AWI** however we get

$$q^\lambda T^{\text{phys}}_{\mu\nu\lambda} = \lim_{M\to\infty} q^\lambda T^{\text{reg}}_{\mu\nu\lambda} = 2mT_{\mu\nu}(m) - \lim_{M\to\infty} 2MT_{\mu\nu}(M). \tag{4.182}$$

Also here a possible massless term like the second term in equation (4.164) cancels in the PV difference (4.178).

Theorem: The **ABJ anomaly** is generated by the limit

$$\lim_{M\to\infty} 2MT_{\mu\nu}(M) = -\mathcal{A}_{\mu\nu}. \tag{4.183}$$

So the PV method selects a satisfied VWI and anomalous AWI automatically.

Proof. We begin with $T_{\mu\nu}(m)$, equation (4.131). For the denominator we introduce the familiar **Feynman parameter integral** (see e.g. [Pietschmann 1983], [Pokorski 1987])

$$\frac{1}{a_1 a_2 a_3} = 2\int_0^1 dx_1 \int_0^{1-x_1} dx_2 \frac{1}{[a_1 x_2 + a_2(1-x_1-x_2) + a_3 x_1]^3}, \tag{4.184}$$

which gives

$$\begin{aligned} T_{\mu\nu} = &-\int \frac{d^4p}{(2\pi)^4} 2\int_0^1 dx_1 \int_0^{1-x_1} dx_2 \\ &\cdot \frac{\text{tr}\,(\not p + m)\gamma_5(\not p - \not q + m)\gamma_\nu(\not p - \not k_1 + m)\gamma_\mu}{[(p^2-m^2)x_2 + ((p-q)^2 - m^2)(1-x_1-x_2) + ((p-k_1)^2 - m^2)x_1]^3} \\ &+ \begin{pmatrix} k_1 \leftrightarrow k_2 \\ \mu \leftrightarrow \nu \end{pmatrix}. \end{aligned} \tag{4.185}$$

The axial traces with 2, 3 or 5 γ_μ-matrices vanish. The traces containing 4 γ_μ-matrices cancel or vanish due to identity (4.138) except the term

$$m\ \text{tr}\ \gamma_5 \not q \gamma_\nu \not k_1 \gamma_\mu = m\ 4i\varepsilon_{\beta\nu\alpha\mu} k_2^\beta k_1^\alpha + \begin{pmatrix} k_1 \leftrightarrow k_2 \\ \mu \leftrightarrow \nu \end{pmatrix} \tag{4.186}$$

so that we get

$$T_{\mu\nu} = \int \frac{d^4p}{(2\pi)^4} 2\int_0^1 dx_1 \int_0^{1-x_1} dx_2 \frac{2m\ 4i\varepsilon_{\mu\nu\alpha\beta} k_1^\alpha k_2^\beta}{[p^2 - 2pk - \bar{m}^2]^3}, \tag{4.187}$$

with

$$
\begin{aligned}
k &= q(1-x_1-x_2)+k_1x_1 \\
\bar{m}^2 &= m^2 - q^2(1-x_1-x_2).
\end{aligned} \tag{4.188}
$$

We have already included the terms from the interchange $k_1 \leftrightarrow k_2$, $\mu \leftrightarrow \nu$ which actually contribute to equation (4.187) after the mass limit $M \to \infty$. The momentum integration is conveniently carried out with the help of the **'t Hooft–Veltman integral formula** ['t Hooft, Veltman 1972]

$$
\int \frac{d^n p}{(p^2 - 2pk - \bar{m}^2)^\alpha} = i^{1-2\alpha}\pi^{n/2}\frac{\Gamma(\alpha - \frac{n}{2})}{\Gamma(\alpha)}\frac{1}{(k^2+\bar{m}^2)^{\alpha - n/2}} =: J_0. \tag{4.189}
$$

In our case $\alpha = 3$ and $n = 4$, then the integral is

$$
J_0 = \frac{\pi^2}{2i}\frac{1}{m^2 + f(x_1,x_2)}, \tag{4.190}
$$

with $f(x_1, x_2)$ a function independent of m. It behaves in the limit of large masses like

$$
\lim_{M\to\infty} J_0(M) = \frac{\pi^2}{2i}\lim_{M\to\infty}\frac{1}{M^2}. \tag{4.191}
$$

So we obtain the relation

$$
\begin{aligned}
&\lim_{M\to\infty} 2MT_{\mu\nu}(M) \\
&= \lim_{M\to\infty}\frac{1}{(2\pi)^4}\frac{\pi^2}{2i}\frac{1}{M^2}\,2M\,2M\,4i\varepsilon_{\mu\nu\alpha\beta}k_1^\alpha k_2^\beta 2\int_0^1 dx_1\int_0^{1-x_1} dx_2 \\
&= \frac{1}{2\pi^2}\varepsilon_{\mu\nu\alpha\beta}k_1^\alpha k_2^\beta = -\mathcal{A}_{\mu\nu}. \qquad \text{Q.E.D.}
\end{aligned} \tag{4.192}
$$

4.3.3 n-dimensional 't Hooft–Veltman regularization

Another attractive regularization procedure has been invented by 't Hooft and Veltman ['t Hooft, Veltman 1972]. There the amplitudes are regulated by the higher dimensions in space–time; the divergences occur as poles at dimension $n = 4$. First all calculations are carried out for $n > 4$ and afterwards in the final result the limit $n \to 4$ is performed. Let us see how the anomaly emerges here (for a review see e.g. [Hořejší 1992a,b]).

The anomaly involves a γ_5-matrix. There is, however, a problem of defining γ_5 correctly in higher dimensions. Choosing

$$
\gamma_5 = i\gamma^0\gamma^1\gamma^2\gamma^3 \tag{4.193}
$$

(note this γ_5-definition [Itzykson, Zuber 1980] differs in sign from the convention we used before in Section 4.1) the **'t Hooft–Veltman rule** is

$$\{\gamma_5, \gamma_\mu\} = 0 \qquad \text{for } \mu = 0, 1, 2, 3 \tag{4.194}$$

$$[\gamma_5, \gamma_\mu] = 0 \qquad \text{for } \mu = 4, 5, \ldots, n-1. \tag{4.195}$$

All other familiar rules for the γ-matrices persist in higher dimensions.

Triangle amplitude: Now we calculate the triangle amplitude (4.130) in n-dimensions. We split the momenta at dimension 4 and we just consider the massless case ($m = 0$) for simplicity

$$T_{\mu\nu\lambda} = -\int \frac{d^4\ell}{(2\pi)^4} \int \frac{d^{n-4}L}{(2\pi)^{n-4}} \operatorname{tr} \frac{1}{\not{\ell} + \not{L}} \gamma_\lambda \gamma_5 \frac{1}{\not{\ell} + \not{L} - \not{q}} \gamma_\nu \frac{1}{\not{\ell} + \not{L} - \not{k}_1} \gamma_\mu + \begin{pmatrix} k_1 \leftrightarrow k_2 \\ \mu \leftrightarrow \nu \end{pmatrix}. \tag{4.196}$$

The internal momenta denote

$$\not{\ell} = \ell_0 \gamma^0 + \ldots + \ell_3 \gamma^3 \tag{4.197}$$

for the first 4 components and

$$\not{L} = L_4 \gamma^4 + \ldots + L_{n-1} \gamma^{n-1} \tag{4.198}$$

for the last $(n-4)$ components. The external momenta

$$q^\nu, k_1^\alpha, k_2^\beta \qquad \nu, \alpha, \beta = 0, 1, 2, 3 \tag{4.199}$$

remain 4-dimensional.

Let us shift the integration momentum for convenience by

$$\ell \to \ell + k_1 \tag{4.200}$$

then the amplitude changes to

$$T_{\mu\nu\lambda} = -\int \frac{d^4\ell}{(2\pi)^4} \int \frac{d^{n-4}L}{(2\pi)^{n-4}} \operatorname{tr} \frac{1}{\not{\ell} + \not{L} + \not{k}_1} \gamma_\lambda \gamma_5 \frac{1}{\not{\ell} + \not{L} - \not{k}_2} \gamma_\nu \frac{1}{\not{\ell} + \not{L}} \gamma_\mu + \begin{pmatrix} k_1 \leftrightarrow k_2 \\ \mu \leftrightarrow \nu \end{pmatrix}. \tag{4.201}$$

Vector Ward identity: The **VWI is satisfied**

$$k_1^\mu T_{\mu\nu\lambda} = k_2^\nu T_{\mu\nu\lambda} = 0. \tag{4.202}$$

This we can verify in the same manner as before in Section 4.3.1, equations (4.165)–(4.168). The analogous equation to formulae (4.167), (4.168) is

$$k_1^\mu T_{\mu\nu\lambda} = -\int \frac{d^4\ell}{(2\pi)^4}\int \frac{d^{n-4}L}{(2\pi)^{n-4}}$$
$$\cdot \operatorname{tr}\left[\gamma_\lambda\gamma_5\frac{1}{\not\ell+\not L-\not k_2}\gamma_\nu\frac{1}{\not\ell+\not L} - \gamma_\lambda\gamma_5\frac{1}{\not\ell+\not L+\not k_1-\not k_2}\gamma_\nu\frac{1}{\not\ell+\not L+\not k_1}\right]$$
$$= \quad -\int\frac{d^4\ell}{(2\pi)^4}[f(\ell-k_1)-f(\ell)] = 0 \qquad (4.203)$$

since the integration shift $\ell \to \ell + k_1$ in the first term is now allowed. Q.E.D.

Axial Ward identity: What about the AWI? Let us study the expression

$$q^\lambda T_{\mu\nu\lambda} \quad = \quad -\int \frac{d^4\ell}{(2\pi)^4}\int\frac{d^{n-4}L}{(2\pi)^{n-4}}$$
$$\cdot\frac{\operatorname{tr}(\not\ell+\not L+\not k_1)\,\not q\gamma_5(\not\ell+\not L-\not k_2)\gamma_\nu(\not\ell+\not L)\gamma_\mu}{[(\ell+k_1)^2-L^2][(\ell-k_2)^2-L^2][\ell^2-L^2]}$$
$$+\begin{pmatrix} k_1\leftrightarrow k_2 \\ \mu\leftrightarrow\nu\end{pmatrix}. \qquad (4.204)$$

Note that

$$\not\ell\,\not\ell = \ell^2, \qquad \not L\,\not L = -L^2, \qquad \not\ell\,\not L + \not L\,\not\ell = 0,$$
$$(\not\ell+\not L)(\not\ell+\not L) = \ell^2 - L^2. \qquad (4.205)$$

Calculating the trace explicitly it happens that all terms cancel or vanish except the terms proportional to $\not L\,\not L$. Therefore we obtain

$$q^\lambda T_{\mu\nu\lambda} \quad = \quad 4\operatorname{tr}\gamma_5\gamma_\mu\gamma_\nu\,\not k_1\,\not k_2\int\frac{d^4\ell}{(2\pi)^4}\int\frac{d^{n-4}L}{(2\pi)^{n-4}}$$
$$\cdot\frac{L^2}{[\ell^2-L^2][(\ell+k_1)^2-L^2)][(\ell-k_2)^2-L^2]}, \qquad (4.206)$$

where we have already included the terms from the interchange $k_1 \leftrightarrow k_2$, $\mu\leftrightarrow\nu$ which actually contribute after the limit $n\to 4$. Next we apply, as usual, the Feynman parameter integral formula (4.184) together with the 't Hooft–Veltman formula (4.189) for the l integration. Then we get

$$q^\lambda T_{\mu\nu\lambda} \quad = \quad 16 i\varepsilon_{\mu\nu\alpha\beta}k_1^\alpha k_2^\beta$$
$$\cdot\frac{1}{(2\pi)^4}\frac{\pi^2}{2i}2\int_0^1 dx_1\int_0^{1-x_1}dx_2\int\frac{d^{n-4}L}{(2\pi)^{n-4}}\frac{L^2}{L^2+f(x_1,x_2)}. \qquad (4.207)$$

For the L-integration we use the **formula**

$$\int \frac{d^N L}{(2\pi)^N} \frac{(L^2)^a}{(L^2+f)^b} = \frac{f^{a-b+N/2}}{(2\sqrt{\pi})^N} \frac{\Gamma(a+N/2)\Gamma(b-a-N/2)}{\Gamma(N/2)\Gamma(b)} \tag{4.208}$$

so that finally in the limit $n \to 4$ the integral

$$\lim_{n\to 4} \int \frac{d^{n-4}L}{(2\pi)^{n-4}} \frac{L^2}{L^2 + f(x_1, x_2)} = -1 \tag{4.209}$$

provides a nonvanishing contribution (from handling $(n-4)/(n-4)$). This is the origin of the anomaly.

Equation (4.207) now yields the familiar **anomalous AWI**

$$q^\lambda T_{\mu\nu\lambda} = -\frac{1}{2\pi^2} \varepsilon_{\mu\nu\alpha\beta} k_1^\alpha k_2^\beta = \mathcal{A}_{\mu\nu}. \tag{4.210}$$

Remark: Although the 't Hooft–Veltman procedure appears totally different to the Pauli–Villars regularization technique where a mass cut-off is introduced, there exists a link between these 2 approaches. Noticing that the momentum quantity L^2 plays the role of a mass squared, Hořejší, Novotný and Zavialov [Hořejší, Novotný, Zavialov 1988] have shown that the dimensional regularization corresponds to a 'continuous superposition' of PV cut-offs.

4.3.4 Singular current operator

Till now we have considered triangle graphs, their regularization creates the anomaly which we have calculated in momentum space. But the same anomaly also occurs in x-space when investigating the current operator.

The canonical commutation relations of field operators teach us that operator products at the same space–time point are singular objects. So the axial current as a product of 2 fermion field operators

$$j_\mu^5(x) = \bar{\psi}(x)\gamma_\mu\gamma_5\psi(x) \tag{4.211}$$

is necessarily singular. The regularization of this current induces the anomaly (see e.g. [Jackiw 1985d], [Adam 1990]).

Point splitting method: To regularize the current operator we use the point splitting method of Schwinger [Schwinger 1962]. There the operators are separated by a small vector ε^μ in the following way

$$j_5^\mu(x,\varepsilon) = \bar{\psi}(x+\varepsilon/2)\gamma^\mu\gamma_5\psi(x-\varepsilon/2) \exp\left[ie \int_{x-\varepsilon/2}^{x+\varepsilon/2} dy^\nu A_\nu(y)\right] \tag{4.212}$$

and

$$j^\mu_{5\,\mathrm{reg}}(x) = \lim_{\varepsilon \to 0} j^\mu_5(x,\varepsilon). \tag{4.213}$$

The exponential is introduced in order to guarantee gauge invariance.

The **rule** is:

i) first all calculations are carried out with $\varepsilon > 0$,

ii) then the limit $\varepsilon \to 0$ is performed in the physical amplitudes.

Now we want to calculate the derivative of the point-split current

$$\begin{aligned}
\partial_\mu j^\mu_5(x,\varepsilon) &= \\
&= \bar{\psi}(x+\varepsilon/2)\, \overleftarrow{\not\partial}\, \gamma_5 \psi(x-\varepsilon/2) \exp\left[ie \int_{x-\varepsilon/2}^{x+\varepsilon/2} dy^\nu A_\nu(y)\right] \\
&\quad - \bar{\psi}(x+\varepsilon/2)\gamma_5 \not\partial\, \psi(x-\varepsilon/2) \exp\left[ie \int_{x-\varepsilon/2}^{x+\varepsilon/2} dy^\nu A_\nu(y)\right] \\
&\quad + \bar{\psi}(x+\varepsilon/2)\gamma^\mu \gamma_5 \psi(x-\varepsilon/2)\partial^x_\mu \exp\left[ie \int_{x-\varepsilon/2}^{x+\varepsilon/2} dy^\nu A_\nu(y)\right].
\end{aligned} \tag{4.214}$$

We use the Dirac equations (4.6) and the definition of the point-split pseudoscalar current

$$P(x,\varepsilon) = \bar{\psi}(x+\varepsilon/2)\gamma_5\psi(x-\varepsilon/2) \exp\left[ie \int_{x-\varepsilon/2}^{x+\varepsilon/2} dy^\nu A_\nu(y)\right] \tag{4.215}$$

then we obtain

$$\begin{aligned}
\partial_\mu j^\mu_5(x,\varepsilon) &= 2imP(x,\varepsilon) \\
&\quad - iej^\mu_5(x,\varepsilon)\left[-\partial^x_\mu \int_{x-\varepsilon/2}^{x+\varepsilon/2} dy^\nu A_\nu(y) + A_\mu(x+\varepsilon/2) - A_\mu(x-\varepsilon/2)\right].
\end{aligned} \tag{4.216}$$

Expanding the gauge potentials yields

$$\partial_\mu j^\mu_5(x,\varepsilon) = 2imP(x,\varepsilon) - iej^\mu_5(x,\varepsilon)\varepsilon^\nu[\partial_\nu A_\mu(x) - \partial_\mu A_\nu(x)]. \tag{4.217}$$

If we naïvely take first the limit $\varepsilon^\nu \to 0$ we find the classical conservation law (4.8) and the corresponding ordinary AWI (4.81) (or PCAC in Section 4.6). However, this procedure is not allowed since the operator $j^\mu_5(x,\varepsilon)$ behaves as a singular like

$$j^\mu_5(x,\varepsilon) \sim \frac{1}{\varepsilon_\mu}. \tag{4.218}$$

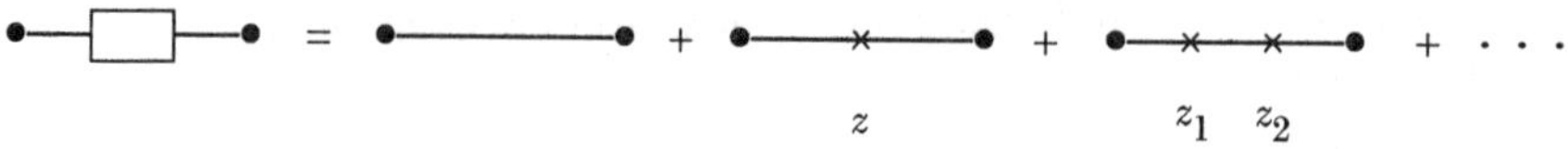

Fig. 4.3. Fermion propagator in the presence of an external field

We have to follow the above rule. First we carry out all calculations with $\varepsilon > 0$ and afterwards we choose $\varepsilon \to 0$. This creates a nonvanishing second term—the anomaly.

Anomaly: Let us consider now the vacuum expectation value of relation (4.217)

$$\begin{aligned}\langle 0|\partial_\mu j_5^\mu(x,\varepsilon)|0\rangle &= 2im\langle 0|P(x,\varepsilon)|0\rangle \\ &\quad - ie\varepsilon^\nu\langle 0|j_5^\mu(x,\varepsilon)|0\rangle[\partial_\nu A_\mu(x) - \partial_\mu A_\nu(x)] \end{aligned} \tag{4.219}$$

and focus on the quantity

$$\begin{aligned}&\varepsilon^\nu\langle 0|j_5^\mu(x,\varepsilon)|0\rangle \\ &= \varepsilon^\nu\langle 0|\bar\psi(x+\varepsilon/2)\gamma^\mu\gamma_5\psi(x-\varepsilon/2)|0\rangle \exp\left[ie\int_{x-\varepsilon/2}^{x+\varepsilon/2} dy^\lambda A_\lambda(y)\right] \\ &= \varepsilon^\nu \text{ tr } \gamma_5\gamma^\mu\tau(x-\varepsilon/2, x+\varepsilon/2)\exp\left[ie\int_{x-\varepsilon/2}^{x+\varepsilon/2} dy^\lambda A_\lambda(y)\right], \end{aligned} \tag{4.220}$$

where we have introduced the fermion propagator in the external field A_λ

$$\tau(x-\varepsilon/2, x+\varepsilon/2) = \langle 0|T\psi(x-\varepsilon/2)\bar\psi(x+\varepsilon/2)|0\rangle. \tag{4.221}$$

We are going to expand this propagator in powers of A_λ

$$\tau(x-\varepsilon/2, x+\varepsilon/2) = \tau_0(-\varepsilon) + ie\int d^4z\,\tau_0(x-\varepsilon/2-z)\,\not{A}(z)\tau_0(z-x-\varepsilon/2) + \dots \tag{4.222}$$

where $\tau_0(x) = iS_F(x)$ denotes the free fermion propagator (3.238). Graphically equation (4.222) is depicted in Figure 4.3. Only the linear term in A_λ is of interest to us since the degree of divergence in the series is decreasing. The free term vanishes in the trace of equation (4.220).

In momentum space the linear term of equation (4.222) has the following representation

$$\tau(x-\varepsilon/2, x+\varepsilon/2) \;\widehat{=}\; ie\int d^4z \int \frac{d^4p}{(2\pi)^4}\frac{d^4q}{(2\pi)^4}\frac{d^4\ell}{(2\pi)^4}e^{-ip(x-\varepsilon/2-z)} \cdot\frac{i}{\not p - m}e^{-iqz}\,\not A(q)e^{-i\ell(z-x-\varepsilon/2)}\frac{i}{\not \ell - m}. \tag{4.223}$$

The z-integral supplies a δ-function (for momentum conservation at the z-vertex, $l = p - q$), which simplifies the ℓ-integration, and now performing the p-integration shift

$$p \to p + q/2 \tag{4.224}$$

gives

$$\tau(x-\varepsilon/2, x+\varepsilon/2) \;\widehat{=}\; -ie\int \frac{d^4p}{(2\pi)^4}\frac{d^4q}{(2\pi)^4}e^{ip\varepsilon}e^{-iqx} \cdot\frac{1}{\not p + \not q/2 - m}\not A(q)\frac{1}{\not p - \not q/2 - m}. \tag{4.225}$$

We need this result (4.225) for the fermion propagator for expression (4.220)

$$\varepsilon^\nu\langle 0|j_5^\mu(x,\varepsilon)|0\rangle = -e\int \frac{d^4p}{(2\pi)^4}\frac{d^4q}{(2\pi)^4}i\varepsilon^\nu e^{ip\varepsilon}e^{-iqx} \cdot\frac{\mathrm{tr}\ \gamma_5\gamma^\mu(\not p + \not q/2 + m)\,\not A(q)(\not p - \not q/2 + m)}{[(p+q/2)^2 - m^2][(p-q/2)^2-m^2]}\exp\left[ie\int_{x-\varepsilon/2}^{x+\varepsilon/2} dy^\lambda A_\lambda(y)\right]. \tag{4.226}$$

In the trace the only surviving term is

$$-\mathrm{tr}\ \gamma_5\gamma^\mu\ \not p\gamma^\lambda\ \not q A_\lambda = -4i\varepsilon^{\mu\alpha\lambda\beta}p_\alpha q_\beta A_\lambda, \tag{4.227}$$

leading to a linear divergent integral (the quadratic divergence disappears because of identity (4.138)). We also observe that the integral shift (4.224)—actually not permissible for linear divergent integrals—does not change our final result because of the special algebraic structure of the trace.

Furthermore we use the identity

$$i\varepsilon^\nu e^{ip\varepsilon} = \frac{\partial}{\partial p_\nu}e^{ip\varepsilon} \tag{4.228}$$

and integrate expression (4.226) by parts

$$\varepsilon^{\nu}\langle 0|j_5^{\mu}(x,\varepsilon)|0\rangle = -e\, 4i\varepsilon^{\mu\alpha\lambda\beta}\int \frac{d^4q}{(2\pi)^4}e^{-iqx} \cdot \int \frac{d^4p}{(2\pi)^4}e^{ip\varepsilon}\frac{\partial}{\partial p_\nu}\frac{p_\alpha q_\beta A_\lambda(q)}{[(p+q/2)^2-m^2][(p-q/2)^2-m^2]} \cdot \exp\left[ie\int_{x-\varepsilon/2}^{x+\varepsilon/2} dy^\lambda A_\lambda(y)\right]. \tag{4.229}$$

Now we may perform the limit $\varepsilon \to 0$ and the remaining linear divergent integral is of type (4.143) which we evaluate via the 'surface' formula (4.146) and limit (4.152)

$$\begin{aligned}\varepsilon^{\nu}\langle 0|j_5^{\mu}(x,\varepsilon)|0\rangle|_{\varepsilon\to 0} &= \frac{e}{8\pi^2}\varepsilon^{\mu\nu\lambda\beta}\int \frac{d^4q}{(2\pi)^4}q_\beta e^{-iqx}A_\lambda(q)\\ &= -\frac{ie}{8\pi^2}\varepsilon^{\mu\nu\beta\lambda}\partial_\beta A_\lambda(x)\end{aligned} \tag{4.230}$$

(we re-expressed the Fourier integral). We finally use the regularized result (4.230) for the second term of equation (4.217) which now provides the **anomaly**

$$\begin{aligned}&-ie\varepsilon^{\nu}\langle 0|j_5^{\mu}(x,\varepsilon)|0\rangle|_{\varepsilon\to 0}\,[\partial_\nu A_\mu(x) - \partial_\mu A_\nu(x)]\\ &= \frac{e^2}{16\pi^2}\varepsilon^{\mu\nu\alpha\beta}F_{\mu\nu}(x)F_{\alpha\beta}(x) = \mathcal{A}(x).\end{aligned} \tag{4.231}$$

So after regularization we obtain an **anomalous divergence of the axial current**

$$\langle 0|\partial_\mu j^{\mu}_{5\,\mathrm{reg}}(x)|0\rangle = 2im\langle 0|P(x)|0\rangle + \mathcal{A}(x) \tag{4.232}$$

or in terms of the operator equation

$$\partial_\mu j^{\mu}_{5\,\mathrm{reg}}(x) = 2imP(x) + \mathcal{A}(x). \tag{4.233}$$

This modification (see also modified PCAC, Section 4.6) is in accordance with our discussion on Ward identities in Section 4.2.1. (Note the pseudoscalar current $P(x,\varepsilon)$ is already regular for $\varepsilon \to 0$.)

4.4 2-dimensional anomaly and dispersion relations

The anomaly originates not only from regularizing the ultraviolet divergences of an amplitude but occurs, in a different approach, in the infrared region as well. This approach works by means of dispersion relations (DR).

The use of dispersion relations to calculate the axial anomaly was proposed originally by Dolgov and Zakharov [Dolgov, Zakharov 1971] (for a review see Hořejší [Hořejší 1992a,b, 1985, 1986] and Huang [Huang 1982]).

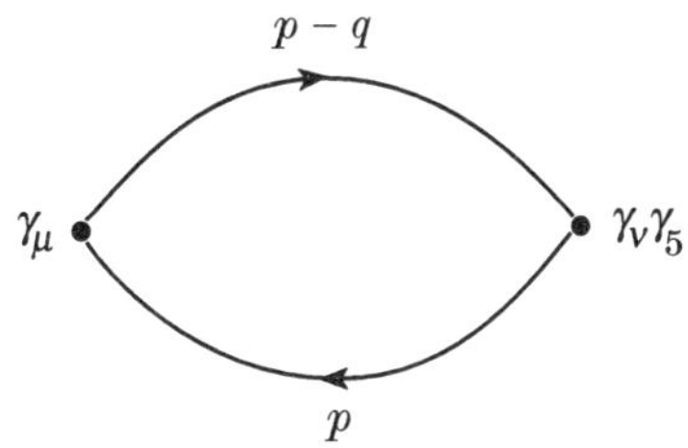

Fig. 4.4. Fermion loop containing a vector and axial current

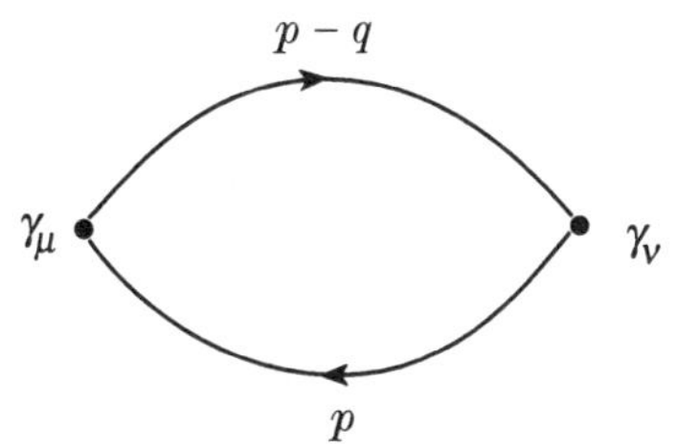

Fig. 4.5. Fermion loop containing a vector and vector current

Even before, Kummer [Kummer 1970] determined the anomaly in a dispersive way relying on a, at that time fashionable, pion–nucleon model. We, however, want to demonstrate the dispersion relation approach in a 2-dimensional example, which already contains all features of the method and is easy to handle. Here we follow closely the work of Adam, Bertlmann and Hofer [Adam, Bertlmann, Hofer 1992, 1993].

4.4.1 Ward identities

For literature specifically on 2-dimensional QFT we refer to [Jackiw 1985d], [Adam, Bertlmann, Hofer 1993], [Abdalla, Abdalla, Rothe 1991], [Dittrich, Reuter 1986], [Ecker 1982], [Grosse 1988], [Löwenstein, Swieca 1971], [Baier, Pilon 1991].

As discussed before, in QFT we work with Green functions which have to satisfy the Ward identities to achieve the renormalizability of the theory. In order to detect the anomaly in two dimensions we consider the 2-point function

$$\langle 0|Tj_\mu(x)j^5_\nu(y)|0\rangle, \tag{4.234}$$

which diagrammatically corresponds to the loop depicted in Figure 4.4. Then we find for the **axial Ward identity** (FT stands for Fourier Transformation)

$$\begin{aligned} q^\nu T^5_{\mu\nu} &= \text{FT } \langle 0|Tj_\mu(x)\partial^\nu_y j^5_\nu(y)|0\rangle \\ &= 2m \text{ FT } i\langle 0|Tj_\mu(x)P(y)|0\rangle \\ &= 2m\, P^5_\mu \end{aligned} \tag{4.235}$$

if we insert the classical result

$$\partial^\nu_y j^5_\nu(y) = 2imP(y) = 2im\bar{\psi}(y)\gamma_5\psi(y). \tag{4.236}$$

Analogously, when considering the 2-point function corresponding to the loop depicted in Figure 4.5

$$\langle 0|Tj_\mu(x)j_\nu(y)|0\rangle \tag{4.237}$$

we get for the **vector Ward identity**

$$\begin{aligned} q^\mu T_{\mu\nu} &= \text{FT } \langle 0|T\partial^\mu_x j_\mu(x)j_\nu(y)|0\rangle = 0 \\ q^\nu T_{\mu\nu} &= \text{FT } \langle 0|Tj_\mu(x)\partial^\nu_y j_\nu(y)|0\rangle = 0 \end{aligned} \tag{4.238}$$

if we rely on the classical conservation law

$$\partial^\mu_x j_\mu(x) = 0. \tag{4.239}$$

But now we will calculate the Green functions explicitly, and we do it in a dispersive way, and we check whether these WI's are satisfied. Dispersion relations in QFT have been first introduced by [Gell-Mann, Goldberger, Thirring 1954].

In two dimensions we have

$$\gamma_\nu\gamma_5 = \varepsilon_{\nu\lambda}\gamma^\lambda, \tag{4.240}$$

which implies for the amplitudes

$$T^5_{\mu\nu} = \varepsilon_{\nu\lambda}T_\mu{}^\lambda. \tag{4.241}$$

In addition we introduce the tensor structure

$$P^5_\mu(q) = \varepsilon_{\mu\nu}q^\nu P(q^2). \tag{4.242}$$

So it is enough to consider just the amplitude $T_{\mu\nu}$, the Fourier transform of the Green function (4.237) containing 2 vector currents. Let us investigate its structure.

Dispersion relations: The general Lorentz decomposition of $T_{\mu\nu}$ is

$$T_{\mu\nu}(q) = q_\mu q_\nu T_1(q^2) - g_{\mu\nu}T_2(q^2). \tag{4.243}$$

All we require is that it vanishes at large q^2

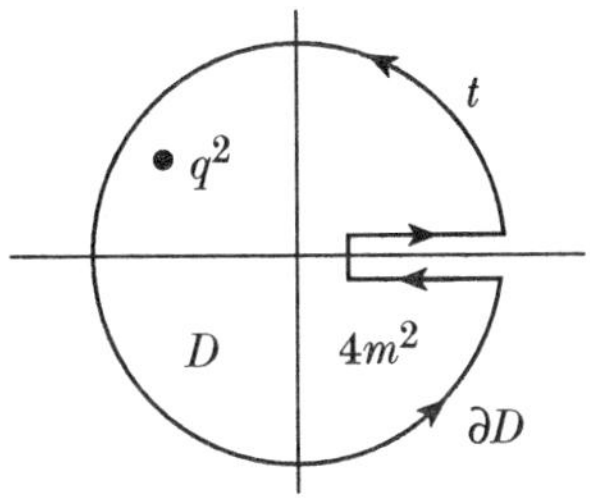

Fig. 4.6. Analyticity domain of the amplitudes $T_i(t)$

$$T_{\mu\nu}(q) \overset{q^2 \to \infty}{\longrightarrow} 0, \tag{4.244}$$

implying

$$\begin{aligned} T_1(q^2) &\overset{q^2 \to \infty}{\longrightarrow} 0 \\ T_2(q^2) &\overset{q^2 \to \infty}{\longrightarrow} \begin{cases} 0 \\ \text{const.} \neq 0. \end{cases} \end{aligned} \tag{4.245}$$

We also know that $T_i(t)$ $(i = 1, 2)$ are analytic functions on the t-plane with a cut starting at $t = 4m^2$ (see Figure 4.6), and along the cut we have

$$T(t + i\epsilon) - T(t - i\epsilon) = T(t + i\epsilon) - T^*(t + i\epsilon) = 2i \,\mathrm{Im}\, T(t + i\epsilon). \tag{4.246}$$

Theorem: Let $T(t)$ be an analytic function on the closure $\bar{D}$ of a domain D, let be $q^2 \in D$, then **Cauchy's integral formula** is valid

$$T(q^2) = \frac{1}{2\pi i} \int_{\partial D} \frac{dt}{t - q^2} T(t). \tag{4.247}$$

So Cauchy's theorem (4.247) provides the dispersion relations for the amplitudes $T_1(q^2)$ and $T_2(q^2)$. The unique choice for the amplitude $T_1(q^2)$ is an **unsubtracted DR** (similarly for $P(q^2)$)

$$T_1(q^2) = \frac{1}{\pi} \int_{4m^2}^{\infty} \frac{dt}{t - q^2} \,\mathrm{Im}\, T_1(t) \tag{4.248}$$

$$P(q^2) = \frac{1}{\pi} \int_{4m^2}^{\infty} \frac{dt}{t - q^2} \,\mathrm{Im}\, P(t). \tag{4.249}$$

Concerning $T_2(q^2)$, however, we have the option to subtract the DR once or not to subtract. The natural choice would again be an **unsubtracted DR**

$$T_2(q^2) = \frac{1}{\pi} \int_{4m^2}^{\infty} \frac{dt}{t - q^2} \, \text{Im } T_2(t) \tag{4.250}$$

since the integral exists.

Then the **AWI is satisfied**

$$\begin{aligned} q^\nu T^5_{\mu\nu}(q) &= q^\nu \varepsilon_{\nu\lambda}(q_\mu q^\lambda T_1(q^2) - g_\mu{}^\lambda T_2(q^2)) \\ &= \varepsilon_{\mu\nu} q^\nu T_2(q^2) = \varepsilon_{\mu\nu} q^\nu 2mP(q^2) = 2mP^5_\mu(q) \end{aligned} \tag{4.251}$$

since the imaginary parts of the amplitudes do fulfil the WI's in any case

$$\begin{aligned} \text{AWI:}\ \text{Im } T_2(t) &= 2m \text{ Im } P(t) \\ \text{VWI:}\ \text{Im } T_2(t) &= t \text{ Im } T_1(t). \end{aligned} \tag{4.252}$$

However, considering the **VWI** we find

$$\begin{aligned} q^\mu T_{\mu\nu} &= q_\nu(q^2 T_1(q^2) - T_2(q^2)) \\ &= q_\nu \left(-\frac{1}{\pi}\right) \int_{4m^2}^{\infty} dt \, \text{Im } T_1(t) \\ &= \mathcal{A} \, q_\nu, \end{aligned} \tag{4.253}$$

with the **anomaly**

$$\mathcal{A} = -\frac{1}{\pi}. \tag{4.254}$$

In the last step we had to do an explicit calculation; we calculated the imaginary part of a fermion loop with help of the Cutkosky rules (see at the end of Section 4.4)

$$t \text{ Im } T_1(t) = \frac{2m^2}{t} \left(1 - \frac{4m^2}{t}\right)^{-1/2}. \tag{4.255}$$

Thus gauge invariance is broken, the vector current is not conserved and the trace of the tensor (4.243)

$$T_\mu{}^\mu(q^2) = q^2 T_1(q^2) - 2T_2(q^2) \tag{4.256}$$

does not vanish in the limit $q^2 \to 0$

$$T_\mu{}^\mu(q^2 = 0) = -2T_2(0) \neq 0. \tag{4.257}$$

Now, for physical reasons, in order to *restore* gauge invariance we have to subtract the DR although the amplitude itself is finite. This is not peculiar for 2 dimensions, the same features also occur in 4 or higher dimensions. So we must use a **once subtracted DR** for the amplitude

$$T_2^R(q^2) = T_2(q^2) - T_2(0) = \frac{q^2}{\pi} \int_{4m^2}^{\infty} \frac{dt}{t(t-q^2)} \text{ Im } T_2(t). \tag{4.258}$$

Then the **subtraction constant represents the anomaly**

$$-T_2(0) = -\frac{1}{\pi} \int_{4m^2}^{\infty} \frac{dt}{t} \text{ Im } T_2(t) = -\frac{1}{\pi} \int_{4m^2}^{\infty} dt \text{ Im } T_1(t) = -\frac{1}{\pi} = \mathcal{A}. \tag{4.259}$$

Furthermore we have

$$T_2^R(q^2) = \frac{q^2}{\pi} \int_{4m^2}^{\infty} \frac{dt}{t-q^2} \text{ Im } T_1(t) = q^2 T_1(q^2) \tag{4.260}$$

and we achieve the familiar gauge invariant tensor structure (with $T_1 \equiv T$)

$$T_{\mu\nu}(q) = (q_\mu q_\nu - q^2 g_{\mu\nu}) T(q^2). \tag{4.261}$$

Now the **VWI is satisfied** and the trace vanishes in the limit $q^2 \to 0$ as it should

$$T_\mu{}^\mu(q^2) = -q^2 T(q^2) \stackrel{q^2 \to 0}{\longrightarrow} 0. \tag{4.262}$$

The **AWI**, on the other hand, contains the **anomaly**

$$\begin{aligned} q^\nu T^5_{\mu\nu} &= q^\nu \varepsilon_{\nu\lambda}(q_\mu q^\lambda T_1(q^2) - g_\mu{}^\lambda T_2^R(q^2)) \\ &= \varepsilon_{\mu\nu} q^\nu (T_2(q^2) - T_2(0)) \\ &= \varepsilon_{\mu\nu} q^\nu (2mP(q^2) + \mathcal{A}). \end{aligned} \tag{4.263}$$

For the massless case $m = 0$ the amplitude $P(q^2)$ vanishes

$$2mP(q^2) \to 0 \qquad \text{for } m \to 0 \tag{4.264}$$

but the anomaly remains

$$q^\nu T^5_{\mu\nu} = \mathcal{A} \varepsilon_{\mu\nu} q^\nu \tag{4.265}$$

so that the axial current is not conserved but provides the well-known **anomaly** result

$$\partial^\nu j^5_\nu = \frac{e}{\pi} \varepsilon_{\nu\mu} \partial^\nu A^\mu = \frac{e}{2\pi} \varepsilon_{\nu\mu} F^{\nu\mu}. \tag{4.266}$$

Remark: Of course, we could also subtract the amplitude $T_2(q^2)$ at some arbitrary point q_a^2; then the anomaly is distributed on both WI's, on the AWI and on the VWI. There is *no* choice to get rid of the anomaly at all (for details see [Adam, Bertlmann, Hofer 1992, 1993]).

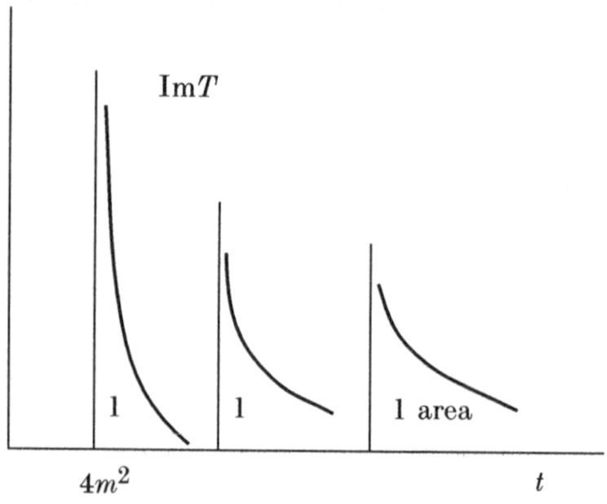

Fig. 4.7. Im $T_1(t)$ approaching a δ-function for $m \to 0$, equation (4.268)

Source of anomaly: The source of the anomaly in this dispersive procedure is the existence of the **superconvergence sum rule**

$$\int_{4m^2}^{\infty} dt \, \mathrm{Im}\, T_1(t) = 1. \tag{4.267}$$

The anomaly corresponds to a threshold singularity of $\mathrm{Im}\, T_1(t)$ at $t = 4m^2$ approaching a δ-function for $m \to 0$ (IR-region)

$$\lim_{m\to 0} \mathrm{Im}\ T_1(t) = \lim_{m\to 0} \frac{2m^2}{t^2}\left(1 - \frac{4m^2}{t}\right)^{-1/2} \theta(t - 4m^2) = \delta(t). \tag{4.268}$$

The limit must be performed in a distributional sense (see Figure 4.7).

This is a surprising feature indeed, but on the other hand it is the DR approach for analytic functions which connects the behaviour of the amplitude at $q^2 \to \infty$ with that at $q^2 \to 0$. A special renormalization procedure—regularizing the other end of the spectrum, the ultraviolet divergences—does not occur at all. For this reason the anomaly is independent of the renormalization procedure, a fact which we have demonstrated already in the previous sections.

4.4.2 n-dimensional regularization procedure

Although the method of DR appears quite different to the methods of regularizations we can reproduce its results in a definite regularization scheme, say, in the n-dimensional regularization procedure of 't Hooft–Veltman ['t Hooft, Veltman 1972]. It is instructive to work it out in more detail.

We calculate the tensor amplitude of the loop containing vector currents

$$T_{\mu\nu}(q) = -i \int \frac{d^2p}{(2\pi)^2} \operatorname{tr} \gamma_\mu \frac{i}{\not p - m} \gamma_\nu \frac{i}{\not p - \not q - m}. \tag{4.269}$$

For the moment it is sufficient to consider the trace (in n dimensions)

$$T_\mu{}^\mu(q^2) = 2^{n/2} i \int \frac{d^n p}{(2\pi)^n} \frac{(2-n)(p^2 - pq) + nm^2}{(p^2 - m^2)[(p-q)^2 - m^2]}. \tag{4.270}$$

The first term containing a logarithmic divergence must be regularized, the second term is already regular.

'Naïve' regularization and dispersion relations: Performing the limit $(2-n) \to 0$ in equation (4.270) before integration we are left with the second term

$$T_\mu^{(1)\mu}(q^2) = 2^{n/2} i \int \frac{d^n p}{(2\pi)^n} \frac{nm^2}{(p^2 - m^2)[(p-q)^2 - m^2]}. \tag{4.271}$$

We follow the standard procedure and insert the **Feynman parameter integral**

$$\frac{1}{ab} = \int_0^1 \frac{dx}{[ax + b(1-x)]^2}, \tag{4.272}$$

together with the 't Hooft–Veltman formula (4.189) where we have (in our case, $\alpha = 2$ and $n = 2$)

$$J_0 = \frac{-i\pi}{q^2 x(1-x) - m^2}. \tag{4.273}$$

Then we obtain

$$T_\mu^{(1)\mu}(q^2) = \frac{1}{\pi} \int_0^1 \frac{m^2 dx}{q^2 x(1-x) - m^2}. \tag{4.274}$$

In the limit $q^2 \to 0$

$$T_\mu^{(1)\mu}(q^2) \to -1/\pi = \mathcal{A}, \tag{4.275}$$

the trace does not vanish but provides precisely the **anomaly** $\mathcal{A}$. We reproduce half of the result (4.257). So we pick up just a $g_{\mu\nu}$-term

$$T_{\mu\nu}^{(1)}(q) = -\frac{1}{2} g_{\mu\nu} T_2(q^2) \tag{4.276}$$

(modulo terms which cancel in the trace, see the 'naïve' regularization below) and we have lost gauge invariance or the VWI (4.238) is broken.

For the amplitude $T_2(q^2)$ we then have an integral representation

$$T_2(q^2) = -\frac{1}{\pi}\int_0^1 \frac{m^2 dx}{q^2 x(1-x) - m^2}, \tag{4.277}$$

which we can easily convert into the familiar DR integral. The two transformations

$$y = 1 - 2x \qquad \text{and} \qquad t = \frac{4m^2}{1-y^2} \tag{4.278}$$

provide the result

$$\begin{aligned} T_2(q^2) &= \frac{1}{\pi}\int_{4m^2}^{\infty} \frac{dt}{t-q^2}\,\frac{2m^2}{t}\left(1-\frac{4m^2}{t}\right)^{-1/2} \\ &= \frac{1}{\pi}\int_{4m^2}^{\infty} \frac{dt}{t-q^2}\ \mathrm{Im}\ T_2(t), \end{aligned} \tag{4.279}$$

which is precisely the unsubtracted DR (4.250) from before.

Thus the 'naïve' kind of regularization corresponds in the DR approach to the case of an unsubtracted DR for the amplitude $T_2(q^2)$. However, we want to explain why we achieve only half of the anomaly result (4.257) after the proper renormalization.

't Hooft–Veltman regularization and dispersion relations: Now we include the divergency term, we renormalize à la 't Hooft–Veltman properly—performing the limit $(2-n) \to 0$ in equation (4.270) after integration—then the explicit calculation gives

$$\begin{aligned} T_{\mu\nu}(q) &= 2^{n/2} i \int \frac{d^n p}{(2\pi)^n}\int_0^1 dx \\ &\cdot \frac{[(p^\alpha p^\beta - p^\alpha q^\beta)(g_{\mu\alpha}g_{\nu\beta} - g_{\mu\nu}g_{\alpha\beta} + g_{\mu\beta}g_{\nu\alpha}) + m^2 g_{\mu\nu}]}{[p^2 - 2pq(1-x) + q^2(1-x) - m^2]^2} \end{aligned} \tag{4.280}$$

when keeping the tensor structure visible. For the p-integration we need the **'t Hooft–Veltman integrals**

$$\int \frac{d^n p p^\mu}{(p^2 - 2pk - m^2)^\alpha} = J_0 k^\mu \tag{4.281}$$

$$\int \frac{d^n p p^\mu p^\nu}{(p^2 - 2pk - m^2)^\alpha} = J_0\left(k^\mu k^\nu + \frac{k^2+m^2}{n+2-2\alpha} g^{\mu\nu}\right), \tag{4.282}$$

with $\alpha = 2$, $n = 2$ and J_0 given by equation (4.273), and we obtain

$$\begin{aligned}
T_{\mu\nu}(q) &= \frac{2^{n/2}\pi}{(2\pi)^n}\int_0^1 \frac{dx}{q^2x(1-x)-m^2} \\
&\cdot\{m^2 g_{\mu\nu} - (1-x)q^\alpha q^\beta(g_{\mu\alpha}g_{\nu\beta} - g_{\mu\nu}g_{\alpha\beta} + g_{\mu\beta}g_{\nu\alpha}) \\
&+ [(1-x)^2 q^\alpha q^\beta + \frac{1}{2-n}(q^2x(1-x)-m^2)g^{\alpha\beta}] \cdot \\
&\cdot(g_{\mu\alpha}g_{\nu\beta} - g_{\mu\nu}g_{\alpha\beta} + g_{\mu\beta}g_{\nu\alpha})\}. \qquad (4.283)
\end{aligned}$$

We observe a cancellation of the divergency $(2-n)^{-1}$ by another factor $(2-n)$ coming from the contraction of the 4γ-matrix trace. It provides the important contribution to finally achieve a gauge invariant amplitude

$$T_{\mu\nu}(q) = (q_\mu q_\nu - q^2 g_{\mu\nu})T(q^2), \qquad (4.284)$$

with

$$T(q^2) = -\frac{1}{\pi}\int_0^1 \frac{dx x(1-x)}{q^2x(1-x)-m^2}. \qquad (4.285)$$

Thus the correct n-dimensional regularization provides the gauge invariant tensor structure automatically, the **VWI is satisfied** and the trace vanishes in the limit $q^2 \to 0$

$$T_\mu{}^\mu(q^2) = -q^2T(q^2) \to -\frac{q^2}{m^2}\frac{1}{6\pi} \to 0. \qquad (4.286)$$

On the other hand, the regularized amplitude (4.284) satisfies the **anomalous AWI** (4.263).

Finally we again convert the integral representation (4.285) into a DR integral by applying the two transformations (4.278) and we find the result

$$\begin{aligned}
q^2T(q^2) &= \frac{q^2}{\pi}\int_{4m^2}^\infty \frac{dt}{t(t-q^2)}\frac{2m^2}{t}\left(1-\frac{4m^2}{t}\right)^{-1/2} \\
&= \frac{q^2}{\pi}\int_{4m^2}^\infty \frac{dt}{(t-q^2)}\ \mathrm{Im}\ T(t) \\
&= \frac{q^2}{\pi}\int_{4m^2}^\infty \frac{dt}{t(t-q^2)}\ \mathrm{Im}\ T_2(t) = T_2^R(q^2), \qquad (4.287)
\end{aligned}$$

which coincides with the unsubtracted DR (4.248) for $T = T_1$ and with the once subtracted DR (4.258) for T_2.

Therefore the proper n-dimensional regularization à la 't Hooft–Veltman corresponds in the DR approach to the case of a once subtracted DR for the amplitude $T_2(q^2)$.

'Naïve' regularization: Now we rediscuss the 'naïve' regularization from before. It exhibits the interesting feature that both Ward identities, the VWI

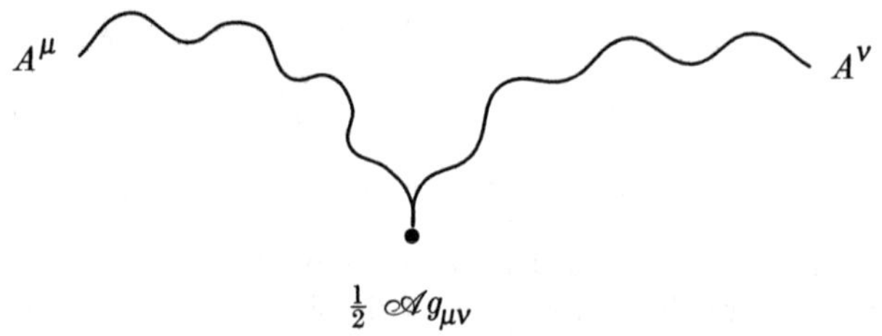

Fig. 4.8. 'Seagull' term

and the AWI, are equally broken. When regularizing the tensor amplitude $T_{\mu\nu}$ (4.283) 'naïvely' we throw away the singular term $(2-n)^{-1}$, in accordance with our 'naïve' procedure from before. Then we have

$$T^{(1)}_{\mu\nu}(q) = \frac{1}{2\pi}\int_0^1 \frac{dx}{q^2x(1-x)-m^2} \cdot\{m^2 g_{\mu\nu} - x(1-x)(2q_\mu q_\nu - q^2 g_{\mu\nu})\}. \qquad (4.288)$$

Identifying the integrals with the amplitudes introduced before we obtain

$$T^{(1)}_{\mu\nu}(q) = -\frac{1}{2}g_{\mu\nu}T_2(q^2) + q_\mu q_\nu T_1(q^2) - \frac{1}{2}g_{\mu\nu}T^R_2(q^2) \qquad (4.289)$$

(note that the second and third terms cancel in the trace and we reproduce equation (4.274)) and we rewrite expression (4.289) in a more convenient fashion by virtue of identities (4.287), (4.259)

$$T^{(1)}_{\mu\nu}(q) = (q_\mu q_\nu - q^2 g_{\mu\nu})T_1(q^2) + \frac{1}{2}\mathcal{A}g_{\mu\nu}. \qquad (4.290)$$

Now we study the Ward identities. The VWI involves only half of the anomaly

$$q^\mu T^{(1)}_{\mu\nu}(q) = \frac{1}{2}\mathcal{A}q_\nu \qquad (4.291)$$

but the AWI contains the other half

$$q^\nu T^5_{\mu\nu}(q) = q^\nu \varepsilon_{\nu\lambda}T^{(1)}{}_\mu{}^\lambda(q) = \varepsilon_{\mu\nu}q^\nu(2mP(q^2) + \frac{1}{2}\mathcal{A}). \qquad (4.292)$$

So the 'naïve' regularization is symmetric in the sense that the anomaly is equally spread in the vector- and axial Ward identity.

'Seagull' term: The term

$$\frac{1}{2}\mathcal{A}g_{\mu\nu} \tag{4.293}$$

is named the **'seagull'** term (see Figure 4.8). It corresponds to a local polynomial in the photon fields which can always be added to the action (recall our discussion of the renormalization ambiguities in Section 4.2.2).

Subtracting the 'seagull' term now we achieve the gauge invariant amplitude

$$T^{(1)}_{\mu\nu}(q) - \frac{1}{2}\mathcal{A}g_{\mu\nu} = (q_\mu q_\nu - q^2 g_{\mu\nu})T_1(q^2), \tag{4.294}$$

with a satisfied VWI and an anomalous AWI.

When adding the 'seagull' (4.293)

$$\begin{aligned} T^{(3)}_{\mu\nu}(q) &:= T^{(1)}_{\mu\nu}(q) + \frac{1}{2}\mathcal{A}g_{\mu\nu} \\ &= (q_\mu q_\nu - q^2 g_{\mu\nu})T_1(q^2) + \mathcal{A}g_{\mu\nu} \end{aligned} \tag{4.295}$$

the VWI gets anomalous

$$q^\mu T^{(3)}_{\mu\nu}(q) = \mathcal{A}q_\nu \tag{4.296}$$

but the AWI is satisfied

$$q^\nu T^5_{\mu\nu} = q^\nu \varepsilon_{\nu\lambda} T^{(3)}{}_\mu{}^\lambda = \varepsilon_{\mu\nu} q^\nu \, 2mP(q^2). \tag{4.297}$$

So by adding or subtracting the 'seagull' (4.293) to the 'naïvely' regularized amplitude $T^{(1)}_{\mu\nu}$ (4.290) we can jump between the 2 extremes of having the anomaly either in the vector- or in the axial Ward identity. But there is no 'seagull' term such that both Ward identities, the VWI *and* AWI, can be satisfied simultaneously. There always remains an anomaly. (For a discussion of the significance of 'seagull' terms see Jackiw [Jackiw 1985d].)

Cutkosky rule: Finally we promised to demonstrate how to calculate the imaginary part of an amplitude. Cutkosky's cutting rule [Cutkosky 1960], which is equivalent to the unitarity relation for the amplitude (see e.g. [Adam, Bertlmann, Hofer 1993]), is the following:

- *Replace in an amplitude each propagator by its discontinuity on mass shell*

$$\frac{1}{p^2 - m^2 \pm i\varepsilon} \to \mp 2\pi i\, \delta(p^2 - m^2)\theta(p^0) \tag{4.298}$$

  then you get the discontinuity—imaginary part times 2i—of the total amplitude.

Example: We need the imaginary part of the tensor $T_{\mu\nu}$ (4.269) with the general decomposition (4.243). Since the imaginary parts satisfy the VWI (4.252) it is enough to consider just the amplitude $T_2(q^2)$. As demonstrated before, the amplitude $T_2(q^2)$ corresponds to the m^2-term in the trace (4.270) of the tensor

$$T_2(q^2) = -T^{(1)}{}_\mu{}^\mu(q^2) \mathrel{\widehat{=}} -T_\mu{}^\mu(q^2). \tag{4.299}$$

Consequently the amplitude $T_2(q^2)$ represents the following integral

$$T_2(q^2) = -i\int \frac{d^2p}{(2\pi)^2}\frac{4m^2}{(p^2-m^2)[(p-q)^2-m^2]}. \tag{4.300}$$

We now calculate the imaginary part according to the above rule

$$\begin{aligned}\text{Im } T_2(q^2) &= 2m^2\int d^2p\,\delta(p^2-m^2)\delta((p-q)^2-m^2)\\ &\quad\cdot\theta(p_0)\theta(q_0-p_0).\end{aligned} \tag{4.301}$$

By using the familiar formula for the δ-functions

$$\delta(f(p)) = \sum_i \frac{\delta(p-p_i)}{|f'(p_i)|} \tag{4.302}$$

(p_i are the roots here; $f(p_i)=0$) we first integrate over p_0

$$\text{Im } T_2(q^2) = 2m^2\int dp_1\frac{1}{2E(p_1)}\delta(q^2-2pq)\theta(q_0-E), \tag{4.303}$$

with

$$E(p_1) = \sqrt{p_1^2+m^2}. \tag{4.304}$$

We conveniently choose the centre of mass system, $q_1 = 0$, and integrate next over p_1, then we get

$$\text{Im } T_2(q_0^2) = \frac{2m^2}{q_0^2}\left(1-\frac{4m^2}{q_0^2}\right)^{-1/2}. \tag{4.305}$$

The Lorentz invariant form is finally (with $T(q^2) = T_1(q^2)$)

$$q^2 \text{ Im } T(q^2) = \text{ Im } T_2(q^2) = \frac{2m^2}{q^2}\left(1-\frac{4m^2}{q^2}\right)^{-1/2}, \tag{4.306}$$

the result which we quoted before in equation (4.255).

Résumé: The chiral anomaly is completely determined by dispersion relations. We have demonstrated this within 2-dimensional QED. The method

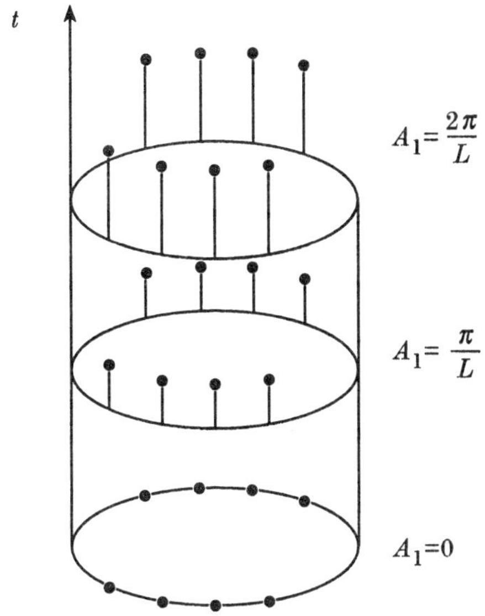

Fig. 4.9. The gauge field values defined on a cylinder

appears rather appealing. All one has to calculate is the imaginary part of an amplitude, which is an easy task.

Working with an unsubtracted DR for the amplitude $T_2(q^2)$ satisfies the AWI but breaks gauge invariance. When restoring gauge invariance we have to use a once subtracted DR for $T_2(q^2)$ which, on the other hand, supplies an anomalous AWI. Comparing this DR approach with the n-dimensional regularization scheme we find that a 'naïve' regularization corresponds to the first case whereas the proper regularization is equivalent to the second. The nature of the anomaly in the DR approach is different in the sense that it originates from a singularity of the amplitude in the infrared region whereas in a regularization procedure it reflects the behaviour in the ultraviolet region.

4.5 The anomaly and the Dirac sea

The Dirac sea simulates the effects of a quantized field theory. In fact, the anomaly can also be understood in terms of the Dirac sea (see e.g. [Jackiw 1985d], [Manton 1985], [Ambjørn, Greensite, Peterson 1983], [Nielson, Ninomiya 1983, 1991], [Shifman 1991], [Adam, Bertlmann, Hofer 1993]). We study the Schwinger model here—QED in 2 dimensions with massless fermions—with x-space compactified on a circle S^1 of length L.

Configuration space: The fields are defined on a cylinder of space and time (see Figure 4.9) and have to obey the boundary conditions

$$\begin{aligned} A_\mu\left(t, x=-\frac{L}{2}\right) &= A_\mu\left(t, x=\frac{L}{2}\right) \\ \psi\left(t, x=-\frac{L}{2}\right) &= -\psi\left(t, x=\frac{L}{2}\right). \end{aligned} \tag{4.307}$$

We choose a gauge where A_1 is independent of x and A_0 (the Coulomb potential) can be neglected. In fact, we treat $A_1(t)$ as an external field which will be switched on adiabatically. Due to the gauge freedom the values $A_1 = 0$ and $A_1 = 2\pi/L$ are gauge equivalent and must be identified. So the true configuration space of the gauge potential is a circle S^1 of length $2\pi/L$.

Conservation laws: The underlying $U(1)$ symmetry (we consider here constant phases) leads to the conservation of the vector current and electric charge

$$\partial^\mu j_\mu = 0 \qquad \text{and} \qquad \dot{Q}(t) = 0, \tag{4.308}$$

with

$$Q(t) = \int dx j_0(t, x). \tag{4.309}$$

The axial $U_A(1)$ symmetry (chiral symmetry) implies the conservation of the axial current and the axial charge

$$\partial^\mu j^5_\mu = 0 \qquad \text{and} \qquad \dot{Q}^5(t) = 0, \tag{4.310}$$

with

$$Q^5(t) = \int dx j^5_0(t, x). \tag{4.311}$$

Introducing the left- and right-handed fermions

$$\psi_L = \begin{pmatrix} \psi_1 \\ 0 \end{pmatrix}, \qquad \psi_R = \begin{pmatrix} 0 \\ \psi_2 \end{pmatrix}, \tag{4.312}$$

with chirality ± 1

$$\gamma_5 \psi_{L,R} = \pm \psi_{L,R}, \tag{4.313}$$

we get the charges

$$Q_{L,R} = \int dx \bar{\psi}_{L,R} \gamma_0 \psi_{L,R} = \int dx \psi^\dagger_{L,R} \psi_{L,R} \tag{4.314}$$

or

$$Q = Q_L + Q_R, \quad Q^5 = Q_L - Q_R. \tag{4.315}$$

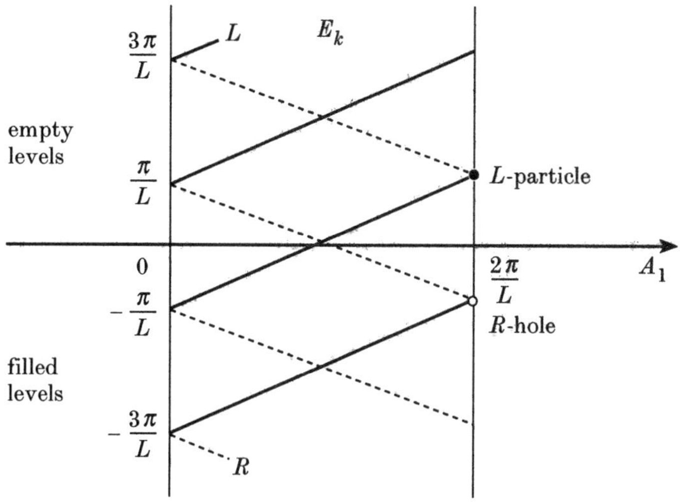

Fig. 4.10. Shift of the energy levels in the Dirac sea. The L-levels increase and the R-levels decrease producing a L-particle and a R-hole

So the axial charge for a L- or R-fermion is

$$Q^5_{L,R} = \int dx \bar{\psi}_{L,R}\gamma_0\gamma_5\psi_{L,R} = \pm \int dx \psi^\dagger_{L,R}\psi_{L,R} = \pm Q_{L,R}. \tag{4.316}$$

Since Q and Q^5 are conserved the numbers of L- and R-fermions are conserved separately.

The laws (4.308) and (4.310) are valid classically. On a quantum level, however, both currents (or charges) (4.308) and (4.310) cannot be conserved simultaneously—an anomaly occurs. How can we observe this phenomenon in a theory à la Dirac?

Energy spectrum: We start from the Dirac equation

$$(i \not{\partial} + \not{A})\psi = 0 \tag{4.317}$$

and we choose the 2-dimensional Dirac matrices in the following way

$$\gamma^0 = \sigma_2, \qquad \gamma^1 = i\sigma_1, \qquad \gamma^5 = \gamma^0\gamma^1 = \sigma_3, \tag{4.318}$$

where σ_i denote the familiar Pauli matrices (see equation (3.281)). Then the Dirac equation is rewritten by

$$\left[i\frac{\partial}{\partial t} + \sigma_3 \left(i\frac{\partial}{\partial x} - A_1\right)\right]\psi = 0. \tag{4.319}$$

According to the boundaries (4.307) the fermion wave function can be expanded into the Fourier series

$$\psi(t,x) = \frac{1}{\sqrt{L}} \sum_k u(k) \exp[-iE_k t] \exp\left[i\frac{2\pi}{L}\left(k+\frac{1}{2}\right)x\right], \tag{4.320}$$

which provides the following energy solutions for the L- and R-fermion eigenstates

$$E_k^L = \frac{2\pi}{L}\left(k+\frac{1}{2}\right) + A_1 \tag{4.321}$$

$$E_k^R = -\frac{2\pi}{L}\left(k+\frac{1}{2}\right) - A_1, \tag{4.322}$$

with $k = 0, \pm 1, \pm 2, \ldots$.

So the energy spectrum is comfortably discrete because of compactification and it depends linearly on A_1. We have plotted its features in Figure 4.10. At $A_1 = 0$ the energy levels for L- and R-fermions are degenerate. If we switch on A_1 the levels split; the L-levels increase whereas the R-levels decrease. At $A_1 = 2\pi/L$ we reproduce exactly the original level structure as it should be for gauge equivalent values. It is precisely this nontrivial restructuring of the infinitely many fermion levels that is the source of the anomaly, as we shall see.

IR-behaviour and anomaly: Now we turn from the 1-particle case to the multiparticle description of QFT by introducing the Dirac sea. We fill up all negative energy levels and keep all positive levels empty. So at $A_1 = 0$ we have a vacuum—the Dirac sea. However, if we increase A_1 from 0 to $2\pi/L$ we produce—by lifting the L-levels and lowering the R-levels—a left-handed particle and a right-handed hole (see Figure 4.10).

What does it imply for the charges? The electric charges of the particle and the hole are opposite. So there is no change in the total electric charge and the vector current is conserved, equation (4.308). The axial charges, however, are identical for both the L-particle and the R-hole, so that the net axial charge changes

$$\Delta Q^5 = 1 + 1 = 2. \tag{4.323}$$

This change compared with the one in the gauge potential A_1 gives

$$\Delta Q^5 = 2 = \frac{L}{\pi}\Delta A_1 \tag{4.324}$$

and per time unit

$$\frac{\Delta Q^5}{\Delta t} = \frac{L}{\pi}\frac{\Delta A_1}{\Delta t}. \tag{4.325}$$

Considering the local change in equation (4.325)

$$\frac{\partial}{\partial t}\int_0^L dx j_0^5(t,x) = \frac{1}{\pi}\frac{\partial}{\partial t}\int_0^L dx A_1(t) \tag{4.326}$$

we obtain the corresponding relation for the axial current

$$\partial_0 j_0^5 = \frac{1}{\pi}\partial_0 A_1 \tag{4.327}$$

or finally written in a Lorentz invariant way we arrive at the **anomaly**

$$\partial^\mu j_\mu^5 = \frac{1}{\pi}\varepsilon_{\mu\nu}\partial^\mu A^\nu, \tag{4.328}$$

which coincides with the result (4.266) determined in the previous section via dispersion relations.

UV-behaviour and anomaly: Alternatively we can also study the behaviour at some UV-cutoff which is more directly related to the perturbative approach in QFT presented before.

In fact, we *have* to regularize when working with the Dirac sea. The total energy, the total charge of the vacuum, the fermionic wave function as superpositions of infinitely many filled states are ill-defined quantities which have to be regularized somehow. A procedure which preserves gauge invariance is the point splitting method of Schwinger [Schwinger 1962] which we have used already in Section 4.3.4. There the gauge invariant regularized currents are

$$j_\mu^{\mathrm{reg}} = \lim_{\varepsilon\to 0}\bar\psi(t,x+\varepsilon)\gamma_\mu\psi(t,x)\exp\left[-i\int_x^{x+\varepsilon} dx A_1\right] \tag{4.329}$$

$$j_\mu^{5\,\mathrm{reg}} = \lim_{\varepsilon\to 0}\bar\psi(t,x+\varepsilon)\gamma_\mu\gamma_5\psi(t,x)\exp\left[-i\int_x^{x+\varepsilon} dx A_1\right]. \tag{4.330}$$

The rule is that the computations are carried out with fixed ε and afterwards the limit $\varepsilon \to 0$ is taken for the physical quantities.

The regularized charges are defined by

$$Q(t) = \int dx j_0^{\mathrm{reg}}(t,x) \tag{4.331}$$

$$Q^5(t) = \int dx j_0^{5\,\mathrm{reg}}(t,x) \tag{4.332}$$

and $Q_{L,R}$ by equation (4.315). Calculating the L- and R-charges explicitly we obtain the sum

$$Q_{L,R} = \sum_k \exp\left\{-i\varepsilon\left[\frac{2\pi}{L}\left(k+\frac{1}{2}\right)+A_1\right]\right\}, \tag{4.333}$$

with the summation

$$\begin{aligned} k &= -1,-2,-3,\ldots \qquad &&\text{for } L \\ k &= 0,1,2,\ldots &&\text{for } R. \end{aligned} \tag{4.334}$$

If we first take the limit $\varepsilon \to 0$ we are back at the unregularized ill-defined quantities

$$Q_{L,R} = \sum_k 1 \tag{4.335}$$

corresponding to the infinitely many filled levels in the Dirac sea.

But summing first and expanding afterwards in ε provides for the L- and R-sea the result

$$Q_L^{\mathrm{vac}} = -\frac{L}{2\pi}\frac{1}{i\varepsilon}+\frac{L}{2\pi}A_1 \tag{4.336}$$

$$Q_R^{\mathrm{vac}} = \frac{L}{2\pi}\frac{1}{i\varepsilon}-\frac{L}{2\pi}A_1. \tag{4.337}$$

Now we observe the following. The electric charge of the vacuum vanishes

$$Q^{\mathrm{vac}} = Q_L^{\mathrm{vac}} + Q_R^{\mathrm{vac}} = 0. \tag{4.338}$$

The electric charge and hence the vector current is conserved. The axial charge, on the other hand, contains a divergent constant $1/\varepsilon$ which will be subtracted to define the regularized quantity; so we are left with a linear dependence on A_1

$$Q^{5\,\mathrm{vac}} = Q_L^{\mathrm{vac}} - Q_R^{\mathrm{vac}} = 2\frac{L}{2\pi}A_1. \tag{4.339}$$

As A_1 increases from 0 to $2\pi/L$ the axial charge changes by 2 units as before, providing the same anomaly result.

Résumé: We consider the Dirac sea, and we switch on a gauge potential. It shifts the energies in this infinite sea; the L-levels increase and the R-levels decrease, there is a spectral flow [Atiyah, Patodi, Singer 1976]. It is this difference of the energy levels—passing through the zero scale ($E = 0$) and

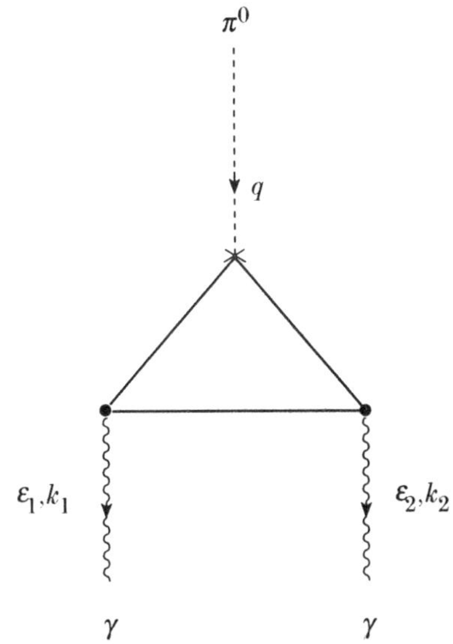

Fig. 4.11. Decay $\pi^0 \to \gamma\gamma$

weighted by the direction of the flow (here +1 and −1)—which corresponds to the index of the Dirac operator (see Chapter 11).

Consequently the L- and R-charges $Q_{L,R}$ change by ± 1 so that the overall axial charge varies by 2 units—an anomaly occurs. On the other hand, when focusing on the energy scale in the UV-region we find the same features.

There is no Feynman diagram of perturbation theory needed here, the anomaly here is a pure topological effect. The Dirac sea is a nice and practicable picture which reveals the anomalous phenomena of the quantum world. It provides an intuitive understanding for the different aspects: the anomaly occurs either as an IR-phenomenon (an index property of the Dirac operator) or as a UV-phenomenon.

4.6 Decay $\pi^0 \to \gamma\gamma$ and PCAC

As already emphasized at the beginning we need anomalies to explain experiments. Anomalies can be experienced in several particle reactions but the most stringent test for the ABJ anomaly is the neutral pion decay into 2 photons. It is also one of the most prominent decays in particle physics which created a puzzling situation for some time, whose solution actually led to the discovery of the anomaly. Let us consider it in more detail; for literature we refer to [Jackiw 1985d], [Cheng, Li 1988], [Itzykson, Zuber 1980], [Adam 1990], [Caprini, Micu, Visinescu 1992].

The **transition matrix element** for the pion decay (see Figure 4.11) is defined by

$$\begin{aligned}&\langle\gamma(\varepsilon_1,k_1),\gamma(\varepsilon_2,k_2)|\pi^0(q)\rangle\\ &= (2\pi)^4\delta^4(q-k_1-k_2)\varepsilon_1^\mu(k_1)\varepsilon_2^\nu(k_2)\Gamma_{\mu\nu}(k_1,k_2,q),\end{aligned} \tag{4.340}$$

where

$$\Gamma_{\mu\nu}(k_1,k_2,q) = e^2 i\int d^4x d^4y\; e^{ik_1x+ik_2y}\,\langle 0|Tj_\mu(x)j_\nu(y)|\pi^0(q)\rangle. \tag{4.341}$$

Because of the negative parity of the π^0 the amplitude (4.341) is a pseudotensor with the general Lorentz structure

$$\Gamma_{\mu\nu}(k_1,k_2,q) = \Gamma(q^2)\varepsilon_{\mu\nu\alpha\beta}k_1^\alpha k_2^\beta. \tag{4.342}$$

LSZ reduction formula: The pion and the photons can be reduced in the transition amplitude (4.340) according to a procedure invented by Lehmann, Symanzik and Zimmermann (LSZ) [Lehmann, Symanzik, Zimmermann 1955].

For scalar fields (analogously for other fields, see e.g. [Itzykson, Zuber 1980], [Lurié 1968]) the S-matrix obeys the following formula:

$$\begin{aligned}S_{f_i} &= \langle cd\ \text{out}|ab\ \text{in}\rangle\\ &= \delta_{f_i} + (-i)^4\int d^4x_1d^4x_2d^4x_3d^4x_4 f_d^*(x_4)f_c^*(x_3)f_b(x_2)f_a(x_1)\\ &\quad\cdot\mathcal{D}(x_1)\mathcal{D}(x_2)\mathcal{D}(x_3)\mathcal{D}(x_4)\langle 0|T\phi(x_1)\phi(x_2)\phi(x_3)\phi(x_4)|0\rangle,\end{aligned} \tag{4.343}$$

with the normalized plane waves

$$f_a(x) = \frac{1}{N}\,e^{-ip_ax}, \qquad N = (2\pi)^{3/2}\sqrt{2p_a^0} \tag{4.344}$$

and the Klein–Gordon operator

$$\mathcal{D}(x) = \Box_x + m^2. \tag{4.345}$$

The fields satisfy the Klein–Gordon equation with a source

$$\mathcal{D}(x)\phi(x) = j(x). \tag{4.346}$$

For the pion amplitude (4.341) then follows

$$\Gamma_{\mu\nu} = e^2 \int d^4y\, e^{ik_2 y} \int d^4z\, e^{-iqz} (\Box_z + m_\pi^2) \cdot \langle 0 | T\phi_\pi(z) j_\nu(y) j_\mu(0) | 0 \rangle. \tag{4.347}$$

The variable x has been transformed away due to energy–momentum conservation and the normalization N is absorbed by the definition of the final scattering matrix element. Extracting the Klein–Gordon operator (integrating by parts) gives

$$\Gamma_{\mu\nu} = e^2(-q^2 + m_\pi^2) \int d^4y d^4z\, e^{ik_2 y - iqz} \cdot \langle 0 | T\phi_\pi(z) j_\nu(y) j_\mu(0) | 0 \rangle. \tag{4.348}$$

We compare this amplitude with the 3-point function $T_{\mu\nu\lambda}$ which we have discussed already in Section 4.2.1 (recall equation (4.79))

$$q^\lambda T_{\mu\nu\lambda} = \int d^4y d^4z\, e^{ik_2 y - iqz} \langle 0 | T \partial_z^\lambda j_\lambda^5(z) j_\nu(y) j_\mu(0) | 0 \rangle. \tag{4.349}$$

In fact, we can find a bridge between these two amplitudes (4.348) and (4.349): the pion field couples to the nonconserved axial current.

Hypothesis: PCAC (partially conserved axial current)

$$\partial^\lambda j_\lambda^{5a} = f_\pi m_\pi^2 \phi_\pi^a(z), \tag{4.350}$$

where $f_\pi = 93$ MeV denotes the pion decay constant measured in the decay $\pi^+ \to \mu^+ \nu_\mu$; and $a = 1, 2, 3$ are the $SU(2)$ isospin indices with $a = 3$ for the neutral pion π^0. For massless pions—Goldstone bosons—the axial current is conserved, the corresponding chiral symmetry is exact.

Using PCAC (4.350) now we find the relation

$$q^\lambda T_{\mu\nu\lambda} = \frac{f_\pi m_\pi^2}{e^2(-q^2 + m_\pi^2)} \Gamma_{\mu\nu}. \tag{4.351}$$

Sutherland–Veltman paradox: Sutherland and Veltman [Sutherland 1967], [Veltman 1967] observed the following behaviour of the amplitudes. Since $T_{\mu\nu\lambda}$ does not contain any poles in the limit $q \to 0$ (there are no physical states between the pion and the vacuum) the l.h.s. of relation (4.351) has to vanish for $q \to 0$. Consequently the pion transition element must vanish too

$$\Gamma(q^2 = 0) = 0. \tag{4.352}$$

So the pion should *not* decay into photons

$$\langle\gamma\gamma|\pi^0\rangle \sim \Gamma(q^2 = m_\pi^2) \approx \Gamma(q^2 = 0) = 0. \tag{4.353}$$

But experimentally the pion decays!

Resolution: This paradoxical situation is resolved by the anomaly. We know already that we have to include the anomaly in the axial Ward identity, equation (4.83), or correspondingly in the divergence of the axial current, equation (4.85).

Adler's proposition:

- **PCAC** *has to be* **modified** *in the presence of gauge fields by the* **anomaly** [Adler 1969, 1970a,b]

$$\partial^\mu j_\mu^{5(3)}(z) = f_\pi m_\pi^2 \phi_\pi^{(3)}(z) + \frac{\alpha}{8\pi}\varepsilon^{\mu\nu\alpha\beta}F_{\mu\nu}F_{\alpha\beta}. \tag{4.354}$$

Note: In this relation the axial current carries $SU(2)$ isospin

$$j_\mu^{5a} = \bar{\psi}\gamma_\mu\gamma_5\frac{\sigma^a}{2}\psi \tag{4.355}$$

and contains a factor 1/2 compared to the usual axial current definition (4.4) (which entered in result (4.86)); accordingly the anomaly coefficient here is $\frac{1}{2}\frac{e^2}{16\pi^2} = \frac{\alpha}{8\pi}$.

Then the **anomaly corrected relation** (4.351) has to be

$$q^\lambda T_{\mu\nu\lambda} = \frac{f_\pi m_\pi^2}{e^2(-q^2+m_\pi^2)}\Gamma_{\mu\nu} - \frac{c}{2\pi^2}\varepsilon_{\mu\nu\alpha\beta}k_1^\alpha k_2^\beta \tag{4.356}$$

and we determine the constant c below in the quark model.

Pion amplitude: In the soft pion limit the pion amplitude is now totally given by the anomaly contribution [Adler 1969, 1970a,b]

$$\lim_{q\to 0}\Gamma_{\mu\nu}(k_1,k_2,q) = \frac{e^2 c}{2\pi^2 f_\pi}\varepsilon_{\mu\nu\alpha\beta}k_1^\alpha k_2^\beta \tag{4.357}$$

and hence

$$\Gamma(q^2=0) = \frac{e^2 c}{2\pi^2 f_\pi}. \tag{4.358}$$

Quark model: Let us calculate the constant c within the simple quark model. For the $SU(3)$ flavour group (up, down, strangeness) the quark spinor denotes

$$\psi(x) = \begin{pmatrix} u(x) \\ d(x) \\ s(x) \end{pmatrix} \tag{4.359}$$

and the currents are defined by

$$j_\mu(x) = \bar{\psi}(x)\gamma_\mu Q\psi(x) \tag{4.360}$$

$$j_\mu^{5(3)}(x) = \bar{\psi}(x)\gamma_\mu\gamma_5\frac{\lambda^3}{2}\psi(x), \tag{4.361}$$

with the matrices

$$Q = \frac{1}{3}\begin{pmatrix} 2 & 0 & 0 \\ 0 & -1 & 0 \\ 0 & 0 & -1 \end{pmatrix}, \qquad \lambda^3 = \begin{pmatrix} 1 & 0 & 0 \\ 0 & -1 & 0 \\ 0 & 0 & 0 \end{pmatrix}. \tag{4.362}$$

The constant c in the anomaly term (4.356) accounts for the matrix couplings in the triangle diagrams, hence we get

$$c = 3 \cdot \frac{1}{2}\ \mathrm{tr}\ \{Q, Q\}\frac{\lambda^3}{2} = 3 \cdot \frac{1}{6} = \frac{1}{2}. \tag{4.363}$$

The factor 3 in front respects the colour degrees of freedom of the quarks.

Decay rate: Finally we have to evaluate numerically the decay rate of the pion. We need the formula for the decay width (see e.g. [Pietschmann 1983]

$$\begin{aligned}\Gamma(\pi^0 \to \gamma\gamma) &= \frac{1}{2m_\pi}\int \frac{d^3k_1}{(2\pi)^3 2k_1^0}\frac{d^3k_2}{(2\pi)^3 2k_2^0}(2\pi)^4\delta^4(q-k_1-k_2) \\ &\quad \cdot \sum_{\text{polar.}} |\varepsilon_1^\mu \varepsilon_2^\nu \varepsilon_{\mu\nu\alpha\beta}k_1^\alpha k_2^\beta \Gamma(q^2)|^2. \end{aligned} \tag{4.364}$$

The calculation of the phase space integral is straightforward. In the soft pion limit with expression (4.358) and with the quark model value $c = 1/2$ (4.363) we obtain

$$\Gamma(\pi^0 \to \gamma\gamma) = \frac{\alpha^2 m_\pi^3}{64\pi^3 f_\pi^2} = 7.63\ \text{eV}. \tag{4.365}$$

This result is in excellent agreement with the experimental rate

$$\Gamma_{\text{experiment}}(\pi^0 \to \gamma\gamma) = 7.37 \pm 1.5\ \text{eV}. \tag{4.366}$$

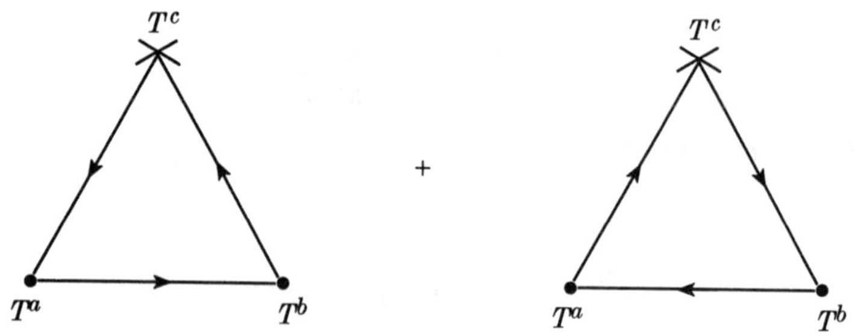

Fig. 4.12. YM coupling matrices in the triangle diagrams

So it is entirely the anomaly which determines the $\pi^0 \to \gamma\gamma$ decay. In addition, the colour factor 3 is necessary to achieve the correct experimental value. Therefore the anomaly also tests the colour degrees of the quarks.

4.7 Singlet anomaly

Next we turn to non-Abelian fields. There also higher loops contribute to the anomaly. Considering first the simplest diagram, the triangle, we have to study the amplitudes

$$T^{abc}_{\mu\nu\lambda}(k_1,k_2) = i\int d^4x d^4y\; e^{ik_1x+ik_2y}\;\langle 0|Tj^a_\mu(x)j^b_\nu(y)j^{5c}_\lambda(0)|0\rangle \tag{4.367}$$

$$T^{abc}_{\mu\nu}(k_1,k_2) = i\int d^4x d^4y\; e^{ik_1x+ik_2y}\;\langle 0|Tj^a_\mu(x)j^b_\nu(y)P^c(0)|0\rangle, \tag{4.368}$$

which now carry internal quantum numbers. We have already defined the non-Abelian currents j^a_μ, j^{5a}_μ, P^a in Section 4.1.2, equation (4.46). Then we know from our discussion of Ward identities (Section 4.2) how the anomaly modifies the **AWI**

$$q^\lambda T^{abc}_{\mu\nu\lambda} = 2mT^{abc}_{\mu\nu} + \mathcal{A}^{abc}_{\mu\nu} + \text{commutator terms}, \tag{4.369}$$

where

$$\mathcal{A}^{abc}_{\mu\nu} = -\frac{c^{abc}}{2\pi^2}\varepsilon_{\mu\nu\alpha\beta}k_1^\alpha k_2^\beta \tag{4.370}$$

denotes the **ABJ anomaly for non-Abelian fields**. The constant c^{abc} accounts for the non-Abelianity of the fields. It is the symmetrized trace of

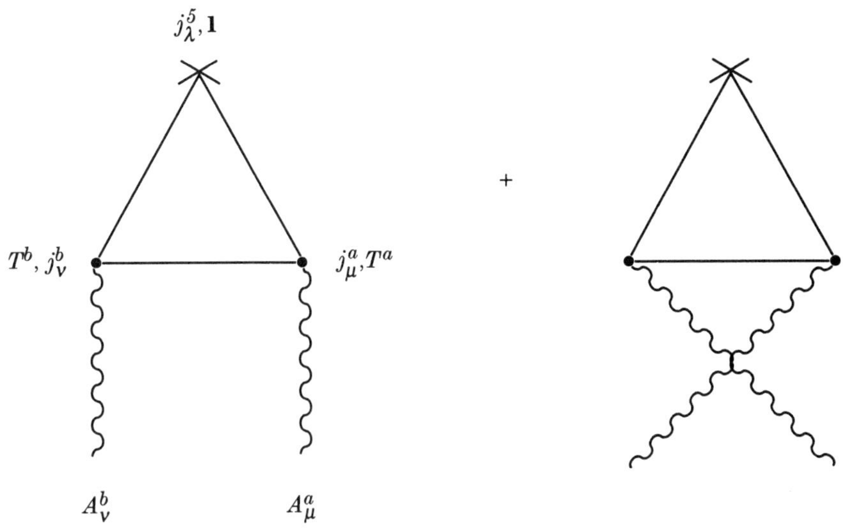

Fig. 4.13. Triangle diagrams for the singlet anomaly

the YM coupling matrices T^a which appear in the triangle diagrams (see Figure 4.12)

$$c^{abc} = \frac{1}{2} \text{ tr } \{T^a, T^b\} T^c \tag{4.371}$$

and which we can express by the symmetric structure constants (3.286) of the $SU(3)$ group

$$c^{abc} = \frac{i}{4} d^{abc}. \tag{4.372}$$

When we keep the axial current j^5_λ Abelian (as a colour singlet) but the gauge fields non-Abelian, so when we choose $T^c = \mathbf{1}$ (see Figure 4.13), we speak of a **singlet anomaly**.

Then the matrix couplings change to

$$c^{abc} \overset{T^c \to \mathbf{1}}{\longrightarrow} c^{ab} = \frac{1}{2} \text{ tr } \{T^a, T^b\} = \text{tr } T^a T^b \tag{4.373}$$

and the **anomalous AWI** is

$$q^\lambda T^{ab}_{\mu\nu\lambda} = 2m T^{ab}_{\mu\nu} + \mathcal{A}^{ab}_{\mu\nu}, \tag{4.374}$$

with the **singlet anomaly** contribution

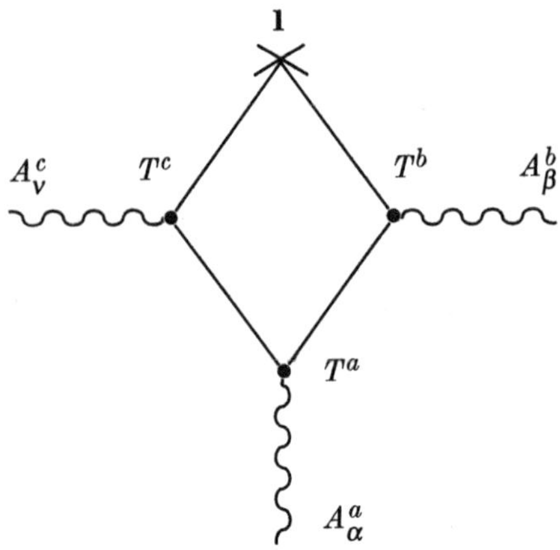

Fig. 4.14. Quadrangle diagram contributing to the singlet anomaly

$$\mathcal{A}^{ab}_{\mu\nu} = -\frac{c^{ab}}{2\pi^2}\varepsilon_{\mu\nu\alpha\beta}k_1^\alpha k_2^\beta. \tag{4.375}$$

As in the Abelian case the expression (4.375) corresponds to an anomalous operator identity for the colour singlet axial current j^5_λ. The only colour singlet operator, which we can construct from the gauge fields, is

$$\partial^\lambda j^5_\lambda = \frac{c^{ab}}{16\pi^2}\varepsilon^{\mu\nu\alpha\beta}F^a_{\mu\nu}F^b_{\alpha\beta}, \tag{4.376}$$

with

$$F^a_{\mu\nu} = \partial_\mu A^a_\nu - \partial_\nu A^a_\mu + [A_\mu, A_\nu]^a \tag{4.377}$$

being the YM field strength. Then the operator identity (4.376) provides the correct triangle result (4.374), (4.375); and we have chosen $m = 0$.

Inserting the coupling (4.373) and using the notation $F_{\mu\nu} = F^a_{\mu\nu}T^a$ we rewrite the **singlet anomaly**

$$\partial^\lambda j^5_\lambda = \frac{1}{16\pi^2}\varepsilon^{\mu\nu\alpha\beta}\ \mathrm{tr}\ F_{\mu\nu}F_{\alpha\beta}, \tag{4.378}$$

where the trace is taken over the gauge group generators T^a.

Equation (4.378) is an obvious generalization to the Abelian case (4.86). But now also nonlinear terms are included in the YM field strength and the singlet anomaly can be re-expressed by the **Chern–Simons term**

$$\partial^\lambda j^5_\lambda = \frac{1}{4\pi^2}\varepsilon^{\mu\nu\alpha\beta}\,\mathrm{tr}\,\partial_\mu\left(A_\nu\partial_\alpha A_\beta + \frac{2}{3}A_\nu A_\alpha A_\beta\right). \tag{4.379}$$

All anomalous WI's follow from the operator identity (4.378). The new feature for YM fields are the selfcoupling terms. In addition, to the linearized version proportional to

$$\partial_\mu A_\nu \partial_\alpha A_\beta, \tag{4.380}$$

which corresponds to the triangle diagram (Figure 4.13) we have a term

$$\partial_\mu A_\nu A_\alpha A_\beta, \tag{4.381}$$

which leads to a quadrangle diagram (see Figure 4.14).

So we get two contributions for the singlet anomaly, the triangle and the quadrangle diagrams. But these are not independent of each other. If, for example, the triangle result vanishes, the quadrangle result is zero too. Hence it is enough to consider just the simple triangle diagram in search of an anomaly.

4.8 Non-Abelian anomaly—Bardeen's result

Finally we discuss the non-Abelian anomaly. The classical conservation laws for non-Abelian fields, which we have derived in Section 4.1.2, are broken in QFT; the anomaly that occurs has been calculated by Bardeen [Bardeen 1969].

Non-Abelian anomaly: For a Lagrangian like equation (4.26) containing vector $\mathcal{V}_\mu$ and axial $\mathcal{A}_\mu$ YM fields the underlying symmetry is $SU_V(3) \times SU_A(3)$ (for $m = 0$). Then the **non-Abelian anomaly** is the following expression (after a laborious work)

$$\begin{aligned}(\mathcal{D}^\mu j^5_\mu)^a &= \frac{1}{16\pi^2}\varepsilon^{\mu\nu\rho\sigma}\,\mathrm{tr}\,T^a\Big[\mathcal{F}^V_{\mu\nu}\mathcal{F}^V_{\rho\sigma} + \frac{1}{3}\mathcal{F}^A_{\mu\nu}\mathcal{F}^A_{\rho\sigma} \\ &\quad - \frac{8}{3}(\mathcal{A}_\mu\mathcal{A}_\nu\mathcal{F}^V_{\rho\sigma} + \mathcal{A}_\mu\mathcal{F}^V_{\nu\rho}\mathcal{A}_\sigma + \mathcal{F}^V_{\mu\nu}\mathcal{A}_\rho\mathcal{A}_\sigma) \\ &\quad + \frac{32}{3}\mathcal{A}_\mu\mathcal{A}_\nu\mathcal{A}_\rho\mathcal{A}_\sigma\Big],\end{aligned} \tag{4.382}$$

with the **Bardeen curvatures**

$$\begin{aligned}\mathcal{F}^V_{\mu\nu} &= \partial_\mu\mathcal{V}_\nu - \partial_\nu\mathcal{V}_\mu + [\mathcal{V}_\mu, \mathcal{V}_\nu] + [\mathcal{A}_\mu, \mathcal{A}_\nu] \\ \mathcal{F}^A_{\mu\nu} &= \partial_\mu\mathcal{A}_\nu - \partial_\nu\mathcal{A}_\mu + [\mathcal{V}_\mu, \mathcal{A}_\nu] + [\mathcal{A}_\mu, \mathcal{V}_\nu].\end{aligned} \tag{4.383}$$

For result (4.382) Bardeen has added suitable counterterms to the action in order to achieve **covariant vector current conservation**

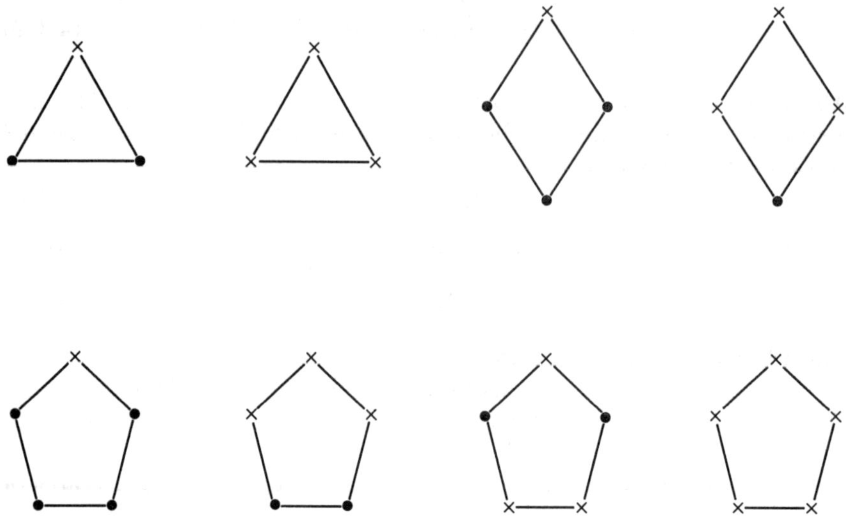

Fig. 4.15. Diagrams contributing to the non-Abelian anomaly in the divergence of the axial current. × depicts the axial current, • the vector current

$$\mathcal{D}_\mu j^\mu = 0. \tag{4.384}$$

If we consider just vector fields ($\mathcal{V}_\mu \equiv A_\mu$, $\mathcal{A}_\mu = 0$) then the **non-Abelian anomaly** is simply

$$(D^\mu j^5_\mu)^a = \frac{1}{16\pi^2}\varepsilon^{\mu\nu\rho\sigma} \text{ tr } T^a F_{\mu\nu} F_{\rho\sigma}. \tag{4.385}$$

This is a straightforward non-Abelian generalization to the singlet case (4.378). The result (4.385) **transforms gauge covariantly** and indeed equals the **covariant anomaly**, which we discuss below.

The anomalous Ward identities corresponding to the operator identity (4.382) involve higher order loop diagrams: the triangle-, quadrangle- and pentagon diagrams (see Figure 4.15). Although the pentagon is a finite diagram it contributes to the anomaly if VWI is required (because of divergent 'seagull' terms to be added; see Bardeen [Bardeen 1969]).

Triangle anomaly: For example, the anomaly corresponding to the triangle diagrams is the following. The triangle which couples to 3 axial fields $\mathcal{A}_\lambda \mathcal{A}_\mu \mathcal{A}_\nu$ amounts to just 1/3 of the triangle which couples to 1 axial- and 2 vector fields $\mathcal{A}_\lambda \mathcal{V}_\mu \mathcal{V}_\nu$. The reason lies in the Bose symmetry of all 3 vertices. Therefore we get

$$\begin{aligned}
&\mathrm{Anom}^a(\Delta) \\
&= \frac{1}{16\pi^2}\varepsilon^{\mu\nu\rho\sigma}\left[\mathrm{tr}\ T^a F^{\mathrm{lin}}_{\mu\nu}(\mathcal{V})F^{\mathrm{lin}}_{\rho\sigma}(\mathcal{V}) + \frac{1}{3}\ \mathrm{tr}\ T^a F^{\mathrm{lin}}_{\mu\nu}(\mathcal{A})F^{\mathrm{lin}}_{\rho\sigma}(\mathcal{A})\right],
\end{aligned} \tag{4.386}$$

where $F^{\mathrm{lin}}_{\mu\nu}$ denotes the linearized version of the field strength (4.383). The triangle result (4.386) is obviously contained in Bardeen's expression (4.382). The several loops which contribute to the anomaly (a detailed discussion can be found in the lecture notes of van Nieuwenhuizen [van Nieuwenhuizen 1988]) are connected with each other. If the triangle contribution vanishes then the other loops, the quadrangle and pentagon, vanish too.

Non-Abelian anomaly and chiral fields: On the other hand, we can consider the Lagrangian in terms of L- and R-handed fields, or positive- and negative chiral fields; recall equation (4.37). The underlying symmetry is $SU_L(3) \times SU_R(3)$ (for $m = 0$). Then the classical conservation laws are broken in a very symmetric way. The **non-Abelian anomaly for L- and R-fields** is (see [Bardeen 1969])

$$\begin{aligned}
-G^a[A^H_\mu] &= (D^H_\mu j^{H\mu})^a \\
&= \eta_H \frac{1}{24\pi^2}\varepsilon^{\mu\nu\rho\sigma}\ \mathrm{tr}\ T^a\partial_\mu\left(A^H_\nu\partial_\rho A^H_\sigma + \frac{1}{2}A^H_\nu A^H_\rho A^H_\sigma\right),
\end{aligned} \tag{4.387}$$

with

$$\eta_H = \left[\begin{array}{l} +1 \\ -1 \end{array}\right. \quad \text{for} \quad H = \left[\begin{array}{l} L \text{ or } + \text{ chirality} \\ R \text{ or } - \text{ chirality} \end{array}\right. . \tag{4.388}$$

So the anomaly just changes sign for L- and R-fields. Therefore we suppress from now on the chirality index $H = L, R$ throughout the book; but we constantly mean L- or R-fields.

The anomaly expression (4.387) is *the* important result which we investigate in this book. It is also called the **consistent anomaly** since it satisfies the **Wess–Zumino consistency condition**. We explain this more precisely in Chapter 8.

But there also exists another type of anomaly, the **covariant anomaly**

$$\widetilde{G}^a[A_\mu] = \pm\frac{1}{32\pi^2}\varepsilon^{\mu\nu\rho\sigma}\ \mathrm{tr}\ T^a F_{\mu\nu}F_{\rho\sigma}, \tag{4.389}$$

with $(+)$ or $(-)$ for R- or L-handed fields. We study its relation to the consistent version (4.387) in a separate chapter (Chapter 10). In the case of vanishing axial fields ($\mathcal{A}_\mu = 0$)

$$A_\mu^{R,L} \to \mathcal{V}_\mu \equiv A_\mu, \qquad j_\mu^{R,L} \to \mp\frac{1}{2}j_\mu^5$$

and

$$\widetilde{G}^a \sim -D_\mu^H j^{H\mu} \to \pm\frac{1}{2}D_\mu j^{5\mu}$$

we reproduce the covariant result (4.385) for the pure axial current.

4.9 Importance of anomalies

The importance of anomalies for physics is twofold:

Anomalies are good for experiment: This is the case if the underlying symmetry is an *external* symmetry. Then the anomalies are responsible for the properties of particles, for the physics of the particle decays or transitions.

Examples:

i) *Decay* $\pi^0 \to \gamma\gamma$.
This is the most prominent decay which is totally determined by the ABJ anomaly [Adler 1969], [Bell, Jackiw 1969]. We have already discussed all details in Section 4.6.

ii) $U(1)$ *problem.*
The QCD Lagrangian (in the limit $m \to 0$) contains a further $U_A(1)$ symmetry, besides the chiral $SU_L(2) \times SU_R(2)$ symmetry and the vector $U_V(1)$ symmetry. This axial symmetry is, however, neither observed in the hadron spectrum (no parity doubling of baryons) nor realized as a Goldstone boson (the η'-meson is just too heavy). The resolution of this problem is found in the existence of the anomaly and of gauge field configurations with nonvanishing topological charge, called instantons. Then the $U_A(1)$ symmetry is spontaneously broken without generating a Goldstone boson ['t Hooft 1976a,b], [Kogut, Susskind 1975].

iii) *Proton spin crisis.*
The $U_A(1)$ axial charge of the proton (nucleon)—the 'spin' of the proton—departs from its expectation in a naïve constituent quark model due to the anomaly [Altarelli, Ross 1988], [Altarelli, Stirling 1989], [Altarelli, Nason, Ridolfi 1994], [Fritzsch 1991], [Shore, Veneziano 1993], [Narison, Shore, Veneziano 1993].

iv) *Reaction* $\gamma\gamma \to \pi^+\pi^-\pi^0$.
This electromagnetic reaction also contains a part given by the non-Abelian anomaly [Wess, Zumino 1971].

v) *Decays* $K^+ \to \pi^+\pi^0\gamma$, $K^+ \to \pi^+\pi^-e^+\nu_e$.
In these nonleptonic Kaon decays there occurs in addition to the weak transition a contribution coming from the non-Abelian anomaly [Ecker, Neufeld, Pich 1992, 1994], [Bijnens, Ecker, Pich 1992], [Wess, Zumino 1971].

Anomalies are bad for theory: More precisely, anomalies are harmful for a quantized perturbative gauge theory. Then the anomalous Ward identities destroy the renormalizability, and thus the consistency, of the gauge theory [Gross, Jackiw 1972]. But also the unitarity of the S-matrix may be spoiled by anomalies [Korthals Altes, Perrottet 1972] (for a discussion see [Hořejší 1994]). This happens if the underlying symmetry is an *internal* symmetry and the gauge fields are internal fields and quantized. Nevertheless, there are attempts to live within anomalous (nonperturbative) theories [Faddeev, Shatashvili 1986], [Jackiw, Rajaraman 1985], [Rajaraman 1985a,b], [Lott, Rajaraman 1985], [Harada, Tsutsui 1987].

In order to guarantee the renormalizability of a theory anomalies must not occur. This implies important restrictions to the physical content of a theory.

Anomaly cancellation conditions:

i) *Vector-like model.*
In a vector-like model where all fermions couple symmetrically in the L- and R-sectors the L-gauge anomalies cancel the R-gauge ones.

ii) *'Safe groups'.*
There are no anomalies if the trace over the gauge group generators vanishes in a certain representation or even for all representations. The condition for it is

$$c^{abc} = \frac{1}{2} \text{ tr } \{T^a, T^b\} T^c = 0 \tag{4.390}$$

resulting from the triangle diagram; then all higher loop contributions vanish too.
There are 'safe' groups in 4 dimensions like $SU(2)$, $SO(2N+1)$ and $SO(4N)$ with $N \geq 2$, or exceptional groups like $E(6)$, $E(8)$ where identity (4.390) is valid for all representations; however, unfortunately not for $SU(N)$ with $N \geq 3$ (see e.g. [Georgi, Glashow 1972]).

iii) *Standard theory.*
The famous anomaly cancellation happens in the standard theory for electroweak interactions with the gauge group $SU(2) \times U(1)$ where the L-handed fermions are arranged in doublets and the R-handed fermions in singlets [Gross, Jackiw 1972], [Bouchiat, Iliopolous, Meyer

1972]. There condition (4.390) is satisfied if (see e.g. [Cheng, Li 1988], [Ecker 1982])

$$\mathrm{tr}\; Q_L = 0, \tag{4.391}$$

with Q_L the charge matrix of the L-fermions. Hence for a lepton doublet we also need a quark doublet with colour factor 3 in order to achieve anomaly cancellation

$$\begin{array}{ll} \begin{pmatrix} e^- \\ \nu_e \end{pmatrix}_L & 3\cdot\begin{pmatrix} u \\ d \end{pmatrix}_L \\ \mathrm{tr}\; Q_L : -1 & 3\cdot\left(\frac{2}{3}-\frac{1}{3}\right) = 1. \end{array} \tag{4.392}$$

This feature repeats itself in every generation

$$\begin{array}{ll} \begin{pmatrix} \mu^- \\ \nu_\mu \end{pmatrix}_L & 3\cdot\begin{pmatrix} c \\ s \end{pmatrix}_L \\ \begin{pmatrix} \tau^- \\ \nu_\tau \end{pmatrix}_L & 3\cdot\begin{pmatrix} t \\ b \end{pmatrix}_L . \end{array} \tag{4.393}$$

So the absence of anomalies restricts the fermionic content of the theory quite severely. It also tests the colour factor 3 of the quarks and it gave very strong support for the existence of the for a long time undiscovered top quark t.

iv) *Nonlocal counterterm.*
When introducing a *nonlocal* counterterm of gauge fields into the action we can compensate the anomaly (see e.g. [Krasnikov 1984, 1985a]). The term is constructed such that a gauge variation just cancels the anomaly contribution.
For example, in the Abelian case the nonlocal polynomial of L-fields is the following

$$\mathcal{S}_{\mathrm{pol}} = \frac{1}{96\pi^2}\varepsilon^{\mu\nu\alpha\beta}\int dxdy\;\partial^\lambda A^L_\lambda(x) D(x-y) F^L_{\mu\nu}F^L_{\alpha\beta}(y), \tag{4.394}$$

where $D(x-y)$ denotes the Green function of the d'Alembert operator. Inserting $\Box^{-1}\Box_x = \mathbf{1}$ we rewrite expression (4.394)

$$\mathcal{S}_{\mathrm{pol}} = \frac{1}{96\pi^2}\varepsilon^{\mu\nu\alpha\beta}\int dx\;\partial^\lambda A^L_\lambda(x)\Box_x^{-1} F^L_{\mu\nu}F^L_{\alpha\beta}(x). \tag{4.395}$$

Performing a gauge variation

$$\delta \mathcal{S}_{\text{pol}} = \frac{1}{96\pi^2} \varepsilon^{\mu\nu\alpha\beta} \int dx \; \partial^\lambda \partial_\lambda \Lambda^L(x) \Box_x^{-1} F^L_{\mu\nu} F^L_{\alpha\beta}(x)$$

we obtain after integrating by parts

$$\begin{aligned} \delta \mathcal{S}_{\text{pol}} &= \frac{1}{96\pi^2} \varepsilon^{\mu\nu\alpha\beta} \int dx \; \Lambda^L(x) F^L_{\mu\nu} F^L_{\alpha\beta}(x) \\ &= \frac{1}{24\pi^2} \varepsilon^{\mu\nu\alpha\beta} \int dx \; \Lambda^L(x) \partial_\mu A^L_\nu(x) \partial_\alpha A^L_\beta(x) \end{aligned} \quad (4.396)$$

which cancels precisely the anomaly (4.111).

v) *Local counterterm with additional fields:*
Analogously a *local* counterterm in the action—but containing additional fields with appropriate gauge transformations—can also compensate the anomaly (see e.g. [Krasnikov 1985b,c]).
A simple Abelian example is

$$\mathcal{S}_{\text{pol}} = \frac{1}{96\pi^2} \varepsilon^{\mu\nu\alpha\beta} \int dx \; \varphi^L(x) F^L_{\mu\nu} F^L_{\alpha\beta}(x), \quad (4.397)$$

with a gauge variation

$$\delta \varphi^L(x) = \Lambda^L(x). \quad (4.398)$$

Then the variation of the polynomial (4.397) leads again to result (4.396) and cancels the anomaly (4.111).
A very sophisticated generalization of this simple example represents the Green–Schwarz anomaly cancellation mechanism in a 10-dimensional supergravity and super Yang–Mills theory ([Green, Schwarz 1984], [Green, Schwarz, West 1985]).

Résumé: There exist two types of chiral anomaly, the singlet- and the non-Abelian anomaly

$$\begin{aligned} \partial^\mu j^5_\mu &= \frac{1}{4\pi^2} \varepsilon^{\mu\nu\rho\sigma} \operatorname{tr} \partial_\mu \left(A_\nu \partial_\rho A_\sigma + \frac{2}{3} A_\nu A_\rho A_\sigma \right) \quad (4.399) \\ (D_\mu j^\mu)^a &= \frac{1}{24\pi^2} \varepsilon^{\mu\nu\rho\sigma} \operatorname{tr} T^a \partial_\mu \left(A_\nu \partial_\rho A_\sigma + \frac{1}{2} A_\nu A_\rho A_\sigma \right). \quad (4.400) \end{aligned}$$

(Note that in the second case the gauge fields and the current are of positive chirality, for negative chirality we get a sign change, the chirality index is suppressed.) Although the two expressions resemble each other very closely they are of different origin. We want to emphasize this different nature of the anomalies. Let us consider the **triangle anomaly** part.

- If the axial current is the generator of the *external* symmetry while the other two vector currents are coupled to gauge bosons—as in the singlet anomaly case—then the anomaly signals the breakdown of this external symmetry in the presence of the gauge fields. Topologically, the singlet anomaly is completely determined by the Atiyah–Singer index theorem in 4 dimensions (see Section 11.1).
- If all three currents (say, L- or R-handed) are coupled to L-, R-gauge fields and generate an *internal* symmetry—as in the case of the non-Abelian anomaly—then the anomaly causes a violation of gauge invariance, an anomalous Ward identity and the ruin of renormalization. Topologically, the non-Abelian anomaly is also determined by an index theorem—but in $(4+2)$ dimensions! (See Section 11.5.3.)
- Nevertheless, the two types of anomaly are related to each other in a very peculiar way. The non-Abelian anomaly in $2n$ dimensions is related to a singlet anomaly in $(2n+2)$ dimensions via the differential geometric Stora–Zumino chain of descent equations (Chapter 9).

5
Path integral and anomaly

As we have discussed already in Chapter 4 the anomaly expresses the breakdown of a classical symmetry by quantum effects. This view of the anomaly arises quite naturally in the path integral formalism. In quantum field theory the basic quantity is the generating functional for the Green functions—a path integral for the classical action—which we have discussed in Chapter 3. How can an anomaly occur when the classical action is invariant under symmetry transformations? It was the important discovery of Fujikawa that the path integral measure—the only quantity which contains the quantum aspect—does not remain invariant under chiral transformations [Fujikawa 1979, 1980]. The associated Jacobian provides precisely the anomaly! In fact, all types of anomaly emerge in this way. This treatment of the anomaly is independent of perturbation theory and is, for this reason, called the nonperturbative approach.

In Section 5.1 we perform a chiral transformation of the path integral and we find the anomalous Ward identity. In Section 5.2 we regularize the transformation Jacobian à la Fujikawa and in this way derive the singlet anomaly; the 2-dimensional case is added in Section 5.3. In Section 5.4 we show the regularization independence of the anomaly and in Section 5.5 we discuss the conflict between gauge- and chiral symmetry in the light of an uncertainty principle. The generalization of the path integral method to non-Abelian fields leading to the non-Abelian anomaly is demonstrated in Section 5.6. Finally, in Section 5.7 we carry out the regularization of the Jacobian by means of the heat kernel method and by the zeta function procedure.

5.1 Fermionic measure and chiral transformation

For the anomaly—and we begin with the singlet anomaly case—we consider quantized Dirac fermions interacting with an external non-Abelian field A_μ^a. The Lagrangian is

$$\mathcal{L} = \bar{\psi}(i \not{D} - m)\psi, \tag{5.1}$$

with the Dirac operator

$$\not{D} = \gamma^\mu D_\mu = \gamma^\mu(\partial_\mu + A_\mu) \tag{5.2}$$

and the gauge potential $A_\mu = A_\mu^a T^a$. The group generators fulfil the commutation relations (3.275). For the actual calculations we perform a Wick rotation into Euclidean space–time

$$\begin{aligned} ix^0 &= x^4 \qquad & \partial_0 &= i\frac{\partial}{\partial x^4} = i\partial_4 \\ i\gamma^0 &= \gamma^4 \qquad & A_0 &= iA_4, \end{aligned} \tag{5.3}$$

with the metric

$$g^{\mu\nu} = -\delta^{\mu\nu}. \tag{5.4}$$

So we get

$$\not{D} = \gamma^\mu D_\mu = g^{\mu\nu}\gamma_\nu D_\mu = -\gamma_\mu D_\mu = -(\gamma_1 D_1 + \gamma_2 D_2 + \gamma_3 D_3 + \gamma_4 D_4). \tag{5.5}$$

We have chosen all γ^μ-matrices anti-Hermitian

$$\gamma^{\mu\dagger} = -\gamma^\mu, \qquad \mu = 1,2,3,4 \tag{5.6}$$

and we recall that the group generators are anti-Hermitian too, $T^{a\dagger} = -T^a$. The γ_5-matrix, on the other hand, remains Hermitian

$$\gamma_5^\dagger = \gamma_5, \tag{5.7}$$

with

$$\gamma_5 = i\gamma^0\gamma^1\gamma^2\gamma^3 = \gamma^4\gamma^1\gamma^2\gamma^3 = -\gamma^1\gamma^2\gamma^3\gamma^4 \tag{5.8}$$

and

$$\{\gamma^\mu, \gamma_5\} = 0. \tag{5.9}$$

Then the Dirac operator turns out to be Hermitian in Euclidean space–time

$$\not{D}^\dagger = \not{D}. \tag{5.10}$$

Chiral transformation: Now we perform a **local chiral transformation**

$$\begin{aligned} \psi &\to \psi' = e^{i\beta(x)\gamma_5}\psi \\ \bar\psi &\to \bar\psi' = \bar\psi e^{i\beta(x)\gamma_5}, \end{aligned} \tag{5.11}$$

where $\beta(x)$ denotes some gauge function. For infinitesimal β the Lagrangian (5.1) changes to

$$\begin{aligned} \mathcal{L} &\to \mathcal{L}' = \bar\psi'(i\not{D} - m)\psi' \\ &= \bar\psi(i\not{D} - m)\psi - (\partial^\mu\beta)\bar\psi\gamma_\mu\gamma_5\psi - 2im\beta\bar\psi\gamma_5\psi \end{aligned} \tag{5.12}$$

and with the usual definitions of the axial current and pseudoscalar density

$$j^5_\mu = \bar{\psi}\gamma_\mu\gamma_5\psi \qquad (5.13)$$

$$P = \bar{\psi}\gamma_5\psi \qquad (5.14)$$

we get

$$\mathcal{L}' = \mathcal{L} - (\partial^\mu \beta) j^5_\mu - 2im\beta P. \qquad (5.15)$$

(Note, for $m = 0$ equation (5.15) is also valid for *finite* β.)

The classical action

$$S' = S + \int dx\; \beta(x)[\partial^\mu j^5_\mu(x) - 2imP(x)], \qquad (5.16)$$

with

$$S = \int dx\; \mathcal{L} \qquad (5.17)$$

remains invariant

$$S' \equiv S \qquad (5.18)$$

iff the classical conservation law for the axial current is satisfied

$$\partial^\mu j^5_\mu = 2imP \qquad (5.19)$$

in accordance with our considerations in Section 4.1.

Path integral: What happens in a quantized field theory? There we have to study the **generating functional** for the Green functions in Euclidean space

$$Z[A_\mu] = \int d\psi d\bar{\psi} \exp\left[\int dx\; \bar{\psi}(i\not{D} - m)\psi\right] = \det(i\not{D} - m) =: \mathcal{N}. \quad (5.20)$$

It is a fermionic path integral, a Grassmann integral, serving as a definition for the determinant of the Dirac operator (recall Section 3.3 for our notations). Performing here the chiral transformation (5.11) Fujikawa discovered that the path integral measure transforms with a Jacobian containing the anomaly [Fujikawa 1979, 1980]. The derivation follows in the next section; here we just quote the result.

Theorem: The **path integral measure** transforms chirally as

$$d\psi' d\bar{\psi}' = d\psi d\bar{\psi}\; J[\beta, A_\mu], \qquad (5.21)$$

where the **transformation Jacobian**

$$J[\beta, A_\mu] = \exp\left[-\int dx\ \beta(x)\mathcal{A}[A_\mu](x)\right] \tag{5.22}$$

contains precisely the **singlet anomaly** in Euclidean space

$$\mathcal{A}[A_\mu](x) = \frac{-i}{16\pi^2}\varepsilon^{\mu\nu\alpha\beta}\ \mathrm{tr}\ F_{\mu\nu}F_{\alpha\beta}. \tag{5.23}$$

The imaginary factor $-i$ is removed in Minkowski space (for $\varepsilon^{0123} = 1$).

Including also the fermionic sources η, $\bar{\eta}$ we now expand the **chirally transformed path integral** (we suppress the source for the A_μ field which we do not need for our purpose)

$$\begin{aligned} Z[\eta, \bar{\eta}, A_\mu, \beta] &= \frac{1}{\mathcal{N}}\int d\psi' d\bar{\psi}' \exp\left\{\int dx[\mathcal{L}' + \bar{\eta}\psi' + \bar{\psi}'\eta]\right\} \\ &= \frac{1}{\mathcal{N}}\int d\psi d\bar{\psi} \exp\left\{\int dx[\mathcal{L} + \bar{\eta}\psi + \bar{\psi}\eta]\right\} \\ &\quad \cdot \exp\left\{\int dx\ \beta(x)[\partial^\mu j^5_\mu - 2imP - \mathcal{A}[A_\mu] + i\bar{\eta}\gamma_5\psi + i\bar{\psi}\gamma_5\eta]\right\} \\ &= \frac{1}{\mathcal{N}}\int d\psi d\bar{\psi} \exp\left\{\int dx[\mathcal{L} + \bar{\eta}\psi + \bar{\psi}\eta]\right\} \\ &\quad \cdot\left\{1 + \int dx\ \beta(x)[\partial^\mu j^5_\mu - 2imP - \mathcal{A}[A_\mu] + i\bar{\eta}\gamma_5\psi + i\bar{\psi}\gamma_5\eta]\right\}. \end{aligned} \tag{5.24}$$

Hence we get the following **functional expansion** (β infinitesimal)

$$Z[\eta, \bar{\eta}, A_\mu, \beta] = Z[\eta, \bar{\eta}, A_\mu, 0] + \int dx\ \beta(x)\frac{\delta}{\delta\beta(x)}Z[\eta, \bar{\eta}, A_\mu, \beta]\bigg|_{\beta=0}, \tag{5.25}$$

with

$$Z[\eta, \bar{\eta}, A_\mu, 0] = \frac{1}{\mathcal{N}}\int d\psi d\bar{\psi} \exp\left\{\int dx[\mathcal{L} + \bar{\eta}\psi + \bar{\psi}\eta]\right\} \tag{5.26}$$

$$\begin{aligned} \frac{\delta}{\delta\beta(x)}Z[\eta, \bar{\eta}, A_\mu, \beta]\bigg|_{\beta=0} &= \frac{1}{\mathcal{N}}\int d\psi d\bar{\psi} \exp\left\{\int dx[\mathcal{L} + \bar{\eta}\psi + \bar{\psi}\eta]\right\} \\ &\quad \cdot[\partial^\mu j^5_\mu - 2imP - \mathcal{A}[A_\mu] + i\bar{\eta}\gamma_5\psi + i\bar{\psi}\gamma_5\eta]. \end{aligned} \tag{5.27}$$

In the above expansion we transformed only the integration variables ψ, $\bar{\psi}$ in the PI. Therefore the generating functional—the 'quantum action'—

must remain invariant under the chiral transformation.

Lemma: Invariance of the 'quantum action'

$$Z[\eta,\bar{\eta},A_\mu,\beta] \equiv Z[\eta,\bar{\eta},A_\mu,0], \tag{5.28}$$

which means that

$$\frac{\delta}{\delta\beta(x)} Z[\eta,\bar{\eta},A_\mu,\beta]\bigg|_{\beta=0} = 0. \tag{5.29}$$

This invariance implies the **anomalous divergence of the axial current** (finally we have to put $\eta = \bar{\eta} = 0$ in condition (5.29))

$$\partial^\mu \langle j^5_\mu \rangle = \langle \partial^\mu j^5_\mu \rangle = 2im\langle P\rangle + \mathcal{A}[A_\mu]. \tag{5.30}$$

The symbol $\langle\ \rangle$ denotes the PI average which equals the vacuum expectation value of the corresponding operator (recall Chapter 3). Thus we reproduce precisely our result (4.232) of Section 4.3.4.

Ward identity: Condition (5.29) determines all (anomalous) Ward identities of the theory by differentiating with respect to the sources. For example, differentiating the functional (5.24)

$$\begin{aligned}\frac{\delta^2 Z[\eta,\bar{\eta},A_\mu,\beta]}{\delta\eta(x_2)\delta\beta(x)}\bigg|_{\beta=0} &= \frac{1}{\mathcal{N}}\int d\psi d\bar{\psi}\exp\left\{\int dx[\mathcal{L}+\bar{\eta}\psi+\bar{\psi}\eta]\right\}\\ &\cdot\{\bar{\psi}(x_2)[\partial^\mu j^5_\mu - 2imP - \mathcal{A}[A_\mu] + i\bar{\eta}\gamma_5\psi + i\bar{\psi}\gamma_5\eta](x)\\ &+ i\bar{\psi}(x)\gamma_5\delta(x-x_2)\}\end{aligned}$$

$$\begin{aligned}\frac{\delta^3 Z[\eta,\bar{\eta},A_\mu,\beta]}{\delta\bar{\eta}(x_1)\delta\eta(x_2)\delta\beta(x)}\bigg|_{\beta=0} &= \frac{1}{\mathcal{N}}\int d\psi d\bar{\psi}\exp\left\{\int dx[\mathcal{L}+\bar{\eta}\psi+\bar{\psi}\eta]\right\}\\ &\cdot\{\psi(x_1)\bar{\psi}(x_2)[\partial^\mu j^5_\mu - 2imP - \mathcal{A}[A_\mu] + i\bar{\eta}\gamma_5\psi + i\bar{\psi}\gamma_5\eta](x)\\ &+ i\psi(x_1)\bar{\psi}(x)\gamma_5\delta(x-x_2) + i\bar{\psi}(x_2)\gamma_5\psi(x)\delta(x-x_1)\}\end{aligned} \tag{5.31}$$

and using the invariance condition (5.29)

$$\frac{\delta^3 Z[\eta,\bar{\eta},A_\mu,\beta]}{\delta\bar{\eta}(x_1)\delta\eta(x_2)\delta\beta(x)}\bigg|_{\beta=0,\eta=0,\bar{\eta}=0} = 0 \tag{5.32}$$

supplies the **anomalous axial WI** (in Euclidean space)

$$\begin{aligned}
&\partial_x^\mu \langle j_\mu^5(x)\psi(x_1)\bar{\psi}(x_2)\rangle \\
&\quad = \; 2im\langle P(x)\psi(x_1)\bar{\psi}(x_2)\rangle + \langle \mathcal{A}[A_\mu](x)\psi(x_1)\bar{\psi}(x_2)\rangle \\
&\qquad - i\langle \gamma_5\psi(x)\bar{\psi}(x_2)\rangle\delta(x-x_1) - i\langle \psi(x_1)\bar{\psi}(x)\gamma_5\rangle\delta(x-x_2),
\end{aligned} \tag{5.33}$$

which in the operator formalism corresponds to the vacuum expectation value of the time-ordered products (in Minkowski space)

$$\begin{aligned}
&\partial_x^\mu \langle 0|Tj_\mu^5(x)\psi(x_1)\bar{\psi}(x_2)|0\rangle \\
&\quad = \; 2im\langle 0|TP(x)\psi(x_1)\bar{\psi}(x_2)|0\rangle + \langle 0|T\mathcal{A}[A_\mu](x)\psi(x_1)\bar{\psi}(x_2)|0\rangle \\
&\qquad - \langle 0|T\gamma_5\psi(x)\bar{\psi}(x_2)|0\rangle\delta(x-x_1) - \langle 0|T\psi(x_1)\bar{\psi}(x)\gamma_5|0\rangle\delta(x-x_2).
\end{aligned} \tag{5.34}$$

In equation (5.33) the factor i in the last two terms and in the anomaly result $\mathcal{A}$ is removed when we rotate back to Minkowski space. This anomalous axial WI (5.34) is the partner equation to the vector WI (4.68) which we have derived in Section 4.2. Of course, we recover result (5.34) when we follow the standard procedure, differentiating the Green function directly and inserting the anomalous divergence $\partial^\mu j_\mu^5 = 2imP + \mathcal{A}[A_\mu]$.

Remark: We want to emphasize that for the singlet anomaly $\mathcal{A}[A_\mu]$—where the axial current represents the generator of an external $U_A(1)$ symmetry—it is the *invariance of the 'quantum action'*

$$\frac{\delta}{\delta\beta(x)} Z[\eta,\bar{\eta},A_\mu,\beta]\bigg|_{\beta=0} = 0 \tag{5.35}$$

under a chiral change of the integration variables which determines all (anomalous) Ward identities.

This is different from the non-Abelian case where all chiral currents are coupled to gauge fields and where both, the fermions *and* the gauge potentials ($A_\mu^\beta = A_\mu + D_\mu\beta$), are chirally transformed. There it is the *noninvariance of the 'quantum action'* under gauge transformations

$$\frac{\delta}{\delta\beta^a(x)} Z[\eta,\bar{\eta},A_\mu^\beta,\beta]\bigg|_{\beta^a=0} = G^a[A_\mu](x) \tag{5.36}$$

—the occurrence of the non-Abelian anomaly $G^a[A_\mu](x)$—that determines all (anomalous) Ward identities.

5.2 Fujikawa's method and singlet anomaly

We certainly have to prove theorem (5.21)–(5.23), we demonstrate now that the Jacobian of the PI measure indeed creates the anomaly. We know already from Chapter 3 that a Grassmann measure transforms with the inverse determinant of the transformation operator. However, this determinant is ill defined in our case. But a correct regularization supplies precisely the anomaly. Let us explain the procedure invented by Fujikawa [Fujikawa 1979], [Fujikawa 1980]; additional literature can be found in [Adam 1990], [Dittrich, Reuter 1986].

Path integral measure: In order to define the PI measure more accurately we decompose the spinors ψ, $\bar{\psi}$ into eigenfunctions of the Dirac operator

$$\begin{aligned}\psi(x) &= \sum_n a_n \varphi_n(x) = \sum_n a_n \langle x|n\rangle \\ \bar{\psi}(x) &= \sum_m \varphi_m^\dagger(x)\bar{b}_m = \sum_m \langle m|x\rangle \bar{b}_m,\end{aligned} \tag{5.37}$$

where the coefficients a_n, $\bar{b}_m$ denote independent Grassmann elements. The Hermitian Dirac operator $\not{D}$ has real eigenvalues λ_n

$$\not{D}\varphi_n(x) = \lambda_n \varphi_n(x) \tag{5.38}$$

and the set of eigenfunctions $\{\varphi_n(x)\}$ is orthonormal and complete

$$\int dx \varphi_m^\dagger(x)\varphi_n(x) = \int dx \langle m|x\rangle\langle x|n\rangle = \langle m|n\rangle = \delta_{mn} \tag{5.39}$$

$$\sum_n \varphi_n(y)\varphi_n^\dagger(x) = \sum_n \langle y|n\rangle\langle n|x\rangle = \langle y|x\rangle = \delta(y-x). \tag{5.40}$$

(Note that in Euclidean space we may choose $\bar{\varphi} = \varphi^\dagger$ so that the group invariant scalar product is positive definite.) So the PI measure and the action can be re-expressed by

$$\begin{aligned}d\psi d\bar{\psi} &= [\det\langle m|x\rangle \det\langle x|n\rangle]^{-1} \prod_n da_n \prod_m d\bar{b}_m \\ &= [\det\langle n|n\rangle]^{-1} \prod_n da_n d\bar{b}_n \\ &= \prod_n da_n d\bar{b}_n\end{aligned} \tag{5.41}$$

$$\int dx\, \bar{\psi}(i\not{D}-m)\psi = \sum_n (i\lambda_n - m)\bar{b}_n a_n \qquad (5.42)$$

and the PI itself, equation (5.20), is formally defined by the following Grassmann integral

$$\begin{aligned}\det(i\not{D}-m) &= \int d\psi d\bar{\psi}\exp\left[\left[\int dx\, \bar{\psi}(i\not{D}-m)\psi\right]\right.\\ &= \int \prod_n da_n d\bar{b}_n \exp\left[\sum_n (i\lambda_n - m)\bar{b}_n a_n\right]\\ &= \prod_n (i\lambda_n - m). \end{aligned} \qquad (5.43)$$

Chiral transformation: How do the Grassmann elements a_n, $\bar{b}_n$ transform under an infinitesimal chiral rotation (5.11)? Let us consider the rotated spinor

$$\psi'(x) = \sum_n a'_n \varphi_n(x), \qquad (5.44)$$

which is, on the other hand,

$$\psi'(x) = (\mathbf{1} + i\beta(x)\gamma_5)\psi(x) = (\mathbf{1} + i\beta(x)\gamma_5)\sum_m a_m \varphi_m(x). \qquad (5.45)$$

Due to the orthonormality of the eigenfunctions, equation (5.39), we obtain

$$a'_n = \sum_m C_{nm} a_m, \qquad (5.46)$$

with the transformation matrix

$$C_{nm} = \delta_{nm} + i\int dx\, \beta(x)\varphi_n^\dagger(x)\gamma_5\varphi_m(x). \qquad (5.47)$$

Analogously we get

$$\bar{b}'_m = \sum_n C_{nm}\bar{b}_n \qquad (5.48)$$

from the rotated adjoint spinor $\bar{\psi}$.

Since the Grassmann measure transforms with the inverse determinant

$$\begin{aligned}\prod_n da'_n &= (\det C)^{-1}\prod_n da_n\\ \prod_m d\bar{b}'_m &= (\det C)^{-1}\prod_m d\bar{b}_m \end{aligned} \qquad (5.49)$$

we find for the change in the PI measure

$$d\psi' d\bar{\psi}' = (\det C)^{-2} d\psi d\bar{\psi} \equiv J[\beta] d\psi d\bar{\psi}. \tag{5.50}$$

Using the formula

$$\det C = \exp[\mathrm{Tr}\ \ln C] \tag{5.51}$$

and expanding the logarithm

$$\ln(1+\beta) = \beta + O(\beta^2) \tag{5.52}$$

we re-express the **Jacobian** in the following way

$$\begin{aligned} J[\beta] &= (\det C)^{-2} = \exp[-2\ \mathrm{Tr}\ \ln C] \\ &= \exp\left[-2\ \mathrm{Tr}\ \ln(\delta_{nm} + i \int dx\ \beta(x)\varphi_n^\dagger(x)\gamma_5\varphi_m(x))\right] \\ &= \exp\left[-2\ \mathrm{Tr}\ i \int dx \beta(x)\varphi_n^\dagger(x)\gamma_5\varphi_m(x)\right] \end{aligned} \tag{5.53}$$

so that

$$J[\beta] = \exp\left[-2i \int dx\ \beta(x) \sum_n \varphi_n^\dagger(x)\gamma_5\varphi_n(x)\right]. \tag{5.54}$$

The sum in the exponential, however, is not well defined

$$\sum_n \varphi_n^\dagger(x)\gamma_5\varphi_n(x) = \mathrm{tr}\ \gamma_5 \cdot \delta(0), \tag{5.55}$$

where we used the completeness (5.40) of the eigenfunctions.

Regularization: Fujikawa's idea was to regularize the sum by introducing a Gaussian cut-off

$$\begin{aligned} \sum_n \varphi_n^\dagger(x)\gamma_5\varphi_n(x) &= \lim_{M\to\infty} \sum_n \varphi_n^\dagger(x)\gamma_5 \exp\left[-\frac{\lambda_n^2}{M^2}\right]\varphi_n(x) \\ &= \lim_{M\to\infty} \sum_n \varphi_n^\dagger(x)\gamma_5 \exp\left[-\frac{\not{D}^2}{M^2}\right]\varphi_n(x) \end{aligned} \tag{5.56}$$

damping the contributions from the large eigenvalues and performing the limit $M \to \infty$ afterwards. We now evaluate the regularized sum in Fourier space. Introducing the Fourier components

$$\varphi_n(x) = \int \frac{d^4k}{(2\pi)^2} e^{ikx} \widetilde{\varphi}_n(k), \tag{5.57}$$

using again the completeness of the eigenfunctions

$$\sum_n \widetilde{\varphi}_n^\dagger(\ell)\Gamma\widetilde{\varphi}_n(k) = \mathrm{Tr}\,\Gamma\cdot\delta(\ell-k) \tag{5.58}$$

and integrating over the momentum ℓ, we obtain

$$\begin{aligned}\sum_n \varphi_n^\dagger(x)\gamma_5\varphi_n(x) &= \\ &= \lim_{M\to\infty}\int\frac{d^4\ell d^4k}{(2\pi)^4}\sum_n \widetilde{\varphi}_n^\dagger(\ell)e^{-i\ell x}\gamma_5\exp\left[-\frac{\not{D}^2}{M^2}\right]e^{ikx}\widetilde{\varphi}_n(k)\\ &= \lim_{M\to\infty}\int\frac{d^4k}{(2\pi)^4}\,\mathrm{Tr}\,e^{-ikx}\gamma_5\exp\left[-\frac{\not{D}^2}{M^2}\right]e^{ikx}.\end{aligned} \tag{5.59}$$

Here the trace has to be taken over the Dirac matrices *and* over the group generators T^a which we denote by Tr.

Next we decompose the Dirac operator

$$\begin{aligned}\not{D}^2 &= \gamma^\mu\gamma^\nu D_\mu D_\nu\\ &= \left(\frac{1}{2}\{\gamma^\mu,\gamma^\nu\}+\frac{1}{2}[\gamma^\mu,\gamma^\nu]\right)D_\mu D_\nu\\ &= D_\mu D^\mu + \frac{1}{4}[\gamma^\mu,\gamma^\nu]F_{\mu\nu},\end{aligned} \tag{5.60}$$

where we used the relation $[D_\mu, D_\nu] = F_{\mu\nu}$. When we insert decomposition (5.60) into the expression (5.59) we are aware that the plane waves shift the differential operator

$$e^{-ikx}f(\partial_\mu)e^{ikx} = f(\partial_\mu + ik_\mu) \tag{5.61}$$

so that we get

$$\begin{aligned}\sum_n \varphi_n^\dagger(x)\gamma_5\varphi_n(x) &= \\ &= \lim_{M\to\infty}\int\frac{d^4k}{(2\pi)^4}\,\mathrm{Tr}\,\gamma_5\exp\left[-\frac{(D_\mu+ik_\mu)(D^\mu+ik^\mu)}{M^2}-\frac{\gamma^\mu\gamma^\nu F_{\mu\nu}}{2M^2}\right]\\ &= \lim_{M\to\infty}M^4\int\frac{d^4k}{(2\pi)^4}e^{k_\mu k^\mu}\,\mathrm{Tr}\,\gamma_5\exp\left[-\frac{2ik_\mu D^\mu}{M}-\frac{D_\mu D^\mu}{M^2}-\frac{\gamma^\mu\gamma^\nu F_{\mu\nu}}{2M^2}\right]\end{aligned} \tag{5.62}$$

after rescaling the momentum $k_\mu \to Mk_\mu$.

Now we expand the exponential and we use the trace properties of the Dirac matrices

$$\text{tr } \gamma_5 = \text{tr } \gamma_5 \gamma^\mu \gamma^\nu = 0 \tag{5.63}$$

$$\text{tr } \gamma_5 \gamma^\mu \gamma^\nu \gamma^\alpha \gamma^\beta = -4\varepsilon^{\mu\nu\alpha\beta}, \tag{5.64}$$

with $\varepsilon^{1234} = \varepsilon^{1230} = 1$. Then only the quadratic term in $\gamma^\mu \gamma^\nu F_{\mu\nu}$ remains in the limit $M \to \infty$ and expression (5.62) becomes

$$\sum_n \varphi_n^\dagger(x)\gamma_5 \varphi_n(x) =$$

$$= \lim_{M\to\infty} \frac{1}{2!} \frac{M^4}{4M^4} \int \frac{d^4k}{(2\pi)^4} e^{-k_\mu k_\mu} \text{ Tr } \gamma_5 \gamma^\mu \gamma^\nu \gamma^\alpha \gamma^\beta F_{\mu\nu} F_{\alpha\beta}. \tag{5.65}$$

We note that our Euclidean convention is

$$k_\mu k^\mu = k_\mu g^{\mu\nu} k_\nu = -k_\mu \delta^{\mu\nu} k_\nu = -k_\mu k_\mu. \tag{5.66}$$

Next we carry out the Gauss integration

$$\int_{-\infty}^{\infty} d^{2n}k e^{-k_\mu k_\mu} = \pi^{2n/2} \tag{5.67}$$

and take the trace over the Dirac matrices. Then the **regularized sum** à la Fujikawa provides the result

$$\sum_n \varphi_n^\dagger(x)\gamma_5 \varphi_n(x) = -\frac{1}{32\pi^2} \varepsilon^{\mu\nu\alpha\beta} \text{ tr } F_{\mu\nu} F_{\alpha\beta}, \tag{5.68}$$

where the trace over the group generators in $F_{\mu\nu} = F_{\mu\nu}{}^a T^a$ remains.

Jacobian: So finally for the **Jacobian** (5.54) of the PI measure we find

$$J[\beta] = \exp\left[-\int dx \; \beta(x) \frac{-i}{16\pi^2} \varepsilon^{\mu\nu\alpha\beta} \text{ tr } F_{\mu\nu} F_{\alpha\beta}\right] \tag{5.69}$$

and recalling our definition (5.22)

$$J[\beta] = \exp\left[-\int dx \; \beta(x) \mathcal{A}[A_\mu](x)\right] \tag{5.70}$$

we obtain the result for the **singlet anomaly**

$$\mathcal{A}[A_\mu] = \frac{-i}{16\pi^2} \varepsilon^{\mu\nu\alpha\beta} \text{ tr } F_{\mu\nu} F_{\alpha\beta}. \tag{5.71}$$

In Minkowski space (with the choice $\varepsilon^{0123} = 1$) we have to remove the factor $-i$ due to the Minkowskian change of the trace (5.64). Then the result (5.71) agrees with our perturbative calculations (4.378) or (4.231) in Chapter 4.

5.3 2-dimensional anomaly

In 2 dimensions the calculations turn out particularly simple. For the Jacobian we have to regularize the sum

$$\begin{aligned}\sum_n \varphi_n^\dagger(x)\gamma_5\varphi_n(x) &= \lim_{M\to\infty}\sum_n \varphi_n^\dagger(x)\gamma_5 \exp\left[-\frac{\lambda_n^2}{M^2}\right]\varphi_n(x)\\ &= \lim_{M\to\infty}\sum_n \varphi_n^\dagger(x)\gamma_5 \exp\left[-\frac{\not{D}^2}{M^2}\right]\varphi_n(x)\\ &= \lim_{M\to\infty}\int\frac{d^2k}{(2\pi)^2}\,\mathrm{tr}\; e^{-ikx}\gamma_5\exp\left[-\frac{\not{D}^2}{M^2}\right]e^{ikx},\end{aligned} \tag{5.72}$$

where our Dirac operator here is (recall equation (4.317))

$$\not{D} = \not{\partial} - i\not{A}. \tag{5.73}$$

A_μ denotes the Abelian photon field and as before $\not{D}$ is chosen Hermitian in Euclidean space, $\not{D} = \not{D}^\dagger$. Decomposing the squared operator

$$\not{D}^2 = D_\mu D^\mu - \frac{i}{4}[\gamma^\mu,\gamma^\nu]F_{\mu\nu} \tag{5.74}$$

we obtain for the sum

$$\begin{aligned}&\sum_n \varphi_n^\dagger(x)\gamma_5\varphi_n(x)\\ &= \lim_{M\to\infty} M^2\int\frac{d^2k}{(2\pi)^2}e^{-k_\mu k_\mu}\;\mathrm{tr}\;\gamma_5\exp\left[-\frac{2ik_\mu D^\mu}{M} - \frac{D_\mu D^\mu}{M^2} + \frac{i\gamma^\mu\gamma^\nu F_{\mu\nu}}{2M^2}\right]\end{aligned} \tag{5.75}$$

after shifting the differential operator (see equation (5.61)) and rescaling the momentum $k_\mu \to Mk_\mu$. We expand the exponential, take the trace of the Dirac matrices, then only the term $\gamma^\mu\gamma^\nu F_{\mu\nu}$ remains in the limit $M\to\infty$. The Gaussian integral gives a factor π (equation (5.67)).

In 2 dimensions our conventions in Minkowski space are (recall Section 4.5, equation (4.318))

$$\begin{aligned}&\gamma^0 = \sigma_2, \qquad \gamma^1 = i\sigma_1, \qquad \gamma_5 = \gamma^0\gamma^1 = \sigma_3,\\ &g_{\mu\nu} = \begin{pmatrix}1 & 0\\ 0 & -1\end{pmatrix}, \qquad \varepsilon_{01} = 1\end{aligned} \tag{5.76}$$

and in Euclidean space

$$\gamma^0 = i\gamma^4, \qquad \gamma_5 = \gamma^0\gamma^1 = i\gamma^4\gamma^1,$$
$$g_{\mu\nu} = -\delta_{\mu\nu}, \qquad \varepsilon_{41} = i\varepsilon_{01} = i. \tag{5.77}$$

Then the relations

$$\gamma_\mu\gamma_5 = \varepsilon_{\mu\nu}\gamma^\nu, \qquad \text{tr}\ \gamma_5\gamma_\mu\gamma_\nu = -2\varepsilon_{\mu\nu} \tag{5.78}$$

are valid in *both* spaces, which is convenient for our calculations.

So we find for the **regularized sum**

$$\begin{aligned}\sum_n \varphi_n^\dagger(x)\gamma_5\varphi_n(x) &= \lim_{M\to\infty} M^2\frac{1}{2M^2}\frac{\pi}{(2\pi)^2}(-2i)\varepsilon_{\mu\nu}F^{\mu\nu} \\ &= -\frac{i}{4\pi}\varepsilon_{\mu\nu}F^{\mu\nu},\end{aligned} \tag{5.79}$$

for the **Jacobian** of the PI measure (equation (5.54))

$$J[\beta] = \exp\left[-\int dx\ \beta(x)\frac{1}{2\pi}\varepsilon_{\mu\nu}F^{\mu\nu}\right] \tag{5.80}$$

and finally for the **anomaly** (recall equation (5.70))

$$\mathcal{A} = \frac{1}{2\pi}\varepsilon_{\mu\nu}F^{\mu\nu}. \tag{5.81}$$

With our definition of the ε-tensor, equations (5.76), (5.77), the expression (5.81) is valid in this form in Euclidean *and* in Minkowski space. Result (5.81) is in accordance with our calculations in Sections 4.4 and 4.5.

5.4 Regularization independence of the anomaly

As Fujikawa [Fujikawa 1980] emphasized the anomaly result is *independent* of the chosen regularization for the large eigenvalue contributions in the sum (5.56). Instead of the exponential damping we could also choose some other function which is smooth and decreasing sufficiently rapidly at infinity

$$f\left(\frac{\lambda_n^2}{M^2}\right), \tag{5.82}$$

with

$$\begin{aligned} f(\infty) &= f'(\infty) = f''(\infty) = \ldots = 0 \\ f(0) &= 1. \end{aligned} \tag{5.83}$$

Replacing exp by f we get

$$\begin{aligned}\sum_n \varphi_n^\dagger(x)\gamma_5\varphi_n(x) &= \lim_{M\to\infty}\sum_n \varphi_n^\dagger(x)\gamma_5 f\left(\frac{\lambda_n^2}{M^2}\right)\varphi_n(x)\\ &= \lim_{M\to\infty}\sum_n \varphi_n^\dagger(x)\gamma_5 f\left(\frac{\not{D}^2}{M^2}\right)\varphi_n(x)\\ &= \lim_{M\to\infty}\int\frac{d^4k}{(2\pi)^4}\,\mathrm{Tr}\,\gamma_5 f\left(\frac{k_\mu k_\mu}{M^2}+\frac{2ik_\mu D^\mu + D_\mu D^\mu}{M^2}+\frac{\gamma^\mu\gamma^\nu F_{\mu\nu}}{2M^2}\right).\end{aligned} \tag{5.84}$$

We expand the function at the point $k_\mu k_\mu/M^2 =: k^2/M^2$ into a Taylor series and we use the trace properties of the Dirac matrices, then only the second derivative contributes to the limit $M\to\infty$

$$\begin{aligned}&\sum_n \varphi_n^\dagger(x)\gamma_5\varphi_n(x)\\ &= \lim_{M\to\infty}\frac{1}{2!}\frac{1}{(2M^2)^2}\int\frac{d^4k}{(2\pi)^4}f''\left(\frac{k^2}{M^2}\right)\,\mathrm{Tr}\,\gamma_5\gamma^\mu\gamma^\nu\gamma^\alpha\gamma^\beta F_{\mu\nu}F_{\alpha\beta}.\end{aligned} \tag{5.85}$$

Inserting the Dirac trace (5.64) and rescaling the momentum $k_\mu \to Mk_\mu$ gives

$$\begin{aligned}&\sum_n \varphi_n^\dagger(x)\gamma_5\varphi_n(x)\\ &= \lim_{M\to\infty} -\frac{1}{2}\int\frac{d^4k}{(2\pi)^4}f''(k^2)\varepsilon^{\mu\nu\alpha\beta}\,\mathrm{tr}\,F_{\mu\nu}F_{\alpha\beta}.\end{aligned} \tag{5.86}$$

We carry out the integral through integration by parts ($f''(k^2)$ means the second derivative with respect to the argument)

$$\begin{aligned}\int\frac{d^4k}{(2\pi)^4}f''(k^2) &= \int_0^\infty\frac{\pi^2k^2dk^2}{(2\pi)^4}f''(k^2) = -\frac{1}{16\pi^2}\int_0^\infty dk^2 f'(k^2)\\ &= -\frac{1}{16\pi^2}f(k^2)\Big|_0^\infty = \frac{1}{16\pi^2},\end{aligned} \tag{5.87}$$

and we recover the desired result (5.68)

$$\sum_n \varphi_n^\dagger(x)\gamma_5\varphi_n(x) = -\frac{1}{32\pi^2}\varepsilon^{\mu\nu\alpha\beta}\,\mathrm{tr}\,F_{\mu\nu}F_{\alpha\beta}. \tag{5.88}$$

For instance, choosing the cut-off function as

$$f\left(\frac{\lambda_n^2}{M^2}\right) = \frac{1}{1 + \lambda_n^2/M^2} \tag{5.89}$$

then Fujikawa's regularization corrresponds to the regularization of Pauli–Villars (Section 4.3.2). For further details we refer to the literature [Fujikawa 1980, 1984, 1985], [Adam 1990], [Delbourgo, Thompson 1985].

5.5 Fujikawa's uncertainty principle

The sum in the Jacobian

$$\sum_n \varphi_n^\dagger(x)\gamma_5\varphi_n(x) = \lim_{y\to x} \text{tr } \gamma_5 \cdot \delta(x-y) \tag{5.90}$$

can be regarded as a *conditionally* convergent series in an infinite dimensional functional space, very much like $+1+1-1-1+1+1-1-1+\ldots$. It has a definite value whenever we specify how to sum the series. Fujikawa's Gaussian cut-off (5.56)—and any regularizer with properties (5.82), (5.83)—corresponds to a summation that imposes *gauge invariance*.

Then it happens that the diagonalization of the Dirac operator $\not{D}$ implies a chiral asymmetry and the chiral trace becomes anomalous, tr $\gamma_5 \neq 0$—in contrast to the 'naive' expectation tr $\gamma_5 = 0$ which would correspond to a summation $(+1+1-1-1)+(+1+1-1-1)+\ldots = 0$.

Fujikawa [Fujikawa 1980, 1989] realized that in QFT the operators γ_5 and $\not{D}$ satisfy a kind of uncertainty principle, analogous to the one we have in nonrelativistic quantum mechanics between the operators x and p.

Proposition: Uncertainty principle

- *We cannot impose both symmetries, the gauge symmetry and the chiral symmetry simultaneously!*
- *The two operators $\not{D}$ and γ_5 do not commute, their commutator expectation value gives rise precisely to the anomaly*

$$\langle\bar{\psi}(x)[\not{D},\gamma_5]\psi(x)\rangle = \mathcal{A}[A_\mu](x). \tag{5.91}$$

Hence $\not{D}$ and γ_5 cannot be diagonalized simultaneously. If we diagonalize the Dirac operator $\not{D}$—imposing gauge symmetry in the trace (5.90)—then the conflicting symmetry—the chirality γ_5—becomes anomalous. So the chiral or axial, or singlet anomaly arises from the conflict of two competing symmetries: the chiral symmetry and the gauge symmetry.

Proof. Let us calculate the expectation value of the commutator (5.91). Within the PI formalism we first evaluate

$$\begin{aligned}
&\langle\bar{\psi}(x)2i\gamma_5\, \not{D}\psi(x)\rangle \\
&= \frac{1}{\mathcal{N}}\int d\psi d\bar{\psi}\; \bar{\psi}(x)2i\gamma_5\, \not{D}\psi(x)\exp\left[\int dx\; \bar{\psi}(i\,\not{D}-m)\psi\right],
\end{aligned} \tag{5.92}$$

with the normalization

$$\mathcal{N} = \int d\psi d\bar{\psi}\exp\left[\int dx\; \bar{\psi}(i\,\not{D}-m)\psi\right] = \det(i\,\not{D}-m). \tag{5.93}$$

Expanding the Dirac spinors ψ, $\bar{\psi}$ in terms of eigenfunctions of $\not{D}$, equation (5.37), we get

$$\begin{aligned}
&\langle\bar{\psi}(x)2i\gamma_5\, \not{D}\psi(x)\rangle \\
&= \frac{1}{\mathcal{N}}\int\prod_i da_i d\bar{b}_i \sum_{m,n}\bar{b}_m a_n \varphi_m^\dagger(x)2i\gamma_5\, \not{D}\varphi_n(x) \\
&\quad \cdot\exp\left[\sum_k (i\lambda_k - m)\bar{b}_k a_k\right].
\end{aligned} \tag{5.94}$$

Respecting now the rules for Grassmann integration, particularly,

$$e^c = 1 + c, \qquad \int dc = 0, \qquad \int dc\, c = 1 \tag{5.95}$$

(see Section 3.3.1), the only contribution which remains is the following

$$\begin{aligned}
&\langle\bar{\psi}(x)2i\gamma_5\, \not{D}\psi(x)\rangle = \\
&= \frac{1}{\mathcal{N}}\int\prod_i da_i d\bar{b}_i \sum_n \bar{b}_n a_n \varphi_n^\dagger(x)2i\gamma_5\, \not{D}\varphi_n(x)\prod_{k\neq n}(i\lambda_k - m)\bar{b}_k a_k \\
&= \frac{1}{\mathcal{N}}\int\prod_i da_i d\bar{b}_i \sum_n \frac{2i\varphi_n^\dagger(x)\gamma_5\, \not{D}\varphi_n(x)}{i(\lambda_n + im)}\prod_k (i\lambda_k - m)\bar{b}_k a_k \\
&= 2\sum_n \frac{\varphi_n^\dagger(x)\gamma_5\, \not{D}\varphi_n(x)}{\lambda_n + im}
\end{aligned} \tag{5.96}$$

since

$$\prod_k (i\lambda_k - m) = \det(i\,\not{D} - m) = \mathcal{N}. \tag{5.97}$$

For the commutator we then find

$$
\begin{aligned}
\langle\bar{\psi}(x)i[\gamma_5, \not{D}]\psi(x)\rangle &= \langle\bar{\psi}(x)2i\gamma_5 \not{D}\psi(x)\rangle \\
&= 2\sum_n \frac{\varphi_n^\dagger(x)\gamma_5 \not{D}\varphi_n(x)}{\lambda_n + im} \\
&= 2\sum_n \frac{\varphi_n^\dagger(x)\gamma_5(\not{D} + im)\varphi_n(x)}{\lambda_n + im} - 2m\sum_n \frac{\varphi_n^\dagger(x)i\gamma_5\varphi_n(x)}{\lambda_n + im} \\
&= -\frac{1}{16\pi^2}\varepsilon^{\mu\nu\alpha\beta} \text{ tr } F_{\mu\nu}F_{\alpha\beta} + 2m\langle\bar{\psi}(x)\gamma_5\psi(x)\rangle,
\end{aligned}
\tag{5.98}
$$

where we used the regularization result (5.68). So we get the anomaly and the symmetry explicitly breaking mass term

$$
\langle\bar{\psi}(x)[\not{D}, \gamma_5]\psi(x)\rangle = \mathcal{A}[A_\mu](x) + 2im\langle P(x)\rangle. \qquad \text{Q.E.D.} \tag{5.99}
$$

5.6 Non-Abelian anomaly

So far we have treated the singlet (or Abelian) anomaly case. But also the non-Abelian anomaly is determined by the PI measure when performing a non-Abelian gauge transformation.

We use the following non-Abelian Lagrangian

$$
\mathcal{L}_{na} = \bar{\psi}i \not{D}\psi, \tag{5.100}
$$

where the Dirac operator

$$
\not{D} = \not{\partial} + \not{\mathcal{V}} + \not{\mathcal{A}}\gamma_5 \tag{5.101}
$$

now contains a vector- and an axial gauge potential

$$
\mathcal{V}_\mu = \mathcal{V}_\mu^a T^a, \qquad \mathcal{A}_\mu = \mathcal{A}_\mu^a T^a. \tag{5.102}
$$

We consider the massless case for simplicity, $m = 0$. This Dirac operator, however, is *not* Hermitian in Euclidean space (we described our notations in Section 5.1)

$$
\not{D}^\dagger = \not{\partial} + \not{\mathcal{V}} - \not{\mathcal{A}}\gamma_5 \neq \not{D}, \tag{5.103}
$$

hence

$$
\not{D}^\dagger(\mathcal{V}, \mathcal{A}) = \not{D}(\mathcal{V}, -\mathcal{A}). \tag{5.104}
$$

So the Dirac operator has no well-defined eigenvalue problem and we cannot use $\not{D}^2$ for the regularization procedure.

One way out is to work with Laplacian operators

$$
\begin{aligned}
\not{D}^\dagger \not{D} &: \not{D}^\dagger \not{D}\varphi_n = \lambda_n^2\varphi_n \\
\not{D} \not{D}^\dagger &: \not{D} \not{D}^\dagger\phi_n = \lambda_n^2\phi_n,
\end{aligned}
\tag{5.105}
$$

where

$$\not{D}\varphi_n = \lambda_n \phi_n \qquad \text{and} \qquad \not{D}^\dagger \phi_n = \lambda_n \varphi_n. \tag{5.106}$$

They are Hermitian and have well-defined eigenstates. The regularization is performed in a gauge covariant way and thus the regularized Jacobian supplies exactly the **covariant anomaly**. Again, we describe Fujikawa's procedure [Fujikawa 1984, 1985]; for further literature see [Adam 1990], [Banerjee, Banerjee, Mitra 1986].

Path integral measure: We expand the spinor fields with respect to the eigenfunctions of the Laplacians

$$\begin{aligned} \psi(x) &= \sum_n a_n \varphi_n(x) = \sum_n a_n \langle x|\varphi_n\rangle \\ \bar{\psi}(x) &= \sum_m \phi_m^\dagger(x) \bar{b}_m = \sum_m \langle \phi_m|x\rangle \bar{b}_m. \end{aligned} \tag{5.107}$$

The function sets $\{\varphi_n\}$, $\{\phi_m\}$ form a complete orthonormal system and, as before, the coefficients a_n, $\bar{b}_m$ are independent Grassmann elements.

Then the PI measure and the action change to

$$d\psi d\bar{\psi} = [\det\langle\phi_m|x\rangle \det\langle x|\varphi_n\rangle]^{-1} \prod_n da_n \prod_m d\bar{b}_m \tag{5.108}$$

$$\int dx\, \bar{\psi} i \not{D} \psi = \sum_n i\lambda_n \bar{b}_n a_n \tag{5.109}$$

so that the PI—the generating functional—or the determinant of the Dirac operator is formally defined by the Grassmann integral

$$\begin{aligned} \det i\not{D} &= \int d\psi d\bar{\psi} \exp\left[\int dx \bar{\psi} i \not{D}\psi\right] \\ &= [\det \phi^\dagger \cdot \det\varphi]^{-1} \int \prod_n da_n d\bar{b}_n \exp\left[\sum_n i\lambda_n \bar{b}_n a_n\right] \\ &= [\det \phi^\dagger \cdot \det\varphi]^{-1} \prod_n i\lambda_n. \end{aligned} \tag{5.110}$$

The quantities φ and $\phi^\dagger$ denote the (infinite dimensional) transformation matrices

$$\varphi := \langle x|\varphi_n\rangle, \qquad \phi^\dagger := \langle\phi_n|x\rangle. \tag{5.111}$$

Their determinants $\det\varphi$, $\det\phi^\dagger$ in front of the Grassmann integral play no role in the calculation of the anomaly.

Chiral transformation: Now we perform a **non-Abelian axial gauge transformation**—a local chiral rotation

$$\begin{aligned}\psi'(x) &= e^{-\beta(x)\gamma_5}\psi(x) \\ \bar{\psi}'(x) &= \bar{\psi}(x)e^{-\beta(x)\gamma_5},\end{aligned} \tag{5.112}$$

where $\beta(x) = \beta^a(x)T^a$ denotes a non-Abelian gauge element and we choose β infinitesimal. Then the Grassmann expansion coefficients transform as

$$\begin{aligned}a'_n &= \sum_m C_{nm}a_m \\ \bar{b}'_m &= \sum_n D_{nm}\bar{b}_n,\end{aligned} \tag{5.113}$$

with

$$\begin{aligned}C_{nm} &= \delta_{nm} - \int dx\ \varphi_n^\dagger(x)\beta(x)\gamma_5\varphi_m(x) \\ D_{nm} &= \delta_{nm} - \int dx\ \phi_n^\dagger(x)\beta(x)\gamma_5\phi_m(x).\end{aligned} \tag{5.114}$$

Note that $\beta = \beta^a T^a$ contains the gauge group generator T^a, a matrix acting on the eigenfunctions φ_m or ϕ_m. Thus β is sandwiched between the eigenfunctions.

The Grassmann measure transforms with the inverse determinant

$$\begin{aligned}\prod_n da'_n &= [\det C]^{-1}\prod_n da_n \\ \prod_m d\bar{b}'_m &= [\det D]^{-1}\prod_m d\bar{b}_m\end{aligned} \tag{5.115}$$

and we re-express the inverse determinants in the familiar way

$$\begin{aligned}[\det C]^{-1} &= \exp\left[-\ \mathrm{tr}\ \ln\left(\delta_{nm} - \int dx\ \varphi_n^\dagger(x)\beta(x)\gamma_5\varphi_m(x)\right)\right] \\ &= \exp\left[\int dx \sum_n \varphi_n^\dagger(x)\beta(x)\gamma_5\varphi_n(x)\right]\end{aligned} \tag{5.116}$$

$$[\det D]^{-1} = \exp\left[\int dx \sum_n \phi_n^\dagger(x)\beta(x)\gamma_5\phi_n(x)\right]. \tag{5.117}$$

So we get a change in the **PI measure**

$$\begin{aligned} d\psi' d\bar{\psi}' &= [\det C \cdot \det D]^{-1} d\psi d\bar{\psi} \\ &=: J[\beta] d\psi d\bar{\psi}, \end{aligned} \tag{5.118}$$

with the **Jacobian**

$$\begin{aligned} J[\beta] &= [\det C \cdot \det D]^{-1} \\ &= \exp\left[\int dx \sum_n (\varphi_n^\dagger(x)\beta(x)\gamma_5\varphi_n(x) + \phi_n^\dagger(x)\beta(x)\gamma_5\phi_n(x))\right]. \end{aligned} \tag{5.119}$$

Regularization: We regularize the Jacobian, which is still ill-defined, with a Gaussian cut-off (analogous to the procedure we described before in Section 5.2)

$$\begin{aligned} &\sum_n (\varphi_n^\dagger(x)\beta(x)\gamma_5\varphi_n(x) + \phi_n^\dagger(x)\beta(x)\gamma_5\phi_n(x)) \\ &= \lim_{M\to\infty} \sum_n \exp\left[-\frac{\lambda_n^2}{M^2}\right] (\varphi_n^\dagger \beta\gamma_5\varphi_n + \phi_n^\dagger\beta\gamma_5\phi_n) \\ &= \lim_{M\to\infty} \sum_n \left(\varphi_n^\dagger\beta\gamma_5 \exp\left[-\frac{\not{D}^\dagger \not{D}}{M^2}\right]\varphi_n + \phi_n^\dagger\beta\gamma_5\exp\left[-\frac{\not{D}\not{D}^\dagger}{M^2}\right]\phi_n\right) \\ &= \lim_{M\to\infty} \int \frac{d^4k}{(2\pi)^4} \operatorname{Tr}\left[\beta e^{-ikx}\gamma_5\left(\exp\left[-\frac{\not{D}^\dagger\not{D}}{M^2}\right] + \exp\left[-\frac{\not{D}\not{D}^\dagger}{M^2}\right]\right)e^{ikx}\right]. \end{aligned} \tag{5.120}$$

Chirality split: Now the trick is to split the Laplacians into positive and negative chirality parts. We will recall the chiral decompositions which we introduced in Section 4.1.2:

projection operator

$$P_\pm = \frac{1}{2}(\mathbf{1} \pm \gamma_5) \qquad \text{with } P_\pm^2 = P_\pm, \quad P_+P_- = 0, \quad P_+ + P_- = \mathbf{1}, \tag{5.121}$$

chiral spinors

$$\psi_\pm = P_\pm\psi \qquad \text{with } \gamma_5\psi_\pm = \pm\psi_\pm, \tag{5.122}$$

chiral gauge fields (or L-, R-handed)

$$A_\mu^+ = \mathcal{V}_\mu + \mathcal{A}_\mu, \qquad A_\mu^- = \mathcal{V}_\mu - \mathcal{A}_\mu, \tag{5.123}$$

Lagrangian

$$\begin{aligned} \mathcal{L}_{na} &= \bar{\psi} i \not{D} \psi = \bar{\psi} i (\not{\partial} + \not{\mathcal{V}} + \not{\mathcal{A}} \gamma_5) \psi \\ &\equiv \bar{\psi}_+ i \not{D}_+ \psi_+ + \bar{\psi}_- i \not{D}_- \psi_-, \end{aligned} \tag{5.124}$$

operators

$$\not{D}_\pm = \not{\partial} + \not{A}_\pm . \tag{5.125}$$

Using the properties

$$\begin{aligned} P_+ (\not{\mathcal{V}} + \not{\mathcal{A}} \gamma_5) &= (\not{\mathcal{V}} - \not{\mathcal{A}}) P_- = \not{A}_- P_- \\ P_- (\not{\mathcal{V}} + \not{\mathcal{A}} \gamma_5) &= (\not{\mathcal{V}} + \not{\mathcal{A}}) P_+ = \not{A}_+ P_+ \end{aligned} \tag{5.126}$$

we split the Dirac operator and its adjoint into + and − chiral fields

$$\not{D} = \not{\partial} + (P_+ + P_-)(\not{\mathcal{V}} + \not{\mathcal{A}} \gamma_5) = \not{\partial} + \not{A}_+ P_+ + \not{A}_- P_- \tag{5.127}$$

$$\not{D}^\dagger (\mathcal{V}, \mathcal{A}) = \not{D}(\mathcal{V}, -\mathcal{A}) = \not{\partial} + \not{A}_- P_+ + \not{A}_+ P_- . \tag{5.128}$$

Then we find for the Laplacians

$$\not{D}^\dagger \not{D} = \not{D}_+^2 P_+ + \not{D}_-^2 P_- \tag{5.129}$$

$$\not{D} \not{D}^\dagger = \not{D}_+^2 P_- + \not{D}_-^2 P_+ \tag{5.130}$$

and for the exponentials

$$\begin{aligned} &\exp\left[-\frac{\not{D}^\dagger \not{D}}{M^2}\right] + \exp\left[-\frac{\not{D} \not{D}^\dagger}{M^2}\right] \\ &= (P_+ + P_-) \exp\left[-\frac{\not{D}_+^2}{M^2}\right] + (P_+ + P_-) \exp\left[-\frac{\not{D}_-^2}{M^2}\right] \\ &= \exp\left[-\frac{\not{D}_+^2}{M^2}\right] + \exp\left[-\frac{\not{D}_-^2}{M^2}\right], \end{aligned} \tag{5.131}$$

where we have made extensive use of the projection operator properties (5.121).

Returning to the regularized sum (5.120), with this separation (5.131) into + and − chiral operators we obtain

$$\begin{aligned} &\sum_n (\varphi_n^\dagger(x) \beta(x) \gamma_5 \varphi_n(x) + \phi_n^\dagger(x) \beta(x) \gamma_5 \phi_n(x)) \\ &= \lim_{M \to \infty} \int \frac{d^4 k}{(2\pi)^4} \operatorname{Tr} \left[\beta e^{-ikx} \gamma_5 \left(\exp\left[-\frac{\not{D}_+^2}{M^2}\right] + \exp\left[-\frac{\not{D}_-^2}{M^2}\right] \right) e^{ikx} \right]. \end{aligned} \tag{5.132}$$

We have reduced the evaluation of the sum to that of the singlet anomaly for a given chiral field. We have described this evaluation already in Section 5.2 and we just have to take over the result (5.68) which represents the **covariant anomaly**

$$\begin{aligned}\sum_n &(\varphi_n^\dagger(x)\beta(x)\gamma_5\varphi_n(x) + \phi_n^\dagger(x)\beta(x)\gamma_5\phi_n(x)) \\ &= -\frac{1}{32\pi^2}\varepsilon^{\mu\nu\alpha\beta}\,\mathrm{tr}\,\beta(F^+_{\mu\nu}F^+_{\alpha\beta} + F^-_{\mu\nu}F^-_{\alpha\beta}) \qquad (5.133)\end{aligned}$$

(with $\varepsilon^{1234} = \varepsilon^{1230} = 1$). $F^\pm_{\mu\nu}$ denotes the field strength corresponding to the chiral gauge field $A^\pm_\mu$. We will discuss the covariant anomaly in detail in Chapter 10.

Gauge transformation: However, the covariant anomaly also occurs in the vector current. Its calculation proceeds quite similarly. We perform a **non-Abelian vector gauge transformation**

$$\begin{aligned}\psi'(x) &= e^{-\alpha(x)}\psi(x) \\ \bar\psi(x) &= \bar\psi(x)e^{\alpha(x)}, \qquad (5.134)\end{aligned}$$

with $\alpha(x) = \alpha^a(x)T^a$ (remember $T^{a\dagger} = -T^a$, thus $\alpha^\dagger = -\alpha$). Then the Grassmann expansion coefficients transform with the matrix

$$\begin{aligned}C_{nm} &= \delta_{nm} - \int dx\, \varphi_n^\dagger(x)\alpha(x)\varphi_m(x) \\ D_{nm} &= \delta_{nm} + \int dx\, \phi_n^\dagger(x)\alpha(x)\phi_m(x), \qquad (5.135)\end{aligned}$$

which leads to the **Jacobian**

$$\begin{aligned}J[\alpha] &= [\det C \cdot \det D]^{-1} \\ &= \exp\left[\int dx \sum_n (\varphi_n^\dagger(x)\alpha(x)\varphi_n(x) - \phi_n^\dagger(x)\alpha(x)\phi_n(x))\right]. \qquad (5.136)\end{aligned}$$

Regularization: We regularize the sum as before

$$\begin{aligned}\sum_n &(\varphi_n^\dagger(x)\alpha(x)\varphi_n(x) - \phi_n^\dagger(x)\alpha(x)\phi_n(x)) \\ &= \lim_{M\to\infty}\int\frac{d^4k}{(2\pi)^4}\,\mathrm{Tr}\left[\alpha e^{-ikx}\left(\exp\left[-\frac{\not{D}^\dagger\not{D}}{M^2}\right] - \exp\left[-\frac{\not{D}\not{D}^\dagger}{M^2}\right]\right)e^{ikx}\right] \qquad (5.137)\end{aligned}$$

and we separate the Laplacians into positive and negative chirality parts (recall equations (5.129), (5.130))

$$\begin{aligned}
&\exp\left[-\frac{\not{D}^\dagger\not{D}}{M^2}\right]-\exp\left[-\frac{\not{D}\not{D}^\dagger}{M^2}\right]\\
&= (P_+-P_-)\exp\left[-\frac{\not{D}_+^2}{M^2}\right]+(P_--P_+)\exp\left[-\frac{\not{D}_-^2}{M^2}\right]\\
&= \gamma_5\left(\exp\left[-\frac{\not{D}_+^2}{M^2}\right]-\exp\left[-\frac{\not{D}_-^2}{M^2}\right]\right).
\end{aligned} \tag{5.138}$$

Altogether we find for the regularized sum

$$\begin{aligned}
&\sum_n(\varphi_n^\dagger\alpha\varphi_n-\phi_n^\dagger\alpha\phi_n)\\
&= \lim_{M\to\infty}\int\frac{d^4k}{(2\pi)^4}\,\mathrm{Tr}\left[\alpha e^{-ikx}\gamma_5\left(\exp\left[-\frac{\not{D}_+^2}{M^2}\right]-\exp\left[-\frac{\not{D}_-^2}{M^2}\right]\right)e^{ikx}\right]
\end{aligned} \tag{5.139}$$

and the evaluation provides the **covariant anomaly** in the vector current

$$\sum_n(\varphi_n^\dagger\alpha\varphi_n-\phi_n^\dagger\alpha\phi_n^\dagger)=-\frac{1}{32\pi^2}\varepsilon^{\mu\nu\alpha\beta}\,\mathrm{tr}\,\alpha(F_{\mu\nu}^+F_{\alpha\beta}^+-F_{\mu\nu}^-F_{\alpha\beta}^-). \tag{5.140}$$

Covariant anomaly: If on the other hand we define

$$J[\Lambda]=\exp\left[\int dx\;\Lambda^a(x)\widetilde{G}^a[A_\mu](x)\right] \tag{5.141}$$

then the **covariant anomaly** $\Lambda^a\widetilde{G}^a[A_\mu]$ is given by expression (5.133) or (5.140).

We get the result for a given chirality—the case we will consider in Chapter 10—by setting the other component to zero. Accordingly, we find the **covariant anomaly for positive or negative chiral fields**

$$\Lambda_+^a\widetilde{G}^a[A_\mu^+] = -\frac{1}{32\pi^2}\varepsilon^{\mu\nu\alpha\beta}\,\mathrm{tr}\,\Lambda_+F_{\mu\nu}^+F_{\alpha\beta}^+ \tag{5.142}$$

$$\Lambda_-^a\widetilde{G}^a[A_\mu^-] = \frac{1}{32\pi^2}\varepsilon^{\mu\nu\alpha\beta}\,\mathrm{tr}\,\Lambda_-F_{\mu\nu}^-F_{\alpha\beta}^-, \tag{5.143}$$

with

$$\Lambda_+=\alpha+\beta,\qquad \Lambda_-=\alpha-\beta. \tag{5.144}$$

This result is also the correct expression in Minkowski space for the usual anomaly definition (see Chapter 10)

$$\widetilde{G}^a[A_\mu^\pm] = -D_\pm^{ba\mu} \langle j_\mu^{\pm b} \rangle_{\text{covar. reg.}} \tag{5.145}$$

since in Minkowski space the current definition changes by a factor i.

Consistent anomaly: We certainly can also find the **consistent anomaly** within the PI procedure but then we have to regularize the Jacobian in a different way.

For example, achieving a Hermitian Dirac operator by analytic continuation of $\mathcal{A} \to i\mathcal{A}$

$$\not{D} \to \not{D} = \not{\partial} + \not{V} + i \not{A}\gamma_5 \equiv \not{D}^\dagger$$

one can use as a Jacobian regulator

$$\exp\left[-\frac{\not{D}^2}{M^2}\right].$$

The authors [Andrianov, Bonora 1984a,b], [Balachandran, Marmo, Nair, Trahern 1982], [Einhorn, Jones 1984], [Gamboa Saraví, Muschietti, Schaposnik, Solomin 1984a], [Gipson 1986] essentially follow this route.

Another possibility suggested by Alvarez-Gaumé and Ginsparg [Alvarez-Gaumé, Ginsparg 1985] is to start with a different differential operator in the Lagrangian

$$i\not{D}_+ \to i\,\widehat{\not{D}} = i(\not{D}_+ + \not{\partial}_-).$$

This operator does not change the gauge theory (up to a normalization factor for the generating functional) and has a well-defined eigenvalue problem. Then the Jacobian is regularized by

$$\exp\left[-\frac{(i\,\widehat{\not{D}})^2}{M^2}\right].$$

The calculations, however, in both cases turn out to be rather tedious, but as a result the consistent anomaly in the familiar Bardeen form (4.387) occurs—up to a trivial solution (which is the gauge variation of a local polynomial in the gauge potentials and can be absorbed by redefining the vacuum functional). We will present a detailed calculation for the latter case in Section 11.5.

5.7 Heat kernel and zeta function regularization

When working with the fermionic PI there always occurs the problem of regularization. Fujikawa's Gaussian cut-off procedure is one possibi-

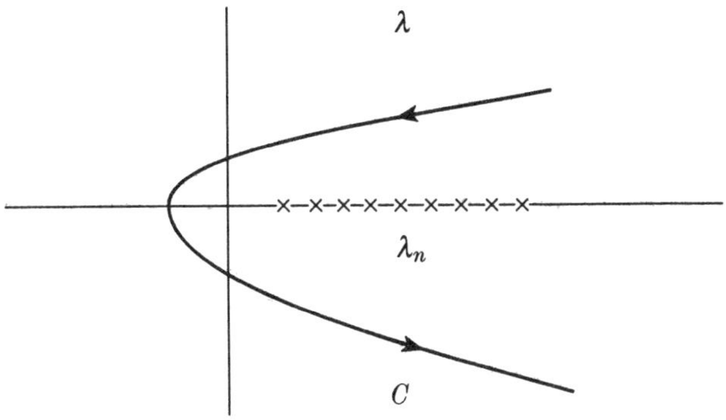

Fig. 5.1. The curve C in the complex λ-plane defining a function of the operator Δ with spectrum $\{\lambda_n\}$

lity but there also exist other techniques—the heat kernel and zeta function regularization—which are elegant and rest mathematically on sound grounds. Their connection to Fujikawa's approach is evident. These regularization techniques have been frequently used in QFT by several authors [Leutwyler 1985a,b, 1986b], [Gasser, Leutwyler 1984], [Gamboa Saraví, Muschietti, Schaposnik, Solomin 1983, 1984a,b], [Bonora, Bregola, Pasti 1985], [Adam 1990], [Ball 1989], [Reuter 1985], [Bertlmann, Launer, de Rafael 1985], [Bertlmann, Dominguez, Loewe, Perrottet, de Rafael 1988], [Hawking 1977]. For the mathematical foundations see [Gilkey 1984], [Berline, Getzler, Vergne 1992]. We shall explain these methods for the example of the anomaly in 2 dimensions where all relevant features occur and the calculations are surveyable. We also follow the work of [Adam, Bertlmann, Hofer 1993] here.

5.7.1 Heat kernel regularization

Let Δ be a **Hermitian, positive, semidefinite operator** with a well-posed eigenvalue problem

$$\Delta\varphi_n(x) = \lambda_n\varphi_n(x). \tag{5.146}$$

The eigenfunctions $\{\varphi_n(x)\}$ are orthonormal and complete

$$\int dx\ \varphi_n^*(x)\varphi_m(x) = \delta_{nm} \tag{5.147}$$

$$\sum_n \varphi_n(x)\varphi_n^*(y) = \delta(x-y) \tag{5.148}$$

and the eigenvalues λ_n are real.

Definition: A **function of the operator** Δ is defined by [Gilkey 1984]

$$f(\Delta) = \frac{1}{2\pi i}\int_C \frac{f(\lambda)}{\lambda - \Delta} d\lambda, \tag{5.149}$$

where the curve C in the complex λ-plane encloses all eigenvalues $\{\lambda_n\}$ and the origin, see Figure 5.1. Specifically, we have

$$e^{-\tau\Delta} = \frac{1}{2\pi i}\int \frac{e^{-\tau\lambda}}{\lambda - \Delta} d\lambda. \tag{5.150}$$

Heat kernel: Let us take some squared integrable function $\varphi(x) \in L^2$, then the expansion into eigenfunctions of Δ gives

$$\begin{aligned} e^{-\tau\Delta}\varphi(x) &= \int dy\, e^{-\tau\Delta}\sum_n \varphi_n(x)\varphi_n^*(y)\varphi(y) \\ &= \int dy \sum_n e^{-\lambda_n\tau}\varphi_n(x)\varphi_n^*(y)\varphi(y) \\ &=: \int dy\, G_\Delta(x,y,\tau)\varphi(y). \end{aligned} \tag{5.151}$$

The operator $e^{-\tau\Delta}$ has a kernel function—the **heat kernel**

$$G_\Delta(x,y,\tau) = \sum_n e^{-\lambda_n\tau}\varphi_n(x)\varphi_n^*(y), \tag{5.152}$$

which can also be expressed by the propagator

$$G_\Delta(x,y,\tau) = \langle x|e^{-\tau\Delta}|y\rangle. \tag{5.153}$$

The initial condition at $\tau = 0$ is

$$G_\Delta(x,y,\tau=0) = \delta(x-y) \tag{5.154}$$

due to the completeness of the eigenfunctions, equation (5.148). The heat kernel satisfies the so-called **heat equation**

$$\Delta G_\Delta(x,y,\tau) = -\frac{\partial}{\partial\tau}G_\Delta(x,y,\tau), \tag{5.155}$$

which can be quickly checked by remembering the eigenvalue equation (5.146).

We also assume that the operator Δ can generally be decomposed into (in Euclidean space)

$$\Delta = -D^2 + X, \qquad D_\mu = \partial_\mu + A_\mu, \qquad D^2 = D_\mu D_\mu, \tag{5.156}$$

where A_μ denotes the gauge potential, D_μ the covariant derivative and X some matrix valued C^∞-function which is positive Hermitian. So Δ is Hermitian and positive as required.

For the special case of

$$\Delta_0 = -\partial^2 \tag{5.157}$$

the heat equation (5.155) has the solution of the **free propagator**

$$G_{\Delta_0}(x, y, \tau) = \langle x|e^{-\tau\Delta_0}|y\rangle = \frac{1}{(4\pi\tau)^{d/2}} \exp\left[-\frac{(x-y)^2}{4\tau}\right], \tag{5.158}$$

with d the dimension of space–time (we recover the nonrelativistic result (3.21) for imaginary times $\tau = it$ in Section 3.1).

We are interested in the behaviour of the heat kernel for small τ; there the following expansion [Seeley 1967] is valid.

Theorem: Heat kernel expansion

$$G_\Delta(x, y, \tau) \overset{\tau \to 0}{=} \frac{1}{(4\pi\tau)^{d/2}} \exp\left[-\frac{(x-y)^2}{4\tau}\right] \sum_n a_n(x, y)\tau^n. \tag{5.159}$$

We shall work in 2 dimensions, $d = 2$. The coefficients $a_n(x, y)$ are called **Seeley coefficients** and they have the property

$$a_n^\dagger(x, y) = a_n(y, x) \tag{5.160}$$

since Δ is Hermitian. They certainly depend on the operator Δ and are fixed by the heat equation. The coefficient a_0 is normalized to unity

$$a_0(x, y) = 1 \tag{5.161}$$

due to the initial condition (5.154). When we substitute the expansion (5.159) into the heat equation (5.155) we find:

Lemma: Recursion relation for the diagonal Seeley coefficients ($x = y$)

$$-\Delta\, a_{n-1}(x, x) = n\, a_n(x, x). \tag{5.162}$$

This is immediately evident when we remember that the Gaussian in expansion (5.159) approaches a δ-function for $\tau \to 0$.

Jacobian: Considering now the path integral we want to regularize the Jacobian (5.54). Corresponding to Fujikawa's procedure here we choose as the heat kernel operator

$$\Delta = \not{D}^2, \tag{5.163}$$

with

$$\not{D} = \not{\partial} - i \not{A} \tag{5.164}$$

for the 2-dimensional case ($\not{D}$ is already Hermitian, $\not{D}^\dagger = \not{D}$). Then we get for the **Jacobian**

$$\begin{aligned} -\ln J[\beta] &= 2i \int dx\ \beta(x) \sum_n \varphi_n^\dagger(x)\gamma_5\varphi_n(x) \\ &= \lim_{\tau\to 0} 2i \int dx\ \beta(x) \sum_n \varphi_n^\dagger(x)\gamma_5 e^{-\lambda_n\tau}\varphi_n(x) \\ &= \lim_{\tau\to 0} 2i \int dx\ \beta(x)\ \mathrm{tr}\ \gamma_5 G_{\not{D}^2}(x,x,\tau) \\ &= \lim_{\tau\to 0} 2i\ \mathrm{Tr}\ \beta\gamma_5 G_{\not{D}^2}, \end{aligned} \tag{5.165}$$

which is just the trace of the heat kernel. Here the notation Tr means the trace with respect to the full operator structure, whereas tr denotes the trace over the Dirac matrices.

Using expansion (5.159) (with $d = 2$) the Jacobian is determined by the Seeley coefficient a_1

$$-\ln J[\beta] = 2i \int dx\ \beta(x)\frac{1}{4\pi}\ \mathrm{tr}\ \gamma_5 a_1(x,x). \tag{5.166}$$

On the other hand, we can calculate the Seeley coefficient a_1 via the recursion relation

$$a_1(x,x) = -\not{D}^2 a_0(x,x) = -\not{D}^2. \tag{5.167}$$

Inserting the Dirac operator (5.164) we have (recall relation (5.74))

$$a_1(x,x) = -D_\mu D^\mu + \frac{i}{4}[\gamma^\mu,\gamma^\nu]F_{\mu\nu}, \tag{5.168}$$

where only the second term contributes to the trace via the formula

$$\mathrm{tr}\ \gamma_5\gamma_\mu\gamma_\nu = -2\varepsilon_{\mu\nu}. \tag{5.169}$$

Then with our definition of the anomaly

$$-\ln J[\beta] = \int dx\ \beta(x)\mathcal{A}[A_\mu](x) \tag{5.170}$$

we find that the Seeley coefficient a_1 already supplies the **anomaly** result

$$\begin{aligned}\mathcal{A}[A_\mu](x) &= \frac{i}{2\pi}\,\mathrm{tr}\;\gamma_5 a_1(x,x)\\ &= \frac{1}{2\pi}\varepsilon_{\mu\nu}F^{\mu\nu}\end{aligned} \tag{5.171}$$

in accordance with our calculations à la Fujikawa in Section 5.3.

5.7.2 Zeta function regularization

Another method to give meaning to the fermionic path integral—the determinant of the Dirac operator—is the zeta function approach. The determinant

$$\det i\,\not{D} = \int d\psi d\bar{\psi} \exp\left[\int dx\;\bar{\psi} i\,\not{D}\psi\right] = \prod_n i\lambda_n \tag{5.172}$$

diverges because of the unbounded product of increasing eigenvalues λ_n of the operator $\not{D}$. So we have to regularize the determinant, the zeta function serves as a convenient technique.

Let Δ again be a Hermitian, positive, semidefinite operator with eigenvalues λ_n and orthonormal, complete eigenfunctions $\{\varphi_n\}$, equations (5.146)–(5.148). But this restriction can also be relaxed. What we actually require is that Δ is an elliptic and invertible operator (see Section 11.2) defined on a compact boundaryless manifold. (So we always assume the fields of the theory to behave at infinity such that a compactification of $\mathbf{R}^m$ is possible, e.g. to S^m.)

Definition: The **zeta function of the operator** Δ is defined by the following sum over the eigenvalues

$$\zeta_\Delta(s) = \sum_n \frac{1}{\lambda_n^s}. \tag{5.173}$$

The sum converges only for sufficiently large positive real part Re s, otherwise it is defined by analytic continuation into the complex s-plane. In particular, $\zeta_\Delta(s)$ and its derivative are regular at $s = 0$. Formally we have

$$\begin{aligned}-\frac{d}{ds}\zeta_\Delta(s)\bigg|_{s=0} &= \sum_n \ln\lambda_n e^{-s\ln\lambda_n}\bigg|_{s=0} = \sum_n \ln\lambda_n\\ &= \ln\prod_n \lambda_n = \ln\det\Delta\end{aligned} \tag{5.174}$$

so that we can define the **regularized determinant** or the **regularized path integral**

$$\int d\psi d\bar{\psi} \exp\left[\left|\int dx\, \bar{\psi} i \not{D} \psi\right|\right] = \det{}^R i \not{D} \equiv \exp\left[-\zeta'_{i \not{D}}(0)\right], \qquad (5.175)$$

where $\zeta'_\Delta(s) := \frac{d}{ds}\zeta_\Delta(s)$.

Generally, we do not know the spectrum $\{\lambda_n\}$ explicitly, hence we cannot work with the sum (5.173). But we can construct another analytic expression for $\zeta_\Delta(s)$ which we are allowed to continue to $s = 0$ in order to achieve the regularization of the determinant (5.175).

Laplace transform: Let us search for the Laplace transform of $\zeta_\Delta(s)$, generally, for the transform

$$f(\lambda) = \int_0^\infty d\tau\, F(\tau) e^{-\lambda\tau}. \qquad (5.176)$$

We remember the integral representation of the Γ-function

$$\Gamma(s) = \int_0^\infty d\tau\, \tau^{s-1} e^{-\tau} \qquad (5.177)$$

and we shift the integration variable $\tau \to \lambda\tau$

$$\Gamma(s) = \lambda^s \int_0^\infty d\tau\, \tau^{s-1} e^{-\lambda\tau}, \qquad \mathrm{Re}\, \lambda,\ \mathrm{Re}\, s > 0. \qquad (5.178)$$

Then the function

$$f(\lambda) = \frac{1}{\lambda^s} \qquad (5.179)$$

has the Laplace transform

$$F(\tau) = \frac{1}{\Gamma(s)} \tau^{s-1} \qquad (5.180)$$

since

$$\frac{1}{\lambda^s} = \frac{1}{\Gamma(s)} \int_0^\infty d\tau\, \tau^{s-1} e^{-\lambda\tau}. \qquad (5.181)$$

For the ζ-function we obtain

$$\zeta_\Delta(s) = \frac{1}{\Gamma(s)} \int_0^\infty d\tau\, \tau^{s-1} \sum_n e^{-\lambda_n \tau}. \qquad (5.182)$$

The summation, on the other hand, is nothing but the **trace of the operator**

$$\mathrm{Tr}\, e^{-\tau\Delta} = \sum_n \langle n| e^{-\tau\Delta} |n\rangle = \sum_n e^{-\lambda_n \tau}, \qquad (5.183)$$

with $|n\rangle$ the complete, orthonormal eigenstates of Δ

$$\Delta|n\rangle = \lambda_n|n\rangle. \tag{5.184}$$

In x-space we have to integrate over the diagonal heat kernel

$$\begin{aligned}\mathrm{Tr}\; e^{-\tau\Delta} &= \int dx\, \langle x|e^{-\tau\Delta}|x\rangle \\ &= \int dx \sum_n \langle x|e^{-\tau\Delta}|n\rangle\langle n|x\rangle = \int dx \sum_n e^{-\lambda_n\tau}\varphi_n(x)\varphi_n^*(x) \\ &= \int dx\; G_\Delta(x,x,\tau),\end{aligned} \tag{5.185}$$

which again is clearly the sum (5.183) due to the normalization (5.147). Altogether we find for the ζ-function the following **Laplace transform**

$$\begin{aligned}\zeta_\Delta(s) &= \frac{1}{\Gamma(s)}\int_0^\infty d\tau\; \tau^{s-1}\; \mathrm{Tr}\; e^{-\tau\Delta} \\ &= \frac{1}{\Gamma(s)}\int_0^\infty d\tau\; \tau^{s-1}\int dx\; G_\Delta(x,x,\tau).\end{aligned} \tag{5.186}$$

Trace: We can also formulate the ζ-function by the **trace of the Δ operator**

$$\zeta_\Delta(s) = \sum_n \lambda_n^{-s} = \mathrm{Tr}\; \Delta^{-s}. \tag{5.187}$$

For complex powers of s the operator is defined by (see equation (5.149))

$$\Delta^s = \frac{1}{2\pi i}\int_C \frac{\lambda^s}{\lambda-\Delta}d\lambda. \tag{5.188}$$

The functional trace then means

$$\begin{aligned}\mathrm{Tr}\;\Delta &= \int dx \sum_n \lambda_n\varphi_n(x)\varphi_n^*(x) =: \int dx\; K_\Delta(x,x,s=1) \\ \mathrm{Tr}\;\Delta^{-s} &= \int dx \sum_n \lambda_n^{-s}\varphi_n(x)\varphi_n^*(x) =: \int dx\; K_\Delta(x,x,-s).\end{aligned} \tag{5.189}$$

For an infinitesimally varied operator we may also **expand the trace** in the following way

$$\mathrm{Tr}\;(\Delta+\delta\Delta)^{-s} = \mathrm{Tr}\;\Delta^{-s} - s\;\mathrm{Tr}\;\Delta^{-s-1}\delta\Delta + O((\delta\Delta)^2). \tag{5.190}$$

Jacobian: Now we return to the path integral and we perform a local chiral transformation

$$\begin{aligned} \psi' &= e^{i\beta(x)\gamma_5}\psi \\ \bar{\psi}' &= \bar{\psi}e^{i\beta(x)\gamma_5}. \end{aligned} \tag{5.191}$$

Then the determinants are related by

$$\begin{aligned} \det i\not{D} &= \int d\psi' d\bar{\psi}' \exp\left[\int dx\ \bar{\psi}' i\not{D}\psi'\right] \\ &= \int J[\beta] d\psi d\bar{\psi} \exp\left[\int dx\ \bar{\psi}e^{i\beta(x)\gamma_5} i\not{D}e^{i\beta(x)\gamma_5}\psi\right] \\ &= J[\beta] \det\left[e^{i\beta(x)\gamma_5} i\not{D}e^{i\beta(x)\gamma_5}\right], \end{aligned} \tag{5.192}$$

with $J[\beta]$ the Jacobian associated to the chiral rotation (5.191).

In the ζ-function approach we work with the regularized determinants according to definition (5.175). So in this case the chiral **Jacobian** represents the ratio of two regularized determinants

$$J_{i\not{D}}[\beta] = \frac{\det{}^R i\not{D}}{\det{}^R e^{i\beta\gamma_5} i\not{D}e^{i\beta\gamma_5}} \tag{5.193}$$

and for β infinitesimally small

$$J_{i\not{D}}[\beta] = \frac{\det{}^R i\not{D}}{\det{}^R i(\not{D} + \{i\beta\gamma_5, \not{D}\})}. \tag{5.194}$$

The Jacobian is for this reason a finite quantity. In Fujikawa's procedure, on the other hand, we just had to regularize an ill-defined Jacobian (and there several regularization choices were possible).

Zeta function and Fujikawa procedure: In order to establish the bridge between the ζ-function and the Fujikawa procedure we have to work with the Hermitian, positive semidefinite operator $\not{D}^2$.

Generally, for a non-Hermitian but elliptic operator D we can regularize the determinant as

$$\det{}^R D = \exp\left[-\frac{1}{2}\zeta'_{D^\dagger D}(0)\right], \tag{5.195}$$

which we infer from conjugate squaring the determinants

$$\det{}^R D (\det{}^R D)^* = \det{}^R D^\dagger D = \exp\left[-\zeta'_{D^\dagger D}(0)\right]. \tag{5.196}$$

The operator $D^\dagger D$ is Hermitian and semipositive, which we need to trace back the ζ-function to the heat kernel, equation (5.186).

In our case we have $D = i\,\not{D}$ (anti-Hermitian) and the determinant is

$$\det{}^R i\,\not{D} = \exp\left[-\frac{1}{2}\zeta'_{\not{D}^2}(0)\right]. \tag{5.197}$$

Then the logarithm of the Jacobian (5.194) arising from the chiral rotation becomes

$$\begin{aligned} \ln J_{i\,\not{D}}[\beta] &= \ln\det{}^R i\,\not{D} - \ln\det{}^R i(\not{D} + \{i\beta\gamma_5, \not{D}\}) \\ &= -\frac{d}{ds}\bigg|_{s=0} \frac{1}{2}[\zeta_{\not{D}^2}(s) - \zeta_{\not{D}^2+\bar{\delta}\,\not{D}^2}(s)], \end{aligned} \tag{5.198}$$

with

$$\bar{\delta}\,\not{D}^2 = \{i\beta\gamma_5, \not{D}^2\} + 2\,\not{D}i\beta\gamma_5\,\not{D} \tag{5.199}$$

a small correction. Expressing the ζ-function by the trace of the operator and expanding the trace according to equation (5.190) we find

$$\begin{aligned} \ln J_{i\,\not{D}}[\beta] &= -\frac{d}{ds}\bigg|_{s=0} \frac{1}{2}[\mathrm{Tr}\,(\not{D}^2)^{-s} - \mathrm{Tr}\,(\not{D}^2 + \bar{\delta}\,\not{D}^2)^{-s}] \\ &= -\frac{d}{ds}\bigg|_{s=0} \frac{1}{2}s\ \mathrm{Tr}\,(\not{D}^2)^{-s-1}\bar{\delta}\,\not{D}^2 \\ &= -\frac{d}{ds}\bigg|_{s=0} 2is\ \mathrm{Tr}\,(\not{D}^2)^{-s}\beta\gamma_5, \end{aligned} \tag{5.200}$$

where we have used the cyclic property of the trace.

Next we use the equality of the following Jacobians.

Lemma:

$$J_{i\,\not{D}}[\beta] \equiv J_{\not{D}^2}[\beta]. \tag{5.201}$$

The Jacobian $J_{\not{D}^2}[\beta]$ means chirally rotated with respect to $\not{D}^2$

$$J_{\not{D}^2}[\beta] = \frac{\det{}^R\,\not{D}^2}{\det{}^R(\not{D}^2 + \{i\beta\gamma_5, \not{D}^2\})}, \tag{5.202}$$

with the determinants regularized according to definition (5.175).

Proof: Denoting

$$\delta\,\not{D}^2 = \{i\beta\gamma_5, \not{D}^2\} \tag{5.203}$$

we calculate

$$\begin{aligned}
\ln J_{\not{D}^2}[\beta] &= \ln \det{}^R \not{D}^2 - \ln \det{}^R(\not{D}^2 + \delta \not{D}^2) \\
&= -\frac{d}{ds}\bigg|_{s=0} [\zeta_{\not{D}^2}(s) - \zeta_{\not{D}^2+\delta \not{D}^2}(s)] \\
&= -\frac{d}{ds}\bigg|_{s=0} [\mathrm{Tr}\ (\not{D}^2)^{-s} - \mathrm{Tr}\ (\not{D}^2 + \delta \not{D}^2)^{-s}] \\
&= -\frac{d}{ds}\bigg|_{s=0} 2is\ \mathrm{Tr}\ (\not{D}^2)^{-s}\beta\gamma_5, \qquad (5.204)
\end{aligned}$$

which coincides with equation (5.200). Q.E.D.

The trace (5.200), (5.204) explicitly means

$$\begin{aligned}
\mathrm{Tr}\ (\not{D}^2)^{-s}\beta\gamma_5 &= \int dx\ \mathrm{tr} \sum_n \lambda_n^{-s}\varphi_n(x)\varphi_n^\dagger(x)\beta(x)\gamma_5 \\
&= \int dx\ \mathrm{tr}\ K_{\not{D}^2}(x,x,-s)\beta(x)\gamma_5. \qquad (5.205)
\end{aligned}$$

Since the trace of the kernel $K_{\not{D}^2}(x,x,-s)$ is regular at $s=0$ [Seeley 1967] we also get

$$-\frac{d}{ds}\bigg|_{s=0} 2is\ \mathrm{Tr}\ (\not{D}^2)^{-s}\beta\gamma_5 = -2i\ \mathrm{Tr}\ (\not{D}^2)^{-s}\beta\gamma_5\big|_{s=0}. \qquad (5.206)$$

Now we relate the trace (5.200), (5.204) to the heat kernel. We insert the Laplace transform (5.181) for λ_n^{-s} into expression (5.205) and we obtain

$$\begin{aligned}
&\mathrm{Tr}\ (\not{D}^2)^{-s}\beta\gamma_5 \\
&= \frac{1}{\Gamma(s)}\int dx \int_0^\infty d\tau\ \tau^{s-1}\ \mathrm{tr} \sum_n e^{-\lambda_n\tau}\varphi_n(x)\varphi_n^\dagger(x)\beta(x)\gamma_5 \\
&= \frac{1}{\Gamma(s)}\int dx \int_0^\infty d\tau\ \tau^{s-1}\ \mathrm{tr}\ G_{\not{D}^2}(x,x,\tau)\beta(x)\gamma_5. \qquad (5.207)
\end{aligned}$$

The heat kernel $G_{\not{D}^2}(x,x,\tau)$ is falling off exponentially for large τ. So we may cut off the integral at some finite value $\tau = T$ since the rest-integral vanishes in the limit $s \to 0$ where we have an overall factor $1/\Gamma(s) = s + O(s^2)$. In fact, only the lower boundary of the integral will contribute in the limit $s \to 0$.

Inserting the heat kernel expansion (5.159) we find for the Jacobian

$$
\begin{aligned}
-\ln J_{\not{D}^2}[\beta] \\
= \quad 2is \int dx \int_0^T d\tau\, \tau^{s-1} \operatorname{tr} \frac{1}{4\pi\tau} \sum_n a_n(x,x)\tau^n \beta(x)\gamma_5 \bigg|_{s=0} .
\end{aligned}
\tag{5.208}
$$

Clearly only the Seeley coefficient a_1 contributes. The a_0-term does not appear since its trace vanishes

$$
\operatorname{tr}\, a_0(x,x)\gamma_5 = \operatorname{tr}\, \gamma_5 = 0.
$$

The higher terms with $n > 1$ are of order $O(s/(s+n-1))$ and vanish too. The cut-off T may be chosen arbitrarily small so that the expansion (5.159) is valid. Finally we obtain for the **Jacobian**

$$
\begin{aligned}
-\ln J_{\not{D}^2}[\beta] \\
&= 2is \int dx \frac{1}{s} T^s \bigg|_{s=0} \operatorname{tr} \frac{1}{4\pi} a_1(x,x)\beta(x)\gamma_5 \\
&= 2i \int dx\, \beta(x) \frac{1}{4\pi} \operatorname{tr}\, \gamma_5 a_1(x,x).
\end{aligned}
\tag{5.209}
$$

This outcome coincides with the result (5.166) of the previous heat kernel procedure which is equivalent to Fujikawa's approach. Of course, result (5.209) represents the **anomaly** (recall equations (5.167)–(5.171))

$$
\mathcal{A}[A_\mu](x) = \frac{i}{2\pi} \operatorname{tr}\, \gamma_5 a_1(x,x) = \frac{1}{2\pi} \varepsilon_{\mu\nu} F^{\mu\nu}. \tag{5.210}
$$

Non-Hermitian zeta function procedure: Now we discuss a ζ-function procedure which is valid more generally. It does not rely on positivity or hermiticity of the associated operator, we only assume ellipticity and invertibility. In this case we calculate the Jacobian corresponding to the Dirac operator $i\not{D}$

$$
J_{i\not{D}}[\beta] = \frac{\det{}^R i\not{D}}{\det{}^R i(\not{D} + \{i\beta\gamma_5, \not{D}\})} \tag{5.211}
$$

and regularize the determinants according to definition (5.175). We demonstrate the procedure with our Dirac operator $\not{D} = \not{\partial} - i\not{A}$ which is chosen Hermitian in Euclidean space, so $i\not{D}$ is anti-Hermitian (see Section 5.3). But the method also works for a *non-Hermitian* $\not{D}$ containing, for example, axial fields as we had in Section 5.6 where we considered the non-Abelian anomaly.

We begin the procedure as before. We represent the determinants by the

ζ-functions associated to the operator $i \not{D}$ and the ζ-functions we express by the operator traces. Then the logarithm of the Jacobian becomes (recall equations (5.204))

$$\begin{aligned} \ln J_{i\not{D}}[\beta] &= -\frac{d}{ds}\bigg|_{s=0} s \operatorname{Tr}(i \not{D})^{-s-1}\delta(i \not{D}) \\ &= -\frac{d}{ds}\bigg|_{s=0} 2is \operatorname{Tr}(i \not{D})^{-s}\beta\gamma_5 \\ &= -2i \operatorname{Tr}(i \not{D})^{-s}\beta\gamma_5\big|_{s=0}, \end{aligned} \tag{5.212}$$

where $\delta(i \not{D})$ is a small correction

$$\delta(i \not{D}) = \{i\beta\gamma_5, i \not{D}\}. \tag{5.213}$$

To evaluate the trace (5.212) we now use the integral representation (5.188) of the operator $(i\not{D})^{-s}$. In this step we do not depend on hermiticity; this step holds for any elliptic, invertible operator. We insert plane waves as a complete system and we get for the trace

$$\begin{aligned} &\operatorname{Tr}(i \not{D})^{-s}\beta\gamma_5 \\ &= \frac{1}{2\pi i}\int_C d\lambda\, \lambda^{-s}\int dx \int \frac{d^2k}{(2\pi)^2} \operatorname{tr} \beta(x)\gamma_5 e^{-ikx}\frac{1}{\lambda - i\not{D}}e^{ikx} \\ &= \frac{1}{2\pi i}\int_C d\lambda\, \lambda^{-s}\int dx \int \frac{d^2k}{(2\pi)^2} \operatorname{tr} \beta(x)\gamma_5 \frac{1}{\lambda + \not{k} - i\not{D}}, \end{aligned} \tag{5.214}$$

where the plane waves have shifted the differential operator (recall equation (5.61)).

Next we expand the denominator according to the formula

$$\frac{1}{A-B} = \frac{1}{A} + \frac{1}{A}B\frac{1}{A} + \frac{1}{A}B\frac{1}{A}B\frac{1}{A} + \dots, \tag{5.215}$$

which gives for the trace

$$\begin{aligned} \operatorname{Tr}(i \not{D})^{-s}\beta\gamma_5 = \frac{1}{2\pi i}\int_C d\lambda\, \lambda^{-s}\int dx \int \frac{d^2k}{(2\pi)^2} \\ \cdot \operatorname{tr} \beta(x)\gamma_5 \left[\frac{1}{\lambda + \not{k}} + \frac{1}{\lambda + \not{k}} i\not{D}\frac{1}{\lambda + \not{k}} \right. \\ \left. + \frac{1}{\lambda + \not{k}} i\not{D}\frac{1}{\lambda + \not{k}} i\not{D}\frac{1}{\lambda + \not{k}} + \dots\right] \end{aligned}$$

$$= \frac{1}{2\pi i}\int_C d\lambda\, \lambda^{-s}\int dx \int \frac{d^2k}{(2\pi)^2}$$
$$\cdot \operatorname{tr}\, \beta(x)\gamma_5\lambda^2\left[\lambda^{-1}\frac{1}{1+\not{k}} + \lambda^{-2}\frac{1}{1+\not{k}}i\,\not{D}\frac{1}{1+\not{k}}\right.$$
$$\left. +\,\lambda^{-3}\frac{1}{1+\not{k}}i\,\not{D}\frac{1}{1+\not{k}}i\,\not{D}\frac{1}{1+\not{k}} + \ldots\right] \tag{5.216}$$

after rescaling the momentum $k_\mu \to \lambda k_\mu$. This corresponds to an expansion of the Dirac operator into Seeley coefficients [Seeley 1967]. We, however, want to evaluate the sum in another, more direct way which turns out to be quite simple. This method is due to Bonora, Bregola and Pasti [Bonora, Bregola, Pasti 1985].

We integrate in the complex λ-plane. Then in the limit $s \to 0$ only the third term of the expansion contributes as a residual integral. We observe that the whole expression is regular at $s = 0$ as claimed above. Using the familiar formula

$$\frac{1}{2\pi i}\int \frac{d\lambda}{\lambda} = 1 \tag{5.217}$$

we obtain

$$\begin{aligned}
&\operatorname{Tr}\,(i\,\not{D})^{-s}\beta\gamma_5\big|_{s=0} \\
&= \int dx \int \frac{d^2k}{(2\pi)^2}\operatorname{tr}\,\beta(x)\gamma_5\frac{1-\not{k}}{1-\not{k}^2}i\,\not{D}\frac{1-\not{k}}{1-\not{k}^2}i\,\not{D}\frac{1-\not{k}}{1-\not{k}^2} \\
&= \int dx \int \frac{d^2k}{(2\pi)^2}\frac{-1}{(1+k^2)^3}\operatorname{tr}\,\beta(x)\gamma_5(\not{D}\,\not{D} + k^2\,\not{D}\,\not{D}) \\
&= \int dx \int \frac{d^2k}{(2\pi)^2}\frac{-1}{(1+k^2)^2}\operatorname{tr}\,\beta(x)\gamma_5\,\not{D}^2.
\end{aligned} \tag{5.218}$$

We carry out the k-integration (note that with our Euclidean metric (5.77) we have $\not{k}\,\not{k} = -k_\mu k_\mu =: -k^2$)

$$\int \frac{d^2k}{(1+k^2)^2} = \int_0^{2\pi} d\varphi\,\frac{1}{2}\int_0^\infty \frac{dk^2}{(1+k^2)^2} = \pi \tag{5.219}$$

and we find for the trace

$$\operatorname{Tr}\,(i\,\not{D})^{-s}\beta\gamma_5\big|_{s=0} = -\int dx\,\beta(x)\,\frac{1}{4\pi}\operatorname{tr}\,\gamma_5\,\not{D}^2. \tag{5.220}$$

For the **Jacobian** (5.212), which we are finally interested in, we then have

$$\begin{aligned} -\ln J_{i\not{D}}[\beta] &= 2i \text{ Tr } (i\not{D})^{-s}\beta\gamma_5\big|_{s=0} \\ &= -2i \int dx\ \beta(x)\ \frac{1}{4\pi}\ \text{tr } \gamma_5\ \not{D}^2. \end{aligned} \tag{5.221}$$

As before, we insert the decomposition (5.74) for $\not{D}^2$

$$-\ln J_{i\not{D}}[\beta] = -\frac{1}{2}\int dx\ \beta(x)\frac{1}{2\pi}\ \text{tr } \gamma_5\gamma_\mu\gamma_\nu F^{\mu\nu} \tag{5.222}$$

and with the trace property (5.78) we finally find the **anomaly** result

$$-\ln J_{i\not{D}}[\beta] = \int dx\ \beta(x)\frac{1}{2\pi}\varepsilon_{\mu\nu}F^{\mu\nu} \tag{5.223}$$

and

$$\mathcal{A}[A_\mu] = \frac{1}{2\pi}\varepsilon_{\mu\nu}F^{\mu\nu} \tag{5.224}$$

in accordance with our previous calculations.

6

Physics in terms of differential forms

In this chapter we shall apply the mathematical formalism developed in Chapter 2—essentially differential geometry and topology—to specific examples in physics that we are familiar with. The geometry of fibre bundles describes the physics of gauge theories in a quite natural way. We shall appreciate the compactness and elegance of the whole formalism.

We begin with the familiar example of electrodynamics (Section 6.1), turn subsequently to Yang–Mills theory (Section 6.2) and formulate the anomalies in terms of differential forms (Section 6.3). We discuss the Dirac monopole (Section 6.4), the Aharonov–Bohm effect (Section 6.5) and the instantons (Section 6.6) and demonstrate how they arise as effects of a nontrivial topology of the fields.

6.1 Abelian fields, electrodynamics

Electrodynamics is certainly the most studied and utilized theory. Here we can follow the development of the theory in the course of history nicely. Maxwell's original equations formulated for the components of the electromagnetic fields were simplified at the end of the last century by introducing a curl and a divergence. At the beginning of this century the equations were already formulated in a beautiful relativistically covariant way. But in the last decades it was differential geometry that supplied us the most concise and elegant formulation of the theory. A modern treatment of electrodynamics can be found in [Thirring 1992], [Bamberg, Sternberg 1990], [Curtis, Miller 1985].

Maxwell's equations: In the standard notation the **Maxwell equations** are

$$\vec{\nabla} \times \vec{E} + \frac{\partial}{\partial t}\vec{B} = 0 \qquad \vec{\nabla}\vec{B} = 0 \qquad \text{homogeneous} \tag{6.1}$$

$$\vec{\nabla} \times \vec{B} - \frac{\partial}{\partial t}\vec{E} = \vec{\jmath} \qquad \vec{\nabla}\vec{E} = \rho \qquad \text{inhomogeneous} \tag{6.2}$$

with $\vec{E}$ and $\vec{B}$ being the electric and the magnetic field, and $\vec{\jmath}$ and ρ the current and the charge density. We introduce the relativistically covariant **field strength tensor**

$$F_{\mu\nu} = \begin{pmatrix} 0 & E_1 & E_2 & E_3 \\ -E_1 & 0 & -B_3 & B_2 \\ -E_2 & B_3 & 0 & -B_1 \\ -E_3 & -B_2 & B_1 & 0 \end{pmatrix} \tag{6.3}$$

$$F^{\mu\nu} = \begin{pmatrix} 0 & -E_1 & -E_2 & -E_3 \\ E_1 & 0 & -B_3 & B_2 \\ E_2 & B_3 & 0 & -B_1 \\ E_3 & -B_2 & B_1 & 0 \end{pmatrix}, \tag{6.4}$$

where the covariant $F_{\mu\nu}$ and contravariant $F^{\mu\nu}$ tensors are related by

$$F_{\mu\nu} = g_{\mu\alpha}F^{\alpha\beta}g_{\beta\nu} \tag{6.5}$$

and we have chosen the metric $g_{\mu\nu}$ to be Minkowskian (2.146). Next we define the **dual field strength tensor**

$$*F_{\mu\nu} = \frac{1}{2}\varepsilon_{\mu\nu\alpha\beta}F^{\alpha\beta} \tag{6.6}$$

in accordance with our dual Hodge star definition (2.148) of the differential forms

$$F = \frac{1}{2}F_{\alpha\beta}dx^{\alpha}dx^{\beta} \tag{6.7}$$

$$*F = \frac{1}{2}F_{\alpha\beta} * dx^{\alpha}dx^{\beta} \equiv \frac{1}{2} * F_{\mu\nu}dx^{\mu}dx^{\nu}. \tag{6.8}$$

In components we find (with $\varepsilon_{0123} = 1$)

$$*F_{\mu\nu} = \begin{pmatrix} 0 & -B_1 & -B_2 & -B_3 \\ B_1 & 0 & -E_3 & E_2 \\ B_2 & E_3 & 0 & -E_1 \\ B_3 & -E_2 & E_1 & 0 \end{pmatrix} \tag{6.9}$$

and we notice that the substitution

$$\vec{B} \to \vec{E}, \qquad \vec{E} \to -\vec{B} \tag{6.10}$$

implies

$$F_{\mu\nu} \to *F_{\mu\nu} \qquad \text{and} \qquad F \to *F. \tag{6.11}$$

Then the **Maxwell equations** can be rewritten in a **relativistically covariant** way

$$\partial_\mu * F^{\mu\nu} = 0 \qquad \text{homogeneous} \tag{6.12}$$

$$\partial_\mu F^{\mu\nu} = j^\nu \qquad \text{inhomogeneous,} \tag{6.13}$$

with the 4-current $j^\nu = (\rho, \vec{\jmath})$.

Differential geometry: Now we turn to the differential geometric formulation. We first express the 2-form F (6.7) in terms of the electric E_i and magnetic field B_i by inserting the explicit tensor (6.3)

$$F = E_i dx^0 dx^i - \frac{1}{2} B_i \varepsilon_{ijk} dx^j dx^k. \tag{6.14}$$

Next we apply the exterior derivative

$$dF = \left(\partial_k E_i - \frac{1}{2}\partial_0 B_j \varepsilon_{jik}\right) dx^0 dx^i dx^k - \frac{1}{2}\partial_\ell B_i \varepsilon_{ijk} dx^\ell dx^j dx^k, \tag{6.15}$$

using $dx^\ell dx^j dx^k = \varepsilon^{\ell jk} dx^1 dx^2 dx^3$ and $\varepsilon_{ijk}\varepsilon^{\ell jk} = 2\delta_i{}^\ell$ we obtain

$$dF = \frac{1}{2}(\partial_k E_i - \partial_i E_k - \partial_0 B_j \varepsilon_{jik}) dx^0 dx^i dx^k - \partial_i B_i dx^1 dx^2 dx^3. \tag{6.16}$$

We compare expression (6.16) with the homogeneous Maxwell equations (6.1) in component form

$$\varepsilon_{j\ell m}\partial_\ell E_m + \partial_0 B_j = 0 \qquad \text{and} \qquad \partial_i B_i = 0, \tag{6.17}$$

which we rewrite by multiplying with ε_{jik} and using

$$\varepsilon_{j\ell m}\varepsilon_{jik} = \delta_{\ell i}\delta_{mk} - \delta_{\ell k}\delta_{mi} \tag{6.18}$$

so that we get

$$\partial_k E_i - \partial_i E_k - \partial_0 B_j \varepsilon_{jik} = 0 \qquad \text{and} \qquad \partial_i B_i = 0. \tag{6.19}$$

Then the **homogeneous Maxwell equation in terms of differential forms** is

$$dF = 0. \tag{6.20}$$

The previous two equations (6.1) combine into one.

For the inhomogeneous equations we consider the dual field strength and its derivative which we obtain from expressions (6.14) and (6.16) by substituting the fields (equation 6.10))

$$*F = -B_i dx^0 dx^i - \frac{1}{2} E_i \varepsilon_{ijk} dx^j dx^k \tag{6.21}$$

$$d*F = \frac{1}{2}(\partial_i B_k - \partial_k B_i - \partial_0 E_j \varepsilon_{jik}) dx^0 dx^i dx^k - \partial_i E_i dx^1 dx^2 dx^3. \tag{6.22}$$

On the other hand, we have the inhomogeneous Maxwell equations (6.2) which we rewrite as before

$$\partial_i B_k - \partial_k B_i - \partial_0 E_j \varepsilon_{jik} = j_j \varepsilon_{jik} \qquad \text{and} \qquad \partial_i E_i = \rho. \tag{6.23}$$

When comparing the two equations (6.22) and (6.23) we need a 3-form—the dual current—which we calculate explicitly by remembering the dual basis (2.158) in 4-dimensional Minkowski space:

current 1-form

$$j = j_\mu dx^\mu = j_0 dx^0 - j_i dx^i \tag{6.24}$$

dual current 3-form

$$\begin{aligned} *j &= j_\mu * dx^\mu = j_0 * dx^0 - j_i * dx^i \\ &= \rho dx^1 dx^2 dx^3 - \frac{1}{2} j_j \varepsilon_{jik} dx^0 dx^i dx^k, \end{aligned} \tag{6.25}$$

with $j_0 \equiv \rho$.

Then we find

$$d*F = -*j. \tag{6.26}$$

Taking again the dual of equation (6.26), noticing that $**j = j$, and introducing the **coderivative** in 4-dimensional Minkowski space

$$\delta = *d* \qquad \text{with } \delta^2 = 0 \tag{6.27}$$

(it differs by a minus sign compared to the Euclidean formula (2.169), (2.172)) leads to the **inhomogeneous Maxwell equation in terms of differential forms**

$$\delta F = -j. \tag{6.28}$$

So both equations (6.2) are represented by the brief equation (6.28).

Gauge potential: Since F is a closed form, $dF = 0$, **Poincaré's lemma** (Sections 2.4 and 7.3) implies that, at least *locally*,

$$\exists \text{ 1-form } A \text{ such that } F = dA. \tag{6.29}$$

In components we have

$$F_{\mu\nu} = \partial_\mu A_\nu - \partial_\nu A_\mu. \tag{6.30}$$

A is the **gauge potential**. It is not uniquely defined; we have the freedom of **gauge transformation** (since $d^2 = 0$)

$$A^\Lambda = A + d\Lambda. \tag{6.31}$$

We may choose the **gauge function** Λ for a given A such that

$$\Delta\Lambda = -\delta A, \tag{6.32}$$

where $\Delta = d\delta + \delta d$ represents the Laplace operator (2.174). Then

$$\delta A^\Lambda = \delta A + \delta d\Lambda = \delta A + \Delta\Lambda = 0 \tag{6.33}$$

is known as the **Lorentz gauge**. In this case equation (6.28) gives

$$-j = \delta F = \delta dA = (d\delta + \delta d)A = \Delta A. \tag{6.34}$$

So in the Lorentz gauge the inhomogeneous Maxwell equation (6.28) corresponds to the set of equations

$$\delta A = 0 \qquad \text{and} \qquad \Delta A = -j, \tag{6.35}$$

which the gauge potential A has to fulfil.

Let us calculate the covariant components (recall equations (2.154) and (2.158)):

$$\begin{aligned}\delta A &= *d*A_\mu dx^\mu = *\partial_\sigma A^\mu \frac{1}{3!}\varepsilon_{\mu\nu\alpha\beta}dx^\sigma dx^\nu dx^\alpha dx^\beta \\ &= \partial_\lambda A^\lambda * \frac{1}{4!}\varepsilon_{\mu\nu\alpha\beta}dx^\mu dx^\nu dx^\alpha dx^\beta = -\partial_\lambda A^\lambda.\end{aligned} \tag{6.36}$$

On the other hand, we get

$$\begin{aligned}\Delta A &= (d\delta + \delta d)A = \delta dA = *d*\partial_\mu A_\nu dx^\mu dx^\nu \\ &= \frac{1}{2}\partial_\sigma\partial_\mu A_\nu \varepsilon^{\mu\nu}{}_{\alpha\beta}\varepsilon^{\sigma\alpha\beta}{}_\rho dx^\rho = -\partial_\sigma\partial^\sigma A_\rho dx^\rho,\end{aligned} \tag{6.37}$$

where we have used the formula

$$\varepsilon^{\mu\nu}{}_{\alpha\beta}\varepsilon^\sigma{}_\rho{}^{\alpha\beta} = -2(g^{\mu\sigma}g^\nu{}_\rho - g^\mu{}_\rho g^{\nu\sigma}). \tag{6.38}$$

So the differential geometric equations (6.35) correspond to the familiar set of equations in the covariant Lorentz gauge (with $\Box = \partial_\sigma\partial^\sigma$)

$$\partial_\lambda A^\lambda = 0 \qquad \text{and} \qquad \Box A_\rho = j_\rho. \tag{6.39}$$

Current conservation: Due to $\delta^2 = 0$ the inhomogeneous Maxwell equation $\delta F = -j$ implies **current conservation**

$$\delta j = 0. \tag{6.40}$$

It corresponds to the components

$$\partial_\lambda j^\lambda = 0 \qquad \text{or} \qquad \frac{\partial}{\partial t}\rho + \vec{\nabla}\vec{\jmath} = 0 \tag{6.41}$$

since the explicit calculation analogous to equation (6.36) gives

$$\delta j = -\partial_\lambda j^\lambda. \tag{6.42}$$

Action: Searching 4-forms with appropriate components we find

$$\begin{aligned} A * j &= A_\rho \frac{1}{3!} j^\mu \varepsilon_{\mu\nu\alpha\beta} dx^\rho dx^\nu dx^\alpha dx^\beta \\ &= A_\lambda j^\lambda dx^0 dx^1 dx^2 dx^3 \end{aligned} \tag{6.43}$$

and

$$\begin{aligned} F * F &= \frac{1}{2}\frac{1}{2}\frac{1}{2} F_{\mu\nu} \varepsilon_{\alpha\beta\rho\sigma} F^{\rho\sigma} dx^\mu dx^\nu dx^\alpha dx^\beta \\ &= \frac{1}{2} F_{\mu\nu} F^{\mu\nu} dx^0 dx^1 dx^2 dx^3 \end{aligned} \tag{6.44}$$

so that we can establish the **Lagrange function of electrodynamics** in terms of differential forms

$$\mathcal{L} = -\frac{1}{2} F * F - A * j \tag{6.45}$$

giving the **action** ($dx^0 dx^1 dx^3 dx^4 = d^4x$ is the oriented volume element)

$$\begin{aligned} S[A] &= \int \left(-\frac{1}{2} F * F - A * j \right) \\ &= \int d^4x \left(-\frac{1}{4} F_{\mu\nu} F^{\mu\nu} - A_\lambda j^\lambda \right). \end{aligned} \tag{6.46}$$

The **gauge invariance** of the action

$$S[A + d\Lambda] - S[A] = -\int d\Lambda * j = \int \Lambda \, d * j = 0 \tag{6.47}$$

is equivalent to **current conservation**

$$d * j = 0 \qquad \text{or} \qquad \partial_\lambda j^\lambda = 0 \tag{6.48}$$

where

$$d * j = \partial_\lambda j^\lambda \frac{1}{4!} \varepsilon_{\mu\nu\alpha\beta} dx^\mu dx^\nu dx^\alpha dx^\beta = \partial_\lambda j^\lambda dx^0 dx^1 dx^2 dx^3. \tag{6.49}$$

Note that we have ignored the boundary term in equation (6.47).

Fibre bundle: Finally, if we want to construct a fibre bundle we have for the **structure group** the Abelian $U(1)$, the gauge group of electromagnetism; for the **base space** we take the 4-dimensional Minkowski space–time and for the **fibre** the set of all possible $U(1)$ group elements. We find that the **principal bundle** is just **trivial**, $P = \mathbf{R}^4 \times U(1)$, since the base space is contractible; the base can be covered with a single trivialization. The **connection** (pulled back to the base) is the gauge potential A and the **curvature** is the field strength F but both multiplied with the Lie algebra factor i.

6.2 Non-Abelian fields, Yang–Mills theory

In the case of non-Abelian fields we have the quantities (recall our introduction of YM theory in Section 3.5.1, and [Sexl, Urbantke 1983])

$$\begin{aligned} A_\mu &= A^a_\mu T^a \\ F_{\mu\nu} &= F^a_{\mu\nu} T^a = \partial_\mu A_\nu - \partial_\nu A_\mu + [A_\mu, A_\nu] \\ D_\mu &= D^a_\mu T^a = \partial_\mu + [A_\mu, \ ], \end{aligned} \tag{6.50}$$

with T^a the generators of the gauge group satisfying the commutator relation (3.275). In differential forms we get:

gauge potential 1-form

$$A = A_\mu dx^\mu \in \text{Lie } G \otimes \Lambda^1(M), \tag{6.51}$$

field strength 2-form

$$\begin{aligned} F &= \frac{1}{2} F_{\mu\nu} dx^\mu dx^\nu \\ &= dA + \frac{1}{2}[A, A] = dA + A^2 \in \text{Lie } G \otimes \Lambda^2(M), \end{aligned} \tag{6.52}$$

covariant derivative

$$D = d + [A, \ ] \tag{6.53}$$

adjoint covariant derivative

$$D^\dagger = \begin{cases} -*D* & \text{Euclidean} \\ *D* & \text{Minkowskian.} \end{cases} \tag{6.54}$$

The covariant adjoint is defined like the simple adjoint (recall equation (2.169); we consider even dimensional manifolds, compact and without boundaries).

Theorem:

$$(D\alpha_{p-1}, \beta_p) = (\alpha_{p-1}, D^\dagger \beta_p). \tag{6.55}$$

Theorem:

$$D^2 = [F, \ ]. \tag{6.56}$$

Proof. Let $\alpha_p \in \text{Lie } G \otimes \Lambda^p(M)$ be a Lie algebra valued p-form, then

$$\begin{aligned} D^2\alpha_p &= D(d\alpha_p + [A, \alpha_p]) \\ &= d^2\alpha_p + [dA, \alpha_p] - [A, d\alpha_p] + [A, d\alpha_p + [A, \alpha_p]] \\ &= [dA, \alpha_p] + [A, [A, \alpha_p]] \\ &= [dA + A^2, \alpha_p] = [F, \alpha_p] \end{aligned} \tag{6.57}$$

since by definition of the commutator

$$[A, [A, \alpha_p]] = [A^2, \alpha_p]. \qquad \text{Q.E.D.} \tag{6.58}$$

However, applied to F we get

$$D^2F = [F, F] = 0. \tag{6.59}$$

Clearly, theorems (3.308) and (6.56) correspond to each other when multiplying the commutator (3.308) with the wedge product $\frac{1}{2}dx^\mu dx^\nu$.

Note: The **Jacobian identity**, when containing forms, also picks up additional sign factors, e.g.

$$[A, [A, \alpha_p]] + (-)^{p+1}[A, [\alpha_p, A]] + [\alpha_p, [A, A]] = 0. \tag{6.60}$$

Theorem:

$$(D^\dagger)^2 = *D**D* = (\pm) \cdot (-)^p * D^2* = (\pm) \cdot (-)^p * [F, *], \tag{6.61}$$

where $(+)$ corresponds to Euclidean and $(-)$ to Minkowski space.

For some p-form α_p we have

$$(D^\dagger)^2\alpha_p = (\pm)\cdot(-)^p * [F, *\alpha_p] \neq 0$$

but specifically for F the result is

$$(D^\dagger)^2 F = (\pm)\cdot(-)^p * [F, *F] = 0 \tag{6.62}$$

since $[F, *F] = 0$.

YM field equation: Analogous to the Maxwell case (6.28) the **YM field equation in terms of differential forms** is

$$D^\dagger F = -j, \tag{6.63}$$

where here

$$j = j^a_\mu T^a dx^\mu \tag{6.64}$$

denotes the **non-Abelian current 1-form**. If we reduce equation (6.63) to its components we get

$$\begin{aligned} D^\dagger F &= *D * \frac{1}{2} F_{\mu\nu} dx^\mu dx^\nu \\ &= -D^\mu F_{\mu\rho} dx^\rho \end{aligned} \tag{6.65}$$

So the differential geometric equation (6.63) corresponds to the familiar YM field equation in covariant components

$$D_\mu F^{\mu\nu} = j^\nu. \tag{6.66}$$

Due to $(D^\dagger)^2 F = 0$ the field equation (6.63) implies **covariant current conservation**

$$D^\dagger j = 0 \quad \text{or} \quad D_\lambda j^\lambda = 0. \tag{6.67}$$

For the **Bianchi identity** we find

$$DF = 0 \tag{6.68}$$

which follows by direct calculation and corresponds to the familiar expression (3.322) when we multiply by the wedge product $\frac{1}{2} dx^\lambda dx^\mu dx^\nu$.

Finally the **YM action** is given by

$$S[A] = \int \text{tr } F * F, \tag{6.69}$$

with the generator normalization (3.277).

6.3 Anomalies

Now we turn to the anomalies which we are actually interested in and we write them in differential geometric notation.

Singlet anomaly: From Section 4.7 we know the anomaly result (recall equation (4.378))

$$\partial^\lambda j_\lambda^5 = -\frac{1}{16\pi^2}\varepsilon^{\mu\nu\alpha\beta}\ \mathrm{tr}\ F_{\mu\nu}F_{\alpha\beta}. \tag{6.70}$$

(We have a sign change since we work here with definition $\varepsilon_{0123} = 1$ instead of $\varepsilon^{0123} = 1$ for equation (4.378).) Defining the **axial current 1-form**

$$j^5 = j_\mu^5 dx^\mu \tag{6.71}$$

its **dual** is

$$*j^5 = \frac{1}{3!}j^{5\mu}\varepsilon_{\mu\nu\alpha\beta}dx^\nu dx^\alpha dx^\beta \tag{6.72}$$

and for the exterior derivative we obtain

$$d*j^5 = \partial^\lambda j_\lambda^5 dx^0 dx^1 dx^2 dx^3. \tag{6.73}$$

Multiplying equation (6.70) now with the wedge product $dx^0dx^1dx^2dx^3$ and using relation (2.154) we immediately find the **singlet anomaly within differential forms**

$$d*j^5 = \frac{1}{4\pi^2}\ \mathrm{tr}\ FF. \tag{6.74}$$

We can re-express equation (6.74),

$$\begin{aligned}\mathrm{tr}\ FF &= \mathrm{tr}\ (dA+A^2)(dA+A^2)\\ &= d\ \mathrm{tr}\left(AdA+\frac{2}{3}A^3\right)\end{aligned} \tag{6.75}$$

(note that $\mathrm{tr}\ A^4 = 0$ by virtue of the wedge product) and we get for the anomaly

$$d*j^5 = \frac{1}{4\pi^2}dQ_3, \tag{6.76}$$

where Q_3 is called the **Chern–Simons form**

$$Q_3 = \mathrm{tr}\left(AdA+\frac{2}{3}A^3\right). \tag{6.77}$$

Non-Abelian anomaly—Bardeen's result: Taking Bardeen's result for L- and R-handed fields (equation (4.387), (4.388) of Section 4.8)

$$\begin{aligned} G^a[A_\mu] &= -(D_\mu j^\mu)^a \\ &= \mp\frac{1}{24\pi^2}\varepsilon^{\mu\nu\rho\sigma} \operatorname{tr} T^a \partial_\mu \left(A_\nu \partial_\rho A_\sigma + \frac{1}{2} A_\nu A_\rho A_\sigma \right) \end{aligned} \quad (6.78)$$

(also here we have a sign change due to our present definition $\varepsilon_{0123} = 1$) we find, in the same way as before, the **non-Abelian anomaly within differential forms**

$$G^a[A] = -(D * j)^a = \pm\frac{1}{24\pi^2} \operatorname{tr} T^a d \left(AdA + \frac{1}{2}A^3 \right). \quad (6.79)$$

(+) for R- and (−) for L-handed fields.

Covariant anomaly: Considering, finally, the **covariant anomaly** the previous result (4.389) immediately gives

$$\widetilde{G}^a[A] = \pm\frac{1}{8\pi^2} \operatorname{tr} T^a FF. \quad (6.80)$$

6.4 Dirac monopole

The Dirac monopole is a simple, but topologically nontrivial example of a gauge theory that casts light on the fascinating relationship between physics and geometry. However, in spite of its beauty, the monopole does not seem to be realized in nature. Nevertheless, its detailed study is well worthwhile; certain features in non-Abelian gauge theories occur in an analogous way (see e.g. Section 6.6).

In classical electrodynamics there is a strong symmetry between the electric field $\vec{E}$ and the magnetic field $\vec{B}$, but still a magnetic charge—**magnetic monopole**—is absent in the Maxwell equations (6.1). Let us now assume the existence of such a magnetic charge with density ρ_M, then we would have

$$\vec{\nabla}\vec{B} = \rho_M \neq 0 \quad (6.81)$$

and the symmetry between $\vec{E}$ and $\vec{B}$ would be perfect. It was Dirac [Dirac 1931] who discovered that, as a consequence, quantum mechanics predicts:

Proposition:

- *If a magnetic monopole exists then its magnetic charge is quantized in units of $1/2|e|$!*

This is a very remarkable outcome of quantum physics. We want to understand this phenomenon first within physics, then by means of modern mathematical techniques—differential geometry and topology. Wu and Yang [Wu, Yang 1975] were the first to reformulate the Dirac monopole in terms of charts and fibre bundles; additional literature can be found in [Sakurai 1985], [Höhne 1990], [Eguchi, Gilkey, Hanson 1980], [Felsager 1981], [Sexl, Urbantke 1983].

Physics: Suppose there is a magnetic monopole with strength g (and no electric field $\vec{E} = 0$) located at the origin

$$\rho_M(\vec{x}) = 4\pi g\, \delta^3(\vec{x}), \tag{6.82}$$

then the **static magnetic field** has the form of a hedgehog

$$\vec{B} = g\frac{\widehat{r}}{r^2} \qquad \text{with } \widehat{r} = \frac{\vec{x}}{r}. \tag{6.83}$$

We might think that the corresponding **vector potential** is given by

$$\vec{A} = g\frac{1-\cos\theta}{r\sin\theta}\widehat{\varphi} \equiv A_\varphi \widehat{\varphi}, \tag{6.84}$$

where we have expressed $\vec{A}$ in terms of spherical coordinates. Generally we have

$$\vec{A} = A_r\widehat{r} + A_\varphi\widehat{\varphi} + A_\theta\widehat{\theta}, \tag{6.85}$$

with $\widehat{r}, \widehat{\varphi}, \widehat{\theta}$ being the unit vectors in the direction of the radius and the angles φ, θ.

The curl of $\vec{A}$ (6.84) provides the desired magnetic field $\vec{B}$ (6.83)

$$\begin{aligned} \vec{B} &= \vec{\nabla}\times\vec{A} = \frac{1}{r^2\sin\theta}\begin{vmatrix} \widehat{r} & r\widehat{\theta} & r\sin\theta\widehat{\varphi} \\ \dfrac{\partial}{\partial r} & \dfrac{\partial}{\partial\theta} & \dfrac{\partial}{\partial\varphi} \\ A_r & rA_\theta & r\sin\theta A_\varphi \end{vmatrix} \\ &= \frac{\widehat{r}}{r^2\sin\theta}\frac{\partial}{\partial\theta}g(1-\cos\theta) = g\frac{\widehat{r}}{r^2}. \end{aligned} \tag{6.86}$$

(Note a sign change in the constant 1 does not affect $\vec{B}$.)

Problem: However, there is a problem. The vector potential (6.84) is singular on the negative z-axis; so for $\theta = \pi$ $\vec{A}$ is not defined. In fact, it turns out that it is impossible to construct a single singularity-free potential which is valid everywhere on $\mathbf{R}^3 \setminus \{0\}$. Why? Let us see.

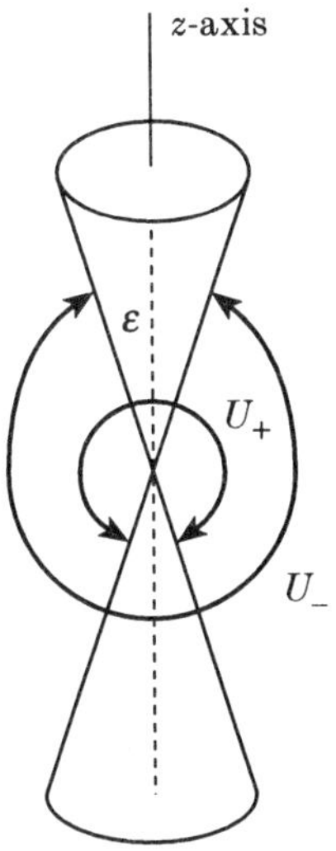

Fig. 6.1. The vector potential $\vec{A}_+$ is valid on U_+, $\vec{A}_-$ on U_-

Integrating the magnetic monopole over a closed surface, a sphere S^2, gives via the Gauss law

$$\int_{S^2} d\vec{f}\vec{B} = \int_{\text{ball}} d^3x \vec{\nabla}\vec{B} = \int_{\text{ball}} d^3x \rho_M(\vec{x}) = 4\pi g \neq 0. \tag{6.87}$$

On the other hand, if $\vec{A}$ were nonsingular on $\mathbf{R}^3 \setminus \{0\}$ we would have via Stokes' theorem

$$\int_{S^2} d\vec{f}\vec{B} = \int_{\partial S^2} d\vec{x}\vec{A} = 0 \tag{6.88}$$

(since $\partial S^2 = \emptyset$) in contradiction to the above monopole result $4\pi g \neq 0$.

Resolution: When considering the total space we need at least two expressions for the vector potential. So we construct a pair of potentials which are valid outside cones (see Figure 6.1)

$$\begin{aligned} \vec{A}_+ &= g\frac{1-\cos\theta}{r\sin\theta}\widehat{\varphi} \qquad \text{for } \theta < \pi - \varepsilon : U_+ \\ \vec{A}_- &= g\frac{-1-\cos\theta}{r\sin\theta}\widehat{\varphi} \qquad \text{for } \theta > \varepsilon : U_-. \end{aligned} \tag{6.89}$$

The potential $\vec{A}_+$ is valid on U_+ and $\vec{A}_-$ on U_-; together they lead to the correct magnetic field $\vec{B}$ everywhere.

What happens now in the overlap region $U_+ \cap U_-$ where both potentials are legitimate? Since both potentials $\vec{A}_+$ and $\vec{A}_-$ correspond to the same magnetic field $\vec{B}$ they must be related by a **gauge transformation**

$$\vec{A}_- = \vec{A}_+ + \vec{\nabla}\Lambda. \tag{6.90}$$

We can calculate explicitly the appropriate **gauge function** Λ. Using the gradient in spherical coordinates

$$\vec{\nabla} = \widehat{r}\frac{\partial}{\partial r} + \widehat{\varphi}\frac{1}{r\sin\theta}\frac{\partial}{\partial\varphi} + \widehat{\theta}\frac{1}{r}\frac{\partial}{\partial\theta} \tag{6.91}$$

we deduce from the difference

$$\begin{aligned}
\vec{A}_- - \vec{A}_+ &= \vec{\nabla}\Lambda \\
-\frac{2g}{r\sin\theta}\widehat{\varphi} &= \widehat{\varphi}\frac{1}{r\sin\theta}\frac{\partial}{\partial\varphi}\Lambda \\
\frac{\partial\Lambda}{\partial\varphi} &= -2g
\end{aligned} \tag{6.92}$$

the result

$$\Lambda = -2g\varphi. \tag{6.93}$$

Quantum mechanics: Next we turn to quantum mechanics. A particle with electric charge e subjected to the magnetic monopole field $\vec{B}$ is described by the **Schrödinger equation**

$$\left[\frac{1}{2m}\left(\vec{p} - \frac{e}{c}\vec{A}\right)^2 + e\phi\right]\psi = i\hbar\frac{\partial}{\partial t}\psi \tag{6.94}$$

where ϕ represents the scalar potential. This equation is *invariant* under the $U(1)$ **gauge transformations**

$$\begin{aligned}
\vec{A}' &= \vec{A} + \vec{\nabla}\Lambda \\
\phi' &= \phi - \frac{1}{c}\frac{\partial}{\partial t}\Lambda \\
\psi' &= \exp\left[i\frac{e}{\hbar c}\Lambda\right]\psi.
\end{aligned} \tag{6.95}$$

So a particular wave function ψ clearly depends on the gauge chosen. In the overlap region $U_+ \cap U_-$ we may use either $\vec{A}_+$ or $\vec{A}_-$ and the corresponding wave functions are connected via gauge transformation (6.95)

$$\psi_- = \exp\left[-i\frac{2eg}{\hbar c}\varphi\right]\psi_+. \tag{6.96}$$

We have already inserted the explicit solution (6.93).

Now let us study the behaviour of the wave function. Each ψ_- and ψ_+ must be *single-valued* if it belongs to the domain of definition of Schrödinger operators. Considering the equator $\theta = \pi/2$ we increase the angle φ from 0 to 2π, then both ψ_- and ψ_+ must return to their original values

$$\varphi = 0 \quad : \quad \psi_- = \psi_+ \tag{6.97}$$

$$\varphi = 2\pi \quad : \quad \psi_- = \exp\left[-i\frac{2eg}{\hbar c}2\pi\right]\psi_+ \equiv \psi_+. \tag{6.98}$$

This implies the **quantization condition**

$$\frac{2eg}{\hbar c} = \pm n, \qquad n = 0, 1, 2, \ldots, \tag{6.99}$$

hence the **magnetic charge must be quantized**

$$g = n\frac{\hbar c}{2|e|} \tag{6.100}$$

or in units where $-e = |e| = \hbar = c = 1$

$$g = \frac{n}{2}. \tag{6.101}$$

Differential geometry: We can consider the gradient $\vec{\nabla}$ (6.91) on the basis $\left\{\frac{\partial}{\partial r}, \frac{\partial}{\partial \varphi}, \frac{\partial}{\partial \theta}\right\}$ of the tangent space $T(M)$. For differential forms, however, we need the dual space, the cotangent space $T^*(M)$, with basis $\{dr, d\varphi, d\theta\}$

$$\vec{\nabla} = dr\frac{\partial}{\partial r} + d\varphi\frac{\partial}{\partial \varphi} + d\theta\frac{\partial}{\partial \theta}. \tag{6.102}$$

So we can identify

$$\begin{aligned} \widehat{r} &\equiv dr \\ \widehat{\varphi} &\equiv r\sin\theta d\varphi \\ \widehat{\theta} &\equiv r d\theta \end{aligned} \tag{6.103}$$

and we obtain from expressions (6.89) the **gauge potential 1-forms**

$$A_\pm = \frac{n}{2}(\pm 1 - \cos\theta)d\varphi, \tag{6.104}$$

which are related by the **gauge transformation**

$$A_+ = A_- + n d\varphi. \tag{6.105}$$

Applying the exterior derivative d which is given by equation (6.102) we find the **field strength 2-form**

$$F = dA_\pm \tag{6.106}$$

$$F = \frac{n}{2}\sin\theta d\theta d\varphi. \tag{6.107}$$

(Note that $d\theta d\varphi$ means a wedge product.)

It is customary to work with Cartesian coordinates. For that we diminish the definition domains ($\varepsilon \to \pi/2$) of the potentials

$$\begin{aligned} U_+ &= \mathbf{R}^3_+ \setminus \{0\} \\ U_- &= \mathbf{R}^3_- \setminus \{0\}, \end{aligned} \tag{6.108}$$

which in the overlap represent the plane minus the origin where the monopole is situated

$$U_+ \cap U_- = \mathbf{R}^2 \setminus \{0\}. \tag{6.109}$$

With the help of the familiar transformation

$$\begin{aligned} x &= r\sin\theta\cos\varphi \\ y &= r\sin\theta\sin\varphi \\ z &= r\cos\theta \end{aligned} \tag{6.110}$$

it is straightforward to re-express the forms (6.104)–(6.107) by Cartesian coordinates:

gauge potential 1-forms

$$A_\pm = \frac{n}{2}\frac{1}{r(z \pm r)}(-ydx + xdy), \tag{6.111}$$

with **gauge transformation**

$$A_+ = A_- + nd\arctan\frac{y}{x} \tag{6.112}$$

and (recall example (2.187)–(2.189) of Section 2.4)

$$d\arctan\frac{y}{x} = \frac{1}{x^2+y^2}(-ydx + xdy) \tag{6.113}$$

field strength 2-form

$$F = dA_\pm \tag{6.114}$$

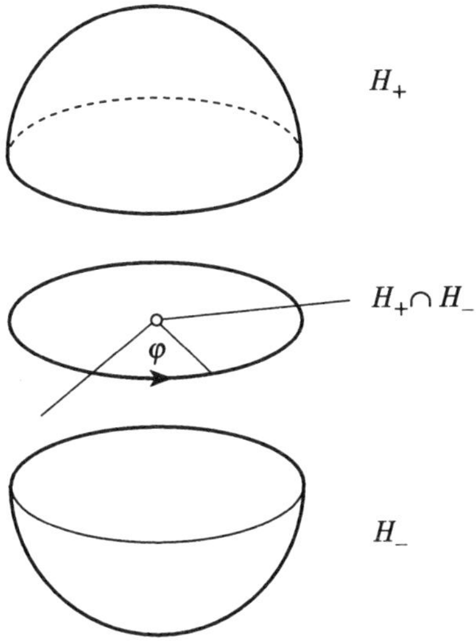

Fig. 6.2. The sphere S^2 is covered by two hemispheres H_+, H_- with the equator $H_+ \cap H_- = S^1$

$$F = \frac{n}{2}\frac{1}{r^3}(zdxdy + xdydz + ydzdx). \tag{6.115}$$

Clearly F is closed

$$dF = d^2 A_\pm = 0 \tag{6.116}$$

but *not* exact since dA_+ is defined only *locally* on U_+ and dA_- on U_-.

Fibre-bundle topology: Now we are going to construct a fibre bundle (recall Section 2.7) which mirrors totally the physics of the Dirac monopole.

The space where we defined the Dirac monopole is $\mathbf{R}^3$ with the origin $\{0\}$ extracted. The monopole is sitting there. Topologically this space is equivalent—*homotopic*—to the sphere S^2. Then the **base manifold** for our bundle will be the sphere S^2

$$M = \mathbf{R}^3 \setminus \{0\} \sim S^2, \tag{6.117}$$

which we cover by two hemispheres H_+, H_-, the **charts** (see Figure 6.2).

The gauge group $U(1)$ of electrodynamics supplies the **structure group** $U(1)$ for the bundle. The set of all possible $U(1)$ gauge symmetry elements establishes the **fibre**. The fibres are identical to the structure group itself, so we construct a **principal bundle** $P(S^2, U(1))$. The wave function $\psi(x)$ corresponds to a **section** of the associated vector bundle—a line bundle. **Locally** the bundle looks like

$$\begin{matrix} H_+ \times U(1) \\ H_- \times U(1) \end{matrix} \quad \text{with bundle coordinates} \quad \begin{matrix} (\theta, \varphi, e^{i\alpha_+}) \\ (\theta, \varphi, e^{i\alpha_-}) \end{matrix} \tag{6.118}$$

and with $0 \leq \theta < \pi$, $0 \leq \varphi < 2\pi$.

The **transition functions** h connect the fibres from one chart to another

$$\begin{aligned} e^{i\alpha_-} &= h_{-+}e^{i\alpha_+} \\ \psi_- &= h_{-+}\psi_+ \end{aligned} \tag{6.119}$$

or they relate the different sections of the bundle. They represent elements of the $U(1)$ gauge group and are functions of φ defined along the equator $(\theta = \pi/2)$ $H_+ \cap H_- = S^1$, $h_{+-}(\varphi) \in U(1)$, and they must glue the fibres together exactly when we complete a full turn around the equator. Then the sole choice is

$$h_{-+} = e^{in\varphi}, \tag{6.120}$$

with n being an integer number. This is the topological origin for Dirac's quantization condition (6.101) of the magnetic charge $(-e = |e| = \hbar = c = 1)$.

The Lie algebra valued **connection 1-form** of the bundle is locally represented by the gauge potentials (recall Section 2.7.2)

$$\mathcal{A}_\pm = iA_\pm, \tag{6.121}$$

which are defined on the hemispheres $H_\pm$. In our definitions we have a Lie algebra factor i when $A_\pm$ is given by equation (6.104).

The **compatibility condition**

$$\mathcal{A}_+ = h_{-+}^{-1}\mathcal{A}_-h_{-+} + h_{-+}^{-1}dh_{-+} \tag{6.122}$$

expresses the gauge transformation law and gives for the Abelian case $U(1)$

$$\mathcal{A}_+ = \mathcal{A}_- + ind\varphi \tag{6.123}$$

in correspondence with equation (6.105).

The Lie algebra valued **curvature 2-form** of the bundle represents the gauge field strength

$$\mathcal{F} = iF, \tag{6.124}$$

with F being given explicitly by equation (6.107). In terms of the gauge potential we have

$$\mathcal{F} = d\mathcal{A}_\pm. \tag{6.125}$$

So $\mathcal{F}$ is closed, $d\mathcal{F} = 0$, but not exact; the Abelian $\mathcal{F}$ is gauge invariant.

Winding number and monopole charge: Consider again the transition functions of the bundle $h_{-+}(\varphi) \in U(1)$, equation (6.120). They define a map from the equator of the base manifold into the structure group

$$h_{-+} : H_+ \cap H_- = S^1 \to U(1). \tag{6.126}$$

These maps are classified by the first homotopy group

$$\Pi_1(U(1)) \simeq \Pi_1(S^1) \simeq \{n\} = \mathbf{Z}. \tag{6.127}$$

The integers n are the **winding numbers** characterizing the different homotopy classes which the bundle belongs to (recall Section 2.2). When φ varies from $0 \to 2\pi$ the group $U(1)$ is wound around n times. The winding number n can be expressed by the integral (recall equation (2.20))

$$\begin{aligned} n &= \frac{1}{2\pi i}\int_0^{2\pi} d\varphi\, h_{-+}^{-1}(\varphi)\frac{d}{d\varphi}h_{-+}(\varphi) \\ &= \frac{1}{2\pi i}\int_0^{2\pi} d\varphi \cdot in, \end{aligned} \tag{6.128}$$

where here n represents the quantized magnetic monopole charge $g = n/2$ (in units $-e = |e| = \hbar = c = 1$).

These winding numbers—the monopole charges—are typical numbers of the topology of the bundle and are also determined by the so-called **first Chern number** or by the **Atiyah–Singer index theorem**—which we shall discuss again later on (see Section 11.4)

$$C_1 \equiv \text{index } D_+ = \frac{i}{2\pi}\int_{S^2} \text{tr } \mathcal{F}. \tag{6.129}$$

The explicit evaluation gives

$$\begin{aligned} -C_1 &= \frac{1}{2\pi}\int_{S^2} F = \frac{1}{2\pi}\left[\int_{H_+} dA_+ + \int_{H_-} dA_-\right] \\ &= \frac{1}{2\pi}\left[\int_{\partial H_+} A_+ + \int_{\partial H_-} A_-\right] = \frac{1}{2\pi}\int_{S^1}(A_+ - A_-) \\ &= \frac{1}{2\pi}\int_0^{2\pi} d\varphi \cdot n = n, \end{aligned} \tag{6.130}$$

where we have used Stokes' theorem (2.128).

Note: The integral representation for the Chern number C_1 or for the index turns out to be *independent* of the connection. Only the gauge trans-

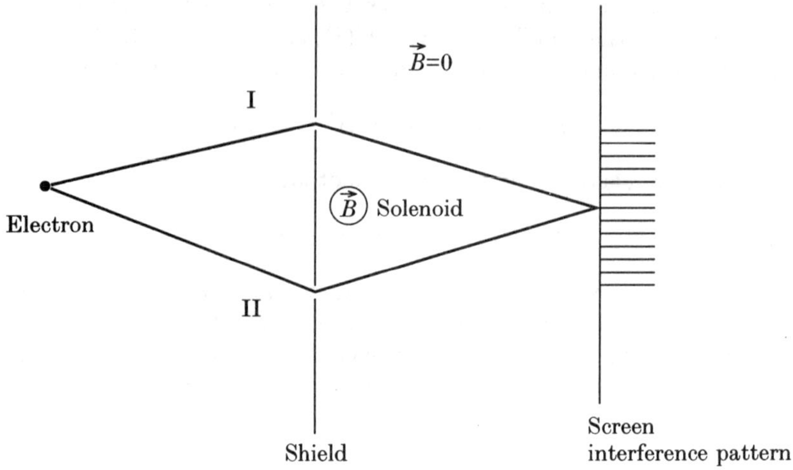

Fig. 6.3. Aharonov–Bohm effect

formation $n d\varphi$ enters into the calculation corresponding to the transition functions $h_{-+}^{-1} dh_{-+}$ which determine the topology of the bundle.

6.5 Aharonov–Bohm effect

In 1959 Bohm and Aharonov wrote a paper investigating the 'significance of electromagnetic potentials in quantum theory' [Aharonov, Bohm 1959]. They found the striking feature of quantum mechanics—the **Aharonov–Bohm effect**—that it is the potential which influences the motion of particles even in regions which are field-free. There is great interest and excitement for the Aharonov–Bohm (AB) effect, especially in the last years, since it touches the foundations of quantum mechanics. The topological description of the effect within fibre bundles has been found by [Wu, Yang 1975] and [Tourrenc 1977] (for a review see [Höhne 1990]). An overview on the theoretical and experimental side has been presented by Peshkin and Tonomura [Peshkin, Tonomura 1989].

Experiment: What precisely is the Aharonov–Bohm effect? We consider here only the magnetic version. Take an electron beam and split it into two parts (see Figure 6.3). The two paths are recombined and form an interference pattern on a screen. A solenoid is placed between the two paths which produces a magnetic field $\vec{B}$ inside. Outside the solenoid the field vanishes everywhere, $\vec{B} = 0$. A shield prevents the electrons from penetrating into the solenoid, so an electron never 'touches' the magnetic field.

Question: Does a change of $\vec{B}$ inside the solenoid influence the interference pattern on the screen?

Answer: Classically, we expect no influence. In fact, there is no AB effect in classical physics since only the Lorentz force determines the motion of the electron and the Lorentz force

$$\vec{K} = \frac{e}{c}\,\dot{\vec{x}} \times \vec{B} = 0 \tag{6.131}$$

vanishes on the paths of the beam (where $\vec{B} = 0$). But quantum mechanically it is the vector potential $\vec{A}$ which enters into the Schrödinger equation (or into the path integral formalism) and the vector potential is nonvanishing outside the solenoid. So we shall get an effect.

Even though there is no magnetic field $\vec{B} = 0$ outside the solenoid, the vector potential is nonvanishing there, $\vec{A} \neq 0$, since the loop integral of $\vec{A}$ around the solenoid equals the magnetic flux ϕ_M. For an infinitesimally thin solenoid we have the vector potential

$$\vec{A}(r) = \frac{\phi_M}{2\pi r^2}(-y, x, 0), \qquad r^2 = x^2 + y^2 \tag{6.132}$$

satisfying

$$\int \vec{B}d\vec{f} = \int (\vec{\nabla} \times \vec{A})d\vec{f} = \phi_M \tag{6.133}$$

and

$$\vec{B} = \vec{\nabla} \times \vec{A} = 0 \qquad \text{for } r \neq 0. \tag{6.134}$$

Quantum mechanics: Now let us consider the quantum mechanical wave function. For $\vec{A} \neq 0$ it is determined by its gauge transformed version (recall equations (6.95))

$$\psi^A(r) = \exp\left[i\frac{e}{\hbar c}\int^r \vec{A}(r)d\vec{x}\right]\psi(r), \tag{6.135}$$

where $\psi(r)$ denotes the wave function corresponding to $\vec{A} = 0$, no magnetic flux inside the solenoid. The total wave function in the interference region of the screen is the superposition of the wave functions along both paths

$$\begin{aligned}
&\psi_I^A(x_0) + \psi_{II}^A(x_0) \\
&\quad = \exp\left[i\frac{e}{\hbar c}\int_I \vec{A}d\vec{x}\right]\psi_I(x_0) + \exp\left[i\frac{e}{\hbar c}\int_{II} \vec{A}d\vec{x}\right]\psi_{II}(x_0) \\
&\quad = \exp\left[i\frac{e}{\hbar c}\int_{II} \vec{A}d\vec{x}\right]\left\{\exp\left[i\frac{e}{\hbar c}\oint \vec{A}d\vec{x}\right]\psi_I(x_0) + \psi_{II}(x_0)\right\},
\end{aligned} \tag{6.136}$$

with x_0 a point on the screen. So we get a relative **phase factor**

$$\exp\left[i\frac{e}{\hbar c}\oint \vec{A}d\vec{x}\right] \tag{6.137}$$

contributing to the interference term.

The relative **phase shift** is

$$\Delta S = \frac{e}{\hbar c}\oint \vec{A}d\vec{x} \tag{6.138}$$

and depends via Stokes' theorem (2.128) on the magnetic flux ϕ_M inside the solenoid

$$\begin{aligned}
\Delta S &= \frac{e}{\hbar c}\oint \vec{A}d\vec{x} = \frac{e}{\hbar c}\int_{\text{area}} (\vec{\nabla}\times\vec{A})d\vec{f} = \frac{e}{\hbar c}\int_{\text{area}} \vec{B}d\vec{f} \\
&= \frac{e}{\hbar c}\phi_M.
\end{aligned} \tag{6.139}$$

The probability of finding the electron on the screen depends on the modulus squared of the total wave function and hence on the phase difference ΔS (6.138), (6.139). If we now vary the magnetic field inside the solenoid we obtain a periodic (sinusoidal) component in the probability of observing the electron on the screen. The period is given by the *magnetic flux unit* $2\pi\hbar c/|e|$. If two values of the magnetic fluxes ϕ_M^1, ϕ_M^2 differ by

$$\frac{e}{\hbar c}(\phi_M^1 - \phi_M^2) = 2\pi n, \qquad n \in \mathbf{Z} \tag{6.140}$$

we find the same interference pattern.

Résumé: We obtain an interference pattern on the screen which arises on pure quantum mechanical grounds. It has the striking feature of depending on the magnetic field *inside* the impenetrable solenoid. The phase factor (6.137) describes the AB effect *completely* and depends on the gauge potential $\vec{A}$ in a *nonlocal* way. So in quantum mechanics it is the vector potential $\vec{A}$ that is more fundamental than the magnetic field $\vec{B}$ itself.

A series of beautiful experiments has been performed to test this impor-

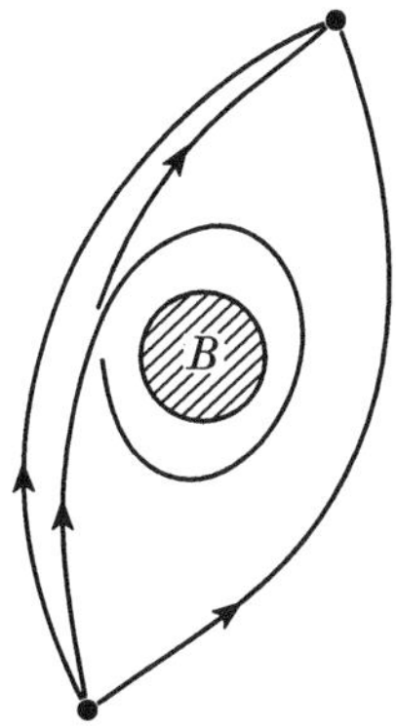

Fig. 6.4. Homotopically different paths in $M = \mathbf{R}^2 \setminus \{D^2\}$

ant quantum effect [Chambers 1960], [Möllenstedt, Bayh 1962], [Tonomura et al. 1982]; overall they confirm the AB predictions nicely. For a review see [Peshkin, Tonomura 1989].

Fibre-bundle topology: Now we turn to the topological description of the AB effect.

The **base manifold** M we have to consider here is the 2-dimensional plane with the origin $\{0\}$ extracted (or cutting out a disc $\{D^2\}$ corresponding to the diameter of the solenoid)

$$M = \mathbf{R}^2 \setminus \{0\}. \tag{6.141}$$

The gauge group $U(1)$ supplies the **structure group** $U(1)$ and all possible gauge group elements build up the **fibre.** So we construct a **principal bundle** $P(M, U(1))$. The wave function $\psi(x)$ corresponds to a **section** of the associated vector bundle—a line bundle.

We rewrite the gauge potential (6.132) as a 1-form (recall $A_0 = 0$ in our pure magnetic case)

$$A = A_\mu dx^\mu = \frac{\phi_M}{2\pi r^2}(-ydx + xdy) \tag{6.142}$$

or in polar coordinates

$$A = \frac{\phi_M}{2\pi} d\varphi. \tag{6.143}$$

It represents the local **connection 1-form** of the bundle which determines the parallel transport of the fibre

$$\mathcal{A} = iA. \tag{6.144}$$

The field strength $F = dA$ corresponds to the local **curvature 2-form**

$$\mathcal{F} = iF = d\mathcal{A}. \tag{6.145}$$

In the AB case we have $\mathcal{F} = 0$, so $\mathcal{A}$ is closed but not exact (recall the example (2.187) of Section 2.4). Thus we cannot gauge away the potential on the whole manifold.

The base space (6.141) is a multiply connected manifold which allows homotopically different paths (see Figure 6.4). They are characterized by the **first homotopy group** $\Pi_1(M) \simeq \Pi_1(S^1) \simeq \{n\} = \mathbf{Z}$, where the integers n represent the **winding numbers** associated with the paths. All these paths contribute to the phase factor and hence to the interference pattern (for a search of higher winding numbers see [Inomata 1986]).

Holonomy: Generally, a fibre point is **parallel transported** along a curve σ in the base M by the phase factor containing the connection (we choose $-e = |e| = \hbar = c = 1$)

$$\exp\left[-\int_\sigma \mathcal{A}\right] \overset{\text{closed curve in } M}{\longrightarrow} \exp\left[-i\oint A\right] = \exp\left[-i\phi_M\right]. \tag{6.146}$$

For a closed curve in M (6.141)—a circle around the origin—the lifted curve in the fibre space $U(1)$ is not closed anymore. We have a **holonomy** in the fibre space. Classifying the curves in M by the winding numbers we obtain the **holonomy group**

$$H = \left\{e^{-in\phi_M}\right\}, \qquad n \in \mathbf{Z}. \tag{6.147}$$

It characterizes a map h of the closed curves in M into the fibre space $U(1)$

$$\begin{aligned} h : \Pi_1(S^1) &\rightarrow U(1) \\ n &\mapsto e^{-in\phi_M}. \end{aligned} \tag{6.148}$$

In other terms, we find this holonomy by considering the parallel transport of the wave function—a fibre point—with respect to the local connection $\mathcal{A}$ (see e.g. [Nakahara 1990]). Then we have

$$\mathcal{D}\psi = (d + \mathcal{A})\psi = 0, \tag{6.149}$$

where $\mathcal{D} = d + \mathcal{A}$ is the covariant derivative. Calculating with polar coordinates (6.143)

$$\mathcal{D}\psi = \left(d + i\frac{\phi_M}{2\pi}d\varphi\right)\psi = 0 \tag{6.150}$$

we easily get the solution

$$\psi(\varphi) = \text{const. } \exp\left[-i\phi_M \frac{\varphi}{2\pi}\right]. \tag{6.151}$$

Along a circle S^1 in the base given by $\varphi : 0 \to 2\pi$ (winding number $n = +1$) the wave function—a fibre point—is transported as

$$\psi(0) \mapsto \psi(0)e^{-i\phi_M}. \tag{6.152}$$

The corresponding fibre curve does not close—a **holonomy** occurs. So geometrically the AB effect corresponds to a holonomy in the fibre space.

Remark: We can view the AB effect as a simple example of a nontrivial topology of the vacuum in a gauge theory. As we shall see in the next section gauge theories allow for a vacuum which exhibits a rich structure—instantons—with physical consequences. Here in the AB case there are no electromagnetic fields ($\vec{B} = 0$, $\vec{E} = 0$) outside the solenoid. So the energy density vanishes and we have a vacuum. But still the gauge potential $A \neq 0$ is nonzero, the vacuum has a *structure*. In fact, $A = \frac{\phi_M}{2\pi} d\varphi$ is a gauge transform—a pure gauge—of the 'pure' vacuum $A = 0$.

6.6 Instantons

Instantons are classical solutions of a non-Abelian field theory which exhibit a nontrivial topology. They are responsible for the vacuum structure of QCD. Their effect on physics (confinement, mass contributions, particle decays ...) has been a hotly debated topic in the last decade; for literature we refer to [Jackiw 1977], [Actor 1979], [Rajaraman 1982], [Shuryak 1988], [Felsager 1981], [Shifman 1994]. Here we discuss some basic features which can be easily understood within the formalism developed so far.

Action: In field theory the fundamental quantity is the vacuum-to-vacuum transition amplitude, the generating functional from which all Green functions can be calculated (recall Chapter 3). It is actually well defined only in *Euclidean* space ($ix^0 = x^4$, $-i\partial_0 = \partial_4$, $-iA_0^a = A_4^a$ and the metric $g_E = -\mathbf{1}$)

$$Z_E = e^{-W_E} = \int dA_\mu e^{-S_E[A]}. \tag{6.153}$$

We have formally absorbed the FP ghost part, the gauge treatment of the fields, into the integration measure dA_μ. For the actual evaluation of expression (6.153) we need the local minima of the Euclidean action, around which the quantum fluctuations are calculated.

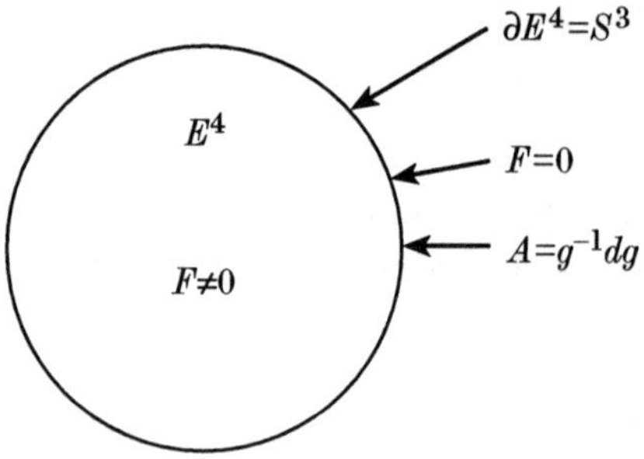

Fig. 6.5. Situation for an instanton

We consider an $SU(2)$ gauge theory which is defined on 4-dimensional Euclidean space $E^4 = \mathbf{R}^4$. The gauge potential and field strength in terms of differential forms are

$$\begin{aligned} A &= A^a_\mu(x)T^a dx^\mu \\ F &= \frac{1}{2}F^a_{\mu\nu}(x)T^a dx^\mu dx^\nu, \qquad T^a = \frac{\sigma^a}{2i} \end{aligned} \tag{6.154}$$

and for the Euclidean action we have

$$\begin{aligned} S_E &= \int d^4x_E\, \frac{1}{4}F^a_{\mu\nu}F^{a\mu\nu} = -\frac{1}{2}\int d^4x_E \text{ tr } F_{\mu\nu}F^{\mu\nu} \\ &= -\int \text{tr } F * F. \end{aligned} \tag{6.155}$$

In the path integral (6.153) the Euclidean action S_E (6.155) has to remain *finite*. This severely restricts the gauge field configurations which contribute. Clearly the field strength must vanish at infinity

$$F \overset{|x|\to\infty}{\longrightarrow} 0 \tag{6.156}$$

but *not* necessarily the gauge potential. $A \to 0$ for $|x| \to \infty$ is the trivial solution, the 'pure' vacuum. As we already know condition (6.156) is achieved for a pure gauge (recall equation (2.510))

$$A(x) \overset{|x|\to\infty}{\longrightarrow} g^{-1}(x)dg(x). \tag{6.157}$$

So we have the situation—which we sketch in Figure 6.5—that the field strength $F \neq 0$ is nonzero inside a volume E^4, but vanishes, $F = 0$, on the boundary $\partial E^4 = S^3$, a 3-dimensional sphere, where the gauge potential approaches a pure gauge. This is the basis for a nontrivial vacuum.

The gauge elements $g(x) \in SU(2)$, $x \in S^3$ represent mappings

$$g : S^3 \to SU(2) \simeq S^3 \tag{6.158}$$

from the sphere S^3 into the gauge group $SU(2)$, which is again isomorphic to a sphere S^3. These mappings are classified according to the homotopy classes which are determined by the topological winding number n (recall Section 2.2)

$$\Pi_3(SU(2)) \simeq \Pi_3(S^3) \simeq \{n\} = \mathbf{Z}. \tag{6.159}$$

Such solutions, discovered by Belavin, Polyakov, Schwartz and Tyupkin [Belavin, Polyakov, Schwartz, Tyupkin 1975], are named the **BPST instanton.**

Topological charge: Let us first introduce a quantity called, in this context, the **Pontrjagin index**

$$q = -\frac{1}{16\pi^2}\int d^4x \ \mathrm{tr}\ F_{\mu\nu} * F^{\mu\nu} = -\frac{1}{8\pi^2}\int_{S^4} \mathrm{tr}\ F^2. \tag{6.160}$$

We have compactified the Euclidean space $E^4 \to S^4$ to a 4-dimensional sphere. The Pontrjagin index is an integer number

$$q = \pm n = 0, \pm 1, \pm 2, \ldots$$

and—as we shall demonstrate—represents the **winding number** of the map $S^3 \to SU(2)$. The integral (6.160) has the remarkable property that it is *only* determined by the topology of the underlying fibre bundle; for this reason q is also called the **topological charge**. The **topological charge density** or the **Pontrjagin density**

$$Q \doteq -\frac{1}{8\pi^2}\ \mathrm{tr}\ F^2 \tag{6.161}$$

is a closed form $dQ = 0$ because of the Bianchi identity $DF = 0$

$$d\ \mathrm{tr}\ F^2 = 2\ \mathrm{tr}\ DFF = 0. \tag{6.162}$$

By virtue of Poincaré's theorem (Section 7.3) it is also locally exact, say on the upper hemisphere H_+ of S^4

$$Q = dK. \tag{6.163}$$

There we have

$$q = -\frac{1}{8\pi^2}\int_{H_+} \mathrm{tr}\ F^2 = \int_{H_+} dK = \int_{S^3} K \tag{6.164}$$

where we applied Stokes' theorem (2.128). K is called the **topological current**; it is the 3-form

$$K = -\frac{1}{8\pi^2}\,\mathrm{tr}\,\left(AdA + \frac{2}{3}A^3\right). \tag{6.165}$$

In Q we recover the singlet anomaly $d * j_5$ result (6.74), in K the Chern–Simons form $Q_3(A, F)$ (6.77). More precisely, we find

$$d * j_5 = -2Q. \tag{6.166}$$

So we can construct a new current $*j_{\mathrm{new}}$, a combination of the axial- and topological current, which is conserved

$$*j_{\mathrm{comb}} = *j_5 + 2K \qquad \text{with } d * j_{\mathrm{comb}} = 0. \tag{6.167}$$

Instanton solution: How do we find explicitly a solution A satisfying the boundary condition (6.157) and classified by a winding number? We start from the **positivity** in Euclidean space

$$\int d^4x (F^a_{\mu\nu} \pm *F^a_{\mu\nu})^2 \geq 0. \tag{6.168}$$

Since

$$(F^a_{\mu\nu} \pm *F^a_{\mu\nu})^2 = 2(F^a_{\mu\nu}F^{a\mu\nu} \pm F^a_{\mu\nu} * F^{a\mu\nu}) \tag{6.169}$$

(note that $*F^a_{\mu\nu} * F^{a\mu\nu} = F^a_{\mu\nu}F^{a\mu\nu}$) we obtain

$$\int d^4x (F^a_{\mu\nu}F^{a\mu\nu} \pm F^a_{\mu\nu} * F^{a\mu\nu}) \geq 0. \tag{6.170}$$

Inserting the Euclidean action (6.155) and the topological charge (6.160) we find

$$S_E \geq 8\pi^2 |q|. \tag{6.171}$$

(Note, if we keep the strong coupling constant g_s in the definition of $A = g_s A^a T^a$ the bound factor $8\pi^2$ becomes $8\pi^2/g_s^2$.) Clearly the inequality (6.168) and consequently the lower bound (6.171) for the action is saturated (equality is achieved) if

$$F = \pm * F. \tag{6.172}$$

This means F being **selfdual** (+ sign) or **antiselfdual** (− sign). Then the YM field equation (6.63) is obeyed automatically due to the Bianchi identity $DF = 0$

$$D^\dagger F = *D * F = \pm * DF = 0. \tag{6.173}$$

The topological charge is positive $q > 0$ for a selfdual field strength, $F = *F$, and negative $q < 0$ for an antiselfdual F, $F = - * F$. We will consider just the selfdual solutions in the following.

Now we construct a field configuration A explicitly which obeys the boundary condition (6.157) and the selfduality (6.172) of F. We remember that the gauge element $g \in SU(2)$ is an element of the group of unitary unimodular two-by-two matrices. Such an element g can be written in terms of the basis $(\mathbf{1}, \vec{\sigma})$, where $\vec{\sigma} = (\sigma^1, \sigma^2, \sigma^3)$ are the Pauli matrices (3.281)

$$g(x) = \frac{1}{r}(x_4 + i\vec{x}\vec{\sigma}) \qquad \text{with } x_4^2 + \vec{x}^2 = r^2. \tag{6.174}$$

This map from $S^3 \to SU(2) \simeq S^3$—the **identity map**—corresponds to winding number $n = 1$. It is the $SU(2)$ analogue of the $U(1)$ case (2.21) which we discussed in Section 2.2. We obtain **maps of higher winding numbers** n by substituting $g \to g^n$. Clearly the inverse element is given by

$$g^{-1}(x) = \frac{1}{r}(x_4 - i\vec{x}\vec{\sigma}) \equiv g^\dagger(x). \tag{6.175}$$

For a solution of equation (6.172) we make the general ansatz

$$A(x) = f(r^2)g^{-1}(x)dg(x), \tag{6.176}$$

with g, g^{-1} expressed by equations (6.174), (6.175). Respecting the boundary condition (6.157) for $r \to \infty$ and avoiding a singularity at origin $r \to 0$ the function $f(r^2)$ has to satisfy the limits

$$\begin{aligned} f(r^2) &\overset{r\to\infty}{\longrightarrow} 1 \\ f(r^2) &\overset{r\to 0}{\longrightarrow} \text{const. } r^2. \end{aligned} \tag{6.177}$$

The simplest function which achieves this is

$$f(r^2) = \frac{r^2}{r^2 + \lambda^2}, \tag{6.178}$$

where the parameter λ specifies the **instanton size**. So the **BPST instanton solution** is

$$A(x) = \frac{r^2}{r^2 + \lambda^2} g^{-1}(x)dg(x), \tag{6.179}$$

with $g(x)$ given by equation (6.174). This field configuration minimizes the action for a fixed winding number, also called the **instanton number** (here $n = 1$)—dominating the functional integral (6.153)—and it is a solution of the field equation.

Topology of instantons: Now we establish the appropriate fibre-bundle set-up to describe the topological features of an instanton (see e.g. [Eguchi, Gilkey, Hanson 1980], [Nakahara 1990]).

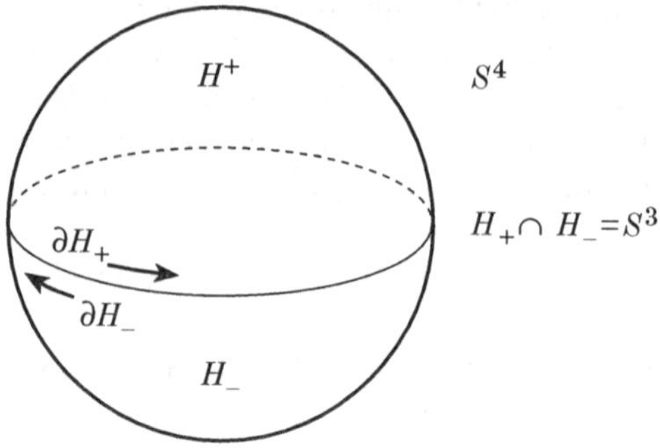

Fig. 6.6. Compactified base manifold S^4 for an instanton

We obtain the **base manifold** by compactifying the Euclidean space $E^4 = \mathbf{R}^4$ to a 4-dimensional sphere S^4 (one-point compactification: the north pole represents the origin, the south pole the points at infinity). We divide the sphere S^4 into two hemispheres H_+ and H_- having the boundary $\partial H_+ = S^3$ and $\partial H_- = -S^3$, a 3-dimensional sphere S^3 (the minus sign reflects the induced orientation). So the intersection, the 'equator', is $H_+ \cap H_- = S^3$ (see Figure 6.6).

The gauge group $SU(2)$ of the non-Abelian field theory provides the **structure group** $SU(2)$ for the bundle. For the **fibres** we also take the elements of the group $SU(2)$, so we construct a **principal bundle** $P(S^4, SU(2))$. The **local bundle patches** look like

$$\begin{matrix} H_+ \times SU(2) \\ H_- \times SU(2) \end{matrix} \quad \text{with bundle coordinates} \quad \begin{matrix} (x, f_+) \\ (x, f_-). \end{matrix} \tag{6.180}$$

The **transition functions** h_{-+} connect the fibres f between the two patches

$$f_- = h_{-+} f_+ \text{ along the 'equator' } H_+ \cap H_- = S^3, \tag{6.181}$$

they 'twist' the neighbouring fibres. A transition function corresponds to a gauge element which in our case is given by the map (6.174)

$$h_{-+}(x) = g(x) \in SU(2), \qquad x \in H_+ \cap H_- = S^3. \tag{6.182}$$

Running along the 'equator' $H_+ \cap H_- = S^3$ we wind once around the group $SU(2) \simeq S^3$. Generally, for higher windings, for instanton number n we have

$$h_{-+}(x) = g^n(x) \in SU(2), \qquad x \in H_+ \cap H_- = S^3 \tag{6.183}$$

(note that n must be integer due to $SU(2)$-matrix multiplication). For $n = 0$, the constant map

$$g^0 : x \in S^3 \mapsto \mathbf{1} \in SU(2) \tag{6.184}$$

we obtain a *trivial* fibre bundle.

The Lie algebra valued **connection 1-form** of the bundle is locally represented by the gauge potentials A_+, A_- which are defined on the hemispheres H_+, H_-. The **compatibility condition**

$$A_+ = h_{-+}^{-1} A_- h_{-+} + h_{-+}^{-1} dh_{-+} \qquad \text{on } H_+ \cap H_- = S^3 \tag{6.185}$$

describes the gauge transformation law for the YM potentials.

The Lie algebra valued **curvature 2-form** of the bundle is locally given by the field strengths F_+, F_- which are expressed by the potentials

$$F_\pm = dA_\pm + A_\pm^2 \tag{6.186}$$

on the hemispheres H_+, H_-. Again, the compatibility condition

$$F_+ = h_{-+}^{-1} F_- h_{-+} \tag{6.187}$$

corresponds to the gauge transformation of the YM field strength.

Lemma:

$$\int_{S^4} \mathrm{tr} F^2 = -\frac{1}{3} \int_{S^3} \mathrm{tr}(h_{-+}^{-1} dh_{-+})^3. \tag{6.188}$$

So the integral defining the Pontryagin index (6.160) is *independent* of the connection A. It depends *only* on the transition functions h_{-+} of the bundle which determine the 'glueings' of the fibres, and thus the topology of the bundle.

Proof:

$$\begin{aligned}
\int_{S^4} \mathrm{tr}\, F^2 &= \int_{H_+} d\, \mathrm{tr}\, \left(F_+ A_+ - \frac{1}{3} A_+^3\right) + \int_{H_-} d\, \mathrm{tr}\, \left(F_- A_- - \frac{1}{3} A_-^3\right) \\
&= \int_{S^3} \mathrm{tr}\, \left(F_+ A_+ - \frac{1}{3} A_+^3 - F_- A_- + \frac{1}{3} A_-^3\right) \\
&= \int_{S^3} \left[-\frac{1}{3}\, \mathrm{tr}\, (h_{-+}^{-1} dh_{-+})^3 + d\, \mathrm{tr}\, A_- dh_{-+} h_{-+}^{-1}\right].
\end{aligned} \tag{6.189}$$

We applied Stokes' theorem (2.128) ($\partial H_+ = S^3$, $\partial H_- = -S^3$) and we inserted the compatibility conditions (6.185), (6.187). The second term vanishes when using Stokes' theorem again, the remaining provides the Lemma (6.188). Q.E.D.

Note: Of course, we could also choose a vanishing connection on the southern hemisphere, $A_- = 0$ on H_-. Then all topological content about the bundle is carried by A_+ on the 'equator' S^3, $A_+ = h_{-+}^{-1} dh_{-+}$ on $H_+ \cap H_- = S^3$, and Lemma (6.188) follows immediately from the term $-\frac{1}{3}A_+^3$ in the second equation of the proof. (We derive a generalization of Lemma (6.188) to any even dimension within Chern–Simons forms in Section 7.5; see equations (7.114)–(7.117).)

Winding number—instanton number: Next we calculate the integral (6.188) when the transition functions are chosen à la BPST, $h_{-+} = g$, equation (6.174)

$$h_{-+}^{-1} dh_{-+} = \frac{1}{r}(x^4 - i\vec{x}\vec{\sigma}) d\frac{1}{r}(x^4 + i\vec{x}\vec{\sigma}) \qquad \text{and} \qquad x_4^2 + \vec{x}^2 = r^2. \quad (6.190)$$

The integral over S^3 simplifies comfortably when choosing as a reference point on S^3 the north pole ($x^4 = 1$, $\vec{x} = 0$) of the unit sphere, which we may do since h_{-+} maps S^3 uniformly onto $SU(2) \simeq S^3$. Then we have

$$h_{-+}^{-1} dh_{-+} = i\sigma_k dx^k \quad (6.191)$$

and

$$\begin{aligned} \text{tr } (h_{-+}^{-1} dh_{-+})^3 &= i^3 \text{ tr } \sigma_i \sigma_j \sigma_k dx^i dx^j dx^k \\ &= 2\varepsilon_{ijk} dx^i dx^j dx^k = 12 dx^1 dx^2 dx^3, \end{aligned} \quad (6.192)$$

where the wedge product $dx^1 dx^2 dx^3$ is an 'area' element at the north pole of the 3-dimensional sphere S^3, and consequently the integral is

$$\int_{S^3} \text{tr } (h_{-+}^{-1} dh_{-+})^3 = 12 \int_{S^3} dx^1 dx^2 dx^3 = 12 \cdot 2\pi^2 = 24\pi^2 \quad (6.193)$$

since the area of the unit sphere S^3 is $2\pi^2$ (generally, the area of S^n is $a(S^n) = 2\pi^{(n+1)/2} R^n / \Gamma\left(\frac{n+1}{2}\right)$).

Altogether we obtain for the **Pontrjagin index—the topological charge**

$$q = -\frac{1}{8\pi^2} \int_{S^4} \text{tr } F^2 = \frac{1}{24\pi^2} \int_{S^3} \text{tr } (h_{-+}^{-1} dh_{-+})^3 = 1, \quad (6.194)$$

a value which agrees with the winding number 1 of the map h_{-+} or with the instanton number 1.

Finally, for maps with higher winding numbers like

$$h_{-+} = g^2 : S^3 \to SU(2) \simeq S^3 \quad (6.195)$$

we split S^3 into hemispheres H_+, H_-. We may always deform homotopically the map g with winding number 1 as

$$g \to \begin{cases} g_+ & \text{on } H_+ \\ \mathbf{1} & \text{on } H_- \end{cases} \qquad \text{or} \qquad g \to \begin{cases} \mathbf{1} & \text{on } H_+ \\ g_- & \text{on } H_- \end{cases},$$

where g_+, g_- have winding number 1. Then we get for the map (6.195)

$$h_{-+} = g \cdot g \to \begin{cases} g_+ \cdot \mathbf{1} & \text{on } H_+ \\ \mathbf{1} \cdot g_- & \text{on } H_-. \end{cases} \tag{6.196}$$

Consequently we find (by the homotopic invariance of the winding number)

$$\begin{aligned} \frac{1}{24\pi^2} \int_{S^3} \text{tr } (g^{-2} dg^2)^3 &= \frac{1}{24\pi^2} \left[\int_{H_+} \text{tr } (g_+^{-1} dg_+)^3 + \int_{H_-} \text{tr } (g_-^{-1} dg_-)^3 \right] \\ &= \frac{1}{24\pi^2} \cdot 24\pi^2 [1+1] = 2 \end{aligned} \tag{6.197}$$

and iteratively all higher winding numbers.

Proposition:

- *The Pontrjagin index—the topological charge—is a topological invariant*

$$q = -\frac{1}{8\pi^2} \int_{S^4} \text{tr } F^2 = \frac{1}{24\pi^2} \int_{S^3} \text{tr } (h_{-+}^{-1} dh_{-+})^3 = n \tag{6.198}$$

determined only by the transition functions h_{-+} of the fibre bundle and their homotopy classes. It represents the instanton number or winding number of the map

$$h_{-+} : S^3 \to SU(2) \simeq S^3 \qquad \text{with } h_{-+} = g^n \tag{6.199}$$

and g given by the identity map (6.174). It counts the number of times the group $SU(2)$ covers the sphere S^3 under the mapping h_{-+}.

Last, but not least, the Pontrjagin index corresponds to the **second Chern number** C_2 or to the **Atiyah–Singer index theorem**, which we will discuss much later (see Section 11.1)

$$q = -C_2 = \text{index } D_+. \tag{6.200}$$

Résumé: There is a remarkable coherence between physics and geometry. In fact, fibre bundles appear as the natural geometric set-up for describing the physics of gauge theories. Assuming some internal symmetry group—gauge symmetry—the fibre bundle automatically supplies a *pure geometric* interpretation for the concepts of gauge theory, specifically for the gauge potential, the field strength, the gauge element and gauge transformations. In addition it provides a geometric-topological reason for the quantized charge of the magnetic monopole and of the instanton, or, as we shall see later on, for the anomaly of quantum field theory.

This elegant geometric topology is important for our understanding of gauge theories, and we also use it to formulate further aspects—the anomalies—in terms of this fibre bundle technique. Finally, we also find it very helpful to compose a brief *aide-memoire* in form of a list which summarizes the corresponding concepts in physics and geometry.

Aide-memoire	
Physics	**Geometry**
gauge theory	principal fibre bundle
space–time	base manifold
set of all possible gauge symmetry elements	fibre
gauge group	structure group
gauge element	transition function
gauge potential	connection
gauge field strength	curvature
gauge transformations	compatibility conditions
wave function	section of fibre bundle
phase factor of wave function	parallel transport of fibre
Aharonov–Bohm effect	holonomy in fibre space
magnetic monopole charge	winding number classified according to homotopy group $\Pi_1(U(1))$ or Chern number C_1, Atiyah–Singer index theorem
instanton number	winding number according to $\Pi_3(SU(2))$ or C_2, Pontrjagin index, Atiyah–Singer index theorem, topological charge
anomaly of QFT	Atiyah–Singer index theorem, Bismut index theorem, determinant bundle

7
Chern–Simons form, homotopy operator and anomaly

The Chern–Simons form and the homotopy operator play an important role in connection with anomalies. In fact, we can calculate the anomaly on pure algebraic grounds from a variation of the Chern–Simons form by using a homotopy operator.

First we need to discuss a symmetric invariant polynomial of fields (Section 7.1) which is the starting point for the derivation of the Chern–Simons form and 'transgression formula' (Section 7.2). Next we prove the important Poincaré lemma and introduce in this connection a homotopy operator (Section 7.3). A generalization of the 'transgression'—the Cartan homotopy formula—follows in Section 7.4. Then we come to the important part of this chapter; we apply the homotopy formula to a Chern–Simons form with gauge transformed fields and are able to derive in this way the non-Abelian anomaly (Section 7.5). A general formula for the variation of the Chern–Simons form—which expresses the anomaly—is presented at the end (Section 7.6).

7.1 Invariant polynomials

We discuss the properties of a polynomial function of some matrices. For literature we refer to [Eguchi, Gilkey, Hanson 1980], [Nakahara 1990], [Alvarez-Gaumé, Ginsparg 1985].

Definition: Let $\{\alpha_i\}$ be a set of complex $k \times k$ matrices and consider the vector space of **C**-valued functions which are linear and symmetric in each pair α_i

$$P(\alpha_1, \ldots, \alpha_i, \ldots, \alpha_j, \ldots, \alpha_n) = P(\alpha_1, \ldots, \alpha_j, \ldots, \alpha_i, \ldots, \alpha_n), \qquad (7.1)$$

with $1 \leq i, j \leq n$. Choose some group G, in practice $G = GL(k, \mathbf{C})$, $SU(k)$, $U(k)$, and Lie G denotes its Lie algebra; then P is called a **symmetric invariant polynomial** or **characteristic polynomial** if

$$P(g^{-1}\alpha_1 g, \dots, g^{-1}\alpha_n g) = P(\alpha_1, \dots, \alpha_n) \tag{7.2}$$

for all $g \in G$ and $\alpha_i \in$ Lie G.

We now use the invariance of such a polynomial under transformations

$$\alpha_i \to g^{-1}\alpha_i g \tag{7.3}$$

by choosing

$$g \equiv g_t = e^{t\beta}, \tag{7.4}$$

with $\beta \in$ Lie G and $t \in \mathbf{R}$ some parameter, and we differentiate equation (7.2) with respect to t

$$P(g_t^{-1}(-\beta\alpha_1)g_t, \dots, g_t^{-1}\alpha_n g_t) + P(g_t^{-1}(\alpha_1\beta)g_t, \dots, g_t^{-1}\alpha_n g_t) + \dots$$
$$P(g_t^{-1}\alpha_1 g_t, \dots, g_t^{-1}(-\beta\alpha_n)g_t) + P(g_t^{-1}\alpha_1 g_t, \dots, g_t^{-1}(\alpha_n\beta)g_t) = 0. \tag{7.5}$$

For $t = 0$ we obtain the following equation

$$\sum_{i=1}^{n} P(\alpha_1, \dots, \alpha_{i-1}, [\beta, \alpha_i], \alpha_{i+1}, \dots, \alpha_n) = 0. \tag{7.6}$$

Definition: If all α_i are equal then P is an **invariant polynomial of degree** n with notation

$$P(\alpha, \dots, \alpha) \equiv P_n(\alpha) \equiv P(\alpha^n). \tag{7.7}$$

Example: The symmetrized trace str represents such a symmetric invariant polynomial

$$P(\alpha_1, \dots, \alpha_n) = \text{str } (\alpha_1, \dots, \alpha_n) := \sum_{\text{Perm}} \frac{1}{n!} \text{ tr } \alpha_{p_1} \dots \alpha_{p_n}. \tag{7.8}$$

For $n = 3$ we get

$$\begin{aligned} P(\alpha_1, \alpha_2, \alpha_3) &= \text{str } (\alpha_1, \alpha_2, \alpha_3) \\ &= \frac{1}{2}(\text{tr } \alpha_1\alpha_2\alpha_3 + \text{tr } \alpha_1\alpha_3\alpha_2) \end{aligned} \tag{7.9}$$

when using the cyclicity of the trace. If all α_i are equal then a polynomial (7.7) is given solely by the trace

$$P_n(\alpha) = P(\alpha^n) = \text{tr } \alpha^n. \tag{7.10}$$

Next we extend the definition of an invariant polynomial of matrices to that of matrix valued (elements of Lie G) differential forms on a manifold M. Let α_i be a differential form of order d_i

$$\alpha_i = \alpha_{d_i} E^{d_i} := \alpha_{\mu_1 \dots \mu_{d_i}}(x) \frac{1}{d_i!} dx^{\mu_1} \dots dx^{\mu_{d_i}} \tag{7.11}$$

where α_{d_i} and E^{d_i} are obvious short-hand notations for the tensor field and the wedge product. The tensor α_{d_i} is an element of the Lie algebra of G via the generators T^a

$$\alpha_{d_i} = \alpha^a_{d_i} T^a \in \text{Lie } G. \tag{7.12}$$

Definition: An **invariant polynomial** for Lie algebra valued d_i-forms α_i is defined by

$$P(\alpha_1, \dots, \alpha_n) = P(\alpha_{d_1}, \dots, \alpha_{d_n}) E^{d_1} \dots E^{d_n}, \tag{7.13}$$

together with equations (7.1), (7.2).

Note: The product $E^{d_1} \dots E^{d_n}$ denotes again a wedge product. Thus we obtain the following as a rule: extract the form part and treat the matrix part as an invariant polynomial (7.2).

Let β be a Lie algebra valued 1-form

$$\beta = \beta_\mu(x) dx^\mu, \qquad \beta_\mu = \beta^a_\mu T^a \in \text{Lie } G, \tag{7.14}$$

then we can write the following quantity according to the above rules

$$\begin{aligned} &P(\alpha_1, \dots, \alpha_{i-1}, [\beta, \alpha_i], \alpha_{i+1}, \dots, \alpha_n) \\ &= P(\alpha_{d_1}, \dots, \alpha_{d_{i-1}}, [\beta_\mu, \alpha_{d_i}], \alpha_{d_{i+1}}, \dots, \alpha_{d_n})(-)^{d_1 + \dots d_{i-1}} dx^\mu E^{d_1} \dots E^{d_n}, \end{aligned} \tag{7.15}$$

where the dx^μ has been pulled through the wedge products providing the corresponding minus factor. Applying equation (7.6) to the matrix part we obtain the identity

$$\sum_{i=1}^n (-)^{d_1 + \dots + d_{i-1}} P(\alpha_1, \dots, \alpha_{i-1}, [\beta, \alpha_i], \alpha_{i+1}, \dots, \alpha_n) = 0. \tag{7.16}$$

Choosing a p-form

$$\eta = \eta_p(x) E^p, \qquad \eta_p = \eta^a_p T^a \in \text{Lie } G \tag{7.17}$$

instead of the 1-form β just changes the minus factor of the identity (7.16) in an obvious way

$$\sum_{i=1}^{n}(-)^{p(d_1+\ldots+d_{i-1})}P(\alpha_1,\ldots,\alpha_{i-1},[\eta,\alpha_i],\alpha_{i+1},\ldots,\alpha_n)=0. \tag{7.18}$$

Now we apply the exterior derivative d to the invariant polynomial of differential forms

$$dP(\alpha_1,\ldots,\alpha_n)=\sum_{i=1}^{n}(-)^{d_1+\ldots+d_{i-1}}P(\alpha_1,\ldots,\alpha_{i-1},d\alpha_i,\alpha_{i+1},\ldots,\alpha_n) \tag{7.19}$$

and we add the identity (7.16) for $\beta = A$ being a 1-form connection; then we gain the following frequently used formula.

Formula:

$$dP(\alpha_1,\ldots,\alpha_n)=\sum_{i=1}^{n}(-)^{d_1+\ldots d_{i-1}}P(\alpha_1,\ldots,\alpha_{i-1},D\alpha_i,\alpha_{i+1},\ldots,\alpha_n), \tag{7.20}$$

where D expresses the covariant derivative $D = d + [A,\ ]$.

Example: Let us choose the symmetrized trace str

$$\begin{aligned} P(\alpha_1,\ldots,\alpha_n) &= \text{str }(\alpha_1,\ldots,\alpha_n)=\text{str }(\alpha_{d_1},\ldots,\alpha_{d_n})E^{d_1}\ldots E^{d_n} \\ &= \sum_{\text{Perm}}\frac{1}{n!}\text{ tr }\alpha_{p(d_1)}\cdots\alpha_{p(d_n)}\cdot E^{d_1}\ldots E^{d_n}. \end{aligned} \tag{7.21}$$

For instance, for $n = 3$ we get

$$\text{str }(\alpha_1,\alpha_2,\alpha_3)=(-)^{d_2d_3}\text{ str }(\alpha_1,\alpha_3,\alpha_2) \tag{7.22}$$

and on the other hand

$$\begin{aligned} \text{str }(\alpha_1,\alpha_2,\alpha_3) &= \frac{1}{2}(\text{tr }\alpha_{d_1}\alpha_{d_2}\alpha_{d_3}+\text{tr }\alpha_{d_1}\alpha_{d_3}\alpha_{d_2})E^{d_1}E^{d_2}E^{d_3} \\ &= \frac{1}{2}(\text{tr }\alpha_1\alpha_2\alpha_3+(-)^{d_2d_3}\text{ tr }\alpha_1\alpha_3\alpha_2). \end{aligned} \tag{7.23}$$

Definition: If all differential forms are equal $\alpha_i \equiv \alpha$ and α represents a Lie algebra valued p-form $\alpha = \alpha_p E^p$ then P is of degree n with notation

$$P(\alpha,\ldots,\alpha)=P(\alpha_p,\ldots,\alpha_p)E^p\ldots E^p\equiv P_n(\alpha)\equiv P(\alpha^n) \tag{7.24}$$

and, for example, is just the trace

$$P(\alpha^n)=\text{tr }\alpha^n. \tag{7.25}$$

7.2 Transgression formula and Chern–Simons form

Next we are going to construct a certain polynomial in the gauge connection and curvature—the Chern–Simons form.

Let us consider an invariant polynomial of degree n in the curvature 2-form $\Omega \in \text{Lie}\, G \otimes \Lambda^2(M)$ which satisfies

$$P_n(g^{-1}\Omega g) = P_n(\Omega) \tag{7.26}$$

for all gauge elements $g \in G$. The symbol Ω will denote either the gauge field strength tensor $\Omega \to F$ of the Yang–Mills theory or the curvature $\Omega \to R$ of gravitation. The group G is then either the YM gauge group or the frame rotation group. The corresponding connection ω satisfying

$$\Omega = d\omega + \omega^2 \tag{7.27}$$

expresses respectively the gauge potential $\omega \to A$ or the spin connection.

The invariant polynomial (7.26) has the following important properties [Stora 1977], [Eguchi, Gilkey, Hanson 1980], [Stora 1984], [Zumino 1984], [Alvarez-Gaumé, Ginsparg 1985]:

Theorem: Let $P_n(\Omega)$ be an invariant polynomial, then

i) $P_n(\Omega)$ is closed, $dP_n(\Omega) = 0$,

ii) $P_n(\Omega)$ has topologically invariant integrals, or, equivalently, the difference of two invariant polynomials is exact—**'transgression'**

$$P_n(\Omega_1) - P_n(\Omega_0) = dQ_{2n-1}(\omega_1, \omega_0) \tag{7.28}$$

$$Q_{2n-1}(\omega_1, \omega_0) = n \int_0^1 dt P(\omega_1 - \omega_0, \Omega_t^{n-1}), \tag{7.29}$$

with

$$\Omega_t = d\omega_t + \omega_t^2, \quad \omega_t = \omega_0 + t(\omega_1 - \omega_0). \tag{7.30}$$

In fact, any invariant polynomial $P(\Omega)$ being a sum of products of $P_n(\Omega)$ satisfies properties i) and ii). The index $2n-1$ indicates the form degree of Q_{2n-1}.

Proof.

i) By formula (7.20), choosing all $\alpha_i = \Omega$, we have

$$dP_n(\Omega) = \sum_{i=1}^{n} P(\Omega, \ldots, \Omega, \left.D\Omega\right|_i, \Omega, \ldots, \Omega), \tag{7.31}$$

where $D\Omega$ stands on the i-th place, and using the Bianchi identity $D\Omega = 0$ (recall Section 6.2) we have proved $dP_n(\Omega) = 0$.

ii) Let ω_0, ω_1 be two connections on a bundle referred to the same system of local trivializations and Ω_0, Ω_1 denote the corresponding curvatures. As usual we consider a principal bundle (or its associated vector bundle). We define a homotopic connection

$$\omega_t = \omega_0 + t\beta, \qquad \beta = \omega_1 - \omega_0, \tag{7.32}$$

with $t \in [0, 1]$; the homotopic curvature is then given by

$$\Omega_t = d\omega_t + \omega_t^2 = \Omega_0 + tD\beta + t^2\beta^2, \tag{7.33}$$

where D expresses the covariant derivative containing ω_0

$$D = d + [\omega_0, \;]. \tag{7.34}$$

The homotopies continuously deform the fields between

$$\omega_{t=0} = \omega_0, \qquad \Omega_{t=0} = \Omega_0 = d\omega_0 + \omega_0^2 \tag{7.35}$$

and

$$\omega_{t=1} = \omega_1, \qquad \Omega_{t=1} = \Omega_1 = d\omega_1 + \omega_1^2. \tag{7.36}$$

We differentiate Ω_t with respect to t

$$\begin{aligned} \frac{d}{dt}\Omega_t &= \frac{d}{dt}(\Omega_0 + tD\beta + t^2\beta^2) = D\beta + 2t\beta^2 \\ &= d\beta + [\omega_t, \beta] = D_t\beta, \end{aligned} \tag{7.37}$$

where D_t denotes the covariant derivative including ω_t

$$D_t = d + [\omega_t, \;]. \tag{7.38}$$

Next we consider the invariant polynomial $P_n(\Omega_t)$ and differentiate with respect to t

$$\begin{aligned} \frac{d}{dt}P_n(\Omega_t) &= nP(\frac{d}{dt}\Omega_t, \Omega_t, \ldots, \Omega_t) \\ &= nP(D_t\beta, \Omega_t^{n-1}). \end{aligned} \tag{7.39}$$

The formula (7.20) together with the Bianchi identity $D_t\Omega_t = 0$ provides

$$\frac{d}{dt}P_n(\Omega_t) = ndP(\beta, \Omega_t^{n-1}), \tag{7.40}$$

and integrating from $t = 0$ to $t = 1$ we obtain

$$P_n(\Omega_1) - P_n(\Omega_0) = dQ_{2n-1}(\omega_1, \omega_0), \tag{7.41}$$

with

$$Q_{2n-1}(\omega_1, \omega_0) = n\int_0^1 dtP(\omega_1 - \omega_0, \Omega_t^{n-1}). \qquad \text{Q.E.D.} \tag{7.42}$$

The **'transgression formula'** (7.28)–(7.30) is a basic relation for our discussions in the following chapters. It is—as we shall see in Section 7.4—a special case of Cartan's homotopy formula.

Integrating equation (7.41) over a $2n$-dimensional compact manifold M_{2n} without boundary ($\partial M_{2n} = \emptyset$) provides

$$\begin{aligned}\int_{M_{2n}} P_n(\Omega_1) - \int_{M_{2n}} P_n(\Omega_0) &= \int_{M_{2n}} dQ_{2n-1}(\omega_1, \omega_0) \\ &= \int_{\partial M_{2n}} Q_{2n-1}(\omega_1, \omega_0) = 0 \end{aligned} \tag{7.43}$$

by virtue of Stokes' theorem (2.128) and $\partial M_{2n} = \emptyset$. Thus the integrals of invariant polynomials are *invariant* under changes of the connection ω; they represent characteristic numbers of the bundle—Chern numbers—and depend only on the transition functions of the bundle (recall our topology discussions in Chapter 6). They are also invariant under deformations of the manifold M_{2n}.

Chern–Simons form: Choosing finally in formula (7.41) and (7.42) $\omega_0 = 0$ and consequently $\Omega_0 = 0$—which we are allowed to do *only* if the bundle is trivial, or on a local chart where the bundle is trivial—we find

$$P_n(\Omega) = dQ_{2n-1}(\omega, \Omega), \tag{7.44}$$

together with the **Chern–Simons form**

$$Q_{2n-1}(\omega, \Omega) = n\int_0^1 dtP(\omega, \Omega_t^{n-1}) \tag{7.45}$$

and

$$\omega_t = t\omega, \quad \Omega_t = d\omega_t + \omega_t^2 = t\Omega + (t^2 - t)\omega^2. \tag{7.46}$$

Then $\omega_t = t\omega$ is not a connection defined globally on a nontrivial bundle.

Note: Equation (7.44) also follows from Poincaré's lemma, which states that a closed form like $P_n(\Omega)$ is *locally* always exact (see Section (7.3)).

Examples: We illustrate formula (7.45) by some examples. For instance, for YM fields and $n = 2$ we recover the formula for the singlet anomaly (6.75) in 4 dimensions

$$\text{tr } F^2 = dQ_3, \tag{7.47}$$

with

$$\begin{aligned} Q_3(A,F) &= 2\int_0^1 dt \text{ tr } A[tF + (t^2 - t)A^2] \\ &= \text{tr } [AF - \frac{1}{3}A^3] = \text{tr } [AdA + \frac{2}{3}A^3]. \end{aligned} \tag{7.48}$$

For $n = 3$ we obtain the analoguous result in 6 dimensions

$$\text{tr } F^3 = dQ_5, \tag{7.49}$$

with

$$\begin{aligned} Q_5(A,F) &= \text{tr } [AF^2 - \frac{1}{2}A^3F + \frac{1}{10}A^5] \\ &= \text{tr } [A(dA)^2 + \frac{3}{2}A^3dA + \frac{3}{5}A^5]. \end{aligned} \tag{7.50}$$

Remark: As noticed by Stora [Stora 1977] formulae (7.41), (7.42) can be lifted to the principal bundle $P(M,G)$ and one also arrives at formulae like (7.44)–(7.46). But then the formulae contain for $\omega \to \omega_{global}$ the *global* connection, for $\Omega \to \Omega_{global}$ the *global* curvature on $P(M,G)$ and for $d \to \Delta$ the exterior derivative on the bundle with $\Omega_{global} = \Delta\omega_{global} + \omega^2_{global}$.

7.3 Poincaré lemma and homotopy operator

We have to discuss an important proposition—**Poincaré's lemma**. It is instructive to introduce in this connection the **homotopy operator** to gain more insight into its action on differential forms. Therefore we want to elaborate the following proof in detail (for literature we refer to e.g. [Flanders 1963], [Singer, Thorpe 1967], [Curtis, Miller 1985], [Nakahara 1990]).

Proposition: Poincaré's lemma

- *If a domain U in a manifold M is contractible to a point and ω is a p-form on $U \subset M$ with $d\omega = 0$ then there exists a $(p-1)$-form α on U such that $\omega = d\alpha$.*

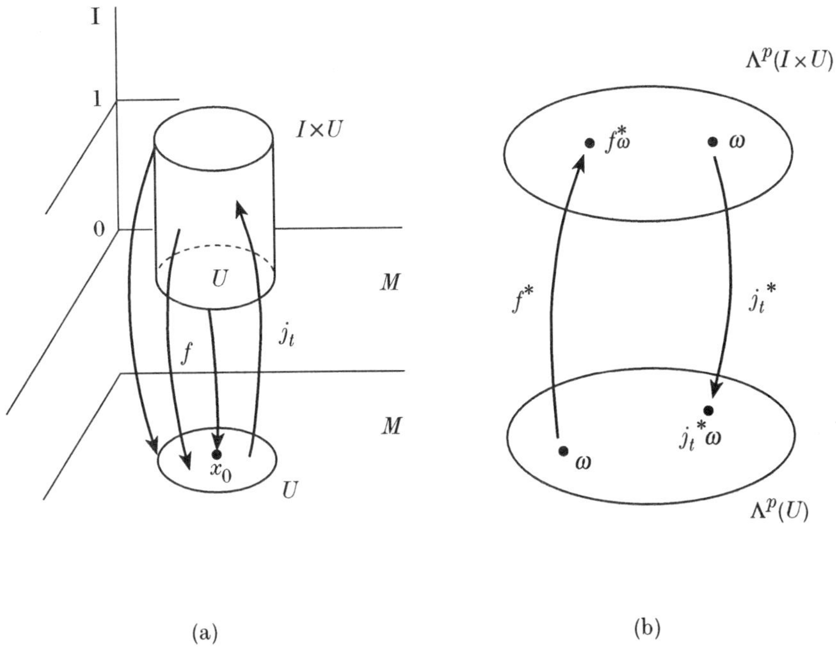

Fig. 7.1. a) Illustration of the maps $j_t(x)$ and $f(t,x)$. b) Pullbacks under these maps in the spaces of the p-forms

Briefly, any closed form ($d\omega = 0$) on M is locally always exact ($\omega = d\alpha$), where locally refers to a coordinate neighbourhood which is contractible to a point.

Proof. We first specify the contraction of a domain to a point. A topological space U is **contractible** if there exists a homotopy (recall Section 2.2)

$$f : I \times U \to U \tag{7.51}$$

such that

$$f(1,x) = x = id_U(x) \quad \text{and} \quad f(0,x) = x_0 \in U, \quad \forall x \in U. \tag{7.52}$$

We construct a cylinder, the product space (see Figure 7.1a))

$$I \times U = \{(t,x)\} \quad \text{with} \quad t \in I = [0,1] \tag{7.53}$$

and let U be contractible according to our presumption. We also define a map from the domain into the cylinder

$$j_t : U \to I \times U \quad \text{with} \quad j_t(x) := (t, x). \tag{7.54}$$

The two maps $j_1(x) = (1, x)$ and $j_0(x) = (0, x)$ identify the domain U with the top and bottom of the cylinder (see Figure 7.1a)).

Let us consider a p-form on $I \times U$ such as

$$\omega = a_p(t, x)E^p + b_{p-1}(t, x)dtE^{p-1} \tag{7.55}$$

by working with the short-hand notation of Section 7.1 (equation (7.11)). Then the pullback (recall Section 2.6.2) under the map (7.54) is (see Figure 7.1b))

$$j_t^* : \Lambda^p(I \times U) \to \Lambda^p(U)$$

$$j_t^*\omega = \omega(t, x)|_{t=const.} = a_p(t, x)E^p \in \Lambda^p(U). \tag{7.56}$$

Definition: Next we define an **operator** K

$$K : \Lambda^p(I \times U) \to \Lambda^{p-1}(U) \tag{7.57}$$

by

$$Ka_p(t, x)E^p = 0 \tag{7.58}$$

$$Kb_{p-1}(t, x)dtE^{p-1} = \left(\int_0^1 b_{p-1}(t, x)dt\right) \cdot E^{p-1}. \tag{7.59}$$

Therefore we have

$$K\omega = \left(\int_0^1 b_{p-1}(t, x)dt\right) \cdot E^{p-1}. \tag{7.60}$$

Lemma:

$$K(d\omega) + d(K\omega) = j_1^*\omega - j_0^*\omega \tag{7.61}$$

valid for any $\omega \in \Lambda^p(I \times U)$.

Proof. We prove equation (7.61) by direct calculation

$$\begin{aligned} K(d\omega) &= K(\partial_j a_p(t, x)dx^j E^p + \frac{\partial}{\partial t}a_p(t, x)dtE^p) \\ &\quad + K\partial_j b_{p-1}(t, x)dx^j dtE^{p-1}, \end{aligned}$$

$$K(d\omega) = \left(\int_0^1 \frac{\partial}{\partial t}a_p(t, x)dt\right) \cdot E^p - \left(\int_0^1 \partial_j b_{p-1}(t, x)dt\right) \cdot dx^j E^{p-1}. \tag{7.62}$$

On the other hand, equation (7.60) gives

$$d(K\omega) = \left(\int_0^1 \partial_j b_{p-1}(t,x)dt\right) \cdot dx^j\, E^{p-1} \tag{7.63}$$

so that altogether we obtain

$$\begin{aligned} K(d\omega) + d(K\omega) &= \left(\int_0^1 \frac{\partial}{\partial t} a_p(t,x)dt\right) \cdot E^p \\ &= j_1^*\omega - j_0^*\omega. \qquad \text{Q.E.D.} \end{aligned} \tag{7.64}$$

We also notice by recalling the contractibility of the domain U that the product map $f \cdot j_t : U \to U$ via $U \xrightarrow{j_t} I \times U \xrightarrow{f} U$ has the boundary conditions (see Figure 7.1a)) $f \cdot j_1 = id_U$, $f \cdot j_0 = x_0$, which implies for the pullbacks of a p-form $\omega \in \Lambda^p(U)$

$$\begin{aligned} j_1^* \cdot f^*\omega &= (f \cdot j_1)^*\omega = \omega \\ j_0^* \cdot f^*\omega &= (f \cdot j_0)^*\omega = 0. \end{aligned} \tag{7.65}$$

Now we are prepared to perform the actual proof of Poincaré's lemma and we prove it by explicit construction. The $(p-1)$-form

$$\alpha = Kf^*\omega \tag{7.66}$$

will be the desired result such that the closed form ω is also exact on the contractible domain U

$$\omega = d\alpha. \tag{7.67}$$

We start from relation (7.61) and replace $\omega \in \Lambda^p(I \times U) \to f^*\omega \in \Lambda^p(I \times U)$. Then we obtain

$$\begin{aligned} K(df^*\omega) + d(Kf^*\omega) &= j_1^* \cdot f^*\omega - j_0^* \cdot f^*\omega \\ &= (f \cdot j_1)^*\omega - (f \cdot j_0)^*\omega. \end{aligned} \tag{7.68}$$

On the left-hand side the first term vanishes since ω is closed ($df^*\omega = f^*d\omega = 0$). On the right-hand side ω remains due to the pullback properties (7.65). Thus we are left with

$$\omega = d(Kf^*\omega) \quad \text{and} \quad \alpha = Kf^*\omega \quad \text{on} \quad U. \qquad \text{Q.E.D.} \tag{7.69}$$

Of course the solution α is not unique, we always can add an arbitrary exact form

$$\alpha \to \alpha + d\beta. \tag{7.70}$$

Homotopy operator: In proving Poincaré's lemma we introduced the operator K (7.57)–(7.59) which acts on p-forms defined on $I \times U$. Usually, however, we work with p-forms $\omega \in \Lambda^p(U)$ defined on U. So we lift ω to the cylinder $I \times U$ via the pullback f^* (see Figure 7.1b)) and apply the operator K there.

Definition: The **homotopy operator** is defined by

$$k = Kf^*, \tag{7.71}$$

with K given by equations (7.57)–(7.59).

Lemma:

$$kd + dk = \mathbf{1}. \tag{7.72}$$

Proof. Starting with relation (7.68) for the K operator, interchanging $df^* = f^*d$, and using the pullback properties (7.65) we find

$$\begin{aligned} Kf^*d\omega + dKf^*\omega &= \omega \\ kd\omega + dk\omega &= \omega \end{aligned} \tag{7.73}$$

for any $\omega \in \Lambda^p(U)$. Q.E.D.

Operator $kd + dk$ thus shows that a closed form ω ($d\omega = 0$) on a contractible space is an exact form ($\omega = d\alpha$)

$$(kd + dk)\omega = \omega, \qquad d(k\omega) = \omega \quad \text{and} \quad \alpha = k\omega \tag{7.74}$$

demonstrating Poincaré's lemma. Note that the application of k reduces the form degree of ω by one unit (because of property (7.59)).

Theorem: The homotopy operator k is nilpotent

$$k^2 = 0. \tag{7.75}$$

Proof. We start with a p-form on a contractible domain U, $\omega = a_p(x)E^p \in \Lambda^p(U)$, and we choose for simplicity the contraction to the zero-point $f(t,x) = tx$. Then the homotopy operator k acts as

$$\begin{aligned} k\omega &= Kf^*a_p(x)E^p \\ &= Kt^p a_p(tx)E^p + Kt^{p-1}x^j a_{ji_1\ldots i_{p-1}}(tx)dtE^{p-1} \\ &= x^j\left(\int_0^1 t^{p-1}a_{ji_1\ldots i_{p-1}}(tx)dt\right)E^{p-1}. \end{aligned} \tag{7.76}$$

Repeating the same procedure gives

$$
\begin{aligned}
k^2\omega &= Kf^*x^j \left(\int_0^1 t^{p-1} a_{ji_1\ldots i_{p-1}}(tx)dt \right) E^{p-1} \\
&= x^j x^\ell \left(\int_0^1 dt \int_0^1 ds\, t^{p-1} s^{p-1} a_{j\ell i_1\ldots i_{p-2}}(stx) \right) E^{p-2} \\
&= 0. \qquad \text{Q.E.D.}
\end{aligned} \tag{7.77}
$$

(Note: $x^j x^l$ is symmetric, whereas $a_{jli_1\ldots i_{p-2}}$ is antisymmetric.)

Algebraic Poincaré lemma: There is also a generalization of Poincaré's lemma, given by Stora [Stora 1977, 1984], which we shall use frequently in this book.

Proposition: Algebraic Poincaré lemma

- *Let $P_p[\phi](x)$ and $Q_{p-1}[\phi](x)$ be forms of degree p and $p-1$ which are local functionals in the basic fields—$[\phi](x)$ denotes collectively all fields and their derivatives—then $P_p = dQ_{p-1}$ iff $dP_p = 0$, for $p+1 \leq m = \dim M$, or iff $\int P_p = 0$, for $p = m$.*

7.4 Cartan homotopy formula

In this section we introduce a homotopy operator on pure algebraic grounds and derive the Cartan homotopy formula. For literature we refer to [Stora 1984], [Zumino 1984], [Alvarez-Gaumé, Ginsparg 1985] and [Mañes, Stora, Zumino 1985].

We start again with a homotopic YM connection as in equation (7.32)

$$A_t = A_0 + t(A_1 - A_0), \qquad t \in [0,1] \tag{7.78}$$

and with the corresponding curvature

$$F_t = dA_t + A_t^2. \tag{7.79}$$

Definition: We define an operator ℓ_t—**homotopy derivation**—by

$$\ell_t A_t = 0 \tag{7.80}$$

$$\ell_t F_t = \delta A_t = dt \frac{\partial}{\partial t} A_t = dt(A_1 - A_0) \tag{7.81}$$

acting on polynomials in A_t and F_t; we require the **antiderivation rule**

$$\ell_t(\alpha_p \beta_q) = (\ell_t \alpha_p)\beta_q + (-)^p \alpha_p (\ell_t \beta_q) \tag{7.82}$$

for the p-forms $\alpha_p \in \Lambda^p$ and q-forms $\beta_q \in \Lambda^q$. The operator ℓ_t decreases the form degree in dx^i by one unit and creates a dt. (Here we follow the approach of [Zumino 1984] and [Alvarez-Gaumé, Ginsparg 1985]; in the work of [Mañes, Stora, Zumino 1985] ℓ_t is defined as an even derivation.)

We can verify explicitly

$$(\ell_t d + d\ell_t)A_t = \ell_t dA_t = \ell_t(F_t - A_t^2) = \delta A_t = dt\frac{\partial}{\partial t}A_t \tag{7.83}$$

$$\begin{aligned}(\ell_t d + d\ell_t)F_t &= \ell_t(F_tA_t - A_tF_t) + d\delta A_t \\ &= \delta A_t A_t + A_t\delta A_t + d\delta A_t \\ &= D_t(\delta A_t) = \delta F_t = dt\frac{\partial}{\partial t}F_t,\end{aligned} \tag{7.84}$$

(the anticommutator of 2 antiderivations is a derivation) which implies for any polynomial S in A_t and F_t (possibly noncommuting)

$$(\ell_t d + d\ell_t)S(A_t, F_t) = \delta S(A_t, F_t) = dt\frac{\partial}{\partial t}S(A_t, F_t). \tag{7.85}$$

So we find the following operator equation.

Lemma:

$$\ell_t d + d\ell_t = \delta. \tag{7.86}$$

Definition: We define the **homotopy operator** (the t-integrated version of the derivation ℓ_t) by

$$k_{01} = \int_0^1 \ell_t. \tag{7.87}$$

Integrating equation (7.85) now with respect to t we arrive at:

Theorem: H. Cartan's homotopy formula

$$S(A_1, F_1) - S(A_0, F_0) = (k_{01}d + dk_{01})S(A_t, F_t). \tag{7.88}$$

A generalization can be found in [Mañes, Stora, Zumino 1985].

We can calculate the following term explicitly

$$\begin{aligned}k_{01}S(A_t, F_t) &= \int_0^1 \ell_t S(A_t, F_t) \\ &= \sum_i (-)^{\deg i} \int_0^1 dt\ S(A_t, F_t \to (A_1 - A_0)|_i, F_t),\end{aligned} \tag{7.89}$$

where on the i-th place the F_t is replaced by $A_1 - A_0$ and the minus factor arises from the antiderivation rule depending on the form degree up to the i-th place.

Specializing the arbitrary polynomial now to an invariant polynomial, $S(A_t, F_t) \to P_n(F_t)$, we obtain

$$\begin{aligned} k_{01} P_n(F_t) &= \sum_{i=1}^{n} (-)^2 \int_0^1 dt\ P_n(F_t \to (A_1 - A_0)|_i\,, F_t) \\ &= n \int_0^1 dt P(A_1 - A_0, F_t^{n-1}) = Q_{2n-1}(A_1, A_0). \end{aligned} \tag{7.90}$$

Since the invariant polynomial is closed $dP_n(F_t) = 0$ the Cartan homotopy formula supplies the **'transgression formula'**

$$P_n(F_1) - P_n(F_0) = dQ_{2n-1}(A_1, A_0). \tag{7.91}$$

Finally, choosing the case $A_0 = 0$, $A_1 = A$—the case of a trivial bundle or local chart—the **homotopy operator** is usually denoted by

$$k_{01} \to k = \int_0^1 \ell_t \tag{7.92}$$

and we get from equations (7.86) and (7.80), (7.81) the **properties**

$$kd + dk = \mathbf{1} \qquad \text{and} \qquad k^2 = 0 \tag{7.93}$$

in accordance with our results (7.72) and (7.75) of Section 7.3.

Applying Cartan's homotopy formula to this special case gives

$$\begin{aligned} P_n(F) &= (kd + dk) P_n(F_t) \\ &= dk P_n(F_t) = dQ_{2n-1}(A, F) \end{aligned} \tag{7.94}$$

and equation (7.90) provides the **Chern–Simons form**

$$Q_{2n-1}(A, F) = kP_n(F_t) = n \int_0^1 dt P(A, F_t^{n-1}), \tag{7.95}$$

with

$$A_t = tA, \qquad F_t = dA_t + A_t^2 = tF + (t^2 - t)A^2. \tag{7.96}$$

7.5 Chern–Simons form, gauge transformations and anomaly

An important application of the homotopy operator and Cartan's homotopy formula occurs in gauge transforming the Chern–Simons form. This procedure provides the anomaly. Again, we follow the work of Zumino [Zumino 1984], Stora [Stora 1984] and Alvarez-Gaumé and Ginsparg [Alvarez-Gaumé, Ginsparg 1985].

Gauge transformations: Consider the gauge transformed field

$$A^g = g^{-1}Ag + g^{-1}dg = g^{-1}(A+V)g, \tag{7.97}$$

with

$$V = dgg^{-1} \tag{7.98}$$

and the transformed curvature

$$F^g = dA^g + (A^g)^2 = g^{-1}Fg, \tag{7.99}$$

where $g = g(x)$ denotes the gauge element.

Let us choose a homotopy such as

$$A_t^g = tg^{-1}Ag + g^{-1}dg = g^{-1}(A_t + V)g, \tag{7.100}$$

with

$$A_t = tA, \qquad t \in [0,1]. \tag{7.101}$$

The homotopic curvature then is

$$F_t^g = dA_t^g + (A_t^g)^2 = g^{-1}F_t g, \tag{7.102}$$

with

$$F_t = dA_t + A_t^2 = tF + (t^2 - t)A^2. \tag{7.103}$$

Both homotopies (7.100) and (7.102) interpolate continuously between

$$A_{t=1}^g = A^g, \qquad F_{t=1}^g = F^g \tag{7.104}$$

and

$$A_{t=0}^g = g^{-1}dg = g^{-1}Vg, \qquad F_{t=0}^g = 0. \tag{7.105}$$

Now we apply the Cartan homotopy formula (7.88) to the Chern–Simons form containing the homotopies (7.100), (7.102)

$$S(A_t, F_t) \to Q_{2n-1}(A_t^g, F_t^g)$$

$$Q_{2n-1}(A^g, F^g) - Q_{2n-1}(g^{-1}dg, 0) = (k_{01}d + dk_{01})Q_{2n-1}(A_t^g, F_t^g). \tag{7.106}$$

Calculating the gauge transformed Chern–Simons term we get

$$Q_{2n-1}(A^g, F^g) = n\int_0^1 dt \ \mathrm{tr}\ A^g(\widehat{F}_t^g)^{n-1}, \tag{7.107}$$

with

$$\widehat{F}_t^g = tF^g + (t^2 - t)(A^g)^2 = g^{-1}\widehat{F}_t g \tag{7.108}$$

and

$$\widehat{F}_t = tF + (t^2 - t)(A + V)^2 \tag{7.109}$$

so that

$$Q_{2n-1}(A^g, F^g) = n \int_0^1 dt \ \mathrm{tr}\ (A + V)\widehat{F}_t^{n-1} = Q_{2n-1}(A + V, F). \tag{7.110}$$

Analogously we can calculate

$$\begin{aligned} Q_{2n-1}(A_t^g, F_t^g) &= n \int_0^1 dt' \ \mathrm{tr}\ (A_t + V) \left[t' F_t + (t'^2 - t')(A_t + V)^2\right]^{n-1} \\ &= Q_{2n-1}(A_t + V, F_t). \end{aligned} \tag{7.111}$$

On the other hand we know that (recall equations (7.90), (7.95))

$$k_{01} dQ_{2n-1}(A_t^g, F_t^g) = k_{01} \mathrm{tr} F_t^n = Q_{2n-1}(A, F) \tag{7.112}$$

and finally we define a $(2n-2)$-form

$$\alpha_{2n-2} := k_{01} Q_{2n-1}(A_t^g, F_t^g) = k_{01} Q_{2n-1}(A_t + V, F_t). \tag{7.113}$$

Then we can rewrite Cartan's homotopy formula (7.106) and we obtain the following **gauge transformation for the Chern–Simons form**

$$Q_{2n-1}(A^g, F^g) = Q_{2n-1}(A, F) + Q_{2n-1}(g^{-1}dg, 0) + d\alpha_{2n-2} \tag{7.114}$$

$$Q_{2n-1}(A + V, F) = Q_{2n-1}(A, F) + Q_{2n-1}(V, 0) + d\alpha_{2n-2}. \tag{7.115}$$

These are important relations satisfied by the Chern–Simons terms. We shall use relation (7.114) again, for instance, for calculating Chern classes (see Section 11.5.3).

Considering the term $Q_{2n-1}(V, 0)$ we can carry out the t-integration explicitly. Formula (7.110) with equation (7.109) supplies

$$Q_{2n-1}(V, 0) = (-)^{n-1} n \int_0^1 dt t^{n-1} (1 - t)^{n-1} \ \mathrm{tr}\ V^{2n-1}. \tag{7.116}$$

The integral represents the familiar beta-function $B(n, n)$ (see equation (9.135)) and we obtain

$$Q_{2n-1}(V, 0) = (-)^{n-1} \frac{(n-1)!n!}{(2n-1)!} \ \mathrm{tr}\ V^{2n-1}, \tag{7.117}$$

with $V = dgg^{-1}$. Note that $Q_{2n-1}(V = dgg^{-1}, 0) = Q_{2n-1}(g^{-1}dg, 0)$ because of the cyclicity of the trace. The form $Q_{2n-1}(V, 0)$ is closed

$$dQ_{2n-1}(V,0) = const.\ \mathrm{tr}\ dVV^{2n-2} = const.\ \mathrm{tr}\ V^{2n} = 0, \tag{7.118}$$

where we have used $dV = V^2$, which can be verified by direct calculation.

Anomaly: Finally we turn to the term α_{2n-2}. As Zumino [Zumino 1984] realized, the term $d\alpha_{2n-2}$ in equations (7.114), (7.115) arising from a gauge variation of the Chern–Simons form contains already the anomaly.

For example, for $n = 2$ we calculate

$$\begin{aligned} \alpha_2 &= k_{01}Q_3(A_t + V, F_t) \\ &= \int_0^1 \ell_t\ \mathrm{tr}\ [(A_t + V)F_t - \frac{1}{3}(A_t + V)^3] \end{aligned} \tag{7.119}$$

when carrying out the t'-integration in formula (7.111). Applying the homotopy derivation ℓ_t with properties (7.80)–(7.82) we get

$$\alpha_2 = \int_0^1 dt\ \mathrm{tr}\ [-tA^2 - VA] = -\mathrm{tr}\ VA \tag{7.120}$$

since $\mathrm{tr}\ A^2 = 0$. Then the derivative of α_2 is

$$d\alpha_2 \doteq \mathrm{tr}\ VdA \tag{7.121}$$

when keeping only the term linear in V which we denote by $\doteq$ (recall that $dV = V^2$). Equation (7.121) expresses the **non-Abelian anomaly in two dimensions** apart from the normalization.

Analogously we calculate the case $n = 3$

$$\begin{aligned} \alpha_4 &= k_{01}Q_5(A_t + V, F_t) \\ &= \int_0^1 \ell_t\ \mathrm{tr}\left[(A_t + V)F_t^2 - \frac{1}{2}(A_t + V)^3 F_t + \frac{1}{10}(A_t + V)^5\right] \\ &= \mathrm{tr}\left[-\frac{1}{2}V(AF + FA) + \frac{1}{2}VA^3 + \frac{1}{4}VAVA + \frac{1}{2}V^3A\right]. \end{aligned} \tag{7.122}$$

For the derivative of α_4, when keeping only linear terms in V, there follows

$$\begin{aligned} d\alpha_4 &\doteq d\ \mathrm{tr}\left[-\frac{1}{2}V(AdA + dAA) - \frac{1}{2}VA^3\right] \\ &\doteq \mathrm{tr}\ Vd\left(AdA + \frac{1}{2}A^3\right). \end{aligned} \tag{7.123}$$

Equation (7.123) represents the **non-Abelian anomaly in 4 dimensions,** Bardeen's result (6.79), apart from the normalization.

Résumé: A gauge variation of the Chern–Simons form, more precisely the linear term in V of the shift $A+V$, already represents the anomaly which we are interested in. This is a remarkable result. In the following chapters we shall elaborate in more detail the relation of the anomaly to gauge transformations.

7.6 Chern–Simons form, variations and anomaly

Alvarez-Gaumé and Ginsparg [Alvarez-Gaumé, Ginsparg 1985] have calculated a general formula for the first-order variation of the Chern–Simons form by working with the homotopy derivation. Their procedure is as follows.

Variation: Consider a variation in the gauge connection

$$A \to A + \delta A, \tag{7.124}$$

implying for the corresponding curvature

$$F \to F + \delta F \quad \text{with} \quad \delta F = d\delta A + \delta A A + A\delta A = D(\delta A). \tag{7.125}$$

Definition: We define a **homotopy derivation** ℓ analogous to equations (7.80), (7.81) by

$$\ell A = 0 \tag{7.126}$$

$$\ell F = \delta A, \tag{7.127}$$

obeying the antiderivation rule (7.82).

Lemma:

$$\ell d + d\ell = \delta. \tag{7.128}$$

Now we study an infinitesimal variation of the Chern–Simons form

$$\begin{aligned}
&Q_{2n-1}(A+\delta A, F+\delta F) - Q_{2n-1}(A,F) \\
&\quad = \delta Q_{2n-1}(A,F) = (\ell d + d\ell) Q_{2n-1}(A,F) \\
&\quad = \ell P(F^n) + d\ell n \int_0^1 dt P(A, F_t^{n-1}) \\
&\quad = nP(\delta A, F^{n-1}) + n(n-1) d \int_0^1 dt P(\delta A, A_t, F_t^{n-2}).
\end{aligned} \tag{7.129}$$

(Note that $\ell F_t = t\delta A$ and $A_t = tA$.) If we next choose a variation where $\delta F = D(\delta A) = 0$ and draw the derivative d into the polynomial (recall formula (7.20)) we obtain

$$\begin{aligned} Q^1_{2n-1}(\delta A, A, F) &:= Q_{2n-1}(A+\delta A, F) - Q_{2n-1}(A, F) \\ &= nP(\delta A, F^{n-1}) - n(n-1)\int_0^1 dt P(\delta A, D(A_t, F_t^{n-2})). \quad (7.130) \end{aligned}$$

The upper index of the polynomial $Q^1_{2n-1}(\delta A, A, F)$ indicates that it is linear in δA whereas the lower index denotes the form degree. This is the important expression for Q^1_{2n-1} and from experience with equation (7.115) of the preceding section we can already infer that Q^1_{2n-1} will turn out as the anomaly in $(2n-2)$ dimensions!

Formula: Relations (7.129) and (7.130) can be re-expressed in a very compact way. Starting with

$$\begin{aligned} nP(\delta A, F^{n-1}) &= n\int_0^1 dt \frac{d}{dt} P(\delta A, F_t^{n-1}) \\ &= n(n-1)\int_0^1 dt P(\delta A, \frac{d}{dt}F_t, F_t^{n-2}), \quad (7.131) \end{aligned}$$

inserting

$$\frac{d}{dt}F_t = F + (2t-1)A^2 = dA + [A_t, A] = D_t A \quad (7.132)$$

we get

$$nP(\delta A, F^{n-1}) = n(n-1)\int_0^1 dt P(\delta A, D_t(A, F_t^{n-2})) \quad (7.133)$$

by virtue of the Bianchi identity $D_t F_t = 0$. Using formula (7.20) again for the derivative of invariant polynomials

$$dP(\delta A, A, F_t^{n-2}) = P(D_t(\delta A), A, F_t^{n-2}) - P(\delta A, D_t(A, F_t^{n-2})) \quad (7.134)$$

gives

$$nP(\delta A, F^{n-1}) = n(n-1)\int_0^1 dt [P(D_t(\delta A), A, F_t^{n-2}) - dP(\delta A, A, F_t^{n-2})]. \quad (7.135)$$

Now we insert expression (7.135) into relation (7.129)

$$\begin{aligned} &Q_{2n-1}(A+\delta A, F+\delta F) - Q_{2n-1}(A,F) \\ &= n(n-1)\int_0^1 dt[P((d\delta A + [tA, \delta A]), A, F_t^{n-2}) \\ &\quad + (t-1)dP(\delta A, A, F_t^{n-2})] \end{aligned} \tag{7.136}$$

and we obtain the **formula** of Alvarez-Gaumé and Ginsparg [Alvarez-Gaumé, Ginsparg 1985]

$$\begin{aligned} &Q_{2n-1}(A+\delta A, F+\delta F) - Q_{2n-1}(A,F) \\ &= n(n-1)\int_0^1 dt(1-t)P(\delta A, d(A, F_t^{n-2})) \\ &\quad + n(n-1)\int_0^1 dttP(D(\delta A), A, F_t^{n-2}). \end{aligned} \tag{7.137}$$

Specializing further to the case where $\delta F = D(\delta A) = 0$ provides the **formula**

$$Q^1_{2n-1}(\delta A, A, F) = n(n-1)\int_0^1 dt(1-t)P(\delta A, d(A, F_t^{n-2})). \tag{7.138}$$

Anomaly: This expression coincides with the formula for the anomaly originally discovered by Zumino [Zumino 1984, 1985a] when δA is shifted by the Faddeev–Popov ghost v (see Chapter 9, equation (9.122)). Substituting the 1-form δA by the 0-form v (which anticommutes with any odd-form, see Chapter 8) we obtain an **anomaly formula**

$$\begin{aligned} &Q^1_{2n-1}(\delta A, A, F) \stackrel{\delta A \to v}{\longrightarrow} Q^1_{2n-2}(v, A, F) \\ &= n(n-1)\int_0^1 dt(1-t)P(v, d(A, F_t^{n-2})). \end{aligned} \tag{7.139}$$

For $n=2$ and $n=3$

$$Q^1_2 = \mathrm{tr}\ vdA \tag{7.140}$$

$$Q^1_4 = \mathrm{tr}\ vd(AdA + \frac{1}{2}A^3) \tag{7.141}$$

we recover the non-Abelian anomaly in two and four dimensions up to the normalization.

The polynomial Q^1_{2n-2} is one of a whole series of terms Q^k_{2n-1-k}, $k = 0, 1, \ldots, 2n-1$, which are linked together by a set of equations—the chain of descent equations—and is therefore called the **chain term**. We will discuss this topic in Chapter 9.

8
Consistent anomaly

In Chapter 4 of this book we introduced the anomaly as the breakdown of a classical conservation law. Now we want to investigate the anomaly further in the light of gauge transformations.

We first introduce an infinitesimal gauge operator, generalize to a BRS operator and find its representation in the functional space of the gauge potentials and Faddeev–Popov ghosts (Section 8.1). Next we discuss the anomalous Ward identity in terms of functional derivatives and subsequently we derive the equation which determines the anomaly—the Wess–Zumino consistency condition—in the gauge transformation variant and in the compact BRS variant (Section 8.2). Different aspects of the anomaly equation, like the algebra-, the cocycle- and the cohomology aspect, are presented at the end (Section 8.3).

8.1 Infinitesimal gauge operator, BRS and geometry

For the calculation of physical processes we need to know the Green functions of the theory which follow from the generating functional (see Chapter 3; here we ignore the sources)

$$Z[A_\mu] = e^{-W[A_\mu]} = \int d\bar{\psi} d\psi \exp[-\int dx \mathcal{L}(A_\mu, \bar{\psi}, \psi)], \tag{8.1}$$

with

$$\mathcal{L} = \bar{\psi} i \not{D} \psi, \qquad \not{D} = \gamma^\mu(\partial_\mu + A_\mu). \tag{8.2}$$

For the anomaly we only consider the fermionic term in the Lagrangian $\mathcal{L}$, and we treat only the massless fermion fields as dynamical variables. The Yang–Mills gauge fields A_μ are regarded as external fields. Therefore we just integrate over the fermion fields and the functional depends on A_μ.

Now we search for an infinitesimal gauge variation of the W-functional in terms of a differential operator. Following Zumino [Zumino 1984] we find its representation in the functional space of the gauge fields A_μ^a and the Faddeev–Popov ghost fields v^a.

Infinitesimal gauge operator: Let us start with the variation of the W-functional with respect to A^a_μ

$$\delta W[A_\mu] = \int dx \delta A^a_\mu(x) \frac{\delta}{\delta A^a_\mu(x)} W[A_\mu]. \tag{8.3}$$

We vary the gauge fields A^a_μ à la BRS (recall Section 3.6)

$$\delta A^a_\mu = D^{ba}_\mu v^b, \tag{8.4}$$

with the covariant derivative

$$D^{ba}_\mu = \partial_\mu \delta^{ba} + f^{bac} A^c_\mu \tag{8.5}$$

as defined in equation (3.306), introducing the Faddeev–Popov ghost v^b. After integrating by parts the variation (8.3) turns into

$$\delta W[A_\mu] = -\int dx v^b D^{ab}_\mu \left.\frac{\delta}{\delta A^a_\mu}\right|_x W[A_\mu]. \tag{8.6}$$

Definition: We define a **gauge operator** represented by a functional differential operator in the functional space of the gauge fields A^a_μ

$$-X^b(x) := D^{ab}_\mu(x) \frac{\delta}{\delta A^a_\mu(x)} = \partial_\mu \frac{\delta}{\delta A^b_\mu(x)} + f^{bca} A^c_\mu(x) \frac{\delta}{\delta A^a_\mu(x)}. \tag{8.7}$$

Theorem: The gauge operators obey the **commutation relation**

$$[X^a(x), X^b(y)] = f^{abc} X^c(x) \delta(x-y). \tag{8.8}$$

Proof.

$$\begin{aligned}
[X^a(x), X^{\bar{a}}(y)] &= X^a(x) X^{\bar{a}}(y) - X^{\bar{a}}(y) X^a(x) \qquad (8.9)\\
&= f^{acb} A^c_\mu \left.\frac{\delta}{\delta A^b_\mu}\right|_x \partial_{\bar{\mu}} \left.\frac{\delta}{\delta A^{\bar{a}}_{\bar{\mu}}}\right|_y - f^{\bar{a}\bar{c}\bar{b}} A^{\bar{c}}_{\bar{\mu}} \left.\frac{\delta}{\delta A^{\bar{b}}_{\bar{\mu}}}\right|_y \partial_\mu \left.\frac{\delta}{\delta A^a_\mu}\right|_x \\
&\quad + \partial_\mu \left.\frac{\delta}{\delta A^a_\mu}\right|_x f^{\bar{a}\bar{c}\bar{b}} A^{\bar{c}}_{\bar{\mu}} \left.\frac{\delta}{\delta A^{\bar{b}}_{\bar{\mu}}}\right|_y + f^{acb} A^c_\mu \left.\frac{\delta}{\delta A^b_\mu}\right|_x f^{\bar{a}\bar{c}\bar{b}} A^{\bar{c}}_{\bar{\mu}} \left.\frac{\delta}{\delta A^{\bar{b}}_{\bar{\mu}}}\right|_y \\
&\quad - \partial_{\bar{\mu}} \left.\frac{\delta}{\delta A^{\bar{a}}_{\bar{\mu}}}\right|_y f^{acb} A^c_\mu \left.\frac{\delta}{\delta A^b_\mu}\right|_x - f^{\bar{a}\bar{c}\bar{b}} A^{\bar{c}}_{\bar{\mu}} \left.\frac{\delta}{\delta A^{\bar{b}}_{\bar{\mu}}}\right|_y f^{acb} A^c_\mu \left.\frac{\delta}{\delta A^b_\mu}\right|_x .
\end{aligned}$$

Applying the functional derivatives, noting that all second order derivatives cancel, we get

$$
\begin{aligned}
&[X^a(x), X^{\bar a}(y)] \\
&= f^{\bar a ab}\left(\partial_\mu^x \delta(x-y)\frac{\delta}{\delta A_\mu^b(y)} + \partial_\mu^y \delta(x-y)\frac{\delta}{\delta A_\mu^b(x)}\right) \\
&\quad - (f^{acb}f^{b\bar a\bar b} + f^{ad\bar b}f^{\bar a cd})\delta(x-y)A_\mu^c(x)\frac{\delta}{\delta A_\mu^{\bar b}(x)}. \qquad (8.10)
\end{aligned}
$$

For the first line we need the relation ($U^\mu(x)$ denotes some functional of x)

$$
\partial_\mu^x \delta(x-y)U^\mu(y) + \partial_\mu^y \delta(x-y)U^\mu(x) = \delta(x-y)\partial_\mu^x U^\mu(x), \qquad (8.11)
$$

which follows from partial integration and by virtue of

$$
\partial_\mu^x \delta(x-y) + \partial_\mu^y \delta(x-y) = 0. \qquad (8.12)
$$

For the second line we use the Jacobi identity for the structure constants

$$
f^{acb}f^{b\bar a\bar b} + f^{ad\bar b}f^{\bar a cd} + f^{a\bar a d}f^{d\bar b c} = 0. \qquad (8.13)
$$

Then expression (8.10) results in

$$
\begin{aligned}
[X^a(x), X^{\bar a}(y)] &= -f^{a\bar a d}\delta(x-y)[\partial_\mu \delta^{bd} + f^{bdc}A_\mu^c(x)]\frac{\delta}{\delta A_\mu^b(x)} \\
&= -f^{a\bar a d}\delta(x-y)D_\mu^{bd}(x)\frac{\delta}{\delta A_\mu^b(x)} \\
&= f^{a\bar a d}X^d(x)\delta(x-y). \qquad \text{Q.E.D.} \qquad (8.14)
\end{aligned}
$$

Definition: We define the **integrated gauge operator** by

$$
v \cdot X := \int dx v^b(x) X^b(x). \qquad (8.15)
$$

Then we rewrite the **gauge variation** (8.6)

$$
\delta W[A_\mu] = \int dx v^b(x) X^b(x) W[A_\mu] = v \cdot X\, W[A_\mu]. \qquad (8.16)
$$

BRS: In the next step we also include the functional space of the Faddeev–Popov ghosts v^a. The variation of a functional with respect to v^a is

$$
\delta W[v] = \int dx \delta v^a(x) \frac{\delta}{\delta v^a(x)} W[v]. \qquad (8.17)
$$

We again vary the FP fields v^a à la BRS (recall Section 3.6)

$$\delta v^a = -\frac{1}{2}[v,v]^a \tag{8.18}$$

so that equation (8.17) turns into

$$\begin{aligned} \delta W[v] &= -\int dx \frac{1}{2}[v,v]^a \left.\frac{\delta}{\delta v^a}\right|_x W[v] \\ &= -\int dx v^b \frac{1}{2} f^{bca} v^c \left.\frac{\delta}{\delta v^a}\right|_x W[v]. \end{aligned} \tag{8.19}$$

Definitions: We define the following **differential FP operator**

$$-X_v^b(x) := \frac{1}{2} f^{bca} v^c(x) \frac{\delta}{\delta v^a(x)} \tag{8.20}$$

and the **integrated FP operator**

$$v \cdot X_v := \int dx v^b(x) X_v^b(x). \tag{8.21}$$

Then we re-express the **FP variation** (8.19) by

$$\delta W[v] = \int dx v^b(x) X_v^b(x) W[v] = v \cdot X_v W[v]. \tag{8.22}$$

Definition: We define a generalized gauge operator—the **BRS operator**

$$\delta \longrightarrow \delta_v = \left\{ \begin{array}{ll} v \cdot X & \text{for } W[A_\mu] \\ v \cdot X_v & \text{for } W[v] \end{array} \right. , \tag{8.23}$$

where we know its functional representation in the gauge fields A_μ^a *and* FP ghosts v^a. We add the index v to indicate that its action creates a FP ghost; δ_v equals $v \cdot X$ when acting on a functional of gauge fields $W[A_\mu]$ and it equals $v \cdot X_v$ when it operates on FP ghosts $W[v]$.

Proposition:

- *The operator δ_v satisfies the* **BRS transformations**

$$\begin{aligned} \delta_v v &= -\frac{1}{2}[v,v] = -v^2 \\ \delta_v A &= -Dv \\ \delta_v F &= [F,v], \end{aligned} \tag{8.24}$$

together with

$$\delta_v^2 = d^2 = 0 \quad , \quad \delta_v d + d\delta_v = 0. \tag{8.25}$$

Proof. i)

$$\begin{aligned}\delta_v v(x) &= v\cdot X_v v(x) = -\int dy v^b(y)\frac{1}{2}f^{bcd}v^c(y)\frac{\delta}{\delta v^d(y)}v^a(x)T^a \\ &= -\frac{1}{2}f^{bca}v^b(x)v^c(x)T^a = -\frac{1}{2}[v,v](x). \end{aligned} \tag{8.26}$$

ii)

$$\delta_v A_\mu(x) = v\cdot XA_\mu(x) = -\int dy v^b(y)D_\nu^{cb}(y)\frac{\delta}{\delta A_\nu^c(y)}A_\mu^a(x)T^a;$$

integrating by parts gives

$$\delta_v A = \delta_v A_\mu(x)dx^\mu = D_\mu^{ba}v^b(x)T^a dx^\mu = -Dv. \tag{8.27}$$

iii) The third equation of (8.24) is a consequence of the second. Of course, one can demonstrate explicitly the action of the functional derivatives.

iv) Applying δ_v twice to a functional of the gauge fields $W[A_\mu]$ gives

$$\begin{aligned}\delta_v\delta_v &= \left\{\begin{array}{c} v\cdot X \\ v\cdot X_v\end{array}\right\} v\cdot X = v\cdot Xv\cdot X + v\cdot X_v v\cdot X \\ &= v\cdot Xv\cdot X - \frac{1}{2}\int dx f^{acb}v^a(x)v^c(x)X^b(x) \\ &= \frac{1}{2}[v,v]\cdot X - \frac{1}{2}[v,v]\cdot X = 0, \end{aligned} \tag{8.28}$$

where we have used the integrated version of algebra (8.8)

$$2\, v\cdot Xv\cdot X = [v,v]\cdot X. \tag{8.29}$$

v) Repeated application to a functional of FP ghosts $W[v]$

$$\delta_v\delta_v = \left\{\begin{array}{c} v\cdot X \\ v\cdot X_v\end{array}\right\} v\cdot X_v = v\cdot X_v v\cdot X_v \tag{8.30}$$

$$= \frac{1}{2}\int dx f^{bca}f^{a\bar{c}\bar{a}}v^b v^c v^{\bar{c}}\left.\frac{\delta}{\delta v^{\bar{a}}}\right|_x = 0. \qquad \text{Q.E.D.} \tag{8.31}$$

The last line vanishes because of the identity

$$f^{bca}f^{a\bar{c}\bar{a}}v^b v^c v^{\bar{c}} = 0. \tag{8.32}$$

Remark: The FP operator $X_v^a(x)$ (8.20) obeys a commutation relation with a factor $\frac{1}{2}$ on the right-hand side implying the nilpotency of δ_v

$$[X_v^a(x), X_v^b(y)] = \frac{1}{2} f^{abc} X_v^c(x) \delta(x-y). \tag{8.33}$$

Functional variation and geometry: We want to equip the functional variation with a differential geometric set-up intended to interpret the BRS transformations geometrically and to shed more light on Zumino's geometric approach of Section 9.3.

Definitions:

Sp $\mathcal{A}$ ...	The **affine space of all gauge connections** on $M = S^{2n}$. It does not correspond to the *physical* configuration space.
$\mathcal{G}$...	The **gauge group space**, the space of all gauge elements $\mathcal{G} = \{g(x)\}$.
Sp $\mathcal{A}/\mathcal{G}$...	The **moduli space**, where we have identified all gauge connections which are linked together by a gauge transformation. It represents the *physical* space of the gauge theory.

Whereas Sp $\mathcal{A}$ is contractible, hence topologically trivial, the moduli space Sp $\mathcal{A}/\mathcal{G}$ is not. Each gauge connected component of Sp $\mathcal{A}/\mathcal{G}$ is classified by the homotopy of the group; for example by $\Pi_1(U(1)) = \mathbf{Z}$—Dirac monopole bundle (see Section 6.4), by $\Pi_3(SU(2)) = \mathbf{Z}$—instanton bundle (see Section 6.6), or by $\Pi_5(SU(N)) = \mathbf{Z}$ with $N \geq 3$—determinant bundle (see Section 11.5.3).

For a differential geometric setting we recall the functional derivatives with respect to $\mathcal{A}_\mu^a(x)$ and define:

$$\left\{\frac{\delta}{\delta \mathcal{A}_\mu^a(x)}\right\} \quad \text{as a basis for the tangent space } T(\text{Sp } \mathcal{A}) \text{ in Sp } \mathcal{A}$$
$$\{\delta \mathcal{A}_\mu^a(x)\} \quad \text{as a basis for the cotangent space } T^*(\text{Sp } \mathcal{A}). \tag{8.34}$$

Both spaces are dual to each other via the **inner product**

$$\delta \mathcal{A}_\mu^a(x) \frac{\delta}{\delta \mathcal{A}_\nu^b(y)} = \delta^a{}_b \delta_\mu{}^\nu \delta(x-y). \tag{8.35}$$

A **wedge product** is defined by

$$\delta \mathcal{A}_\mu^a(x) \delta \mathcal{A}_\nu^b(y) \equiv \delta \mathcal{A}_\mu^a(x) \wedge \delta \mathcal{A}_\nu^b(y) :=$$
$$\delta \mathcal{A}_\mu^a(x) \otimes \delta \mathcal{A}_\nu^b(y) - \delta \mathcal{A}_\nu^b(y) \otimes \delta \mathcal{A}_\mu^a(x), \tag{8.36}$$

and, analogous to the ordinary exterior derivative (2.103), we also introduce

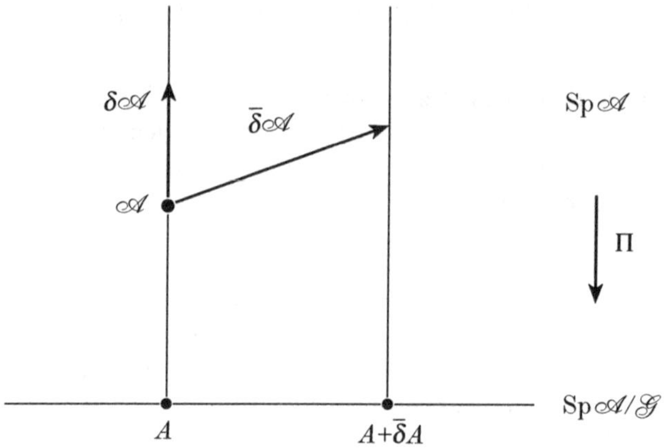

Fig. 8.1. Variations in Sp $\mathcal{A}$

an **exterior derivative in functional space Sp $\mathcal{A}$**

$$\delta = \frac{\delta}{\delta \mathcal{A}_\mu^a} \delta \mathcal{A}_\mu^a = \int dx \frac{\delta}{\delta \mathcal{A}_\mu^a(x)} \delta \mathcal{A}_\mu^a(x). \tag{8.37}$$

A general **vector field V on Sp $\mathcal{A}$** is given by

$$V = \int dx V_\mu^a(x) \frac{\delta}{\delta \mathcal{A}_\mu^a(x)} \tag{8.38}$$

and we describe a ***p*-form B on Sp $\mathcal{A}$** as

$$B = \frac{1}{p!} \int dx_1 \dots dx_p B^{\mu_1 \dots \mu_p}_{a_1 \dots a_p} \delta \mathcal{A}^{a_1}_{\mu_1}(x_1) \dots \delta \mathcal{A}^{a_p}_{\mu_p}(x_p), \tag{8.39}$$

where the $\delta \mathcal{A}^{a_i}_{\mu_i}$ products are wedged.

Note: Here the group indices a_i represent the contravariance (or covariance) in the forms on Sp $\mathcal{A}$ and we raise and lower them for this reason. The Lorentz indices μ_i express the tensor structure on the space–time manifold M.

Bundle: We construct a fibre bundle, a principal bundle with the spaces Sp $\mathcal{A}$ and Sp $\mathcal{A}/\mathcal{G}$ whose fibre is $\mathcal{G}$. Space $\mathcal{A}$ has the projection Π : Sp $\mathcal{A} \to$ Sp $\mathcal{A}/\mathcal{G}$, see Figure 8.1 (recall Section 3.5.2; for further literature see [Nakahara 1990], [Cotta-Ramusino, Reina 1984], [Falqui, Reina 1985]).

Choosing a gauge potential $A(x) = A_\mu^a(x) dx^\mu T^a$ as a representative of the equivalence class $[A] \in$ Sp $\mathcal{A}/\mathcal{G}$ we consider the gauge transformed configuration

$$\mathcal{A}(x,g) = g^{-1}(x)[A(x)+d]g(x) \in \text{Sp } \mathcal{A} \tag{8.40}$$

being defined on $M \times \mathcal{G}$.

We apply the exterior derivative (8.37)—we now use a bar over the letter for reasons which become clear below—representing a functional variation in Sp $\mathcal{A}$ defined on $M \times$ Sp $\mathcal{A}$ (see Figure 8.1) and we obtain

$$\begin{aligned} \bar{\delta}\mathcal{A} &= \bar{\delta}g^{-1}(A+d)g + g^{-1}\bar{\delta}Ag - g^{-1}A\bar{\delta}g + g^{-1}\bar{\delta}dg \\ &= -d(g^{-1}\bar{\delta}g) - \mathcal{A}g^{-1}\bar{\delta}g - g^{-1}\bar{\delta}g\mathcal{A} + g^{-1}\bar{\delta}Ag \\ &= -\mathcal{D}(g^{-1}\bar{\delta}g) + g^{-1}\bar{\delta}Ag. \end{aligned} \tag{8.41}$$

$\mathcal{D}$ denotes the familiar covariant derivative

$$\mathcal{D} = d + [\mathcal{A},\ ]. \tag{8.42}$$

The variation splits into two parts. The first part represents the derivative along the fibre $\mathcal{G}$, whereas the second term corresponds to the derivative along the moduli space $\mathcal{A}/\mathcal{G}$, see Figure 8.1.

Comparison: The comparison with Zumino's geometrical approach shows that the derivative operator δ of Section 9.3 corresponds to restricting the variation $\bar{\delta}$ in Sp $\mathcal{A}$ along the fibre $\mathcal{G}$.

We reproduce the structure equations (9.68)

$$\bar{\delta}\mathcal{A}\big|_{\text{fibre}} = \delta\mathcal{A} = -\mathcal{D}v_{MC}, \tag{8.43}$$

where v_{MC} is the **Maurer–Cartan form on group space**

$$v_{MC} = g^{-1}\ \bar{\delta}\big|_{\text{fibre}}\, g = g^{-1}\delta g \tag{8.44}$$

and the equation (recall Section 2.6.6)

$$\bar{\delta}v_{MC}\big|_{\text{fibre}} = \delta v_{MC} = -g^{-1}\delta g g^{-1}\delta g = -v_{MC}^2 \tag{8.45}$$

is called the **Maurer–Cartan structure equation in group space.**

Tracing back the variation along the fibre to the group parameters of Section 9.3

$$g(x) \to g(x,\theta), \qquad \mathcal{A}(x,g) \to \mathcal{A}(x,\theta), \qquad \delta \to \frac{\partial}{\partial\theta^\alpha}d\theta^\alpha \tag{8.46}$$

we recover

$$\bar{\delta}g\big|_{\text{fibre}} = \delta g = \frac{\partial}{\partial\theta^\alpha}g(x,\theta)d\theta^\alpha \tag{8.47}$$

$$v_{MC} = g^{-1}\delta g = g^{-1}(x,\theta)\frac{\partial}{\partial\theta^\alpha}g(x,\theta)d\theta^\alpha \tag{8.48}$$

$$\bar{\delta}\mathcal{A}\Big|_{\text{fibre}} = \delta\mathcal{A} = \frac{\partial}{\partial\theta^\alpha}\mathcal{A}_\mu(x,\theta)d\theta^\alpha dx^\mu. \tag{8.49}$$

Note that $d\theta^\alpha dx^\mu$ is a wedge product.

Identification: Now we turn to the geometric interpretation of the objects used in field theory, the FP ghost v_{FP} and the BRS operator s, or the operator δ_v. They obey an equation, the first of equations (8.24), which is of the Maurer–Cartan type, (8.45), (9.62). Therefore it is quite natural to identify the FP ghost with the Maurer–Cartan form (MC) on the group

$$v_{FP} = v^a T^a \longleftrightarrow v_{MC} = g^{-1}\delta g = v_\alpha d\theta^\alpha. \tag{8.50}$$

The degree in the 'ghost number' matches the degree of the frame $d\theta^\alpha$ in group space. Thus the FP ghost has a geometric interpretation which explains the 'wrong' statistics of this scalar field.

The generalized gauge operator δ_v—representing by construction the BRS operator s of Section 3.6—obeys the same algebra relation (8.24) as the exterior derivative on the group space, when finally the gauged quantities in (8.43), (9.68) are evaluated at $g(x,\theta=0)=\mathbf{1}$. Consequently we identify this operator (8.23) with the group derivate (9.43), with the variation in Sp $\mathcal{A}$ along the fibre $\mathcal{G}$

$$s, \delta_v = \begin{cases} v\cdot X \\ v\cdot X_v \end{cases} \longleftrightarrow \delta = \frac{\partial}{\partial\theta^\alpha}d\theta^\alpha. \tag{8.51}$$

These identifications have been proposed by Leinaas and Olaussen [Leinaas, Olaussen 1982], Zumino [Zumino 1984], and Stora [Stora 1977]. A rigorous treatment within fibre bundles and cohomology has been presented by Bonora and Cotta-Ramusino [Bonora, Cotta-Ramusino 1983]. Somewhat deviating aspects can be found in [Thierry-Mieg 1980] and [Quirós, de Urries, Hoyos, Mazón, Rodriguez 1981].

In this book we frequently use the FP ghost synonymously with the MC form on the group and the extended gauge operator δ_v, or the BRS operator s synonymously with the exterior derivative on the group space.

8.2 Wess–Zumino consistency condition

Again, we start from the generating functional (8.1) where the dynamical massless fermions are interacting with the external gauge fields. The fields are chosen as in Section 4.1 either left (L)- or right (R)-handed but we avoid labelling the field symbols by L, or R. We assume the theory renormalized in the one-loop approximation and consider the functional $W[A_\mu]$ of the one-particle irreducible Green functions. This is what we need for the discussion of an anomaly which is determined by loops like the ones of Section 4.8.

Anomalous Ward identity: Let us perform an infinitesimal gauge transformation as in equation (8.23)

$$A_\mu \to A_\mu + \delta_v A_\mu \tag{8.52}$$

then on a classical level the action remains invariant

$$\delta_v S_{\text{class}}(A_\mu) = \delta_v \int dx \mathcal{L}(A_\mu, \bar{\psi}, \psi) = 0. \tag{8.53}$$

If we still have gauge invariance on a quantum level—a **Ward identity**

$$X^a W[A_\mu] = 0 \quad , \quad \delta_v W[A_\mu] = 0 \tag{8.54}$$

(considering the differential and the integrated version) the gauge theory remains renormalizable. If, however, gauge invariance (8.54) is lost we find the following proposition (recall Section 4.2).

Proposition: Anomalous Ward identity

- *The gauge transformed 'quantum action' represents precisely Bardeen's non-Abelian anomaly*

$$X^a W[A_\mu] = G^a[A_\mu]$$

$$\delta_v W[A_\mu] = G(v, A) = \int dx v^a(x) G^a[A_\mu](x). \tag{8.55}$$

Proof.

$$\begin{aligned} \frac{\delta}{\delta A^b_\mu} W[A_\mu] &= \frac{\int d\bar{\psi} d\psi \, \bar{\psi}\gamma^\mu i T^b \psi \exp[-\int dy \bar{\psi} i(\partial\!\!\!/ + A\!\!\!/)\psi]}{\int d\bar{\psi} d\psi \exp[-\int dy \bar{\psi} i(\partial\!\!\!/ + A\!\!\!/)\psi]} \\ &= \langle j^{\mu b} \rangle, \end{aligned} \tag{8.56}$$

with

$$j^{\mu b} = \bar{\psi}\gamma^\mu i T^b \psi, \tag{8.57}$$

then

$$\begin{aligned} X^a W[A_\mu] &= -D^{ba}_\mu \frac{\delta}{\delta A^b_\mu} W[A_\mu] \\ &= \langle -D^{ba}_\mu j^{\mu b} \rangle \\ &= G^a[A_\mu]. \qquad \text{Q.E.D.} \end{aligned} \tag{8.58}$$

In the last step we used Bardeen's definition of the non-Abelian anomaly (Section 4.8, equation (4.387)) and the fact that $G^a[A_\mu]$ depends only on the gauge fields A_μ so that the average over the fermion fields ψ, $\bar{\psi}$ drops.

Of course, the 'quantum action' $W[A_\mu]$ is a highly *nonlocal* and non-polynomial functional in A. However, its gauge variation—the anomaly $G(v, A)$—turns out to be *local* and polynomial in A (integrated over the manifold M).

Explicitly, we use Bardeen's result in terms of differential forms (equation (6.79) of Section 6.3)

$$G(v, A) = \frac{1}{24\pi^2} \int_M \operatorname{tr} v d(AdA + \frac{1}{2} A^3), \tag{8.59}$$

which is, on the other hand, also the solution of the chain term Q_4^1 (see equations (7.141) and (9.102)), so that we have:

Proposition:

- *The integrated Q_4^1 chain term equals the anomaly apart from the normalization*

$$G(v, A) = \frac{1}{24\pi^2} \int_{M=S^4} Q_4^1(v, A). \tag{8.60}$$

We discuss this result later on.

Wess–Zumino consistency condition: Previously we defined the anomaly by the infinitesimal gauge variation of the 'quantum action'; now we find a kind of differential equation which the anomaly has to obey, and we derive it in several versions. One follows straight away when we apply the commutation relations (8.8) of the gauge operator X^a to the 'quantum action' $W[A_\mu]$

$$(X^a(x)X^b(y) - X^b(y)X^a(x))W[A_\mu] = f^{abc} X^c(x) W[A_\mu] \delta(x - y) \tag{8.61}$$

and use the anomalous Ward identity (8.55).

Theorem: Wess–Zumino consistency condition

$$X^a(x)G^b[A_\mu](y) - X^b(y)G^a[A_\mu](x) = f^{abc} G^c[A_\mu](x)\delta(x - y) \tag{8.62}$$

([Wess, Zumino 1971]). It describes a requirement which a possible anomaly has to fulfil. There exists, of course, a **trivial solution**

$$G^a_{\text{triv}}[A_\mu](x) = X^a(x)\widehat{G}[A_\mu](x), \tag{8.63}$$

where $\widehat{G}[A_\mu]$ denotes a *local* polynomial in A_μ and its derivatives.

The anomaly condition (8.62) has now been solved in full generality [Becchi, Rouet, Stora 1976a], [Brandt, Dragon, Kreuzer 1989, 1990a,b,c] and it determines the anomaly so that, in reverse, one can use it as a definition for the anomaly.

Definition: Any solution of the Wess–Zumino (WZ) consistency condition which is not trivial—is not a gauge variation of a *local* functional in the basic fields—represents an **anomaly**.

Note: If we know the first term of the Bardeen anomaly (8.59), which follows quickly from the triangle graph calculation (see Chapter 4), the WZ consistency condition (8.62) completely determines the remaining second term of the anomaly (8.59).

For the integrated version of condition (8.62) we first work with ordinary gauge elements.

Let be

$$v_\alpha(x) = v^a_\alpha(x)T^a \tag{8.64}$$

an ordinary gauge element, v^a_α a commuting scalar field. We mark such gauge elements by an additional index below: $\alpha, \beta = 1, 2, \ldots$.

The corresponding **infinitesimal gauge operator**

$$\delta_{v_\alpha} = \int dx v^a_\alpha(x) X^a(x) = v_\alpha \cdot X, \tag{8.65}$$

with X^a defined by equation (8.7) generates ordinary gauge transformations. The gauge operators themselves satisfy the group commutation relations—the integrated version (with $v_1(x)v_2(y)$) of Theorem (8.8).

Theorem: Gauge commutator

$$[\delta_{v_1}, \delta_{v_2}] = \delta_{[v_1,v_2]}. \tag{8.66}$$

Applying these commutation relations to $W[A_\mu]$

$$\delta_{v_1}\delta_{v_2}W[A_\mu] - \delta_{v_2}\delta_{v_1}W[A_\mu] = \delta_{[v_1,v_2]}W[A_\mu] \tag{8.67}$$

and using the **anomalous Ward identity**

$$\delta_{v_\alpha}W[A_\mu] = G(v_\alpha, A), \tag{8.68}$$

we arrive at:

Theorem: Wess–Zumino consistency condition

$$\delta_{v_1} G(v_2, A) - \delta_{v_2} G(v_1, A) = G([v_1, v_2], A). \tag{8.69}$$

In this connection δ_{v_α} is also called the **Ward operator** in the literature. Of course, we could integrate directly the local form of WZ (8.62) with $v_1(x)v_2(y)$ and arrive at the integrated version (8.69).

Descent equation: Now we follow Stora [Stora 1977, 1984] and Zumino [Zumino 1984], [Zumino, Wu, Zee 1984] and construct a descent equation involving the ordinary gauge elements.

Starting with the polynomial

$$P(F^n) = \text{tr } F^n, \tag{8.70}$$

it is closed and gauge invariant

$$d \text{ tr } F^n = 0 \tag{8.71}$$

$$\delta_{v_\alpha} \text{ tr } F^n = 0 \tag{8.72}$$

by virtue of the Bianchi identity $DF = 0$ and the gauge commutator for F.

According to Poincaré's lemma (recall Section 7.3) the closure (8.71) implies the existence of a $(2n-1)$-form Q^0_{2n-1} such that locally

$$\text{tr } F^n = dQ^0_{2n-1}. \tag{8.73}$$

Rewriting gauge invariance (8.72)

$$\delta_{v_\alpha} dQ^0_{2n-1} = d\delta_{v_\alpha} Q^0_{2n-1} = 0 \tag{8.74}$$

(note that d and δ_{v_α} commute) Poincaré again assures a $(2n-2)$-form Q^1_{2n-2} such that locally

$$\delta_{v_\alpha} Q^0_{2n-1}(A) = dQ^1_{2n-2}(v_\alpha, A). \tag{8.75}$$

Equation (8.75) is called a **descent equation**. It relates the Chern–Simons form Q^0_{2n-1} in $(2n-1)$ dimensions to the term Q^1_{2n-2}, the anomaly in $(2n-2)$ dimensions (the upper index denotes the power of v_α). It is one of a whole set of equations (see Chapter 9).

Proposition:

- *The chain term Q^1_{2n-2} represents the anomaly in $(2n-2)$ dimensions*

$$G(v_\alpha, A) = \int dx v^a_\alpha(x) G^a[A_\mu](x) \equiv N \int_{S^{2n-2}} Q^1_{2n-2}(v_\alpha, A). \tag{8.76}$$

We quote the normalization N later on (e.g. in Chapters 9, 11).

Proof. We prove that expression (8.76) obeys the WZ consistency condition (8.69).

We suppose that we can extend the gauge potentials from our chosen $(2n-2)$-dimensional sphere S^{2n-2} to a $(2n-1)$-dimensional ball B^{2n-1} whose boundary is the sphere $\partial B^{2n-1} = S^{2n-2}$.

Consider now the following functional

$$U[A] = N \int_{B^{2n-1}} Q^0_{2n-1}(A), \tag{8.77}$$

its gauge variation is

$$\begin{aligned} \delta_{v_\alpha} U[A] &= N \int_{B^{2n-1}} \delta_{v_\alpha} Q^0_{2n-1}(A) = N \int_{B^{2n-1}} dQ^1_{2n-2}(v_\alpha, A) \\ &= N \int_{S^{2n-2}} Q^1_{2n-2}(v_\alpha, A) = G(v_\alpha, A) \end{aligned} \tag{8.78}$$

by virtue of equation (8.75) and Stokes' theorem (2.128). Then

$$\begin{aligned} &\delta_{v_1} G(v_2, A) - \delta_{v_2} G(v_1, A) = \\ &= \delta_{v_1}\delta_{v_2} U[A] - \delta_{v_2}\delta_{v_1} U[A] = [\delta_{v_1}, \delta_{v_2}] U[A] \\ &= \delta_{[v_1, v_2]} U[A] = G([v_1, v_2], A). \qquad \text{Q.E.D.} \end{aligned} \tag{8.79}$$

BRS: Working finally with the anticommuting FP ghosts $v(x)v(y)$ the integrated version of condition (8.62) becomes

$$2\, v \cdot X\, v \cdot G = [v, v] \cdot G, \tag{8.80}$$

which we use for calculating

$$\begin{aligned} \delta_v G(v, A) &= \delta_v \int dx v^a(x) G^a[A_\mu](x) \\ &= \int dx v \cdot X_v v^a(x) G^a[A_\mu](x) + v \cdot X v \cdot G \\ &= -\frac{1}{2}[v, v] \cdot G + \frac{1}{2}[v, v] \cdot G = 0. \end{aligned} \tag{8.81}$$

Theorem: Wess–Zumino consistency condition

$$\delta_v G(v, A) = 0. \tag{8.82}$$

This is the most compact and elegant form of the WZ consistency condition and was found by Stora [Stora 1977, 1984] and Zumino [Zumino 1984].

Of course, condition (8.82) follows immediately from the anomalous Ward identity (8.55) since δ_v represents the nilpotent BRS operator

$$\delta_v G(v, A) = \delta_v^2 W[A_\mu] = 0. \tag{8.83}$$

For the same argument the expression

$$G_{\text{triv}}(v, A) = \delta_v \widehat{G}[A] \tag{8.84}$$

with $\widehat{G}[A]$ a *local* polynomial in A is just a **trivial solution**. Such a local functional $\widehat{G}[A]$ corresponds to an additional term in the 'quantum action', which just redefines the regularization scheme and does *not* alter the anomalous content of the gauge theory (recall our discussion in Section 4.2).

8.3 Algebra, cocycles and cohomology

8.3.1 Faddeev–Popov ghosts and gauge elements

Let us consider the two formalisms discussed so far:

FP ghost formalism	**gauge element formalism**
v^a FP ghost, anticommuting	v^a_α gauge element, commuting
δ_v BRS operator	δ_{v_α} ordinary gauge operator
$\delta_v^2 = 0$ nilpotency	$[\delta_{v_1}, \delta_{v_2}] = \delta_{[v_1, v_2]}$ commutation relation
$\delta_v G(v, A) = 0$ WZ consistency condition	$\delta_{v_1} G(v_2, A) - \delta_{v_2} G(v_1, A) = G([v_1, v_2], A)$ WZ consistency condition

As noted by [Becchi, Rouet, Stora 1976c] and [Alvarez-Gaumé, Ginsparg 1985] both formalisms are totally equivalent to each other. Either we can work with the FP ghost v and with the BRS operator δ_v or with the ordinary gauge element v_α and its infinitesimal gauge operator δ_{v_α}. Then the role of the commutation relation for the gauge operator δ_{v_α} is taken over by the nilpotency of the BRS operator δ_v and the WZ consistency condition on the right-hand side is given by the concise form on the left-hand side. We can transfer the gauge element formalism directly into the FP ghost formalism. How does it work?

We can construct a FP ghost, which we have already identified with the MC form in group space, by choosing two ordinary gauge functions v_1, v_2 (see Sections 8.1 and 9.3)

$$v = v_1 d\theta^1 + v_2 d\theta^2 = v_\alpha d\theta^\alpha, \tag{8.85}$$

together with

$$v_\alpha = g^{-1}(x,\theta)\frac{\partial}{\partial\theta^\alpha}g(x,\theta). \tag{8.86}$$

The BRS operator δ_v (8.23), which we have identified with the group derivation δ (9.43), creates the BRS transformations (8.24). For instance,

$$\delta_v A = -Dv = d\theta^\alpha D v_\alpha = d\theta^\alpha \delta_{v_\alpha} A \tag{8.87}$$

implies that

$$\delta_v = d\theta^\alpha \delta_{v_\alpha}. \tag{8.88}$$

Then we get for the **anomalous Ward identity**, multiplying equation (8.68) with $d\theta^\alpha$, the result (8.55) in the FP scheme

$$\begin{aligned} d\theta^\alpha \delta_{v_\alpha} W[A_\mu] &= d\theta^\alpha \int_M v_\alpha^a G^a[A] \\ \delta_v W[A_\mu] &= \int_M v^a G^a[A] = G(v,A). \end{aligned} \tag{8.89}$$

But to achieve in equation (8.88) the full significance of the BRS operator δ_v the operator δ_{v_α} also has to act upon the gauge functions v_α. The ordinary behaviour of the gauge operator δ_{v_α} is given up; δ_{v_α} must reproduce the BRS transformation (8.24)

$$\delta_v v = -v^2. \tag{8.90}$$

Inserting the constructions (8.85) and (8.88)

$$\begin{aligned} d\theta^\alpha d\theta^\beta \delta_{v_\alpha} v_\beta &= -d\theta^\alpha d\theta^\beta v_\alpha v_\beta \\ d\theta^1 d\theta^2 (\delta_{v_1} v_2 - \delta_{v_2} v_1) &= -d\theta^1 d\theta^2 [v_1, v_2] \end{aligned} \tag{8.91}$$

then leads to the **BRS-like transformation**

$$\delta_{v_1} v_2 - \delta_{v_2} v_1 = -[v_1, v_2]. \tag{8.92}$$

In this way we have extended the action of the gauge operator into a functional space of gauge fields A *and* gauge elements v_α. As in Section 8.1 we find its **representation**

$$\delta_{v_\alpha} = \begin{cases} v_\alpha \cdot X & \text{for } W[A_\mu] \\ v_\alpha \cdot X_v & \text{for } W[v_\alpha] \end{cases}, \tag{8.93}$$

where the operator X_v (8.20), is rewritten in terms of gauge elements v_β.

Finally we can convert the **WZ consistency condition** (8.69) into its concise form (8.82). We multiply equation (8.69) with $d\theta^1 d\theta^2$ and find

$$\begin{aligned}
d\theta^1 d\theta^2 [\int v_2^a \delta_{v_1} G^a[A] - \int v_1^a \delta_{v_2} G^a[A] - \int [v_1, v_2]^a G^a[A]] &= 0 \\
d\theta^\alpha d\theta^\beta [\int v_\beta^a \delta_{v_\alpha} G^a[A] + \int (\delta_{v_\alpha} v_\beta^a) G^a[A]] &= 0 \\
-\int v^a \delta_v G^a[A] + \int (\delta_v v^a) G^a[A] &= 0 \\
\delta_v G(v, A) = \delta_v \int v^a G^a[A] &= 0. \\
\text{Q.E.D.} \qquad & \qquad (8.94)
\end{aligned}$$

Of course, when we extend the gauge operator δ_{v_α} à la BRS this concise form of the WZ consistency condition follows immediately from the anomalous Ward identity (8.89) by virtue of $\delta_v^2 = d\theta^\alpha d\theta^\beta \delta_{v_\alpha} \delta_{v_\beta} = 0$.

8.3.2 Algebra

The above geometric detour, by identifying the FP ghost with the MC form on group space, is certainly not necessary to show the equivalence between the two representations of the WZ consistency condition. Stora's view in his GIFT lectures [Stora 1986] is purely algebraic.

Stora's view: Stora introduces—in the sense of vector spaces—a **dual algebra *Lie** G to the Lie algebra Lie G of the gauge group G. Defining

$$\begin{aligned}
&\{v_a\} \qquad \text{as a basis for Lie } G \\
&\{v^a\} \qquad \text{as a dual basis for *Lie } G \qquad (8.95)
\end{aligned}$$

both spaces are dual to each other via the inner product

$$(v^a, v_b) = \delta^a{}_b. \qquad (8.96)$$

Stora then regards the FP ghost v as the **Lie algebra valued generator of the dual *Lie** G

$$v = v^a v_a. \qquad (8.97)$$

Whereas the ordinary Lie algebra generators satisfy the usual **commutation relations**

$$[v_a, v_b] = f_{ab}{}^c v_c \qquad (8.98)$$

the duals obey the **BRS transformation**

$$\delta v^a = -\frac{1}{2} f^a{}_{bc} v^b v^c. \qquad (8.99)$$

(For obvious reasons here we raise and lower the indices of the structure constants in our notation.) Multiplying equation (8.99) by v_a gives

$$\delta v = -\frac{1}{2}[v, v]. \tag{8.100}$$

(Recall our MC discussion in Section 2.6.6.)

Remembering the familiar 2-form (2.111), evaluated on the tangent space $T(M)$, $dx^\mu dx^\nu(X, Y) = X^\mu Y^\nu - Y^\mu X^\nu$, we can calculate a **'Faddeev–Popov ghost 2-form'** $v^a v^b$ analogously

$$v^a v^b(\xi, \eta) = \xi^a \eta^b - \eta^a \xi^b \tag{8.101}$$

on the vector fields (gauge elements)

$$\xi, \eta \in \text{Lie } G; \qquad \xi = \xi^a v_a, \qquad \eta = \eta^a v_a, \tag{8.102}$$

where $\{v_a\}$ is the basis for Lie G in the spirit of Stora.

WZ consistency condition: Now we are prepared to reformulate the WZ consistency condition. We start with representation (8.82)

$$\delta G(v, A) = \int dx \delta[v^a(x) G_a[A_\mu](x)] = 0,$$

$$\int dxdy \left[\frac{\delta}{\delta A^b_\mu(y)} \delta A^b_\mu(y) + \frac{\delta}{\delta v^b(y)} \delta v^b(y)\right] v^a(x) G_a[A_\mu](x) = 0, \tag{8.103}$$

where δ (we drop the index v) means a variation of the fields à la BRS

$$\int dxdy \left[(D_\mu v)^b(y) v^a(x) \frac{\delta}{\delta A^b_\mu(y)} G_a[A_\mu](x) \right.$$
$$\left. - \frac{1}{2}[v, v]^b(y) \frac{\delta}{\delta v^b(y)} v^a(x) G_a[A_\mu](x)\right] = 0. \tag{8.104}$$

We integrate by parts and evaluate the 'FP ghost 2-form' $v^a(x)v^b(y)$ on the vector fields ξ, η

$$\int dxdy v^a(x) v^c(y)(\xi, \eta) \left[D_\mu{}^b{}_c(y) \frac{\delta}{\delta A^b_\mu(y)} G_a[A_\mu](x) \right.$$
$$\left. - \frac{1}{2} f_{ac}{}^b G_b[A_\mu](x) \delta(x - y)\right] = 0,$$

$$\int dxdy[\xi^a(x)\eta^c(y) - \eta^a(x)\xi^c(y)] \left[D_\mu{}^b{}_c(y) \frac{\delta}{\delta A^b_\mu(y)} G_a[A_\mu](x) \right.$$
$$\left. - \frac{1}{2} f_{ac}{}^b G_b[A_\mu](x) \delta(x-y) \right] = 0. \tag{8.105}$$

Recalling the ordinary **gauge operator** (8.65)

$$\delta_\eta = -\int dy \eta^c(y) D_\mu{}^b{}_c(y) \frac{\delta}{\delta A^b_\mu(y)} = \eta \cdot X, \tag{8.106}$$

with its commutation relation (8.66) provides the **WZ consistency condition** for the gauge elements ξ, η straight away in accordance with equation (8.69)

$$\delta G(v, A)(\xi, \eta) = \delta_\xi G(\eta, A) - \delta_\eta G(\xi, A) - G([\xi, \eta], A) = 0. \tag{8.107}$$

8.3.3 Cocycle

We can also consider the WZ consistency condition from a geometric point of view leading to a cocycle condition.

We return to the space of all gauge potentials Sp $\mathcal{A}$ (Section 8.1). We denote

$$\left\{ \frac{\delta}{\delta A^a_\mu(x)} \right\} \quad \text{as the basis for } T(\text{Sp } \mathcal{A})$$
$$\{\delta A^a_\mu(x)\} \quad \text{as the basis for } T^*(\text{Sp } \mathcal{A}), \tag{8.108}$$

and introduce vector fields (8.38) and p-forms (8.39). For example, the **fundamental vector field** is described by

$$X_\xi = \int dx (D_\mu \xi)^a(x) \frac{\delta}{\delta A^a_\mu(x)} \tag{8.109}$$

and corresponds, in our previous notations (8.65), (8.106), to the gauge operator δ_ξ

$$X_\xi = \xi \cdot X = \delta_\xi. \tag{8.110}$$

The **exterior derivative in Sp $\mathcal{A}$** is given by

$$\delta = \int dx \frac{\delta}{\delta A^a_\mu(x)} \delta A^a_\mu(x). \tag{8.111}$$

Lie derivative: Let us consider the Lie derivative in Sp $\mathcal{A}$. What does it mean? We calculate its action on a functional $f[A_\mu]$ along a curve which is produced by the fundamental vector field (8.109) (recall Section 2.6)

$$\begin{aligned}\mathcal{L}_{X_\xi} f[A_\mu] &= \left.\frac{d}{dt}\right|_{t=0} f[A^a_\mu + t(D_\mu\xi)^a] \\ &= \lim_{t\to 0}\frac{1}{t}\left\{f[A^a_\mu] + t\int dx (D_\mu\xi)^a(x)\frac{\delta}{\delta A^a_\mu(x)} f[A^a_\mu] - f[A^a_\mu]\right\} \\ &= \int dx (D_\mu\xi)^a(x)\frac{\delta}{\delta A^a_\mu(x)} f[A^a_\mu] \\ &= X_\xi f[A_\mu]. \end{aligned} \tag{8.112}$$

Thus the Lie derivative in Sp $\mathcal{A}$ represents the usual directional derivative $\delta/\delta A^a_\mu(x)$ along $(D_\mu\xi)^a(x)$. It gives the fundamental vector field X_ξ or the familiar gauge operator δ_ξ (8.65), (8.106), $\mathcal{L}_{X_\xi} \to X_\xi, \delta_\xi$.

Analogous to the ordinary space–time manifold M we also have here in Sp $\mathcal{A}$ the **formula**

$$\mathcal{L}_{X_\xi} = i_{X_\xi}\delta + \delta i_{X_\xi}, \tag{8.113}$$

where i_{X_ξ} denotes the **interior product** of a form on Sp $\mathcal{A}$ with the fundamental vector field X_ξ.

Then we can express the **anomalous Ward identity** (8.68) geometrically by

$$\mathcal{L}_{X_\xi} W[A_\mu] = G(\xi, A). \tag{8.114}$$

Cocycle condition: Now let us define some 1-form on Sp $\mathcal{A}$

$$G = \int dx G^\mu_a[A](x)\delta A^a_\mu(x), \tag{8.115}$$

with $G^\mu_a[A]$ some functional of the gauge potential A. The inner product with a vector field X gives a 0-form on Sp $\mathcal{A}$

$$i_X G = G(X) = \int dx G^\mu_a[A](x) X^a_\mu(x). \tag{8.116}$$

Analogous to differential forms on M we can derive here the following formula for the above 1-form on Sp $\mathcal{A}$.

Lemma:

$$\begin{aligned} \delta G(X,Y) &= XG(Y) - YG(X) - G([X,Y]) \\ &= \mathcal{L}_X G(Y) - \mathcal{L}_Y G(X) - G([X,Y]), \end{aligned} \tag{8.117}$$

$$i_Y i_X \delta G = \mathcal{L}_X i_Y G - \mathcal{L}_Y i_X G - i_{[X,Y]} G. \tag{8.118}$$

If the form is closed, $\delta G = 0$, it is also called a **cocycle** in the language of cohomology (Section 2.5.2).

Next we choose G to be the variation—the exterior derivative in Sp $\mathcal{A}$—of the vacuum functional $W[A_\mu]$

$$G = \delta W[A_\mu] = -\delta \ln Z[A_\mu]. \tag{8.119}$$

Then the left-hand side of formula (8.117), (8.118) vanishes

$$\delta G(X_\xi, X_\eta) = i_{X_\eta} i_{X_\xi} \delta G = 0 \tag{8.120}$$

and we find the **WZ consistency condition** written **geometrically in Sp $\mathcal{A}$**

$$\mathcal{L}_{X_\xi} G(\eta, A) - \mathcal{L}_{X_\eta} G(\xi, A) - G([\xi,\eta], A) = 0. \tag{8.121}$$

Proposition:

- *The WZ consistency condition corresponds to a cocycle condition in Sp $\mathcal{A}$!*

Remark:

i) The logarithm in equation (8.119), however, may not be globally well defined on Sp $\mathcal{A}$. As noticed by Falqui and Reina [Falqui, Reina 1985] if there is an anomaly the functional $Z[A_\mu]$ vanishes at some gauge orbits in Sp $\mathcal{A}$. Excluding these points from the domain of $Z[A_\mu]$ the above functional (8.119) is defined. But then the 1-form G (8.119) is just closed (and not exact).

ii) Furthermore the 1-form G (8.119) depends *nonlocally* on the gauge potential A (since the 'quantum action' $W[A_\mu]$ is a nonlocal functional). Only if the derivative is restricted along the fundamental vector field—along the fibre $\mathcal{G}$ in Figure 8.1—

$$\delta \to i_{X_\xi} \delta = \delta_\xi \tag{8.122}$$

the functional emerges locally in the fields A and represents the non-Abelian anomaly. We shall discuss this point further in Chapter 10.

Notation: Let us finally list the several aspects and notations:

$$
\begin{array}{rll}
 & G = \delta W[A_\mu] & \\
\stackrel{i_{X_\xi}}{\longrightarrow} & i_{X_\xi} G = i_{X_\xi}\delta W[A_\mu] & \text{variation along fibre } \mathcal{G}, \text{ 0-form in Sp } \mathcal{A} \\
 & = \delta_\xi W[A_\mu] & \text{ordinary gauge operator} \\
 & = X_\xi W[A_\mu] & \text{fundamental vector field in Sp } \mathcal{A} \\
 & = \mathcal{L}_{X_\xi} W[A_\mu] & \text{Lie derivative in Sp } \mathcal{A} \\
 & = G(\xi, A) & \text{non-Abelian consistent anomaly.}
\end{array}
$$

8.3.4 Cohomology

The above considerations have been generalized mathematically by several authors [Bonora, Cotta-Ramusino 1983, 1986], [Dubois-Violette, Talon, Viallet 1985a,b], [Stora 1986], [Kastler, Stora 1986a,b], [Tröster 1994]. A principal fibre bundle $P(M,G)$ is constructed with M the (compact) space–time manifold, G the gauge group (compact Lie group) and Lie G the Lie algebra of G. Again, we consider Sp $\mathcal{A}$, the space of all connections in $P(M,G)$, and we denote the space of all *local* functionals on Sp $\mathcal{A}$ by $\Gamma_{\text{loc}}(\text{Sp } \mathcal{A})$. It is a representation space for G and Lie G. The representation of the Lie G in $\Gamma_{\text{loc}}(\text{Sp } \mathcal{A})$ is achieved by the Lie derivative

$$\mathcal{L}_{X_\xi} f[A] = \left.\frac{d}{dt}\right|_{t=0} f[A + tD\xi] = X_\xi f[A], \tag{8.123}$$

with $\xi \in$ Lie G and $f[A] \in \Gamma_{\text{loc}}(\text{Sp } \mathcal{A})$.

Definitions: Recalling the cohomology of Section 2.5.2 we define the **vector space** $C^k(\text{Lie } G, \Gamma_{\text{loc}}(\text{Sp } \mathcal{A}))$ **of all** k**-cochains** of Lie G with values in $\Gamma_{\text{loc}}(\text{Sp } \mathcal{A})$ by considering the space of all multilinear skew-symmetric maps (k-forms on Lie G)

$$f_k[A] : \text{Lie } G \underbrace{\times \ldots \times}_{k \text{ factors}} \text{Lie } G \to \Gamma_{\text{loc}}(\text{Sp } \mathcal{A}). \tag{8.124}$$

Then we get a **coboundary operator**

$$\delta : C^k(\text{Lie } G, \Gamma_{\text{loc}}(\text{Sp } \mathcal{A})) \to C^{k+1}(\text{Lie } G, \Gamma_{\text{loc}}(\text{Sp } \mathcal{A})) \tag{8.125}$$

on $C^*(\text{Lie } G, \Gamma_{\text{loc}}(\text{Sp } \mathcal{A})) = \bigoplus_k C^k(\text{Lie } G, \Gamma_{\text{loc}}(\text{Sp } \mathcal{A}))$ with $\delta^2 = 0$ which is defined by

$$\begin{aligned}\delta f_k[A](\xi_1,\ldots,\xi_{k+1}) \\ = & \sum_{i=1}^{k+1}(-)^{i+1}\mathcal{L}_{X_{\xi_i}} f_k[A](\xi_1,\ldots,\widehat{\xi}_i,\ldots,\xi_{k+1}) \\ & +\sum_{i<j}(-)^{i+j} f_k[A]([\xi_i,\xi_j],\xi_1,\ldots,\widehat{\xi}_i,\ldots,\widehat{\xi}_j,\ldots,\xi_{k+1}). \quad (8.126)\end{aligned}$$

The symbol $\widehat{}$ denotes the omission of the corresponding element in Lie G.

As in Section 2.5.2 we specify the **space of cocycles in Lie** G of degree k by

$$Z^k(\text{Lie } G, \Gamma_{\text{loc}}(\text{Sp } \mathcal{A})) = \{f_k | \delta f_k = 0\} \quad (8.127)$$

and the **space of coboundaries in Lie** G by

$$B^k(\text{Lie } G, \Gamma_{\text{loc}}(\text{Sp } \mathcal{A})) = \{f_k | f_k = \delta f_{k-1}, f_{k-1} \in C^{k-1}\}. \quad (8.128)$$

Since $\delta^2 = 0$ we have

$$B^k(\text{Lie } G, \Gamma_{\text{loc}}(\text{Sp } \mathcal{A})) \subset Z^k(\text{Lie } G, \Gamma_{\text{loc}}(\text{Sp } \mathcal{A})) \quad (8.129)$$

and we can define the **cohomology in Lie** G

$$H^k(\text{Lie } G, \Gamma_{\text{loc}}(\text{Sp } \mathcal{A})) = Z^k(\text{Lie } G, \Gamma_{\text{loc}}(\text{Sp } \mathcal{A}))/B^k(\text{Lie } G, \Gamma_{\text{loc}}(\text{Sp } \mathcal{A})). \quad (8.130)$$

The cohomology space of the complex $(C^*(\text{Lie } G, \Gamma_{\text{loc}}(\text{Sp } \mathcal{A})), \delta)$ is called the **cohomology of the Lie algebra of the gauge group**.

Physics: The physical meaning of the objects involved here is:

- the coboundary operator δ represents the BRS operator,
- the cochains $f_k[A]$ are local functionals in the gauge potential A containing k FP ghosts,
- the elements of the cohomology H^k are integrated anomalous terms with k FP ghosts.

For example, in the case of $k = 1$ we obtain

$$\delta f_1[A](\xi_1,\xi_2) = \mathcal{L}_{X_{\xi_1}} f_1[A](\xi_2) - \mathcal{L}_{X_{\xi_2}} f_1[A](\xi_1) - f_1[A]([\xi_1,\xi_2]). \quad (8.131)$$

For $f_1[A] \in H^1(\text{Lie } G, \Gamma_{\text{loc}}(\text{Sp } \mathcal{A}))$ being an element of the first cohomology class the **cocycle condition**

$$\delta f_1[A] = 0 \quad (8.132)$$

or equivalently

$$\mathcal{L}_{X_{\xi_1}} f_1[A](\xi_2) - \mathcal{L}_{X_{\xi_2}} f_1[A](\xi_1) - f_1[A]([\xi_1, \xi_2]) = 0 \tag{8.133}$$

represents the **WZ consistency condition**. The cohomology element

$$f_1[A](\xi) = G(\xi, A) \tag{8.134}$$

expresses the integrated physical **anomaly**. It is determined uniquely modulo trivial solutions

$$G_{\text{triv}} = \delta f_0[A]. \tag{8.135}$$

We shall rediscuss these cohomology solutions in Section 9.3 in terms of the views of [Dubois-Violette, Talon, Viallet 1985a,b].

9

Stora–Zumino chain of descent equations

Now we come to one of the main chapters of the book. We shall show how the singlet anomaly in $2n$ dimensions determines the non-Abelian anomaly in $(2n-2)$ dimensions via a set of equations. These are part of a whole chain of equations, which descend in their form degree, thus called the **Stora–Zumino chain of descent equations** [Stora 1977, 1984], [Zumino 1984]. We are going to derive this chain on pure mathematical grounds—algebra and differential geometry—and discuss its meaning in physics afterwards. We offer the topological aspect of a chain, described by an index theorem, later on in Chapter 11.

We begin with Stora who approaches the chain from the physical side by working with the gauge potential A, the FP ghost v and the BRS operator s (Section 9.1). Then we extend the formalism to nontrivial gauge bundles (Section 9.2) and afterwards present Zumino's pure geometric approach (Section 9.3). Of course, the two approaches are equivalent. Explicit solutions for a whole chain are calculated in Section 9.4 and general formulae for the chain terms are derived in Section 9.5.

9.1 Stora's approach to the chain

Since we are already familiar with the BRS transformation, the anomalous Ward identity and the WZ consistency condition, we are prepared to follow Stora's approach [Stora 1977, 1984].

Let us consider a renormalizable field theory given by the vacuum functional $W[A_\mu]$, equation (8.1), where the chiral fermions are coupled to the external Yang–Mills fields A_μ. The FP ghosts v are introduced via the BRS transformations (8.24), then the **anomalous Ward identity**

$$sW[A_\mu] = G(v, A) \tag{9.1}$$

defines the **anomaly** which we express by the integral

$$G(v, A) = \int_M v^a G^a[A]. \tag{9.2}$$

The 4-form $v^a G^a[A]$ is linear in v, local in A and v, and defined up to a derivative of a local form $d\eta$. In particular $G^a[A]$ denotes Bardeen's anomaly result (6.79) of Section 6.3.

Next we consider the **WZ consistency condition** (following from the anomalous Ward identity (9.1) and $s^2 = 0$)

$$sG(v, A) = s \int_M v^a G^a[A] = 0, \tag{9.3}$$

the equation which determines the anomaly. We rewrite equation (9.3) in a local way by using the algebraic lemma of Poincaré (recall Section 7.3) which guarantees that equation (9.3) is satisfied by any 4-form for which $s(v^a G^a)$ is exact

$$s(v^a G^a[A]) = -dQ_3^2(v, A). \tag{9.4}$$

The term Q_3^2 has to be a polynomial in v and A with FP ghost number 2 and form degree 3. What is the solution of equation (9.4)? It can be found in a very general way.

Now we perform pure mathematics. We consider the symmetric invariant polynomial $P(F^n)$ (recall Section 7.1). Since $P(F^n)$ is closed (due to the Bianchi identity $DF = 0$) it is locally exact (Poincaré lemma, Section 7.3)

$$P(F^n) = dQ_{2n-1}(A, F). \tag{9.5}$$

We know the polynomial Q_{2n-1}, of form degree $(2n-1)$, already; it is the **Chern–Simons form** which is explicitly given by (recall Section 7.2)

$$Q_{2n-1}(A, F) = n \int_0^1 dt P(A, F_t^{n-1}), \tag{9.6}$$

where

$$F_t = tF + (t^2 - t)A^2 \tag{9.7}$$

(geometrically, we consider a trivial fibre bundle). Equation (9.5) together with equations (9.6), (9.7)—also called **'transgression'**—is the starting point of the following mathematical operations.

Stora noticed that a shift in both, in the gauge connection and in the derivate

$$\begin{aligned} A &\rightarrow \widehat{A} = A + v \\ d &\rightarrow \Delta = d + s \end{aligned} \tag{9.8}$$

leaves the curvature invariant. Stora nicknamed this the 'Russian formula' [Stora 1984], [Mañes, Stora, Zumino 1985].

Theorem: 'Russian formula'

$$\widehat{F}(\widehat{A}) = \Delta\widehat{A} + \widehat{A}^2 \equiv dA + A^2 = F(A). \tag{9.9}$$

Proof: Theorem (9.9) is valid by virtue of the BRS equations (8.24)

$$\begin{aligned}\Delta(A+v) &= dA + dv + sA + sv \\ &= dA + A^2 - (A^2 + Av + vA + v^2) \\ &= F - (A+v)^2. \qquad \text{Q.E.D.}\end{aligned} \tag{9.10}$$

Of course, the Bianchi identity for the shifted fields is also satisfied

$$\widehat{D}\widehat{F} = \Delta\widehat{F} + [\widehat{A}, \widehat{F}] = 0 \tag{9.11}$$

so that we obtain the 'transgression formula' for the Stora shift.

Theorem: 'Shifted transgression'

$$P(\widehat{F}^n) = \Delta Q_{2n-1}(\widehat{A}, \widehat{F}), \tag{9.12}$$

with

$$Q_{2n-1}(\widehat{A}, \widehat{F}) = n\int_0^1 dt P(\widehat{A}, \widehat{F}_t^{n-1}) \quad \text{and} \quad \widehat{F}_t = t\widehat{F} + (t^2 - t)\widehat{A}^2. \tag{9.13}$$

Proof. Let us take some infinitesimal variation of the fields

$$\widehat{A} \to \widehat{A} + \delta_t\widehat{A} \quad \Longrightarrow \quad \delta_t\widehat{F} = \widehat{D}(\delta_t\widehat{A}). \tag{9.14}$$

Then we calculate

$$\begin{aligned}\delta_t P(\widehat{F}^n) &= nP(\delta_t\widehat{F}, \widehat{F}^{n-1}) = nP(\widehat{D}(\delta_t\widehat{A}), \widehat{F}^{n-1}) \\ &= n\Delta P(\delta_t\widehat{A}, \widehat{F}^{n-1}).\end{aligned} \tag{9.15}$$

Considering the homotopies $\widehat{A}_t = t\widehat{A}$, $\widehat{F}_t = d\widehat{A}_t + \widehat{A}_t^2$ together with $\widehat{D}_t\widehat{F}_t = \Delta\widehat{F}_t + [\widehat{A}_t, \widehat{F}_t] = 0$, and choosing finally $\delta_t = dt\frac{\partial}{\partial t}$, we find after integration

$$\begin{aligned}\delta_t P(\widehat{F}_t^n) &= n\Delta P(\delta_t\widehat{A}_t, \widehat{F}_t^{n-1}) \\ P(\widehat{F}^n) &= n\Delta\int_0^1 dt P(\widehat{A}, \widehat{F}_t^{n-1}). \qquad \text{Q.E.D.}\end{aligned} \tag{9.16}$$

Applying the 'Russian formula' (9.9) now we can equate the 'transgression' (9.5) with its shifted version (9.12)

$$\Delta Q_{2n-1}(A+v,F) = dQ_{2n-1}(A,F). \tag{9.17}$$

We expand the Chern–Simons form in powers of the FP ghost v

$$\begin{aligned} Q_{2n-1}(A+v,F) &= Q^0_{2n-1}(A,F) + Q^1_{2n-2}(v,A,F) \\ &\quad + Q^2_{2n-3}(v,A,F) + \ldots + Q^{2n-1}_0(v), \end{aligned} \tag{9.18}$$

where the upper index denotes the powers of v and the lower index the form degree. We insert this expansion into formula (9.17)

$$\begin{aligned} &(d+s)Q^0_{2n-1} + (d+s)Q^1_{2n-2} + (d+s)Q^2_{2n-3} \\ &+ \ldots + (d+s)Q^{2n-1}_0 = dQ^0_{2n-1}; \end{aligned} \tag{9.19}$$

we compare the terms of the same form degree and same power in v then we obtain the following set of equations:

Stora–Zumino chain of descent equations:

$$\begin{aligned} P(F^n) - dQ^0_{2n-1} &= 0 \\ sQ^0_{2n-1} + dQ^1_{2n-2} &= 0 \\ sQ^1_{2n-2} + dQ^2_{2n-3} &= 0 \\ \cdots\cdots & \\ sQ^{2n-2}_1 + dQ^{2n-1}_0 &= 0 \\ sQ^{2n-1}_0 &= 0. \end{aligned} \tag{9.20}$$

Comparing now the chain (9.20) with physics we observe that the third equation represents precisely the local version of the WZ consistency condition (9.4). Therefore we may identify the chain term Q^1_4, or in general Q^1_{2n-2}, with the anomaly $v^a G^a[A]$ (which we found in Chapter 8).

Proposition:

- *The chain term Q^1_{2n-2} represents the anomaly*

$$G(v,A) = N \int_{M_{2n-2}} Q^1_{2n-2}(v,A). \tag{9.21}$$

The **normalization**

$$N = -2\pi i \frac{i^n}{(2\pi)^n n!} \cdot (\pm) \tag{9.22}$$

where the factor $(\pm)$ stands for positive or negative chirality fields (L- or R-fields), however, is not fixed by the chain (9.20) and must be calculated by other methods:

i) perturbation theory, Feynman diagrams (Section 4.8),
ii) Fujikawa path integral formalism (Sections 5.6 and 11.5.2),
iii) topological analysis, Atiyah–Singer index theorem (Section 11.5.3).

For instance, when using perturbation theory it is enough to calculate the simplest anomalous loop—the triangle diagram in 4 dimensions—which is an easy task. (Recall that there are no radiative corrections due to the Adler–Bardeen theorem.) Then we have the normalization and the total anomaly expression—which would otherwise involve more complicated calculations—is given by the chain term Q^1_4. This is the practical value of the SZ chain; we can easily solve all chain terms Q^k_{2n-1-k}. We postpone the discussion of the higher order chain terms Q^k_{2n-1-k} to Section 9.3.

Remark: The SZ chain starts with some invariant polynomial. However, which polynomial we have to choose depends on the physical theory (see e.g. gravitation, Section 12.8). Of course, in general, the invariant polynomial may represent a linear combination of trace products like

$$\mathrm{tr}\ F^p \mathrm{tr}\ \widetilde{F}^q \tag{9.23}$$

where $\widetilde{F} = d\widetilde{A} + \widetilde{A}^2$, and $\widetilde{A}$ describes some connection (e.g. $\widetilde{A} \equiv A$, or some gauge connection $\widetilde{A} \not\equiv A$, or $\widetilde{A} \equiv \omega$ the spin connection of gravitation). Then the corresponding chain terms are given by

$$\begin{aligned} Q^0_{2p+2q-1}(A, \widetilde{A}) &= aQ^0_{2p-1}(A)\mathrm{tr}\ \widetilde{F}^q + b\mathrm{tr}\ F^p Q^0_{2q-1}(\widetilde{A}) \\ Q^1_{2p+2q-2}(A, \widetilde{A}) &= aQ^1_{2p-2}(A)\mathrm{tr}\ \widetilde{F}^q + b\mathrm{tr}\ F^p Q^1_{2q-2}(\widetilde{A}), \end{aligned} \tag{9.24}$$

with $a + b = 1$; e.g. $a = \frac{p}{p+q}, b = \frac{q}{p+q}$.

9.2 Chain for nontrivial gauge bundles

In deriving the chain of descent equations, we have considered up till now gauge theories with a trivial fibre-bundle structure. But the algebraic construction of the anomaly and of the whole chain can certainly be extended to a global validity on nontrivial gauge bundles. This generalization has been worked out by Mañes, Stora and Zumino [Mañes, Stora, Zumino 1985] and we briefly summarize their result.

We start with a gauge theory containing a structure group G, a compact Lie group, and consider a principal bundle $P(M, G)$ (recall Section 2.7). The base manifold M is as usual of even dimension, compact and without boundary. The connections on $P(M,G)$ are locally represented by 1-forms, the gauge potentials, with values in Lie G. The transition functions of $P(M,G)$ are locally given by the gauge elements $g(x)$ on M into G with suitable gluing properties. In addition to the gauge potential A a fixed background field A_0 will be introduced whenever the bundle $P(M,G)$ is not trivial. Such a field A_0 also occurs in perturbation theory where one perturbes in the neighbourhood of A_0.

We introduce the BRS transformations (8.24) but we do not transform the field A_0

$$sA_0 = 0. \tag{9.25}$$

The anomaly appears via the **anomalous Ward identity**

$$sW[A, A_0] = G(v, A, A_0) = \int_M v^a G^a[A, A_0], \tag{9.26}$$

and from the nilpotency $s^2 = 0$ follows the **WZ consistency condition**

$$sG(v, A, A_0) = s \int_M v^a G^a[A, A_0] = 0. \tag{9.27}$$

Again, the solution $v^a G^a[A, A_0]$ is ambiguous; the replacement

$$v^a G^a[A, A_0] \to v^a G^a[A, A_0] + s\widehat{G}[A] + d\eta \tag{9.28}$$

also solves equation (9.27) where $\widehat{G}$ and η are local polynomials of the fields.

For deriving the chain we consider now the **'transgression formula'** (7.28)–(7.30), valid for nontrivial bundles

$$\begin{aligned} P(F^n(A)) - P(F^n(A_0)) &= dQ_{2n-1}(A, A_0) \\ &= nd \int_0^1 dt P(A - A_0, F_t^{n-1}), \end{aligned} \tag{9.29}$$

with the homotopies

$$F_t = dA_t + A_t^2 \tag{9.30}$$

and

$$A_t = tA + (1-t)A_0. \tag{9.31}$$

Shifting the gauge field A and the derivative d à la Stora–Zumino (9.8) and applying the 'Russian formula' (9.9) to the 'shifted transgression' we obtain the relation

$$\begin{aligned} dQ_{2n-1}(A, A_0) &= \Delta Q_{2n-1}(A + v, A_0) \\ &= n\Delta \int_0^1 dt P(A + v - A_0, \widehat{F}_t^{n-1}), \end{aligned} \tag{9.32}$$

with the homotopies

$$\widehat{F}_t = \Delta \widehat{A}_t + \widehat{A}_t^2 \tag{9.33}$$

and

$$\widehat{A}_t = t(A + v) + (1 - t)A_0. \tag{9.34}$$

Expanding the Chern–Simons form as usual in powers of v

$$Q_{2n-1}(A + v, A_0) = \sum_{p=0}^{2n-1} Q^p_{2n-1-p}(v, A, A_0) \tag{9.35}$$

the relation (9.32) provides the desired chain of descent equations when comparing terms of the same order:

Chain of descent equations:

$$\begin{aligned} P(F^n(A)) - P(F^n(A_0)) - dQ^0_{2n-1}(A, A_0) &= 0 \\ sQ^p_{2n-1-p}(v, A, A_0) + dQ^{p+1}_{2n-2-p}(v, A, A_0) &= 0 \\ sQ_0^{2n-1}(v) &= 0 \end{aligned} \tag{9.36}$$

and $p = 0, 1, \ldots, 2n - 2$.

In particular the chain term $Q^1_{2n-2}(v, A, A_0)$ solves the WZ consistency condition (9.27) and therefore represents the anomaly up to the normalization

$$NQ^1_{2n-2}(v, A, A_0) = v^a G^a[A, A_0]. \tag{9.37}$$

The set of equations (9.36) containing a fixed background connection A_0 is globally valid on a nontrivial bundle $P(M, G)$. Only for a trivial bundle one may choose $A_0 = 0$ and one recovers the usual local chain formulae which we have considered before.

9.3 Zumino's approach to the chain

In a Yang–Mills gauge theory the element $g(x)$ of a gauge group G is usually just a function of x, the space–time manifold M. Zumino [Zumino 1984] extends this space; the gauge element

$$g = g(x, \theta) \tag{9.38}$$

should depend on additional parameters

$$\theta = \{\theta^\alpha\}, \tag{9.39}$$

variables which form a manifold in a space, the **group parameter space** M_θ with dimension $\dim M_\theta = p$.

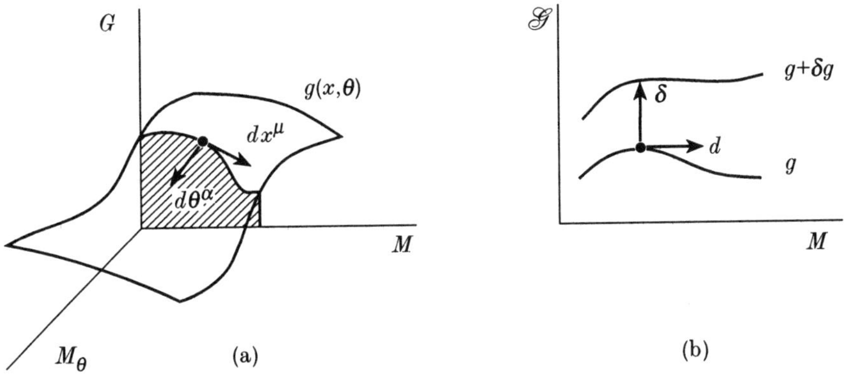

Fig. 9.1. a) Gauge group elements on $M \times M_\theta$. b) Gauge group space $\mathcal{G}$

Definitions: Analogous to the base space M we define:

$$\left\{\frac{\partial}{\partial\theta^\alpha}\right\} \quad \text{as a basis for the tangent space } T(M_\theta) \text{ in } M_\theta,$$
$$\{d\theta^\alpha\} \quad \text{as a basis for the cotangent space } T^*(M_\theta). \tag{9.40}$$

Both spaces are dual to each other via the **inner product**

$$\left(d\theta^\alpha, \frac{\partial}{\partial\theta^\beta}\right) = \delta^\alpha{}_\beta \tag{9.41}$$

and $\dim T^*(M_\theta) = \dim T(M_\theta) = \dim M_\theta$.

A **wedge product** is defined by

$$d\theta^\alpha d\theta^\beta \equiv d\theta^\alpha \wedge d\theta^\beta := d\theta^\alpha \otimes d\theta^\beta - d\theta^\beta \otimes d\theta^\alpha, \tag{9.42}$$

and we also introduce an **exterior derivative in the group parameter space M_θ**

$$\delta = \frac{\partial}{\partial\theta^\alpha} d\theta^\alpha \qquad \text{with} \qquad \delta^2 = 0, \tag{9.43}$$

together with the antiderivative rule.

Now, the x-space M with dimension $\dim M = 2n$ is enlarged by considering the Cartesian product

$$M \times M_\theta \tag{9.44}$$

and again we define wedge products in this enlarged space (9.44), e.g.

$$\begin{aligned} &dx^{\mu_1} \ldots dx^{\mu_{2n}} d\theta^{\alpha_1} \ldots d\theta^{\alpha_p} \equiv \\ &\equiv \quad dx^{\mu_1} \wedge dx^{\mu_2} \wedge \ldots \wedge dx^{\mu_{2n}} \wedge d\theta^{\alpha_1} \wedge d\theta^{\alpha_2} \wedge \ldots \wedge d\theta^{\alpha_p}. \end{aligned} \tag{9.45}$$

An **extended gauge group element** defines the map (see Figure 9.1a)

$$\begin{array}{ccccccc} g(x,\theta): & M & \times & M_\theta & \longrightarrow & G \\ & \downarrow & & \downarrow & & \\ & S^{2n} & \times & S^p & \longrightarrow & G, \end{array} \tag{9.46}$$

where we finally work with compactified manifolds. The parameter dependence of $g(x,\theta)$ actually means that we also vary g at each point $x \in M$; so we consider the space of all gauge elements $\{g(x)\} = \mathcal{G}$, the **gauge group space** (see Figure 9.1b).

An infinitesimal gauge transformation involves an **extended gauge element** v given by

$$v = g^{-1}\delta g, \tag{9.47}$$

or, more explicitly,

$$v = v(x,\theta) = v_\alpha(x,\theta)d\theta^\alpha, \tag{9.48}$$

with

$$v_\alpha(x,\theta) = g^{-1}(x,\theta)\frac{\partial}{\partial\theta^\alpha}g(x,\theta). \tag{9.49}$$

Thus v acts as a 1-form in gauge group space $\mathcal{G}$ (or in M_θ) but as a 0-form on the base manifold M, the x-space, and again as a 1-form in the product space $M \times \mathcal{G}$ (or in $M \times M_\theta$) (see Figs. 7.1a,b).

Gauge transformations: Let $A(x)$ be some Yang–Mills reference gauge potential over the base manifold M. We carry out a gauge transformation over the product space $M \times M_\theta$ (or $M \times \mathcal{G}$)

$$A \xrightarrow{g} A^g(x,\theta) = g^{-1}(x,\theta)[A(x) + d + \delta]g(x,\theta) = g^{-1}[A + \Delta]g, \tag{9.50}$$

where we introduce a new **exterior differential operator** Δ acting in the product space $M \times \mathcal{G}$ (or $M \times M_\theta$)

$$\Delta = d + \delta, \qquad \text{with } \Delta^2 = 0 \tag{9.51}$$

since $d^2 = \delta^2 = d\delta + \delta d = 0$.

Considering, on the other hand, just a θ-family of gauge transformed configurations like

$$\mathcal{A}(x,\theta) = g^{-1}(x,\theta)[A(x) + d]g(x,\theta) \tag{9.52}$$

we rewrite the gauge potential A^g (or connection on $M \times M_\theta$) by

$$A^g = \mathcal{A} + v. \tag{9.53}$$

The corresponding field strengths, or curvatures, are

$$F = dA + A^2 = F(A), \tag{9.54}$$

$$\mathcal{F} = d\mathcal{A} + \mathcal{A}^2 = \mathcal{F}(\mathcal{A}), \tag{9.55}$$

with the gauge transformation

$$\mathcal{F} = g^{-1}Fg; \tag{9.56}$$

and

$$\begin{aligned} F \xrightarrow{g} F^g &= \Delta A^g + (A^g)^2 = F^g(A^g) &(9.57)\\ &= \Delta(\mathcal{A}+v) + (\mathcal{A}+v)^2 =: \widehat{\mathcal{F}}(\mathcal{A}+v), &(9.58)\end{aligned}$$

with the gauge transformation

$$F^g = g^{-1}Fg. \tag{9.59}$$

The gauge transformation formulae (9.56) and (9.59) follow quickly by explicit calculation.

Structure equations: Next we apply the δ-operator to v, $\mathcal{A}$ and $\mathcal{F}$:

$$\delta v = \delta(g^{-1}\delta g) = \delta g^{-1}\delta g, \tag{9.60}$$

by virtue of $\delta^2 = 0$, and using

$$\delta g^{-1}g + g^{-1}\delta g = 0, \tag{9.61}$$

which follows from $g^{-1}g = \mathbf{1}$, we get

$$\delta v = -g^{-1}\delta g g^{-1}\delta g = -v^2 = -\frac{1}{2}[v, v]. \tag{9.62}$$

Equation (9.62) represents the **Maurer–Cartan structure equation in group space** $\mathcal{G}$ and the element $v = g^{-1}\delta g$ is called the **Maurer–Cartan form on group space** $\mathcal{G}$ (recall Section 8.1).

The application of δ on the connection $\mathcal{A}$ is

$$\delta\mathcal{A} = \delta[g^{-1}(A+d)g] = \delta g^{-1}(A+d)g - g^{-1}A\delta g + g^{-1}\delta dg. \tag{9.63}$$

Using equation (9.61) and

$$dv = -g^{-1}dgg^{-1}\delta g - g^{-1}\delta dg \tag{9.64}$$

we obtain

$$\delta\mathcal{A} = -dv - \mathcal{A}v - v\mathcal{A} = -\mathcal{D}v, \tag{9.65}$$

where

$$\mathcal{D} = d + [\mathcal{A},\ ] \tag{9.66}$$

is the **covariant derivative** with respect to the gauge field $\mathcal{A}$.

Clearly, for the curvature $\mathcal{F}$ we find

$$\delta\mathcal{F} = \mathcal{F}v - v\mathcal{F} = [\mathcal{F}, v]. \tag{9.67}$$

Summarizing, we gain the following **structure equations**

$$\begin{aligned} \delta v &= -v^2 \\ \delta\mathcal{A} &= -\mathcal{D}v \\ \delta\mathcal{F} &= [\mathcal{F}, v]. \end{aligned} \tag{9.68}$$

We certainly recognize the **BRS algebra** in equations (9.68) and we identify, as before in Section 8.1, the MC form on the gauge group with the FP ghost and the exterior derivative on the group with the BRS operator

$$v_{MC} = g^{-1}\delta g = v_\alpha d\theta^\alpha \longleftrightarrow v_{FP} = v^a T^a$$

$$\delta = \frac{\partial}{\partial\theta^\alpha} d\theta^\alpha \longleftrightarrow s. \tag{9.69}$$

Chain of descent equations: Now we can derive the chain of descent equations in an analogous manner as we did before.

We start with the **'transgression'**

$$\mathrm{tr}\ \mathcal{F}^n = dQ_{2n-1}(\mathcal{A}, \mathcal{F}), \tag{9.70}$$

where

$$Q_{2n-1}(\mathcal{A}, \mathcal{F}) = n\int_0^1 dt\ \mathrm{tr}\ \mathcal{A}\mathcal{F}_t^{n-1} \quad \text{and} \quad \mathcal{F}_t = t\mathcal{F} + (t^2 - t)\mathcal{A}^2. \tag{9.71}$$

We shift the fields and the operator

$$\mathcal{A} \to \widehat{\mathcal{A}} = \mathcal{A} + v \quad , \quad d \to \Delta = d + \delta \tag{9.72}$$

so that we obtain the **'shifted transgression'**

$$\mathrm{tr}\ \widehat{\mathcal{F}}^n = \Delta Q_{2n-1}(\mathcal{A} + v, \widehat{\mathcal{F}}), \tag{9.73}$$

where

$$Q_{2n-1}(\mathcal{A}+v, \widehat{\mathcal{F}}) = n\int_0^1 dt\ \mathrm{tr}\ (\mathcal{A}+v)\widehat{\mathcal{F}}_t^{n-1} \text{ and } \widehat{\mathcal{F}}_t = t\widehat{\mathcal{F}} + (t^2 - t)(\mathcal{A}+v)^2. \tag{9.74}$$

Now using the **'Russian formula'**

$$\widehat{\mathcal{F}}(\widehat{\mathcal{A}}) = \Delta\widehat{\mathcal{A}} + \widehat{\mathcal{A}}^2 \equiv d\mathcal{A} + \mathcal{A}^2 = \mathcal{F}(\mathcal{A}) \tag{9.75}$$

(notice, this was evaluated before at $\mathcal{A} \to A, g \to \mathbf{1}$; within a fibre-bundle set-up identity (9.75) is also called the **'horizontality condition'**, see our discussion in Section 11.5.3) we find the relation

$$\Delta Q_{2n-1}(\mathcal{A} + v, \mathcal{F}) = dQ_{2n-1}(\mathcal{A}, \mathcal{F}). \tag{9.76}$$

We expand the Chern–Simons form in powers of v

$$\begin{aligned} &Q_{2n-1}(\mathcal{A} + v, \mathcal{F}) \\ &\quad = Q^0_{2n-1}(\mathcal{A}, \mathcal{F}) + Q^1_{2n-2}(v, \mathcal{A}, \mathcal{F}) + Q^2_{2n-3}(v, \mathcal{A}, \mathcal{F}) + \dots \\ &\quad \dots + Q^{2n-2}_1(v, \mathcal{A}) + Q^{2n-1}_0(v). \end{aligned} \tag{9.77}$$

The upper index labels the powers of v, the Maurer–Cartan form on group space—it counts the degree of the form in group space $\mathcal{G}$ (or M_θ)—and the lower index indicates the degree of the form on the space–time manifold M.

We insert the expansion (9.77) into relation (9.76), and compare the terms of the same order in space–time M as well as in group space $\mathcal{G}$—note that d increases the form degree in M by one unit and so does δ in $\mathcal{G}$—then we arrive at the **Stora–Zumino chain of descent equations**

$$\begin{aligned} \text{tr } \mathcal{F}^n - dQ^0_{2n-1} &= 0 \\ \delta Q^0_{2n-1} + dQ^1_{2n-2} &= 0 \\ \delta Q^1_{2n-2} + dQ^2_{2n-3} &= 0 \\ \dots\dots & \\ \delta Q^{2n-2}_1 + dQ^{2n-1}_0 &= 0 \\ \delta Q^{2n-1}_0 &= 0. \end{aligned} \tag{9.78}$$

Before discussing its meaning we observe that the several chain terms Q^k_{2n-1-k} are *not* determined uniquely. For instance, we can replace

$$\begin{aligned} Q^0_{2n-1} &\to Q^0_{2n-1} + d\alpha \\ Q^1_{2n-2} &\to Q^1_{2n-2} + \delta\alpha + d\beta \\ &\dots\dots \end{aligned} \tag{9.79}$$

and the above tower of descent equations (9.78) remains invariant. These different choices for the chain terms—when interpreted in terms of physics—are equivalent.

Again we emphasize that the above chain of equations (9.78) is only valid for gauge fields with vanishing topological number—the trivial bundle case.

Cohomology: What is the meaning of the descent equations? On the mathematical side there is the Dubois-Violette, Talon and Viallet [Dubois-Violette, Talon, Viallet 1985a,b] interpretation that chain (9.78) corresponds to a cohomology problem.

Generally, an **anomalous term** B is defined by its **consistency condition**

$$\delta B = 0. \tag{9.80}$$

B is a polynomial in the fields and their derivatives, integrated over the space–time manifold M. However, solutions like

$$B = \delta \widehat{B}, \tag{9.81}$$

where $\widehat{B}$ represents a local functional in the fields and their derivatives, are **trivial** since $\delta^2 = 0$. Indeed, for $B = G$ anomalies and $B = S$ Schwinger terms these trivial solutions can be absorbed by redefining the 'quantum action'.

For example, in case of the **anomaly** when we set

$$B \to G = N \int_M Q^1_{2n-2} \tag{9.82}$$

the condition $\delta G = 0$ implies, by virtue of the algebraic Poincaré lemma (Section 7.3), the descent equation

$$\delta Q^1_{2n-2} = -dQ^2_{2n-3}. \tag{9.83}$$

This means Q^1_{2n-2} is a **δ-cocycle modulo** d.

For the **trivial solution** $G = \delta \widehat{G}$ when we write

$$\widehat{G} = N \int_M L^0_{2n-2} \tag{9.84}$$

Poincaré again ensures the existence of a polynomial L^1_{2n-3} such that

$$Q^1_{2n-2} = \delta L^0_{2n-2} + dL^1_{2n-3}. \tag{9.85}$$

Then Q^1_{2n-2} is called a **δ-coboundary modulo** d. The task is now to calculate the space of δ-cocycles modulo d quotiented by the δ-coboundaries modulo d. This is $H^1_{2n-2}(\delta/d)$, the **δ-cohomology modulo** d in FP ghost degree 1 and space–time degree $(2n-2)$ (recall Section 8.3.4).

As another example, Faddeev [Faddeev 1984] discovered that certain **Schwinger terms** obey an analogous cohomology problem

$$B \to S = N \int Q^2_{2n-3}. \tag{9.86}$$

The consistency condition $\delta S = 0$ implies the next descent equation

$$\delta Q^2_{2n-3} = -dQ^3_{2n-4}, \tag{9.87}$$

leading to the cohomology $H^2_{2n-3}(\delta/d)$.

Dubois-Violette, Talon and Viallet have calculated all cohomologies $H^k_{2n-1-k}(\delta/d)$ $(k = 0, 1, \ldots, 2n-1)$—all possible anomalous terms—for the BRS algebra of differential forms which are generated by exterior products of A, v and their differentials d, δ. (For further literature see [Kastler, Stora 1986a,b].)

Physics: What about physics? There the research of the last years has come up with fascinating results. For the moment we just summarize some results which we explain further in this book:

- δ, v
  The derivative δ on group space is interpreted as the BRS operator. The Maurer–Cartan form v is identified with the Faddeev–Popov ghost field (Chapter 8).
- tr $\mathcal{F}^n$
  The trace of the curvatures $\mathcal{F}$ expresses the singlet anomaly in $2n$ dimensions. The normalization, found first in perturbation theory (Chapter 4), can also be calculated by Fujikawa's path integral method (Chapter 5) or by the index theorem (Chapter 11).
- Q^0_{2n-1}
  The Chern–Simons form serves as a term to redefine the vacuum functional, the 'quantum action', leading to so-called topological field theories, specifically to the Chern–Simons theories (for literature see e.g. [Deser, Jackiw, Templeton 1984], [Dunne, Jackiw, Trugenberger 1989], [Witten 1989a,b]).
- Q^1_{2n-2}
  The chain term to order v represents the non-Abelian anomaly in $(2n-2)$ dimensions. It is the consistent anomaly obeying the Wess–Zumino consistency condition (Chapter 8). Again, besides perturbation theory (Chapter 4), Fujikawa's method (Chapter 5) or the Atiyah–Singer index theorem (Chapter 11) supplies the correct normalization.
- Q^2_{2n-3}
  The chain term to order v^2 has been identified by Faddeev [Faddeev 1984] and by Mickelsson [Mickelsson 1985b] with the Schwinger term in an equal time commutator of Gauss-law operators for a Yang–Mills theory (for further literature see [Rupp 1998]).
- Q^3_{2n-4}
  The chain term to order v^3 has an interpretation within ordinary quantum mechanics. It is related to the failure of the Jacobi identity for

velocity operators at a magnetic monopole [Jackiw 1985 a,b,c], [Grossman 1985], [Wu, Zee 1985], [Mickelsson 1985a]. For attempts to interpret this term within field theory see [Jo 1985], [Levy 1987], [Zhang 1987].

For the higher order chain terms Q^k_{2n-1-k} with $k \geq 4$ no physical interpretation is known up to now. But there is a common belief that all solutions of the chain will have a counterpart in physics.

Comparison of Stora's with Zumino's approach:

Stora	Zumino
basic fields and operators	
A	$\mathcal{A} = g^{-1}(A + d)g$
F	$\mathcal{F} = d\mathcal{A} + \mathcal{A}^2$
v_{FP}	$v_{MC} = g^{-1}\delta g,\ g = g(x, \theta)$
s	$\delta = \frac{\partial}{\partial\theta^\alpha} d\theta^\alpha$
algebra	
$sv_{FP} = -v_{FP}^2$	$\delta v_{MC} = -v_{MC}^2$
$sA = -Dv_{FP}$	$\delta\mathcal{A} = -\mathcal{D}v_{MC}$
$sF = [F, v_{FP}]$	$\delta\mathcal{F} = [\mathcal{F}, v_{MC}]$
shift	
$A \to A + v_{FP}$	$\mathcal{A} \to \mathcal{A} + v_{MC}$
$d \to d + s$	$d \to d + \delta$
'Russian formula'	
$\widehat{F}(A + v_{FP}) = F(A)$	$\widehat{\mathcal{F}}(\mathcal{A} + v_{MC}) = \mathcal{F}(\mathcal{A})$

chain of descent equations	
$\mathrm{tr}\ F^n - dQ^0_{2n-1} = 0$	$\mathrm{tr}\ \mathcal{F}^n - dQ^0_{2n-1} = 0$
$sQ^0_{2n-1} + dQ^1_{2n-2} = 0$	$\delta Q^0_{2n-1} + dQ^1_{2n-2} = 0$
$sQ^1_{2n-2} + dQ^2_{2n-3} = 0$	$\delta Q^1_{2n-2} + dQ^2_{2n-3} = 0$
......	......
$sQ^{2n-2}_1 + dQ^{2n-1}_0 = 0$	$\delta Q^{2n-2}_1 + dQ^{2n-1}_0 = 0$
$sQ^{2n-1}_0 = 0$	$\delta Q^{2n-1}_0 = 0$

identification		
v_{FP}	$\longleftrightarrow$	v_{MC}
s	$\longleftrightarrow$	δ

anomaly and WZ consistency condition
$NQ^1_{2n-2} = v^a G^a[A]$
$sG(v, A) = s \int v^a G^a[A] = 0$

9.4 Explicit solutions for the chain terms

With the operator rules, we have learned so far, it is possible to solve explicitly the whole chain of descent equations (9.78). Let us demonstrate this term-by-term for the simple case of $n = 2$.

Beginning with the first equation the singlet anomaly in 4 dimensions, we insert $\mathcal{F}$, equation (9.55),

$$\begin{aligned} \mathrm{tr}\ \mathcal{F}^2 &= d\ \mathrm{tr}\ [\mathcal{A}d\mathcal{A} + \frac{2}{3}\mathcal{A}^3] \qquad (9.88)\\ &= dQ^0_3, \qquad (9.89) \end{aligned}$$

so that we can read off the Chern–Simons term in 3 dimensions

$$Q^0_3 = \mathrm{tr}\ [\mathcal{A}d\mathcal{A} + \frac{2}{3}\mathcal{A}^3] = \mathrm{tr}\ [\mathcal{A}\mathcal{F} - \frac{1}{3}\mathcal{A}^3]. \qquad (9.90)$$

Then we apply δ and use the structure equations (9.68)

$$\begin{aligned} \delta Q_3^0 &= \delta \text{ tr } [\mathcal{A}\mathcal{F} - \frac{1}{3}\mathcal{A}^3] \\ &= \text{tr } [\delta\mathcal{A}\mathcal{F} - \mathcal{A}\delta\mathcal{F} - \frac{1}{3}3\delta\mathcal{A}\mathcal{A}^2] \\ &= -d \text{ tr } v(\mathcal{F} - \mathcal{A}^2). \end{aligned} \tag{9.91}$$

Comparison with the chain

$$\delta Q_3^0 = -dQ_2^1 \tag{9.92}$$

gives

$$Q_2^1 = \text{tr } v(\mathcal{F} - \mathcal{A}^2) = \text{tr } vd\mathcal{A}. \tag{9.93}$$

Solution (9.93) represents the non-Abelian anomaly in 2 dimensions (besides a normalization factor).

Next we apply the δ-operator to result (9.93)

$$\delta Q_2^1 = \delta \text{ tr } vd\mathcal{A} = \text{tr } [\delta v d\mathcal{A} + vd\delta\mathcal{A}] = d \text{ tr } v^2\mathcal{A}, \tag{9.94}$$

compare with the chain

$$\delta Q_2^1 = -dQ_1^2 \tag{9.95}$$

and find explicitly

$$Q_1^2 = -\text{tr } v^2\mathcal{A}. \tag{9.96}$$

Solution (9.96) is the result for the 1-dimensional Schwinger term occurring in a $1+1$-dimensional Yang–Mills field theory (note that time is fixed in the corresponding commutator relations).

Finally, using the above recipe for the term Q_1^2 we get

$$\delta Q_1^2 = -\delta\text{tr } v^2\mathcal{A} = \frac{1}{3}d \text{ tr } v^3 \tag{9.97}$$

$$\delta Q_1^2 = -dQ_0^3 \tag{9.98}$$

$$Q_0^3 = -\frac{1}{3} \text{ tr } v^3. \tag{9.99}$$

Clearly the last equation of the chain is satisfied

$$\delta Q_0^3 = -\frac{1}{3}\delta \text{ tr } v^3 = -\text{tr } \delta v v^2 = \text{tr } v^4 = 0. \tag{9.100}$$

Set of solutions:

i) For $n = 2$, we collect all terms

$$Q_3^0 = \text{tr } [\mathcal{A}\mathcal{F} - \frac{1}{3}\mathcal{A}^3] = \text{tr } [\mathcal{A}d\mathcal{A} + \frac{2}{3}\mathcal{A}^3]$$

$$
\begin{aligned}
Q_2^1 &= \text{tr } v(\mathcal{F} - \mathcal{A}^2) = \text{tr } vd\mathcal{A} \\
Q_1^2 &= -\text{tr } v^2\mathcal{A} \\
Q_0^3 &= -\frac{1}{3} \text{ tr } v^3.
\end{aligned} \tag{9.101}
$$

ii) For $n = 3$, the case which we are actually interested in, we gain the following explicit solutions analogously

$$
\begin{aligned}
Q_5^0 &= \text{tr } [\mathcal{A}\mathcal{F}^2 - \frac{1}{2}\mathcal{A}^3\mathcal{F} + \frac{1}{10}\mathcal{A}^5] \\
&= \text{tr } [\mathcal{A}(d\mathcal{A})^2 + \frac{3}{2}\mathcal{A}^3 d\mathcal{A} + \frac{3}{5}\mathcal{A}^5] \\
Q_4^1 &= \text{tr } v[\mathcal{F}^2 - \frac{1}{2}(\mathcal{A}^2\mathcal{F} + \mathcal{F}\mathcal{A}^2 + \mathcal{A}\mathcal{F}\mathcal{A}) + \frac{1}{2}\mathcal{A}^4] \\
&= \text{tr } vd(\mathcal{A}d\mathcal{A} + \frac{1}{2}\mathcal{A}^3) \\
Q_3^2 &= -\frac{1}{2} \text{ tr } [(v^2\mathcal{A} + v\mathcal{A}v + \mathcal{A}v^2)\mathcal{F} - v^2\mathcal{A}^3 - v\mathcal{A}v\mathcal{A}^2] \\
&= -\frac{1}{2} \text{ tr } [(v^2\mathcal{A} + v\mathcal{A}v + \mathcal{A}v^2)d\mathcal{A} + v^2\mathcal{A}^3] \\
Q_2^3 &= \frac{1}{2} \text{ tr } [-v^3\mathcal{F} + v^3\mathcal{A}^2 + \mathcal{A}v\mathcal{A}v^2] \\
&= \frac{1}{2} \text{ tr } [-v^3 d\mathcal{A} + \mathcal{A}v\mathcal{A}v^2] \\
Q_1^4 &= \frac{1}{2} \text{ tr } v^4\mathcal{A} \\
Q_0^5 &= \frac{1}{10} \text{ tr } v^5.
\end{aligned} \tag{9.102}
$$

Term Q_5^0 is the explicit solution of the Chern–Simons form in 5 dimensions—dQ_5^0 is nothing but the singlet anomaly in 6 dimensions—and term Q_4^1 provides the familiar result for the non-Abelian anomaly in 4 dimensions, equation (6.79) (apart from the normalization factor). Q_3^2 represents the 3-dimensional Schwinger term occurring in a $1+3$-dimensional Yang–Mills field theory (see [Faddeev 1984]).

Alternatively, the chain of descent equations can also be solved in a reversed way, beginning with Q_0^{2n-1} and ending with Q_{2n-1}^0 (see [Sorella 1993], [Sorella, Tătaru 1994]).

Although the recipe for finding the explicit solutions of a chain is quite simple the actual calculations can become rather cumbersome in higher dimensions. But fortunately we can find simple, general formulae for all chain terms. We present these in the next section.

9.5 Zumino's formulae for chain terms

As we have discussed in Section 9.3 the chain terms Q^k_{2n-1-k} belonging to the chain of descent equations are not defined uniquely. Zumino [Zumino 1985a] takes advantage of this ambiguity to simplify the explicit solution. His choice for Q^k_{2n-1-k}—the **simplest choice**—is characterized by having

i) the smallest power in the gauge connection $\mathcal{A}$,
ii) the largest power in the number of derivatives dv.

As before, only gauge connections with vanishing topological number are considered which correspond to trivial fibre bundles. In order to create a chain of descent equations we have to start with some kind of 'transgression', which we obtain by considering some homotopy of the fields.

Homotopy: For the chain terms Q^k_{2n-1-k} with Zumino's simplest choice we consider the **homotopic connection**

$$\mathcal{A}_t = t\mathcal{A} + v, \qquad t \in [0,1], \tag{9.103}$$

where

$$\mathcal{A}_1 = \mathcal{A} + v, \qquad \mathcal{A}_0 = v. \tag{9.104}$$

The associated **homotopic curvature** is

$$\widehat{\mathcal{F}}_t = \Delta\mathcal{A}_t + \mathcal{A}_t^2, \qquad \Delta = d + \delta, \tag{9.105}$$

together with

$$\begin{aligned} \widehat{\mathcal{F}}_{t=1} &= \Delta(\mathcal{A}+v) + (\mathcal{A}+v)^2 = \widehat{\mathcal{F}}(\mathcal{A}+v) \equiv \mathcal{F}(\mathcal{A}) \\ \widehat{\mathcal{F}}_{t=0} &= (d+\delta)v + v^2 = dv, \end{aligned} \tag{9.106}$$

where we have also used the 'Russian formula' (9.75).

We rewrite $\widehat{\mathcal{F}}_t$ by inserting again the structure equations (9.68)

$$\begin{aligned} \widehat{\mathcal{F}}_t &= td\mathcal{A} + dv + t\delta\mathcal{A} + \delta v + t^2\mathcal{A}^2 + t(\mathcal{A}v + v\mathcal{A}) + v^2 \\ &= \mathcal{F}_t + (1-t)dv, \end{aligned} \tag{9.107}$$

with

$$\mathcal{F}_t = t\mathcal{F} + (t^2 - t)\mathcal{A}^2. \tag{9.108}$$

From expression (9.107) we also reproduce equations (9.106).

Next we differentiate $\widehat{\mathcal{F}}_t$, equation (9.105), with respect to t

$$\frac{\partial\widehat{\mathcal{F}}_t}{\partial t} = \Delta\frac{\partial\mathcal{A}_t}{\partial t} + \left[\mathcal{A}_t, \frac{\partial\mathcal{A}_t}{\partial t}\right], \tag{9.109}$$

we introduce the **homotopic covariant derivative**

$$\widehat{\mathcal{D}}_t = \Delta + [\mathcal{A}_t, \ ] \tag{9.110}$$

and obtain

$$\frac{\partial \widehat{\mathcal{F}}_t}{\partial t} = \widehat{\mathcal{D}}_t \frac{\partial \mathcal{A}_t}{\partial t}. \tag{9.111}$$

Of course, the Bianchi identity is satisfied

$$\widehat{\mathcal{D}}_t \widehat{\mathcal{F}}_t = 0. \tag{9.112}$$

Considering now a symmetric invariant polynomial $P(\widehat{\mathcal{F}}_t^n)$ we find

$$\begin{aligned}\frac{\partial}{\partial t} P(\widehat{\mathcal{F}}_t^n) &= nP\left(\frac{\partial \widehat{\mathcal{F}}_t}{\partial t}, \widehat{\mathcal{F}}_t^{n-1}\right) \\ &= nP\left(\widehat{\mathcal{D}}_t \frac{\partial \mathcal{A}_t}{\partial t}, \widehat{\mathcal{F}}_t^{n-1}\right) \\ &= n\Delta P\left(\frac{\partial \mathcal{A}_t}{\partial t}, \widehat{\mathcal{F}}_t^{n-1}\right) \end{aligned} \tag{9.113}$$

(we used the property (7.20) for invariant polynomials together with the Bianchi identity (9.112)). The integration in the interval $t \in [0, 1]$ leads to:

Theorem: 'Transgression'

$$P(\mathcal{F}^n) - P((dv)^n) = \Delta Q_{2n-1}, \tag{9.114}$$

where

$$Q_{2n-1} = n \int_0^1 dt P(\mathcal{A}, (\mathcal{F}_t + (1-t)dv)^{n-1}). \tag{9.115}$$

This is the basic relation for creating the desired chain of descent equations.

Chain of descent equations: Analogous to Section 9.3 we expand the Chern–Simons term (9.115) in powers of dv

$$P(\mathcal{A}, (\mathcal{F}_t + (1-t)dv)^{n-1}) = \sum_{k=0}^{n-1} \binom{n-1}{k} (1-t)^k P((dv)^k, \mathcal{A}, \mathcal{F}_t^{n-1-k}), \tag{9.116}$$

with

$$\binom{n-1}{k} = \frac{(n-1)!}{k!(n-k-1)!}.$$

(Note that dv behaving as a 2-form on the underlying manifold $M \times M_\theta$ (or $M \times \mathcal{G}$) can be interchanged freely in the symmetric polynomial P.) Separating each term of a given power of dv with the obvious notation

$$Q_{2n-1} = Q^0_{2n-1} + Q^1_{2n-2} + Q^2_{2n-3} + \ldots + Q^{n-1}_n \tag{9.117}$$

we rewrite relation (9.114)

$$P(\mathcal{F}^n) - P((dv)^n) = (d+\delta)Q^0_{2n-1} + (d+\delta)Q^1_{2n-2} + \ldots + (d+\delta)Q^{n-1}_n. \tag{9.118}$$

We compare all terms of the same order and obtain the first part of the **chain of descent equations**

$$\begin{aligned} P(\mathcal{F}^n) - dQ^0_{2n-1} &= 0 \\ \delta Q^0_{2n-1} + dQ^1_{2n-2} &= 0 \\ \delta Q^1_{2n-2} + dQ^2_{2n-3} &= 0 \\ \ldots\ldots & \\ \delta Q^{n-1}_n + P((dv)^n) &= 0. \end{aligned} \tag{9.119}$$

Formulae: The expansion (9.116) for the Chern–Simons form (9.115) supplies a **formula** to order $(dv)^k$ for each term of the chain (9.119)

$$Q^k_{2n-1-k} = \frac{n!}{k!(n-1-k)!} \int_0^1 dt (1-t)^k P((dv)^k, \mathcal{A}, \mathcal{F}_t^{n-1-k}), \tag{9.120}$$

with $0 \le k \le n-1$.

i) For $k = 0$ we recover result (9.71) for the Chern–Simons form Q^0_{2n-1}.

ii) For $k = 1$

$$Q^1_{2n-2}(dv, \mathcal{A}, \mathcal{F}) = n(n-1) \int_0^1 dt (1-t) P(dv, \mathcal{A}, \mathcal{F}_t^{n-2}) \tag{9.121}$$

$$Q^1_{2n-2}(v, \mathcal{A}, \mathcal{F}) = n(n-1) \int_0^1 dt (1-t) P(v, d(\mathcal{A}, \mathcal{F}_t^{n-2})). \tag{9.122}$$

The second formula is equivalent to the first since Q^1_{2n-2} is defined only up to a total derivative of a local form, which we can neglect. For instance, for $n = 2$ and for $n = 3$ we recover the result we found before (equations (9.101), (9.102)).

Note: In Section 7.6 we also derived a formula for the anomaly with the help of the homotopy operator. We observe that both results (7.138) and (9.122) agree when we substitute $A + \delta A \to \mathcal{A} + v$.

iii) For $k = 2$

$$Q^2_{2n-3}(dv, \mathcal{A}, \mathcal{F}) = \frac{n(n-1)(n-2)}{2!} \int_0^1 dt(1-t)^2 P((dv)^2, \mathcal{A}, \mathcal{F}_t^{n-3}). \tag{9.123}$$

iv) For $k = n-1$

$$\begin{aligned} Q_n^{n-1}(dv, \mathcal{A}) &= \frac{n!}{(n-1)!} \int_0^1 dt(1-t)^{n-1} P((dv)^{n-1}, \mathcal{A}) \\ &= P((dv)^{n-1}, \mathcal{A}). \end{aligned} \tag{9.124}$$

So far we know the chain terms up to $k \leq n-1$ in the power of dv. In order to find expressions for the higher powers we have to consider a different homotopy of the fields.

Homotopy: Zumino's choice is [Zumino 1985a]

$$\mathcal{A}_t = tv, \tag{9.125}$$

where

$$\mathcal{A}_1 = v, \qquad \mathcal{A}_0 = 0, \tag{9.126}$$

together with

$$\widehat{\mathcal{F}}_t = \Delta \mathcal{A}_t + \mathcal{A}_t^2 = t(d+\delta)v + t^2v^2 = tdv + (t^2 - t)v^2, \tag{9.127}$$

where

$$\widehat{\mathcal{F}}_{t=1} = dv, \qquad \widehat{\mathcal{F}}_{t=0} = 0. \tag{9.128}$$

Then equations (9.111)–(9.113)—the basis for the 'transgression'—are also valid for this homotopy choice (9.125) and (9.127) leading to:

Theorem: 'Transgression'

$$P((dv)^n) = \Delta Q_{n-1}, \tag{9.129}$$

where

$$Q_{n-1} = n \int_0^1 dt P(v, (tdv + (t^2 - t)v^2)^{n-1}). \tag{9.130}$$

Chain of descent equations: Again we expand in powers of dv

$$\begin{aligned} &P(v, (tdv + (t^2-t)v^2)^{n-1}) \\ &= \sum_{k=0}^{n-1} \binom{n-1}{k} t^{n-1-k}(t^2-t)^k P((dv)^{n-1-k}, v, (v^2)^k), \end{aligned} \tag{9.131}$$

and separate in expression (9.130) each term of a given power in v

$$Q_{n-1} = Q^n_{n-1} + Q^{n+1}_{n-2} + Q^{n+2}_{n-3} + \ldots + Q^{2n-1}_0. \tag{9.132}$$

Splitting the relation (9.129) according to expansion (9.132)

$$P((dv)^n) = (d+\delta)Q^n_{n-1} + (d+\delta)Q^{n+1}_{n-2} + \ldots + (d+\delta)Q^{2n-1}_0 \tag{9.133}$$

and comparing all terms of the same order we find the second part of the **chain of descent equations**

$$\begin{aligned} P((dv)^n) - dQ^n_{n-1} &= 0 \\ \delta Q^n_{n-1} + dQ^{n+1}_{n-2} &= 0 \\ \ldots\ldots & \\ \delta Q^{2n-2}_1 + dQ^{2n-1}_0 &= 0 \\ \delta Q^{2n-1}_0 &= 0. \end{aligned} \tag{9.134}$$

Formulae: Now we look for a formula for the chain terms. In the expansion (9.131) the t-dependence can be factorized generating integrals—**beta functions**—such as

$$(-)^k B(n, k+1) = (-)^k \int_0^1 dt t^{n-1}(1-t)^k = (-)^k \frac{(n-1)!k!}{(n+k)!} \tag{9.135}$$

so that we find the **formula** to order $(dv)^{n-1-k}$ or v^{n+k}

$$Q^{n+k}_{n-1-k}(v) = (-)^k \frac{n!(n-1)!}{(n-1-k)!(n+k)!} P((dv)^{n-1-k}, v, (v^2)^k), \tag{9.136}$$

with $0 \leq k \leq n-1$.

All these terms (9.136) contain neither the curvature $\mathcal{F}$ nor the connection $\mathcal{A}$ but only v, the Maurer–Cartan form on the group!

For example, for $k = 0$ we have

$$Q^n_{n-1}(v) = P((dv)^{n-1}, v), \tag{9.137}$$

and for $k = n-1$, the last term of the chain

$$Q^{2n-1}_0(v) = (-)^{n-1} \frac{n!(n-1)!}{(2n-1)!} P(v^{2n-1}). \tag{9.138}$$

Formulae (9.120) and (9.136) contain all chain terms. Both formulae have been derived from different homotopies but they belong to one and the same chain. Clearly, both parts (9.119) and (9.134) fit together.

Examples: Finally, we present the examples which we can compare with the solutions of Section 9.4. For the symmetric invariant polynomial P we choose the trace (see Section 7.1). In the case of $n = 2$ formulae (9.120) and (9.136) give

$$\begin{aligned} Q_3^0 &= \operatorname{tr}\,(\mathcal{A}d\mathcal{A} + \frac{2}{3}\mathcal{A}^3) \\ Q_2^1 &= \operatorname{tr}\, dv\mathcal{A} \\ Q_1^2 &= \operatorname{tr}\, dvv \\ Q_0^3 &= -\frac{1}{3}\operatorname{tr}\, v^3. \end{aligned} \tag{9.139}$$

In the anomaly term Q_2^1 the derivative can be shifted from v to the gauge potential $\mathcal{A}$, then it agrees explicitly with (9.101). The difference is just a total differential which can be neglected. Furthermore we observe that the Schwinger term Q_1^2 in its simplest choice is independent of the gauge connection $\mathcal{A}$! Whereas the equivalent solution (9.101) exhibits a linear dependence on $\mathcal{A}$.

For the example $n = 3$ the formulae (9.120) and (9.136) provide:

$$\begin{aligned} Q_5^0 &= \operatorname{tr}\,[\mathcal{A}(d\mathcal{A})^2 + \frac{3}{2}\mathcal{A}^3 d\mathcal{A} + \frac{3}{5}\mathcal{A}^5] \\ Q_4^1 &= \frac{1}{2}\operatorname{tr}\, dv(\mathcal{A}d\mathcal{A} + d\mathcal{A}\mathcal{A} + \mathcal{A}^3) \\ Q_3^2 &= \operatorname{tr}\,(dv)^2\mathcal{A} \\ Q_2^3 &= \operatorname{tr}\,(dv)^2 v \\ Q_1^4 &= -\frac{1}{2}\operatorname{tr}\, dvv^3 \\ Q_0^5 &= \frac{1}{10}\operatorname{tr}\, v^5. \end{aligned} \tag{9.140}$$

In comparison with expressions (9.102) we observe the economical choice of solutions (9.140). Again, the anomaly term Q_4^1 agrees explicitly with expression (9.102) after shifting the derivative to the second factor. The Schwinger term Q_3^2 appears linear in the gauge potential $\mathcal{A}$, and this is the minimal choice. Thus the Schwinger term—as a solution of the chain—*must* contain a dynamical variable like $\mathcal{A}$ in this dimension. The equivalent expression Q_3^2 (9.102), however, is much more complicated; there the dependence on $\mathcal{A}$ is quadratic and cubic. The next terms Q_2^3 and Q_1^4 of the solution set (9.140) are already independent of the gauge connection $\mathcal{A}$, in contrast to the equivalent terms (9.102).

10
Covariant anomaly

The non-Abelian anomaly, Bardeen's result, which we have discussed so far is called the consistent anomaly because it is a solution of the Wess–Zumino consistency condition. However, it is not gauge covariant since the corresponding current does not transform covariantly under gauge transformations. But there exists another type of non-Abelian anomaly—the covariant anomaly—whose corresponding current does transform gauge covariantly and consequently the covariant anomaly itself. These two types of anomaly reflect two different regularization procedures—the ambiguities in the regularization—when calculating anomalous terms. In one case gauge invariance is lost in the regularization, in the other it is maintained no matter whether we work with Feynman diagrams (see e.g. [Leutwyler 1985a,b]) or with the path integral measure (see e.g. [Fujikawa 1984, 1985], [Banerjee, Banerjee, Mitra 1986], in Chapter 5).

In Section 10.1 we introduce the Bardeen–Zumino polynomial which relates the consistent- to the covariant anomaly. We discuss the features of the covariant type. This discussion will be reformulated by using differential forms on the space–time manifold M (Section 10.2), as well as differential forms on Sp $\mathcal{A}$, the space of all gauge connections (Section 10.3). Finally, we extend the anomalies to nonlocal forms over Sp $\mathcal{A}$ (Section 10.4).

10.1 Bardeen–Zumino polynomial and covariant anomaly

Naïvely we expect the non-Abelian current j^a_μ (8.57) to transform covariantly under gauge transformations

$$\delta_\xi j^a_\mu = [j_\mu, \xi]^a. \tag{10.1}$$

However, in the presence of an anomaly this is not so. Let us find the correct gauge transformation law for the consistent current. Here we follow the work of Bardeen and Zumino (BZ) [Bardeen, Zumino 1984].

Definition: We introduce a general **variation** by

$$\delta_Y A = Y, \tag{10.2}$$

together with

$$\delta_Y F = D(\delta_Y A) = DY, \tag{10.3}$$

where D denotes the covariant derivative $D = d + [A, \,]$.

We can view this variation as a derivative in the space of gauge connections Sp $\mathcal{A}$ (recall Section 8.1). The variation δ_Y represents a derivative along an arbitrary (but fixed) vector field Y

$$\delta_Y = i_Y \delta = \int dx Y^a_\mu(x) \frac{\delta}{\delta A^a_\mu(x)}, \tag{10.4}$$

where δ denotes the exterior derivative (8.111). Analogously the gauge variation δ_ξ expresses a derivative along the fundamental vector field (8.109)

$$\delta_\xi = i_{X_\xi} \delta = \int dx (D_\mu \xi)^a(x) \frac{\delta}{\delta A^a_\mu(x)}. \tag{10.5}$$

Theorem: Commutation relation

$$[\delta_Y, \delta_\xi] = \delta_{[Y,\xi]}. \tag{10.6}$$

Proof. Recalling the transformations

$$\delta_\xi A = D\xi, \qquad \delta_\xi F = [F, \xi], \qquad \delta_\xi Y = 0 \tag{10.7}$$

we verify explicitly

$$\begin{aligned} [\delta_Y, \delta_\xi] A &= (\delta_Y \delta_\xi - \delta_\xi \delta_Y) A = \delta_Y (d\xi + [A, \xi]) \\ &= [Y, \xi] = \delta_{[Y,\xi]} A \end{aligned} \tag{10.8}$$

$$\begin{aligned} [\delta_Y, \delta_\xi] F &= (\delta_Y \delta_\xi - \delta_\xi \delta_Y) F = [DY, \xi] - [D\xi, Y] \\ &= D[Y, \xi] = \delta_{[Y,\xi]} F. \end{aligned} \tag{10.9}$$

Relations (10.8) and (10.9) imply validity for any polynomial in A and F so that the operator equation (10.6) is satisfied. Q.E.D.

Gauge transformation law: In order to derive the transformation law for the consistent current we start with the vacuum functional (8.1) and apply the commutator relation (10.6)

$$[\delta_Y, \delta_\xi] W[A_\mu] = \delta_{[Y,\xi]} W[A_\mu]. \tag{10.10}$$

Using the anomalous Ward identity (8.68) we get

$$\delta_Y \int dx \xi^a(x) G_a[A_\mu](x) - \delta_\xi \int dx Y_\mu^a(x) \frac{\delta}{\delta A_\mu^a(x)} W[A_\mu]$$

$$= \int dx [Y_\mu, \xi]^a(x) \frac{\delta}{\delta A_\mu^a(x)} W[A_\mu]. \quad (10.11)$$

Definition: The **consistent current** is defined as a variation of the vacuum functional (recall equation (8.56))

$$\langle j_a^\mu(x) \rangle = \frac{\delta}{\delta A_\mu^a(x)} W[A_\mu]. \quad (10.12)$$

Then we find

$$\int dx Y_\mu^a(x) \delta_\xi \langle j_a^\mu(x) \rangle$$

$$= \int dx Y_\mu^a(x) [\langle j^\mu \rangle, \xi]_a(x) + \int dx \xi^a(x) \delta_Y G_a[A_\mu](x), \quad (10.13)$$

and extracting the vector field Y_μ^a provides the following theorem:

Theorem: Gauge transformation law

$$\delta_\xi \langle j_a^\mu(x) \rangle = [\langle j^\mu \rangle, \xi]_a(x) + \int dy \xi^b(y) \frac{\delta}{\delta A_\mu^a(x)} G_b[A_\mu](y). \quad (10.14)$$

The naïve transformation law (first term) is broken by the occurrence of the anomaly (second term). Only if the anomaly vanishes does the consistent current also transform covariantly.

Bardeen–Zumino polynomial: Bardeen and Zumino [Bardeen, Zumino 1984] discovered that one can construct a gauge covariant current leading to the covariant type of non-Abelian anomaly. The trick is to add a polynomial to the consistent current which gauge transforms in an opposite way in order to cancel the unwanted anomalous term.

Proposition:

- *There exists a local polynomial in the fields A_μ—the* **BZ polynomial** *$\mathcal{P}_a^\mu[A_\mu]$—with the gauge transformation property*

$$\delta_\xi \mathcal{P}_a^\mu[A_\mu](x) = [\mathcal{P}^\mu[A_\mu], \xi]_a(x) - \int dy \xi^b(y) \frac{\delta}{\delta A_\mu^a(x)} G_b[A_\mu](y) \quad (10.15)$$

so that the **covariant current**

$$\widetilde{\jmath}^{\mu}_{a} = j^{\mu}_{a} + \mathcal{P}^{\mu}_{a} \tag{10.16}$$

obeys the correct **covariant transformation law**

$$\delta_{\xi}\widetilde{\jmath}^{\mu}_{a} = [\widetilde{\jmath}^{\mu}, \xi]_{a}. \tag{10.17}$$

Using Bardeen's result (4.387) for the consistent anomaly (for negative chirality) BZ find a unique result for the polynomial

$$\mathcal{P}^{\mu}_{a}[A_{\mu}] = -\frac{1}{48\pi^2}\varepsilon^{\mu\nu\rho\sigma} \text{ tr } T_a(A_\nu F_{\rho\sigma} + F_{\rho\sigma}A_\nu - A_\nu A_\rho A_\sigma). \tag{10.18}$$

Using expression (10.18) one can reproduce equation (10.15).

Definition: Analogous to the consistent anomaly the **covariant anomaly** is defined by the covariant divergence of the covariant current

$$\widetilde{G}_a[A_\mu] = -(D_\mu \widetilde{\jmath}^{\mu})_a. \tag{10.19}$$

Inserting the covariant current (10.16) provides the relation

$$\begin{aligned} \widetilde{G}_a[A_\mu] &= -(D_\mu j^\mu)_a - (D_\mu \mathcal{P}^\mu)_a \\ &= G_a[A_\mu] - (D_\mu \mathcal{P}^\mu)_a. \end{aligned} \tag{10.20}$$

Now we just use the results for the consistent anomaly (4.387) and the BZ polynomial (10.18), then we find for the **covariant anomaly**

$$\widetilde{G}_a[A_\mu] = \frac{1}{32\pi^2}\varepsilon^{\mu\nu\rho\sigma} \text{ tr } T_a F_{\mu\nu} F_{\rho\sigma}. \tag{10.21}$$

For the **integrated covariant anomaly** we obtain

$$\begin{aligned} \widetilde{G}(\xi, A) &= \xi \cdot \widetilde{G} = \int dx \xi^a(x) \widetilde{G}_a[A_\mu](x) \\ &= \frac{1}{32\pi^2}\int dx \varepsilon^{\mu\nu\rho\sigma}\xi^a(x) \text{ tr } T_a F_{\mu\nu} F_{\rho\sigma}(x). \end{aligned} \tag{10.22}$$

Actually the covariant anomaly is expressed only by field strength tensors F, as expected by the covariance. Nevertheless we keep the gauge potentials A in the argument.

Covariant anomaly condition: Whereas the consistent current $\langle j^{\mu}_{a}\rangle$ is given by the variation of the vacuum functional (10.12)—leading to the WZ consistency condition—the covariant current $\langle \widetilde{\jmath}^{\mu}_{a}\rangle$ cannot be obtained as a variation of a functional. So it is no surprise that the covariant anomaly does *not* satisfy the WZ condition. Instead, the following condition can be derived [Adam 1990], [Banerjee, Banerjee, Mitra 1986]:

We start from the **covariant gauge transformation law**

$$\delta_\eta \widetilde{G}(\xi, A) = \widetilde{G}([\eta, \xi], A). \tag{10.23}$$

We give a general derivation in Section 10.2; here we can check equation (10.23) easily by using the explicit expression (10.22).

Then the difference of two gauge variations yields the theorem:

Theorem: Covariant anomaly condition

$$\delta_\xi \widetilde{G}(\eta, A) - \delta_\eta \widetilde{G}(\xi, A) = 2\widetilde{G}([\xi, \eta], A). \tag{10.24}$$

Equation (10.24) differs from the WZ consistency condition (8.69), (8.107) by the factor 2!

Alternatively, we may view the **condition for the covariant anomaly**

$$\delta_\xi \widetilde{G}(\eta, A) - \delta_\eta \widetilde{G}(\xi, A) - \widetilde{G}([\xi, \eta], A) = \widetilde{G}([\xi, \eta], A) \tag{10.25}$$

as a WZ consistency condition broken by an 'inhomogeneity' which is the covariant anomaly itself.

Résumé: The non-Abelian anomalies may occur in two different types—the consistent or the covariant anomaly—corresponding to the two different currents—the consistent or the covariant current—which the anomalies depend on.

Consistent anomaly:

- The consistent current is defined as the variation of the vacuum functional.
- As a result the covariant divergence of this current—the consistent anomaly—satisfies the WZ consistency condition.
- The consistent anomaly is part of the Stora–Zumino chain of descent equations.
- However, the consistent current does not transform covariantly under gauge transformations. The covariant transformation law is broken by the anomaly.

Covariant anomaly:

- By adding the BZ polynomial to the consistent current one can obtain a covariantly transforming current.

- However, the covariant current cannot be obtained from the variation of a vacuum functional.
- For this reason the covariant divergence of the covariant current—the covariant anomaly—does *not* satisfy the WZ consistency condition. The covariant anomaly obeys a modified WZ condition containing an additional anomaly term as an 'inhomogeneity'.
- Consequently the covariant anomaly does not fulfil the Stora–Zumino chain of descent equations but it is part of more complicated algebraic relations (see [Tsutsui 1989], [Abud, Ader, Gieres 1990], [Bonora, Cotta-Ramusino 1986]).
- The anomaly cancellation conditions remain the same for either type—covariant or consistent.
- The covariant anomaly may have its physical significance when the covariant current is coupled to other nongauged external fields.

10.2 Covariant anomaly and differential forms

What we have discussed in the previous section—the BZ polynomial and the covariant anomaly—we can also formulate, which we prefer, in the compact notation of differential forms on the space–time manifold M (dim $M = m$).

Definitions: **Current 1-form** (consistent current)

$$j = j^a_\mu dx^\mu T_a, \quad \text{with} \quad \langle j^\mu_a \rangle = \frac{\delta W}{\delta A^a_\mu}. \tag{10.26}$$

Dual current $(m-1)$-form

$$\langle * j \rangle = \frac{\delta W}{\delta A}, \quad \text{with} \quad * j = \frac{1}{(m-1)!} \varepsilon_{\mu_1 \mu_2 \ldots \mu_m} j^{\mu_1}_a dx^{\mu_2} \ldots dx^{\mu_m} T^a. \tag{10.27}$$

Treating

$$\delta A = \delta A^a_\mu dx^\mu T_a \tag{10.28}$$

as a 1-form on the manifold M we consider

$$\frac{\delta}{\delta A} = \frac{1}{(m-1)!} \varepsilon_{\mu_1 \mu_2 \ldots \mu_m} \frac{\delta}{\delta A^a_{\mu_1}} dx^{\mu_2} \ldots dx^{\mu_m} T^a \tag{10.29}$$

as a dual form. Then the scalar product of such forms can be written as (recall Section 2.4, equation (2.164))

$$
\begin{aligned}
\delta A \cdot \frac{\delta}{\delta A} &= \left(\delta A, \frac{\delta}{\delta A}\right) \\
&= \int_M \delta A^a_\mu \frac{\delta}{\delta A^a_\mu} \frac{1}{m!} \varepsilon_{\mu_1\mu_2\ldots\mu_m} dx^{\mu_1} dx^{\mu_2} \ldots dx^{\mu_m} \\
&= \int dx \delta A^a_\mu \frac{\delta}{\delta A^a_\mu}.
\end{aligned} \tag{10.30}
$$

Gauge transformation law: Now we look for the gauge transformation law of the consistent current by working with differential forms. Applying the commutation relation (10.6) to the vacuum functional again

$$[\delta_Y, \delta_\xi] W[A_\mu] = (\delta_Y \delta_\xi - \delta_\xi \delta_Y) W[A_\mu] = \delta_{[Y,\xi]} W[A_\mu], \tag{10.31}$$

using

$$\delta_Y W = \delta_Y A \cdot \frac{\delta W}{\delta A} = Y \cdot \langle *j \rangle \tag{10.32}$$

and the anomalous Ward identity (8.68) we obtain

$$\delta_Y\, \xi \cdot G[A] - \delta_\xi\, Y \cdot \langle *j \rangle = [Y, \xi] \cdot \langle *j \rangle, \tag{10.33}$$

which is rewritten:

Theorem: Gauge transformation law

$$Y \cdot \delta_\xi \langle *j \rangle = Y \cdot [\langle *j \rangle, \xi] + \delta_Y\, \xi \cdot G[A]. \tag{10.34}$$

In deriving equation (10.34) Y does not change under gauge transformations $\delta_\xi Y = 0$. If instead we also allow Y to transform as

$$\delta_\xi Y = [Y, \xi], \tag{10.35}$$

then we find the theorem more compact.

Theorem: Gauge transformation law

$$\delta_\xi\, Y \cdot \langle *j \rangle = \delta_Y\, \xi \cdot G[A]. \tag{10.36}$$

Here the variations act on all factors to the right but ξ remains unchanged $\delta_Y \xi = 0$. In components both versions (10.36) and (10.34) correspond to equation (10.13).

Bardeen–Zumino polynomial and covariant anomaly: As explained before, we introduce the BZ polynomial in order to obtain a covariantly transforming current.

Proposition:

- *There exists the local* **BZ polynomial** $\mathcal{P}$ *with transformation property*

$$\delta_\xi Y \cdot \mathcal{P} = -\delta_Y \xi \cdot G[A] \tag{10.37}$$

so that the **covariant current**

$$*\widetilde{\jmath} = *j + \mathcal{P} \tag{10.38}$$

transforms as

$$\delta_\xi Y \cdot \langle *\widetilde{\jmath} \rangle = 0. \tag{10.39}$$

The transformation law (10.39) can be regarded as a gauge invariance rather than a covariance. (Note, $\mathcal{P}$ is certainly the dual of a 1-form.)

Definition: The **covariant anomaly** is defined by

$$\widetilde{G}(\xi, A) = -\xi \cdot D * \widetilde{\jmath}. \tag{10.40}$$

Inserting the covariant current (10.38) we get the relation

$$\begin{aligned} \widetilde{G}(\xi, A) &= -\xi \cdot D * j - \xi \cdot D\mathcal{P} \\ &= G(\xi, A) - \xi \cdot D\mathcal{P}. \end{aligned} \tag{10.41}$$

Lemma: The anomaly $\widetilde{G}$—a covariant divergence of a covariant current—also transforms in a covariant way.

Proof. By definition we have (we choose the case $\delta_\xi Y = 0$)

$$\delta_\eta * \widetilde{\jmath} = [*\widetilde{\jmath}, \eta], \tag{10.42}$$

which implies covariance for

$$\begin{aligned} \delta_\eta\, D * \widetilde{\jmath} &= (\delta_\eta D) * \widetilde{\jmath} + D\, \delta_\eta * \widetilde{\jmath} \\ &= [D\eta, *\widetilde{\jmath}] + D[*\widetilde{\jmath}, \eta] = [D * \widetilde{\jmath}, \eta] \end{aligned} \tag{10.43}$$

and consequently covariance for

$$\begin{aligned} \delta_\eta \widetilde{G}(\xi, A) &= -\xi \cdot \delta_\eta\, D * \widetilde{\jmath} = -\xi \cdot [D * \widetilde{\jmath}, \eta] \\ &= -[\eta, \xi] \cdot D * \widetilde{\jmath} = \widetilde{G}([\eta, \xi], A). \qquad \text{Q.E.D.} \end{aligned} \tag{10.44}$$

Formulae: Bardeen and Zumino have found explicit expressions for the BZ polynomial and covariant anomaly in any (even) dimension. The procedure is the following. Since the consistent anomaly is known we first solve equation (10.37) for the polynomial $\mathcal{P}$ and then relation (10.41) for the covariant anomaly. The actual procedure, however, is rather technical (for details see [Bardeen, Zumino 1984]), so we just quote the results given in $(2n-2)$ dimensions:

BZ polynomial

$$Y \cdot \mathcal{P} = -Nn(n-1) \int_0^1 dt\, t \int_M P(Y, A, F_t^{n-2}), \tag{10.45}$$

with $F_t = tF + (t^2 - t)A^2$.

Covariant anomaly

$$\widetilde{G}(v, A) = Nn \int_M P(v, F^{n-1}) \tag{10.46}$$

in components

$$\widetilde{G}_a[A_\mu] = N \frac{n}{2^{n-1}} \varepsilon^{\mu_1 \nu_1 \cdots \mu_{n-1} \nu_{n-1}} \mathrm{tr}\, T_a F_{\mu_1 \nu_1} \cdots F_{\mu_{n-1} \nu_{n-1}}. \tag{10.47}$$

P denotes a symmetric invariant polynomial and we have already quoted the normalization N in Section 9.1, equation (9.22).

Examples:

i) *in 2 dimensions* $(n = 2,\ N = -\frac{i}{4\pi})$
Covariant anomaly

$$\widetilde{G}(v, A) = -\frac{i}{2\pi} \int \mathrm{tr}\, vF \tag{10.48}$$

$$\widetilde{G}_a[A_\mu] = -\frac{i}{4\pi} \varepsilon^{\mu\nu} \mathrm{tr}\, T_a F_{\mu\nu}. \tag{10.49}$$

BZ polynomial
integrating the formula

$$\int \mathrm{tr}\, Y\mathcal{P} = Y \cdot \mathcal{P} = \frac{i}{4\pi} 2 \int_0^1 dt\, t \int \mathrm{str}(Y, A) \tag{10.50}$$

gives

$$\mathcal{P} = \frac{i}{4\pi} A. \tag{10.51}$$

We obtain the components in conformity with notation (10.16) from the dual (recall $\widetilde{\jmath} = \jmath + *\mathcal{P}$)

$$*\mathcal{P} = \varepsilon_{\mu\nu}\mathcal{P}^{\nu}dx^{\mu} \tag{10.52}$$

then

$$\mathcal{P}^{\mu}_a[A_\mu] = \operatorname{tr} T_a \mathcal{P}^{\mu} = \frac{i}{4\pi}\varepsilon^{\mu\nu}\operatorname{tr} T_a A_\nu. \tag{10.53}$$

ii) *in 4 dimensions* ($n = 3$, $N = \frac{1}{24\pi^2}$)
Covariant anomaly

$$\widetilde{G}(v, A) = \frac{1}{8\pi^2}\int \operatorname{tr} vF^2 \tag{10.54}$$

$$\widetilde{G}_a[A_\mu] = \frac{1}{32\pi^2}\varepsilon^{\mu\nu\rho\sigma}\operatorname{tr} T_a F_{\mu\nu}F_{\rho\sigma}. \tag{10.55}$$

BZ polynomial
integrating the formula

$$Y \cdot \mathcal{P} = -\frac{1}{24\pi^2}3\cdot 2\int_0^1 dt\, t\int \operatorname{str}(Y, A, (tF + (t^2 - t)A^2)) \tag{10.56}$$

gives

$$\int \operatorname{tr} Y\mathcal{P} = -\frac{1}{24\pi^2}\int \operatorname{tr} Y(AF + FA - \frac{1}{2}A^3) \tag{10.57}$$

and therefore

$$\mathcal{P} = \frac{1}{3!}\mathcal{P}_{\nu\rho\sigma}dx^\nu dx^\rho dx^\sigma = -\frac{1}{24\pi^2}(AF + FA - \frac{1}{2}A^3). \tag{10.58}$$

Again, we find the components from the dual

$$*\mathcal{P} = \frac{1}{3!}\varepsilon_{\mu\nu\rho\sigma}\mathcal{P}^{\nu\rho\sigma}dx^\mu =: \mathcal{P}_\mu dx^\mu \tag{10.59}$$

then

$$\begin{aligned}\mathcal{P}^{\mu}_a[A_\mu] &= \operatorname{tr} T_a\mathcal{P}^{\mu}\\ &= -\frac{1}{48\pi^2}\varepsilon^{\mu\nu\rho\sigma}\operatorname{tr} T_a(A_\nu F_{\rho\sigma} + F_{\rho\sigma}A_\nu - A_\nu A_\rho A_\sigma),\end{aligned} \tag{10.60}$$

which is the result (10.18) quoted before.

Remark: The chain term $Q^1_{2n-2}(v, A, F)$ represents a polynomial in the variables v (linear), A, F where the leading term in powers of F is

$$Q^1_{2n-2}(v, A, F) = P(v, F^{n-1}) + \ldots. \tag{10.61}$$

For the consistent anomaly this implies to leading order

$$G(v,A) = N\int_M Q^1_{2n-2}(v,A,F) = N\int_M P(v,F^{n-1}) + \ldots . \qquad (10.62)$$

We observe that the covariant anomaly (10.46) is a factor n—a factor 3 ($n=3$) in 4-dimensional YM field theory—larger than the leading term in the consistent anomaly. This reflects the Bose symmetrization factor when calculating the anomaly with Feynman diagrams. It also shows that the anomaly cancellation condition, as discussed in Section 4.9, remains the same for both anomalies, the covariant and the consistent types.

Recipe: We find a recipe to obtain the covariant anomaly from a singlet anomaly:

- *The covariant anomaly in $(2n-2)$ dimensions follows from a singlet anomaly in $2n$ dimensions by substituting $F \to F + v$ and selecting the linear term in v!*

The **singlet anomaly** is given by

$$\begin{aligned} G_{\text{singlet}} &= -2\pi i \int_{M_{2n}} \text{ch } F = -2\pi i \frac{i^n}{(2\pi)^n n!}\int_{M_{2n}} \text{tr } F^n \\ &= N\int_{M_{2n}} \text{tr } F^n, \end{aligned} \qquad (10.63)$$

where $-2\pi i$ is a phase factor arising from topology and will be explained later on in Section 11.5.3; the integral corresponding to an index theorem contains the Chern character in $2n$ dimensions (see Section 11.4) and gives the normalization N (9.22). Then the following substitution provides the **covariant anomaly**

$$\begin{aligned} G_{\text{singlet}} &\overset{F\to F+v}{\longrightarrow} N\int \text{tr } (F+v)^n \\ &\widehat{=} \; Nn\int_{M_{2n-2}} \text{tr } vF^{n-1} = \widetilde{G}(v,A). \end{aligned} \qquad (10.64)$$

This illustrates that a singlet anomaly cancellation in $2n$ dimensions also guarantees a cancellation of non-Abelian anomalies in $(2n-2)$ dimensions.

10.3 Geometry in the space of gauge potentials

Finally we also formulate the anomaly equations as differential forms over Sp $\mathcal{A}$, the space of all gauge connections (see e.g. [Bonora, Cotta-Ramusino 1983, 1986], [Zumino 1985b], [Bao, Nair 1985], [Kelnhofer 1991]). We have already described the differential geometric set-up in Sections 8.1 and 8.3.

Again we start from the vacuum functional (8.1)

$$W[A] = -\ln Z[A]. \tag{10.65}$$

Definition: We define a *nonlocal* **current 1-form on Sp $\mathcal{A}$** by

$$\begin{aligned} \langle J[A] \rangle &= \int dx \langle j_a^\mu[A](x) \rangle \delta A_\mu^a(x) \\ &=: \delta W[A] = -\delta \ln Z[A] = -\frac{1}{Z[A]} \delta Z[A], \end{aligned} \tag{10.66}$$

where

$$\langle j_a^\mu[A](x) \rangle = \frac{\delta}{\delta A_\mu^a(x)} W[A] \tag{10.67}$$

coincides with the usual definition (8.56) for the consistent current.

Anomalous Ward identity: The interior product of the current with respect to the fundamental vector field X_ξ

$$i_{X_\xi} \langle J[A] \rangle = i_{X_\xi} \delta W[A] = (i_{X_\xi} \delta + \delta i_{X_\xi}) W[A] = \mathcal{L}_{X_\xi} W[A] \tag{10.68}$$

provides the Lie derivative on $W[A]$ (note: the interior product of a 0-form vanishes and we used relation (8.113)). Since the Lie derivative $\mathcal{L}_{X_\xi}$ along the fundamental vector field represents the gauge operator δ_ξ geometrically (recall Section 8.3.3), we can formulate the **anomalous Ward identity** by

$$\mathcal{L}_{X_\xi} W[A] = G(\xi, A). \tag{10.69}$$

Then the **consistent anomaly** is given as a restriction of the current 1-form along X_ξ

$$G(\xi, A) = i_{X_\xi} \langle J[A] \rangle. \tag{10.70}$$

In local components we have

$$\int dx \xi_b(x) G^b[A](x) = \int dx \langle j_a^\mu[A](x) \rangle (D_\mu \xi)^a(x). \tag{10.71}$$

Wess–Zumino consistency condition: The consistency condition follows from the **commutator of the Lie derivatives** (see Section 2.6)

$$[\mathcal{L}_{X_\xi}, \mathcal{L}_{X_\eta}] = \mathcal{L}_{[X_\xi, X_\eta]} \tag{10.72}$$

and from the **commutator of the fundamental vector fields** (8.66)

$$[X_\xi, X_\eta] = X_{[\xi, \eta]} \tag{10.73}$$

implying the **identity**

$$\mathcal{L}_{[X_\xi, X_\eta]} = \mathcal{L}_{X_{[\xi,\eta]}}. \tag{10.74}$$

Applying the commutation relation to the vacuum functional

$$([\mathcal{L}_{X_\xi}, \mathcal{L}_{X_\eta}] - \mathcal{L}_{X_{[\xi,\eta]}})W[A] = 0 \tag{10.75}$$

and using the anomalous Ward identity (10.69) we find:

Theorem: WZ consistency condition

$$\mathcal{L}_{X_\xi} G(\eta, A) - \mathcal{L}_{X_\eta} G(\xi, A) - G([\xi, \eta], A) = 0. \tag{10.76}$$

Gauge transformation law: For the gauge transformation of the consistent current we start with the identity

$$\mathcal{L}_{X_\xi} \langle J[A] \rangle = (i_{X_\xi} \delta + \delta i_{X_\xi}) \langle J[A] \rangle. \tag{10.77}$$

The first term on the r.h.s. vanishes because the current 1-form is closed (recall our discussion in Section 8.3.3)

$$\delta \langle J[A] \rangle = 0 \tag{10.78}$$

and for the second term we insert the anomaly definition (10.70). Then we immediately find the following theorem:

Theorem: Gauge transformation law

$$\mathcal{L}_{X_\xi} \langle J[A] \rangle = \delta G(\xi, A). \tag{10.79}$$

The gauge invariance of the current is broken by the anomaly. Equation (10.79) represents the geometric realization in Sp $\mathcal{A}$ of Bardeen–Zumino's formulation of the current transformation law (10.36).

For a strict equivalence to expression (10.36) we have to confine the variations in equation (10.79) along an arbitrary but fixed vector field $Y[A]$

$$i_Y \mathcal{L}_{X_\xi} \langle J[A] \rangle = i_Y \delta G(\xi, A). \tag{10.80}$$

Rewriting equation (10.80) with the help of the commutator property

$$[\mathcal{L}_{X_\xi}, i_Y] = i_{[X_\xi, Y]} \tag{10.81}$$

we obtain:

Theorem: Gauge transformation law

$$\mathcal{L}_{X_\xi} i_Y \langle J[A] \rangle = i_{[X_\xi, Y]} \langle J[A] \rangle + i_Y \delta G(\xi, A). \tag{10.82}$$

This equation describes the anomalous covariant transformation property of the consistent current closest to the spirit of Bardeen–Zumino.

Local components: In local components relation (10.82) coincides with expression (10.14) which we had in Section 10.1. It is instructive to carry out this little exercise. We first calculate the l.h.s. (note that the interior product of a 0-form vanishes)

$$\begin{aligned}\mathcal{L}_{X_\xi} i_Y \langle J[A]\rangle &= (i_{X_\xi}\delta + \delta i_{X_\xi}) i_Y \langle J[A]\rangle \\ &= \int dx (D_\mu \xi)^a(x) \frac{\delta}{\delta A^a_\mu(x)} \int dy \langle j^\nu_b[A](y)\rangle Y^b_\nu[A](y) \\ &= \int dy Y^b_\nu[A](y) \int dx (D_\mu\xi)^a(x) \frac{\delta}{\delta A^a_\mu(x)} \langle j^\nu_b[A](y)\rangle \\ &\quad + \int dx dy \langle j^\nu_b[A](y)\rangle (D_\mu\xi)^a(x) \frac{\delta}{\delta A^a_\mu(x)} Y^b_\nu[A](y).\end{aligned} \tag{10.83}$$

For the r.h.s. we need a Lie bracket in the space of gauge potentials.

Definition: The **Lie bracket in Sp $\mathcal{A}$** is defined by

$$[X,Y] = \int dy [X,Y]^b_\nu(y) \frac{\delta}{\delta A^b_\nu(y)}, \tag{10.84}$$

with

$$[X,Y]^b_\nu(y) = \int dx X^a_\mu(x) \frac{\delta}{\delta A^a_\mu(x)} Y^b_\nu(y) - \int dx Y^a_\mu(x) \frac{\delta}{\delta A^a_\mu(x)} X^b_\nu(y). \tag{10.85}$$

The X, Y are vector fields

$$X = \int dx X^\mu_a(x) \frac{\delta}{\delta A^a_\mu(x)}, \qquad Y = \int dx Y^\mu_a(x) \frac{\delta}{\delta A^a_\mu(x)}. \tag{10.86}$$

Then we can calculate the interior product of the current

$$\begin{aligned} i_{[X_\xi,Y]}\langle J[A]\rangle &= \int dy \langle j^\nu_b[A](y)\rangle [X_\xi, Y]^b_\nu(y) \\ &= \int dx dy \langle j^\nu_b[A](y)\rangle (D_\mu\xi)^a(x) \frac{\delta}{\delta A^a_\mu(x)} Y^b_\nu[A](y) \\ &\quad + \int dy Y^b_\nu[A](y) \int dx [\langle j^\nu[A]\rangle, \xi]_b(x) \delta(x-y).\end{aligned} \tag{10.87}$$

The last term in equation (10.82) gives

$$i_Y \delta G(\xi, A) = \int dy Y^b_\nu[A](y) \frac{\delta}{\delta A^b_\nu(y)} \int dx \xi^a(x) G_a[A](x). \qquad (10.88)$$

Collecting our results (10.83), (10.87) and (10.88) we formulate relation (10.82) in local components

$$\begin{aligned} &\int dy Y^b_\nu[A](y) \int dx (D_\mu \xi)^a(x) \frac{\delta}{\delta A^a_\mu(x)} \langle j^\nu_b[A](y)\rangle \\ &= \int dy Y^b_\nu[A](y) \int dx [\langle j^\nu[A]\rangle, \xi]_b(x) \delta(x-y) \\ &\quad + \int dy Y^b_\nu[A](y) \int dx \xi^a(x) \frac{\delta}{\delta A^b_\nu(y)} G_a[A](x). \end{aligned} \qquad (10.89)$$

Since the vector field $Y^b_\nu[A]$ has been chosen arbitrarily we can drop it here and we recover precisely the local transformation law (10.14).

Bardeen–Zumino polynomial and covariant anomaly: As previously we introduce the BZ polynomial to achieve a gauge invariant current.

Proposition:

- *There exists a local polynomial 1-form in Sp $\mathcal{A}$—the* **BZ polynomial**

$$\mathcal{P}[A] = \int dx \mathcal{P}^\mu_a[A](x) \delta A^a_\mu(x), \qquad (10.90)$$

with transformation property

$$\mathcal{L}_{X_\xi} \mathcal{P}[A] = -\delta G(\xi, A) \qquad (10.91)$$

so that the **covariant current 1-form**

$$\widetilde{J}[A] = J[A] + \mathcal{P}[A] \qquad (10.92)$$

obeys **gauge invariance**

$$\mathcal{L}_{X_\xi} \langle \widetilde{J}[A]\rangle = 0. \qquad (10.93)$$

Restricting equations (10.91) and (10.93) along an arbitrary but fixed vector field $Y[A]$ we obtain the **transformation property**

$$\mathcal{L}_{X_\xi} i_Y \mathcal{P}[A] = i_{[X_\xi, Y]} \mathcal{P}[A] - i_Y \delta G(\xi, A) \qquad (10.94)$$

and the **covariant transformation law**

$$\mathcal{L}_{X_\xi} i_Y \langle \widetilde{J}[A]\rangle = i_{[X_\xi, Y]} \langle \widetilde{J}[A]\rangle. \qquad (10.95)$$

The local components $\mathcal{P}_a^\mu[A](x)$ are exactly the BZ solutions in four dimensions (10.60) or in two dimensions (10.53).

Definition: Analogous to the consistent case (10.70) we define the **covariant anomaly** by restricting the covariant current 1-form along X_ξ

$$\widetilde{G}(\xi, A) = i_{X_\xi}\langle\widetilde{J}[A]\rangle. \tag{10.96}$$

In local components we have

$$\int dx \xi_b(x)\widetilde{G}^b[A](x) = \int dx \langle\widetilde{\mathcal{J}}_a^\mu[A](x)\rangle(D_\mu\xi)^a(x) \tag{10.97}$$

which coincides with definition (10.19) (after partial integration).

Then the two types of anomaly—the covariant and consistent types—just differ by

$$\widetilde{G}(\xi, A) = G(\xi, A) + i_{X_\xi}\mathcal{P}[A], \tag{10.98}$$

where

$$i_{X_\xi}\mathcal{P}[A] = \mathcal{P}[A](X_\xi) \tag{10.99}$$

denotes the BZ polynomial evaluated at the fundamental vector field X_ξ.

10.4 Nonlocal extensions of anomalies in Sp $\mathcal{A}$

We consider again the vacuum functional $W[A]$. It is highly *nonlocal* in A.

$$Z[A] = e^{-W[A]} = \int d\bar{\psi}d\psi \exp[-\int dx\bar{\psi}i\gamma^\mu(\partial_\mu + A_\mu)\psi]. \tag{10.100}$$

Therefore the current 1-form (10.66)

$$\langle J[A]\rangle = \delta W[A] \tag{10.101}$$

also represents a *nonlocal* functional in Sp $\mathcal{A}$. Restricting, however, the current to a gauge orbit—evaluating along a fundamental vector field X_ξ—

$$i_{X_\xi}\langle J[A]\rangle = G(\xi, A) \tag{10.102}$$

yields the anomaly which turns out *locally* in the gauge connection A. This locality is a fact (and a consequence of the anomaly calculation with Feynman diagrams). The same is also true for the covariant case.

On the other hand, the difference of the consistent and covariant current—the BZ polynomial $\mathcal{P}[A]$—does represent a *local* functional of A

$$\langle\widetilde{J}[A]\rangle - \langle J[A]\rangle = \mathcal{P}[A]. \tag{10.103}$$

Since the BZ polynomial 1-form is a local functional in the whole space Sp $\mathcal{A}$, it clearly remains local along a gauge orbit, characterizing the two types of anomaly

$$i_{X_\xi}\mathcal{P}[A] = \widetilde{G}(\xi, A) - G(\xi, A). \tag{10.104}$$

The anomalies themselves (consistent or covariant), however, *cannot* be expressed as *local* forms on Sp $\mathcal{A}$ but, as emphasized by Zumino [Zumino 1985b] and Stora [Stora 1991], *nonlocal extensions* do exist.

Nonlocal extensions: We present two examples in 2 dimensions.

i) *Consistent anomaly*
The **functional 1-form**

$$G = N\int \mathrm{tr}\; dA\Box^{-1}D^\lambda \delta A_\lambda \tag{10.105}$$

represents a **nonlocal extension on Sp $\mathcal{A}$** of the consistent anomaly. The nonlocality is caused by the nonlocal operator $\Box^{-1}$, where $\Box = D^\lambda D_\lambda$. The operator $\Box^{-1}$ means the Green function $\mathcal{D}(x-y)$ defined by

$$\Box_x \mathcal{D}(x-y) = \delta(x-y). \tag{10.106}$$

In components the 1-form (10.105) denotes

$$\begin{aligned} G &= N\int dx^\mu dx^\nu\; \mathrm{tr}\; \partial_\mu A_\nu(x)\Box_x^{-1}D^\lambda T_a \delta A^a_\lambda(x) \\ &= N\int dx dy \varepsilon^{\mu\nu}\; \mathrm{tr}\; \partial_\mu A_\nu(x)\mathcal{D}(x-y)D^\lambda T_a\delta A^a_\lambda(y). \end{aligned} \tag{10.107}$$

The last line shows explicitly the nonlocal correlation of the two terms $\partial_\mu A_\nu(x)$ and $D^\lambda \delta A_\lambda(y)$.
Restricting, however, the nonlocal 1-form (10.105) along a gauge orbit $\delta A_\lambda \to sA_\lambda$ (we use the BRS formalism now)

$$\begin{aligned} G(sA) = i_{sA}G &= N\int \mathrm{tr}\; dA\Box^{-1}D^\lambda sA_\lambda \\ &= N\int \mathrm{tr}\; dA\Box^{-1}D^\lambda D_\lambda v \\ &= N\int \mathrm{tr}\; dAv = G(v, A) \end{aligned} \tag{10.108}$$

supplies us the *local* functional $G(v, A)$—the consistent anomaly.

ii) *Covariant anomaly*
Similarly, the nonlocal **functional 1-form on Sp $\mathcal{A}$**

$$\widetilde{G} = N \cdot 2 \int \mathrm{tr}\ F \Box^{-1} D^{\lambda} \delta A_{\lambda} \tag{10.109}$$

represents a **nonlocal extension** of the covariant anomaly. The restriction along the gauge orbit

$$\begin{aligned} \widetilde{G}(sA) = i_{sA}\widetilde{G} &= N \cdot 2 \int \mathrm{tr}\ F \Box^{-1} D^{\lambda} sA_{\lambda} \\ &= N \cdot 2 \int \mathrm{tr}\ Fv = \widetilde{G}(v, A) \end{aligned} \tag{10.110}$$

reduces to the *local* functional $\widetilde{G}(v, A)$—the covariant anomaly.

Recipe: In order to construct a nonlocal extension of an anomaly into the space of all gauge connections Sp $\mathcal{A}$ we find the following recipe (for the covariant case, for example):

- *Start with the anomaly*

$$\widetilde{G}(v, A) = N \cdot n \int \mathrm{str}\ F^{n-1} v \tag{10.111}$$

insert the identity $\Box^{-1}\Box = \mathbf{1}$ *with* $\Box = D^{\lambda} D_{\lambda}$

$$\widetilde{G}(v, A) = N \cdot n \int \mathrm{str}\ F^{n-1} \Box^{-1} D^{\lambda} D_{\lambda} v \tag{10.112}$$

and use the BRS transformation $sA_{\lambda} = D_{\lambda} v$

$$\widetilde{G}(v, A) = N \cdot n \int \mathrm{str}\ F^{n-1} \Box^{-1} D^{\lambda} sA_{\lambda}. \tag{10.113}$$

Then, by extending the BRS variation $sA_{\lambda} \to \delta A_{\lambda}$ *to a general variation, the* **nonlocal extension on Sp $\mathcal{A}$** *is given by the* **1-form**

$$\widetilde{G} = N \cdot n \int \mathrm{str}\ F^{n-1} \Box^{-1} D^{\lambda} \delta A_{\lambda}. \tag{10.114}$$

Restricting, conversely, the 1-form $\widetilde{G}$ (10.114) to the gauge orbit $\delta A_{\lambda} \to sA_{\lambda}$

$$\widetilde{G}(sA) = i_{sA}\widetilde{G} = \widetilde{G}(v, A) \tag{10.115}$$

brings back the local anomaly $\widetilde{G}(v, A)$.

11
Index and anomaly

In this chapter we study one of the most fascinating aspects of the anomaly—its role in the topology of gauge theories. In Chapters 2 and 6 we discussed the fibre bundles as the proper geometric set-up for the physics of gauge theories. Here we shall see that the anomaly also has a 'natural' explanation; it occurs as an obstruction in certain nontrivial bundles and it is determined completely by a topological quantity—the index.

First in Section 11.1 we discuss the relation of the singlet anomaly to the Atiyah–Singer index theorem and then, after some introduction into index theory (Section 11.2), we describe the geometric–topological character of the non-Abelian anomaly in the context of index theorems. In particular, we show the connection between the index of the Weyl operator and the heat kernel of the Laplacian in Section 11.3, shedding much more light on Fujikawa's regularization procedure. In Section 11.4 we present the Atiyah–Singer index theorem for the case of YM fields. Introducing, in Section 11.5.1, a special Dirac operator which is equivalent to the Weyl operator, we calculate, in Section 11.5.2, the non-Abelian anomaly, Bardeen's result, by the path integral method. Finally, in Section 11.5.3, we explain the procedure of Alvarez-Gaumé and Ginsparg, how to determine the non-Abelian anomaly by a generalized index theorem.

11.1 Singlet anomaly and index

It has been discovered by several authors [Jackiw, Rebbi 1976, 1977], [Nielsen, Schroer 1977], [Nielsen, Römer, Schroer 1977], [Römer 1979, 1981], [Ansourian 1977], [Kiskis 1977], [Brown, Carlitz, Lee 1977] that the singlet anomaly is related to the index of a differential operator. This discovery triggered a new era of topological investigations for the anomaly. We proceed here within our path integral approach of Chapter 5 and study the Jacobian of the PI measure.

As Fujikawa [Fujikawa 1979, 1980] noticed, the evaluation of the Jacobian in the PI corresponds to the calculation of the index of the Dirac operator for a given chirality. Why is this so?

Let us consider the eigenvalue equation for the Dirac operator in Euclidean space

$$\not{D}\varphi_n(x) = \lambda_n \varphi_n(x), \tag{11.1}$$

with $\{\varphi_n(x)\}$ being a complete orthonormal system of eigenfunctions. Then $\gamma_5\varphi_n$ satisfies the equation with negative eigenvalues $(-\lambda_n)$

$$\not{D}\gamma_5\varphi_n(x) = -\lambda_n \gamma_5\varphi_n(x) \tag{11.2}$$

due to $\{\gamma_\mu, \gamma_5\} = 0$.

Lemma: The eigenfunctions φ_n and $\gamma_5\varphi_n$ are orthogonal for $\lambda_n \neq 0$

$$\varphi_n \perp \gamma_5\varphi_n \qquad \text{or} \qquad (\varphi_n, \gamma_5\varphi_n) = \int dx\; \varphi_n^\dagger(x)\gamma_5\varphi_n(x) = 0. \tag{11.3}$$

Proof. The orthogonality holds because of $\not{D}^\dagger = \not{D}$. Let us consider

$$(\gamma_5\varphi_n, \not{D}\varphi_n) = \lambda_n(\gamma_5\varphi_n, \varphi_n),$$

which is, on the other hand,

$$(\gamma_5\varphi_n, \not{D}\varphi_n) = (\not{D}^\dagger\gamma_5\varphi_n, \varphi_n) = -(\gamma_5\, \not{D}\varphi_n, \varphi_n) = -\lambda_n(\gamma_5\varphi_n, \varphi_n)$$

implying

$$(\gamma_5\varphi_n, \varphi_n) = 0 \quad \text{for } \lambda_n \neq 0. \qquad \text{Q.E.D.} \tag{11.4}$$

Zero-modes: However, this is not so for zero-eigenvalue $\lambda_n = 0$. In this case φ_n^0 and $\gamma_5\varphi_n^0$ are eigenfunctions to the same eigenvalue $\lambda_n = 0$

$$\not{D}\varphi_n^0(x) = 0 \qquad \text{and} \qquad \not{D}\gamma_5\varphi_n^0(x) = 0. \tag{11.5}$$

This is also true for positive or negative chirality eigenfunctions

$$\varphi_{n+}^0 = P_+\varphi_n^0, \qquad \varphi_{n-}^0 = P_-\varphi_n^0, \tag{11.6}$$

with $P_\pm = \frac{1}{2}(\mathbf{1} \pm \gamma_5)$. We clearly have

$$\not{D}\varphi_{n\pm}^0 = \frac{1}{2}\not{D}(\mathbf{1} \pm \gamma_5)\varphi_n^0 = 0. \tag{11.7}$$

The zero-modes $\varphi_{n\pm}^0$ are also eigenfunctions of γ_5 with eigenvalues ± 1

$$\begin{aligned} \gamma_5\varphi_{n+}^0 &= \frac{1}{2}(\gamma_5 + \mathbf{1})\varphi_n^0 = \varphi_{n+}^0 \\ \gamma_5\varphi_{n-}^0 &= \frac{1}{2}(\gamma_5 - \mathbf{1})\varphi_n^0 = -\varphi_{n-}^0. \end{aligned} \tag{11.8}$$

Jacobian: Now we return to the **Jacobian** of the PI, equation (5.54), which arises from the infinitesimal chiral transformation (5.11)

$$J[\beta] = \exp\left[-2i\int dx\ \beta(x)\sum_n \varphi_n^\dagger(x)\gamma_5\varphi_n(x)\right]. \tag{11.9}$$

We have evaluated the sum in the exponent and it supplies the **anomaly result** (recall equation (5.68))

$$\sum_n \varphi_n^\dagger(x)\gamma_5\varphi_n(x) = -\frac{1}{32\pi^2}\varepsilon^{\mu\nu\alpha\beta}\ \mathrm{tr}\ F_{\mu\nu}F_{\alpha\beta}, \tag{11.10}$$

with $\varepsilon^{1234} = \varepsilon^{1230} = 1$. Considering, on the other hand, the eigenfunctions—we keep the transformation parameter β constant—then we find that due to the orthogonality condition (11.3) only the zero-modes survive

$$\begin{aligned}\int dx \sum_n \varphi_n^\dagger(x)\gamma_5\varphi_n(x) &= \int dx \sum_n \varphi_n^{0\dagger}(x)\gamma_5\varphi_n^0(x)\\ &= \sum_n \int dx\ \varphi_{n+}^{0\dagger}(x)\varphi_{n+}^0(x) - \sum_n \int dx\ \varphi_{n-}^{0\dagger}(x)\varphi_{n-}^0(x)\\ &= n_+ - n_-.\end{aligned} \tag{11.11}$$

n_+ and n_- denote the number of positive and negative chirality zero-modes.

Index: The difference (11.11)—as we shall learn later on—is a characteristic integer number attached to a differential operator. It is called the **index**. In fact, it represents precisely the **index of the Dirac operator** projected to the positive chirality subspace

$$\text{index } D_+ = n_+ - n_-, \tag{11.12}$$

with

$$D_+ = \not{D} P_+ = \not{D}|_{\{\varphi_{n+}\}}. \tag{11.13}$$

Using the anomaly result (11.10) we obtain an explicit expression for the index

$$\text{index } D_+ = -\frac{1}{32\pi^2}\int dx\ \varepsilon^{\mu\nu\alpha\beta}\ \mathrm{tr}\ F_{\mu\nu}F_{\alpha\beta}, \tag{11.14}$$

which represents the famous Atiyah–Singer index theorem. Rewritten within differential forms we have:

Theorem: Atiyah–Singer index theorem

$$\text{index } D_+ = -\frac{1}{8\pi^2}\int_{S^4} \text{tr } F^2. \tag{11.15}$$

We study this relation in more detail later on, when we work on a compact manifold like the sphere S^4. But also in our Euclidean space $\mathbf{R}^4$—where we are actually working in the PI formalism—the relation (11.14) is true; but then we have to assume that the gauge potential approaches a pure gauge configuration $A_\mu \to g^{-1}\partial_\mu g$ at infinity $x \to \infty$ (as found by [Jackiw, Rebbi 1977]). We discussed this situation in our previous study of instantons (Chapter 6). In fact, we met the integral (11.14) in Section 6.6. It was equation (6.160), named the **Pontrjagin index** or the **topological charge** q. Hence we find the identity

$$n_+ - n_- \equiv q. \tag{11.16}$$

We also remember that the topological charge, and therefore the index (11.12)–(11.15), is determined only by the topology of the bundle (recall equation (6.198)).

The regularized sum of eigenfunctions represents an **index density**

$$dx \cdot \sum_n \varphi_n(x)\gamma_5\varphi_n(x) = -\frac{1}{8\pi^2} \text{ tr } F^2 = \text{Index } D_+. \tag{11.17}$$

So Fujikawa's regularization procedure (Section 5.2) corresponds to a local evaluation of the index.

Anomaly: The **singlet anomaly** $\mathcal{A}$ (or axial-, or chiral anomaly), which is the anomalous divergence of the chiral current, occurs in the Jacobian via

$$J[\beta] = \exp\left[-\int dx\ \beta(x)\mathcal{A}[A_\mu](x)\right], \tag{11.18}$$

therefore we find the following **relation between the anomaly and the index** (in Euclidean space)

$$\begin{aligned}\int dx\ \partial^\mu j^5_\mu(x) &= \int dx\ \mathcal{A}[A_\mu](x) = \int dx\ 2i\sum_n \varphi_n^\dagger(x)\gamma_5\varphi_n(x)\\ &= 2i\cdot(n_+ - n_-) = 2i \cdot \text{index } D_+\end{aligned} \tag{11.19}$$

or in terms of differential forms we have:

Theorem: Anomaly and index

$$d * j^5 = \mathcal{A}[A_\mu] = 2i \cdot \text{Index } D_+. \tag{11.20}$$

11.2 Fredholm and elliptic operators and index theory

We want to give an idea of the concept of an index as it is used in the mathematical literature (see e.g. [Gilkey 1984], [Booss, Bleecker 1985], [Berline, Getzler, Vergne 1992], [Lawson, Michelsohn 1989], [Nash 1991]). Let us begin with some definitions.

Operator: A **linear operator** $T : E \to F$ acting between the Hilbert spaces E, F is said to be **bounded** if the **operator norm**

$$\|T\| := \sup_{x \neq 0} \frac{\|Tx\|}{\|x\|} < \infty \tag{11.21}$$

is finite, where the **norms** in E, F are induced by the inner product

$$\|x\| = (x, x)^{1/2}. \tag{11.22}$$

Kernel: The **kernel**, or null space, of the operator T is defined by

$$\ker T = \{e \in E | Te = 0\}. \tag{11.23}$$

Range: The **range**, or image, of the operator T is given by

$$\text{range } T = \{f \in F | Te = f \text{ for some } e \in E\}. \tag{11.24}$$

Note that $\ker T$ is always closed, whereas range T need not be closed.

Orthogonal complement: Let $A \subset F$, then the **complement** of A is $A^c = \{a^c \in F | a^c \notin A\}$. We have $A \cup A^c = F$, $A \cap A^c = \emptyset$. The orthogonal space—**orthogonal complement**—is defined by $A_\perp = \{a_\perp \in F | (a, a_\perp) = 0 \; \forall \, a \in A\}$. Note that $A_\perp$ is closed.

Adjoint operator: Next we consider the **adjoint operator** $T^* : E \leftarrow F$, adjoint with respect to the inner product

$$(f, Te) = (T^* f, e) \qquad \text{for } e \in E, \; f \in F. \tag{11.25}$$

Then we find for the orthogonal complement of the range:

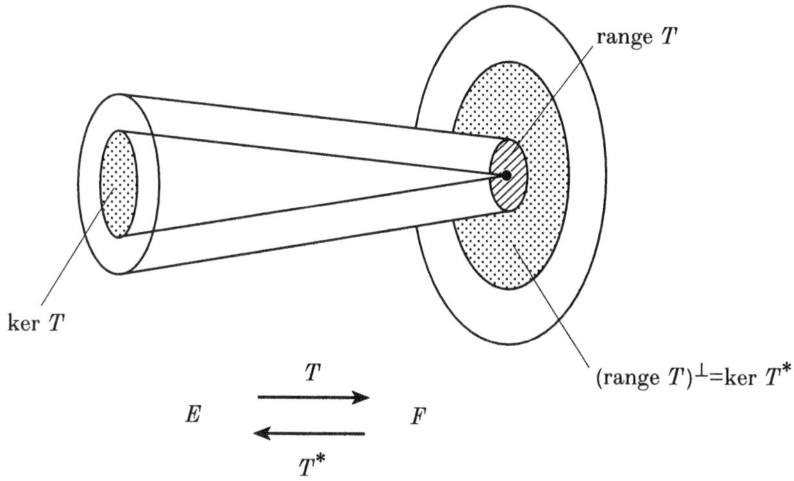

Fig. 11.1. The operator $T : E \to F$, the adjoint $T^* : E \leftarrow F$, $\ker T$ (dotted area), range T (hatched area), $(\text{range } T)^\perp = \ker T^*$ (dotted strip)

Lemma:

$$(\text{range } T)^\perp = \ker T^*. \tag{11.26}$$

Proof. Take the elements $f \in \text{range } T$, $g \in (\text{range } T)^\perp$, then

$$0 = (g, f) = (g, Te) = (T^* g, e) \tag{11.27}$$

$\forall\, e \in \text{dom } T$ (domain of T). Consequently we have $T^* g = 0$ or $g \in \ker T^*$ and vice versa, thus $(\text{range } T)^\perp = \ker T^*$. Q.E.D.

Cokernel: We define the **cokernel** of the operator T

$$\text{coker } T := F \,/\, \text{range } T \simeq (\text{range } T)^\perp = \ker T^*, \tag{11.28}$$

where the symbol $\simeq$ denotes isomorphic.

We have illustrated the above concepts in Figure 11.1.

Fredholm operator: We speak of a **Fredholm operator** $T \in \text{FRED}(E, F)$

$$\text{if } T : E \to F \text{ with} \quad T \text{ bounded}$$
$$\dim \ker T < \infty, \quad \dim \ \text{coker } T < \infty. \tag{11.29}$$

This means that the homogeneous equation $Te = 0$ has a finite number of linear independent solutions and to solve the inhomogeneous equation $Te = f$ the function f has to obey a finite number of conditions $(g, f) = 0$, where $g \in \ker T^*$; thus $T^* g = 0$.

Note that if $T \in \mathrm{FRED}(E, F)$ then range T and range T^* are closed.

Lemma:

$$\text{i)} \qquad \text{if } T \in \mathrm{FRED}(E,F) \Rightarrow T^* \in \mathrm{FRED}(F,E),$$

$$\text{ii)} \qquad \text{if } \begin{matrix} T_1 \in \mathrm{FRED}(E,F) \\ T_2 \in \mathrm{FRED}(F,G) \end{matrix} \Rightarrow T_2 T_1 \in \mathrm{FRED}(E,G). \qquad (11.30)$$

Now we are prepared to define the index of a Fredholm operator.

Index: Let $T \in \mathrm{FRED}(E, F)$, then we define the **index** of T

$$\text{index } T := \dim\ker T - \dim\operatorname{coker} T. \qquad (11.31)$$

Because of the finiteness of the dimensions of the kernel and the cokernel this definition makes sense and, using isomorphism (11.28), we also have

$$\text{index } T = \dim\ker T - \dim\ker T^*. \qquad (11.32)$$

This kind of index is actually named the **analytic index**.

Example: Let $\{\phi_n\}$ be a complete orthonormal basis for L^2, the space of square integrable functions. We define the displacement operator by

$$T\phi_n = \begin{cases} \phi_{n-1} & \text{if } n > 0 \\ 0 & \text{if } n = 0 \\ \phi_n & \text{if } n < 0. \end{cases} \qquad (11.33)$$

Then T is surjective (onto), hence $\ker T^* = (\text{range } T)^\perp = \emptyset$. The space

$$\ker T = \{\phi_0 | T\phi_0 = 0\} \qquad (11.34)$$

is 1-dimensional, therefore we find for the **index**

$$\text{index } T = \dim\ker T - \dim\ker T^* = 1. \qquad (11.35)$$

Considering next the operator squared for the $n = 1$ basis vector we get

$$T^2\phi_1 = T\phi_0 = 0 \qquad (11.36)$$

in addition to

$$T^2\phi_0 = 0. \tag{11.37}$$

Hence the space

$$\ker T^2 = \{\phi_0, \phi_1 | T^2\phi_0 = T^2\phi_1 = 0\} \tag{11.38}$$

is 2-dimensional and the index is

$$\text{index } T^2 = 2. \tag{11.39}$$

Generally we obtain

$$\text{index } T^n = n \in \mathbf{Z}. \tag{11.40}$$

This shows that the index is a (here surjective) map from the space of Fredholm operators FRED into the integer numbers **Z**

$$\text{index: FRED}(E, F) \to \mathbf{Z}. \tag{11.41}$$

Lemma:

$$\begin{aligned} &\text{i)} \quad \text{index } T = -\text{index } T^* \\ &\text{ii)} \quad \text{index } T = 0 \quad \text{if } T = T^* \text{ selfadjoint.} \end{aligned} \tag{11.42}$$

Both are evident from the index definition (11.32).

On the other hand, for a vanishing index we may have

$$\begin{aligned} &\textit{either} \quad \dim\ker T = \dim\ker T^* = 0 \\ &\textit{or} \quad \dim\ker T = \dim\ker T^* \neq 0 \end{aligned} \tag{11.43}$$

then the statement

$$\text{index } T = 0 \tag{11.44}$$

is *equivalent* to the more familiar **Fredholm alternative:**

either the inhomogeneous equation $Te = f$ has a unique solution $e \in E$ for a given $f \in F$;

or the homogeneous equation $Te = 0$ has finitely many, linear independent solutions $\{e_n\}$ and just as many solutions $\{g_n\}$ as the adjoint equation $T^*g = 0$. The inhomogeneous equation is solvable whenever $(g_1, f) = \ldots = (g_n, f) = 0$.

Lemma: Direct sum

Let $T \in \text{FRED}(E, F)$, $S \in \text{FRED}(E', F')$, then

$$\begin{aligned} &\text{i)} \quad S \oplus T \in \text{FRED}(E' \oplus E, F' \oplus F) \\ &\text{ii)} \quad \text{index } (S \oplus T) = \text{index } S + \text{index } T. \end{aligned} \tag{11.45}$$

Proof. Property i) is evident; property ii) is quickly verified:

$$\begin{aligned}\text{index }(S\oplus T) &= \dim\ker(S\oplus T)-\dim\ker(S\oplus T)^* \\ &= \dim\ker S+\dim\ker T-\dim\ker S^*-\dim\ker T^* \\ &= \text{index } S+\text{index } T. \qquad \text{Q.E.D.}\end{aligned} \tag{11.46}$$

Attention! The ordinary sum of two Fredholm operators is *not* in general Fredholm-like again, $S+T\notin$ FRED. For instance, we could set $T=-S$.

Lemma: Product
Let $T\in \text{FRED}(E,F)$, $S\in\text{FRED}(F,G)$, then

$$\text{index } ST=\text{index } S+\text{index } T. \tag{11.47}$$

We postpone the proof until after our homotopy discussion.

Theorem: The index

$$\text{index: FRED}(E,F)\to \mathbf{Z} \tag{11.48}$$

is **locally constant** or **homotopic invariant.**

This homotopic invariance is an important property of the index—which we need when we work with the index theorems later on.

Homotopy: (in analogy with the homotopic functions discussed in Section 2.2)

Two Fredholm operators $T_0, T_1\in\text{FRED}(E,F)$ are **homotopic** (T_0 can be continuously deformed into T_1)

$$T_0\sim T_1 \tag{11.49}$$

if there exists a **continuous family of Fredholm operators** $\mathcal{T}_t$ parametrized over the product space $E\times I$

$$E\times I\xrightarrow{\mathcal{T}_t}F \qquad \text{with } I=[0,1] \tag{11.50}$$

such that

$$\mathcal{T}_{t=0}|_{E\times\{0\}}=T_0 \qquad \text{and} \qquad \mathcal{T}_{t=1}|_{E\times\{1\}}=T_1. \tag{11.51}$$

Example: Let us consider the **group of invertible operators** on a Hilbert space $\mathcal{H}$, specifically the group of complex $N\times N$ matrices $GL(N,\mathbf{C})$. We study the group of invertible operators on the product space $\mathcal{H}\oplus\mathcal{H}$, specifically matrices in $GL(2N, \mathbf{C})$ which can be written as 2×2 block matrices.

We choose two such operators $S, T \in GL(N, \mathbf{C})$, then the following two operators defined on the direct sum space

$$\mathcal{T}_0 = \begin{pmatrix} ST & 0 \\ 0 & \mathbf{1} \end{pmatrix}, \qquad \mathcal{T}_{\pi/2} = \begin{pmatrix} T & 0 \\ 0 & S \end{pmatrix} \tag{11.52}$$

are **homotopic**

$$\mathcal{T}_0 \sim \mathcal{T}_{\pi/2}, \tag{11.53}$$

which means that there exists a continuous family of Fredholm operators $\mathcal{T}_t$

$$GL(2N, \mathbf{C}) \times I \xrightarrow{\mathcal{T}_t} GL(2N, \mathbf{C}) \qquad \text{with } I = [0, \pi/2] \tag{11.54}$$

such that

$$\mathcal{T}_{t=0} = \mathcal{T}_0 \qquad \text{and} \qquad \mathcal{T}_{t=\pi/2} = \mathcal{T}_{\pi/2}. \tag{11.55}$$

The explicit solution—the **homotopy**—is

$$\mathcal{T}_t = U_t^{-1} \begin{pmatrix} S & 0 \\ 0 & \mathbf{1} \end{pmatrix} U_t \begin{pmatrix} T & 0 \\ 0 & \mathbf{1} \end{pmatrix} \tag{11.56}$$

with the rotation matrices

$$U_t = \begin{pmatrix} \cos t & \sin t \\ -\sin t & \cos t \end{pmatrix}, \qquad U_t^{-1} = \begin{pmatrix} \cos t & -\sin t \\ \sin t & \cos t \end{pmatrix}. \tag{11.57}$$

Property (11.55) is easily verified. So the deformation here involves a rotation around an angle $t \in [0, \pi/2]$ of the space where S is operating.

Proof of Lemma (11.47): The homotopy also allows us to prove Lemma (11.47). What we need is the direct sum of the index, lemma (11.45), the homotopic invariance of the index, theorem (11.48), and the example (11.52), (11.53) of two homotopic operators. Then we quickly prove

$$\begin{aligned} \text{index } ST &= \text{index } ST + \text{index } \mathbf{1} \\ &= \text{index } (ST \oplus \mathbf{1}) \\ &= \text{index } (T \oplus S) \\ &= \text{index } T + \text{index } S. \qquad \text{Q.E.D.} \end{aligned} \tag{11.58}$$

Elliptic differential operator: In physics we mainly work with differential operators. Mathematically we can describe a differential operator D in a very general way as acting between the spaces of sections of complex vector bundles E, F over a compact manifold M without boundary. In this sense D is a linear map on the sections Γ of the bundles

$$D : \Gamma(M, E) \to \Gamma(M, F), \tag{11.59}$$

which has the following form

$$D = \sum_{|\alpha| \leq N} a^{\alpha}(x) \frac{\partial^{|\alpha|}}{\partial x^{\alpha}}, \tag{11.60}$$

where α denotes a multi-index

$$\begin{aligned} \alpha &= (\alpha_1, \ldots, \alpha_m), \qquad \alpha_i \in \mathbf{Z} \text{ and } \alpha_i \geq 0 \\ |\alpha| &= \alpha_1 + \ldots + \alpha_m. \end{aligned} \tag{11.61}$$

The derivative means

$$\frac{\partial^{|\alpha|}}{\partial x^{\alpha}} = \left(\frac{\partial}{\partial x^1}\right)^{\alpha_1} \left(\frac{\partial}{\partial x^2}\right)^{\alpha_2} \cdots \left(\frac{\partial}{\partial x^m}\right)^{\alpha_m} \tag{11.62}$$

and N is the order of the differential operator D.

For a particular section $s(x) \in \Gamma(M, E)$ of the bundle E—all sections in dom D are actually smooth functions, $\Gamma(M, E) \equiv C^{\infty}(E)$—the local action of D on s is

$$(Ds(x))_i = \sum_{|\alpha| \leq N} a^{\alpha}_{ij}(x) \frac{\partial^{|\alpha|}}{\partial x^{\alpha}} s_j(x). \tag{11.63}$$

Note that x denotes the point in the manifold M and its coordinate as well. The coefficients $a^{\alpha}_{ij}(x)$ are smooth $k \times k$ matrices depending on x and $1 \leq i, j \leq k$ with $k = \dim E = \dim F$ the bundle dimensions.

We are mainly interested in the case of the Dirac operator, $N = 1$, and in the Laplacian, $N = 2$. For example, if E is the spin bundle over M, the (massless) Dirac operator is

$$D = i\gamma^{\mu}\partial_{\mu} \tag{11.64}$$

and acts on a section $\psi(x)$ of E as

$$(D\psi(x))_i = i(\gamma^{\mu})_{ij}\partial_{\mu}\psi_j(x). \tag{11.65}$$

Symbol: The **symbol** σ of the differential operator D is defined by

$$\sigma[D](x, \xi) = \sum_{|\alpha| \leq N} a^{\alpha}(x)\xi^{\alpha}. \tag{11.66}$$

It is a polynomial of order N in the dual variable ξ, where ξ^α means

$$\xi^\alpha = \xi_1^{\alpha_1}\xi_2^{\alpha_2}\dots\xi_m^{\alpha_m}. \tag{11.67}$$

The pair (x, ξ) is regarded as the coordinates in the cotangent bundle $T^*(M)$.

It is also customary to describe the symbol by the Fourier transform of the operator; in this case it is more convenient to replace

$$\frac{\partial^{|\alpha|}}{\partial x^\alpha} \rightarrow \frac{1}{i^{|\alpha|}}\frac{\partial^{|\alpha|}}{\partial x^\alpha}, \tag{11.68}$$

then we have

$$\begin{aligned} Ds(x) &= \int d\xi \sum_{|\alpha|\leq N} a^\alpha(x)\frac{1}{i^{|\alpha|}}\frac{\partial^{|\alpha|}}{\partial x^\alpha}\widetilde{s}(\xi)e^{ix\xi} \\ &= \int d\xi\ \sigma[D](x,\xi)\widetilde{s}(\xi)e^{ix\xi}. \end{aligned} \tag{11.69}$$

The **leading symbol** is the highest order part of the polynomial (11.66)

$$\sigma_L[D](x,\xi) = \sum_{|\alpha|=N} a^\alpha(x)\xi^\alpha \tag{11.70}$$

and is homogeneous and of order N in ξ.

Ellipticity: The differential operator D is termed **elliptic** if its leading symbol is invertible for *all* (x, ξ) with $\xi \neq 0$.

For example, the **Laplace operator** defined by

$$\Delta := \left(\frac{\partial}{\partial x^1}\right)^2 + \ldots + \left(\frac{\partial}{\partial x^m}\right)^2 \tag{11.71}$$

has a symbol

$$\sigma[\Delta](\xi) = \xi^2 = \xi_1^2 + \ldots + \xi_m^2 \tag{11.72}$$

and is clearly an elliptic operator. More generally, let D be a differential operator of order 2 with real coefficients, acting between two trivial line bundles; then the leading symbol is

$$\sigma_L[D](x,\xi) = a_{11}(x)\xi_1^2 + (a_{12}(x) + a_{21}(x))\xi_1\xi_2 + a_{22}(x)\xi_2^2. \tag{11.73}$$

The operator D is elliptic if $\forall\, x \in M$ the leading symbol $\sigma_L[D]$ is invertible—this is the case if the curve

$$\sigma_L[D](x,\xi) = 1 \tag{11.74}$$

describes an ellipse in the ξ_1, ξ_2 plane $\forall\, x \in M$.

Another example is the Dirac operator (11.64). Its leading symbol is

$$\sigma_L[D](x,\xi) = \gamma^\mu \xi_\mu = \not{\xi}. \tag{11.75}$$

From

$$\not{\xi}^2 = \xi^2 \cdot \mathbf{1} \tag{11.76}$$

we find that $\not{\xi}$ is invertible in Euclidean space

$$\not{\xi}^{-1} = \frac{\not{\xi}}{\xi^2}. \tag{11.77}$$

In contrast, the **d'Alembert operator**

$$\Box := \left(\frac{\partial}{\partial x^1}\right)^2 + \ldots + \left(\frac{\partial}{\partial x^{m-1}}\right)^2 - \left(\frac{\partial}{\partial x^m}\right)^2 \tag{11.78}$$

is not elliptic—but **hyperbolic**—since its (leading) symbol

$$\sigma_L[D](\xi) = \xi_1^2 + \ldots + \xi_{m-1}^2 - \xi_m^2 \tag{11.79}$$

vanishes for

$$\xi_m^2 = \xi_1^2 + \ldots + \xi_{m-1}^2 \tag{11.80}$$

and is therefore not invertible on this cone.

The same is true for the Dirac operator in Minkowski space (see the previous example).

Note: Elliptic differential operators, like the Laplacian (11.71), are clearly unbounded with respect to the usual L^2-norm but are bounded with respect to Sobolev norms (see e.g. [Gilkey 1984]).

Theorem: An elliptic differential operator over a compact boundaryless manifold is a Fredholm operator (in the sense of a suitable Sobolev norm).

This is an important theorem since now the whole previous index discussion is applicable to elliptic operators which we are interested in in physics. The elliptic operators play a central role in the topological view of index theory, which we shall see in the next sections.

Weyl operator: The former Dirac operator $\not{D}$ (5.2) in Euclidean space, equation (5.5), is an elliptic operator. Thus $\not{D}$ has an index but

$$\text{index } \not{D} = 0 \tag{11.81}$$

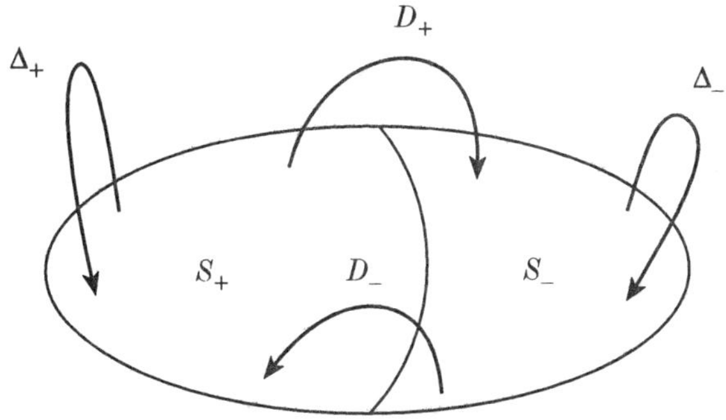

Fig. 11.2. The Weyl operators D_+, D_- and the Laplacians Δ_+, Δ_- acting on the positive and negative chirality spinor spaces S_+, S_-

since $\not{D}$ is selfadjoint, $\not{D} = \not{D}^\dagger$. Therefore $\not{D}$ itself is not interesting for index theory and we have to consider a different operator. We introduce the **Weyl operators**

$$D_+ := \not{D}P_+, \qquad D_- := \not{D}P_-, \tag{11.82}$$

with the chirality projection operators $P_\pm = \frac{1}{2}(\mathbf{1} \pm \gamma_5)$. The **adjoints** are

$$D_+^\dagger = \not{D}P_- = D_-, \qquad D_-^\dagger = \not{D}P_+ = D_+. \tag{11.83}$$

Separating the space of positive and negative chirality spinors

$$S_+ = \{\varphi_+ = P_+\varphi\}, \qquad S_- = \{\varphi_- = P_-\varphi\}, \tag{11.84}$$

where the φ's are the eigenfunctions of $\not{D}$, then the Weyl operators act on the spaces (see Figure 11.2)

$$S_+ \underset{D_-}{\overset{D_+}{\rightleftarrows}} S_- \tag{11.85}$$

since

$$D_+\varphi_+ = \not{D}P_+P_+\varphi = \not{D}P_+\varphi = P_-\,\not{D}\varphi = \lambda P_-\varphi = \lambda\varphi_- \in S_-. \tag{11.86}$$

The **Laplacians**, on the other hand,

$$\begin{aligned} \Delta_+ &= D_+^\dagger D_+ = D_-D_+ \\ \Delta_- &= D_-^\dagger D_- = D_+D_- \end{aligned} \tag{11.87}$$

act on (see Figure 11.2)

$$S_+ \xrightarrow{\Delta_+} S_+, \qquad S_- \xrightarrow{\Delta_-} S_-. \tag{11.88}$$

The **index of the Weyl operator** D_+

$$\begin{aligned} \text{index } D_+ &= \dim\ker D_+ - \dim\ker D_+^\dagger \\ &= \dim\ker D_+ - \dim\ker D_- \\ &= n_+ - n_- \end{aligned} \tag{11.89}$$

is the difference of the number of (linearly independent) zero-modes of positive and negative chirality, which we have already met in the Jacobian of the Fujikawa procedure, equation (11.11).

11.3 Heat kernel and index

There is an interesting connection between the index of a differential operator and the heat kernel of its Laplacian (see e.g. [Gilkey 1984], [Berline, Getzler, Vergne 1992]).

Let D be an **elliptic differential operator** (11.60) of order N and D^* its adjoint

$$C^\infty(E) \underset{D^*}{\overset{D}{\rightleftarrows}} C^\infty(F) \tag{11.90}$$

acting on the sections C^∞ of the vector bundles E, F over a compact, boundaryless manifold M.

Laplacians: We construct two **Laplacians**

$$\begin{aligned} \Delta_+ &:= D^*D \\ \Delta_- &:= DD^* \end{aligned} \tag{11.91}$$

acting on

$$C^\infty(E) \xrightarrow{\Delta_+} C^\infty(E), \qquad C^\infty(F) \xrightarrow{\Delta_-} C^\infty(F). \tag{11.92}$$

They are elliptic too and Hermitian (selfadjoint)

$$\begin{aligned} \Delta_+^* &= (D^*D)^* = D^*D = \Delta_+ \\ \Delta_-^* &= (DD^*)^* = DD^* = \Delta_-. \end{aligned} \tag{11.93}$$

Since the manifold is compact the spectrum of $\Delta_\pm$ is *discrete* and the degeneracy of each eigenvalue is *finite*. So the Laplacians $\Delta_\pm$ have a well-defined eigenvalue problem. Suppose the eigenvalue equation

$$\Delta_+\phi_\lambda = D^*D\phi_\lambda = \lambda\phi_\lambda, \tag{11.94}$$

then

$$D\phi_\lambda = \psi_\lambda \tag{11.95}$$

is an eigenfunction of Δ_- with the same eigenvalue

$$\Delta_-\psi_\lambda = DD^*\psi_\lambda = DD^*D\phi_\lambda = \lambda D\phi_\lambda = \lambda\psi_\lambda. \tag{11.96}$$

The spectrum for the nonzero eigenvalues ($\lambda \neq 0$) of Δ_+ and Δ_- is identical. Denoting the **spaces of the eigenfunctions** of Δ_+ and Δ_- by

$$E_+(\lambda) = \{\phi_\lambda\}, \qquad E_-(\lambda) = \{\psi_\lambda\} \tag{11.97}$$

we have (for $\lambda \neq 0$)

$$\dim E_+(\lambda) = \dim E_-(\lambda) \tag{11.98}$$

and the operators act on

$$E_+(\lambda) \underset{D^*}{\overset{D}{\rightleftarrows}} E_-(\lambda) \tag{11.99}$$

$$E_+(\lambda) \xrightarrow{\Delta_+} E_+(\lambda), \qquad E_-(\lambda) \xrightarrow{\Delta_-} E_-(\lambda). \tag{11.100}$$

However, this is *not* true for the zero-modes, $\lambda = 0$. But the zeros of Δ_+ and D, respectively Δ_- and D^*, are the same.

Lemma:

$$\begin{aligned} \ker\Delta_+ &= \ker D \\ \ker\Delta_- &= \ker D^*. \end{aligned} \tag{11.101}$$

Proof.

$$\begin{aligned} \ker\Delta_+ &= \{f|\Delta_+ f = 0\} \\ \ker D &= \{f|Df = 0\} \end{aligned} \tag{11.102}$$

i) $\Delta_+ f = D^*Df$

$$\text{if } Df = 0 \Rightarrow \Delta_+ f = 0 \tag{11.103}$$

ii) if $f \in \ker\Delta_+$, $\Delta_+ f = D^*Df = 0$

$$\begin{aligned} &\Rightarrow \quad 0 = (D^*Df, f) = (Df, Df) \\ &\Rightarrow \quad Df = 0, \qquad f \in \ker D. \qquad \text{Q.E.D.} \end{aligned} \tag{11.104}$$

Now we are prepared to establish the following important theorem.

Theorem: Index and heat kernel

$$\text{index } D = \text{Tr}_{E_+} e^{-t\Delta_+} - \text{Tr}_{E_-} e^{-t\Delta_-} \quad , \quad \forall t > 0. \tag{11.105}$$

Proof. We compute the traces

$$\begin{aligned}
&\text{Tr}_{E_+} e^{-t\Delta_+} - \text{Tr}_{E_-} e^{-t\Delta_-} \\
&= \sum_{\lambda,\phi_\lambda} \langle \phi_\lambda | e^{-t\Delta_+} | \phi_\lambda \rangle - \sum_{\lambda,\psi_\lambda} \langle \psi_\lambda | e^{-t\Delta_-} | \psi_\lambda \rangle \\
&= \sum_\lambda e^{-t\lambda} \left[\sum_{\phi_\lambda} \langle \phi_\lambda | \phi_\lambda \rangle - \sum_{\psi_\lambda} \langle \psi_\lambda | \psi_\lambda \rangle \right] \\
&= \sum_\lambda e^{-t\lambda} [\dim E_+(\lambda) - \dim E_-(\lambda)].
\end{aligned} \tag{11.106}$$

Due to equality (11.98) only the zero-modes remain, with the property (11.101) leading to the index (11.32)

$$\begin{aligned}
\text{Tr}_{E_+} e^{-t\Delta_+} - \text{Tr}_{E_-} e^{-t\Delta_-} &= e^{t\cdot 0}[\dim E_+(0) - \dim E_-(0)] \\
&= \dim\ker \Delta_+ - \dim\ker \Delta_- \\
&= \dim\ker D - \dim\ker D^* \\
&= \text{index } D. \qquad \text{Q.E.D.}
\end{aligned} \tag{11.107}$$

Weyl operator: Specifically, for our previous Dirac operator projected to a given chirality—the Weyl operator $D_\pm$ (11.82)—we obtain:

Theorem: Index and heat kernel

$$\text{index } D_+ = \text{Tr}_S \gamma_5 \, e^{-t\not{D}^2}, \qquad \forall t > 0. \tag{11.108}$$

Proof.

$$\begin{aligned}
\text{index } D_+ &= \text{Tr}_{S_+} \, e^{-tD_-D_+} - \text{Tr}_{S_-} \, e^{-tD_+D_-} \\
&= \text{Tr}_{S_+} \, e^{-t\not{D}^2} P_+ - \text{Tr}_{S_-} \, e^{-t\not{D}^2} P_- \\
&= \text{Tr}_{S=S_+\oplus S_-} \, e^{-t\not{D}^2}(P_+ - P_-) \\
&= \text{Tr}_S \gamma_5 \, e^{-t\not{D}^2}. \qquad \text{Q.E.D.}
\end{aligned} \tag{11.109}$$

Remembering Fujikawa's anomaly procedure with path integrals (Section 5.2)—where the regularized sum in the PI Jacobian describes the index—we notice that the index–heat kernel formula (11.108) corresponds precisely to Fujikawa's Gaussian regularization choice! So we just have to expand the heat kernel into Seeley coefficients, equation (5.159), pick up the t-independent part and perform the limit $t \to 0$.

Example: In 2 dimensions we get

$$\mathrm{Tr}_S \gamma_5 \, e^{-t\not{D}^2} = \mathrm{Tr}_S \gamma_5 G_{\not{D}^2} = \int dx \; \mathrm{tr} \; \gamma_5 G_{\not{D}^2}(x,x,t)$$

$$= \int dx \frac{1}{4\pi} \; \mathrm{tr} \; \gamma_5 a_1(x,x) = -\frac{1}{4\pi} \int dx \; \varepsilon_{\mu\nu} F^{\mu\nu}, \tag{11.110}$$

where we have used equations (5.167)–(5.169) together with the convential ε-tensor definition $\varepsilon_{14} \equiv \varepsilon_{12} = 1$ for Euclidean space.

So we find

$$\text{index } D_+ = -\frac{1}{4\pi} \int dx \; \varepsilon^{\mu\nu} F_{\mu\nu} = -\frac{1}{2\pi} \int_{S^2} F, \tag{11.111}$$

which represents the **Atiyah–Singer index theorem** in two dimensions. We learn this in the next section.

11.4 Atiyah–Singer index theorem

Now it is time to discuss more precisely Atiyah and Singer's index theorem (AS) [Atiyah, Singer 1968a,b,c], [Atiyah, Singer 1971a,b], which we have met already in the preceding sections. Up to now we have studied the so-called analytic index (11.32) of an elliptic differential operator. Atiyah and Singer have shown that this analytic index equals an index which is fully determined by topology—called the **topological index** for this reason. This topological index is a topological invariant depending only on the fibre bundles considered and on the manifold. Moreover, this topological invariant can be expressed as an integral over certain **characteristic classes**. These characteristic classes can be found explicitly for a given differential operator.

Chern character: In our case of the Dirac operator containing the YM gauge potential $A = A^a_\mu dx^\mu \; T^a$ the characteristic class is given by the **Chern character**

$$\begin{aligned} \mathrm{ch}(F) &= \mathrm{tr} \; \exp\left[\frac{i}{2\pi} F\right] \\ &= r + \frac{i}{2\pi} \; \mathrm{tr} \; F + \frac{1}{2!}\left(\frac{i}{2\pi}\right)^2 \; \mathrm{tr} \; F^2 + \ldots, \end{aligned} \tag{11.112}$$

where r is the dimension of the group and F is the curvature 2-form,

$$F = dA + A^2. \tag{11.113}$$

Theorem: Atiyah–Singer index theorem

$$\text{index } D_+ = \int_{M_{2n}} \text{ch}(F). \tag{11.114}$$

D_+ is the Weyl operator (11.82). The integral is taken over the compact manifold M_{2n} with dimension $2n$ so that we pick up the n-th term only

$$\text{index } D_+ = \frac{1}{n!}\left(\frac{i}{2\pi}\right)^n \int_{M_{2n}} \text{tr } F^n. \tag{11.115}$$

We have given a proof via the heat kernel or Fujikawa procedure before.

So the AS index theorem is an interesting connection between the *local* properties of a differential operator—the local information about the solutions of a differential equation—and its *global* properties on a manifold.

Physics: Let us return to physics.

i) *Instantons*
The YM fields compactified on a sphere S^4—the instantons—correspond geometrically to an $SU(2)$ principal bundle $P(SU(2), S^4)$ over the sphere S^4 (recall our discussion in Section 6.6). The **index of the $P(SU(2), S^4)$ bundle** is then

$$\text{index } D_+ = \frac{1}{2!}\left(\frac{i}{2\pi}\right)^2 \int_{S^4} \text{tr } F^2 = -\frac{1}{8\pi^2}\int_{S^4} \text{tr } F^2 \equiv q. \tag{11.116}$$

We recover the Pontrjagin index or topological charge q (or the instanton number) which we have discussed already in Section 6.6.

ii) *Anomaly*
Recalling the relation between the anomaly and the index, equation (11.20), we obtain the **anomaly from the index density**

$$\begin{aligned} d * j^5 = \mathcal{A}[A_\mu] = 2i \cdot \text{Index } D_+ &= -i\frac{1}{4\pi^2} \text{ tr } F^2 \quad \text{in 4 dim} \\ &= -i\frac{1}{\pi}F \quad \text{in 2 dim.} \end{aligned} \tag{11.117}$$

(The factor $-i$ is removed in Minkowski space.)

iii) *Dirac monopole*
The Dirac monopole that we studied in Section 6.4 corresponds geometrically to a $U(1)$ principal bundle $P(U(1), S^2)$ over the sphere S^2. Remembering that the curvature in the $U(1)$ case is pure imaginary $F \to iF$, the **index of the $P(U(1), S^2)$ bundle** is given by

$$\text{index } D_+ = \frac{i}{2\pi}\int_{S^2} \text{tr } iF = -\frac{1}{2\pi}\int_{S^2} F = -n. \tag{11.118}$$

It represents the quantized monopole charge n.

Remark: We again emphasize the pure topological content of the AS index theorem. The integral (11.115) is independent of the chosen connection A, it depends only—which we have demonstrated in Chapter 6—on the transition functions of the underlying bundle.

We find the remarkable result that basic features of QFT, like anomalies or instanton- and monopole charges, are determined by a global, topological quantity—the AS index theorem.

11.5 Non-Abelian anomaly and generalized index theorem

Again, we return to the fermionic path integral. Alvarez-Gaumé and Ginsparg (AGG) [Alvarez-Gaumé, Ginsparg 1984, 1985] have introduced a special Dirac operator $\widehat{D}$ in order to calculate the non-Abelian anomaly. This operator $\widehat{D}$ can be used either for regularizing the Jacobian of the fermionic PI or—what interests us most—for determining the anomaly from an index theorem.

11.5.1 Differential operator $\widehat{D}$

Let us start with the usual non-Abelian Lagrangian (see e.g. equation (5.124))

$$\mathcal{L}_{na} = \bar{\psi}_+ i \not{D}_+ \psi_+, \tag{11.119}$$

where we consider only the massless ($m = 0$), positive chirality part. The Dirac operator

$$\not{D}_+ = \not{\partial} + \not{A}_+ \tag{11.120}$$

contains only the positive chirality gauge potential (recall equation (5.123)); from now on we suppress the indication + in the gauge potential. We rewrite the Lagrangian (11.119) in terms of the Weyl operator (11.82) $D_+ = \not{D} P_+$

$$\mathcal{L}_{na} = \bar{\psi} P_- i \not{D} P_+ \psi = \bar{\psi} i \not{D} P_+ \psi = \bar{\psi} i D_+ \psi. \tag{11.121}$$

The Euclidean convention we choose here is the following

$$\begin{aligned}
\gamma^{\mu\dagger} &= \gamma^\mu \qquad \text{Hermitian} \\
\gamma_5 &= i^n \prod_{\mu=1}^{2n} \gamma^\mu \\
\gamma_5^\dagger &= \gamma_5, \qquad (\gamma_5)^2 = \mathbf{1}
\end{aligned} \tag{11.122}$$

with the metric

$$g^{\mu\nu} = \delta^{\mu\nu}. \tag{11.123}$$

(Note that this choice differs from our convention in Chapter 5.)

The gauge group generators T^a satisfy, as before, the commutation relations (3.275) and they are anti-Hermitian $T^{a\dagger} = -T^a$. Hence the operator $i\,\not{D}$ becomes Hermitian

$$(i\,\not{D})^\dagger = i\,\not{D} \tag{11.124}$$

and

$$\not{D} = \gamma^\mu D_\mu = g^{\mu\nu}\gamma_\nu D_\mu = \gamma_\mu D_\mu. \tag{11.125}$$

Now we consider the generating functional

$$Z[A_\mu] = e^{-W[A_\mu]} = \int d\bar{\psi}d\psi \exp\left[-\int dx\; \bar{\psi} iD_+ \psi\right] \stackrel{?}{=} \det iD_+. \tag{11.126}$$

The fermionic PI (again, the notation here differs from that in Chapter 5) *formally* gives a determinant. However, this determinant is *not* well-defined; we *cannot* identify the determinant with the product of the eigenvalues

$$\det iD_+ \stackrel{?}{=} \prod i\lambda_n \tag{11.127}$$

since the eigenvalue equation

$$iD_+\psi_n \stackrel{?}{=} \lambda_n \psi_n \tag{11.128}$$

is meaningless. The Weyl operator iD_+ has *no* well-defined eigenvalue problem—in contrast to the Dirac operator $i\,\not{D}$—since the operator iD_+ maps positive chirality spinors ψ_+ into negative chirality spinors ψ_-

$$S_+ \otimes V \xrightarrow{iD_+} S_- \otimes V. \tag{11.129}$$

$S_\pm = \{\psi_\pm\}$ are the sections of the spin bundle and V is a section of a vector bundle associated with the gauge group.

The way out proposed by Alvarez-Gaumé and Ginsparg [Alvarez-Gaumé, Ginsparg 1984, 1985] is to define a new differential operator.

Definition: Differential operator

$$\widehat{D} := \gamma^\mu(\partial_\mu + A_\mu P_+) = \partial_- + D_+ \tag{11.130}$$

or in a γ_5-diagonal basis $\begin{pmatrix} \psi_+ \\ \psi_- \end{pmatrix}$

$$\widehat{D} = \begin{pmatrix} 0 & \partial_- \\ D_+ & 0 \end{pmatrix}, \tag{11.131}$$

with $\partial_\pm = \not\partial P_\pm$.

Then the generating functional, the path integral, defines a determinant

$$\begin{aligned} Z_{\mathrm{AGG}}[A_\mu] &= e^{-W_{\mathrm{AGG}}[A_\mu]} \\ &= \int d\bar{\psi} d\psi \exp\left[-\int dx\, \bar{\psi} i\widehat{D}\psi\right] = \det i\widehat{D}. \end{aligned} \tag{11.132}$$

The operator $i\widehat{D}$ acts on Dirac fermions mapping the combined space of positive and negative chirality spinors into itself

$$(S_+ \oplus S_-) \otimes V \xrightarrow{i\widehat{D}} (S_+ \oplus S_-) \otimes V. \tag{11.133}$$

Hence $i\widehat{D}$ has a well-defined eigenvalue problem and the determinant exists. Furthermore $i\widehat{D} = i(\partial_- + D_+)$ has gauge couplings *only* to positive chirality spinors and it has only positive chirality zero-modes which coincide with those of iD_+. So a gauge theory arising from $i\widehat{D}$ is equivalent to a gauge theory from iD_+—the difference lies in an overall normalization constant that is independent of the gauge potential. In addition, the absolute value of the determinant turns out gauge invariant so that only the *phase* of the determinant may receive an anomalous contribution under gauge variation. This fact will become important when we relate the anomaly to the index, in Section 11.5.3.

Properties: Altogether we find the following $i\widehat{D}$ **operator properties**:

i) well-defined eigenvalue problem,
ii) gauge coupling only to positive chirality spinors,
iii) only positive chirality zero-modes contribute,
iv) $|\det i\widehat{D}|$ is gauge invariant.

Proof. We will prove property iv), which is important for our anomaly discussion. Let us consider

$$|\det i\widehat{D}|^2 = \det i\widehat{D} \det(i\widehat{D})^\dagger = \det(i\widehat{D}(i\widehat{D})^\dagger); \tag{11.134}$$

we decompose the operators as

$$i\widehat{D}(i\widehat{D})^\dagger = \begin{pmatrix} 0 & i\partial_- \\ iD_+ & 0 \end{pmatrix} \begin{pmatrix} 0 & iD_- \\ i\partial_+ & 0 \end{pmatrix} = \begin{pmatrix} i\partial_- i\partial_+ & 0 \\ 0 & iD_+ iD_- \end{pmatrix} \tag{11.135}$$

so that the determinant becomes

$$\begin{aligned}\det(i\widehat{D}(i\widehat{D})^\dagger) &= \det\begin{pmatrix} i\partial_- i\partial_+ & 0 \\ 0 & iD_+ iD_- \end{pmatrix} \\ &= \det(i\partial_- i\partial_+)\det(iD_+ iD_-).\end{aligned} \tag{11.136}$$

We relate the second determinant to the Hermitian Dirac operator (in a γ_5 diagonal basis)

$$i\not{D} = \begin{pmatrix} 0 & iD_- \\ iD_+ & 0 \end{pmatrix} \tag{11.137}$$

then we have

$$\begin{aligned}|\det i\not{D}|^2 &= \det(i\not{D})^2 \\ &= \det\begin{pmatrix} iD_- iD_+ & 0 \\ 0 & iD_+ iD_- \end{pmatrix} = [\det(iD_+ iD_-)]^2 \\ \det i\not{D} &= \det(iD_+ iD_-).\end{aligned} \tag{11.138}$$

Altogether we find

$$\begin{aligned}|\det i\widehat{D}|^2 &= \det(i\partial_- i\partial_+)\det i\not{D} \\ |\det i\widehat{D}| &= \text{const.}\cdot(\det i\not{D})^{1/2},\end{aligned} \tag{11.139}$$

where const. is some (irrelevant) gauge potential independent constant. The Dirac operator $i\not{D}$ has only real gauge invariant eigenvalues λ_n^{Dirac} (see our discussion below) so that its determinant

$$\det i\not{D} = \prod_n \lambda_n^{\text{Dirac}} \tag{11.140}$$

is gauge invariant. This also implies gauge invariance for $|\det i\widehat{D}|$. Q.E.D.

Lemma: The operator $i\widehat{D}$ is elliptic but not Hermitian.

Its eigenvalues are therefore complex satisfying the eigenvalue equations

$$\begin{aligned}i\widehat{D}\varphi_n(x) &= \lambda_n\varphi_n(x) \\ (i\widehat{D})^\dagger\chi_n(x) &= \lambda_n^*\chi_n(x) \quad \text{or} \quad \chi_n^\dagger i\overleftarrow{\widehat{D}} = \lambda_n\chi_n^\dagger.\end{aligned} \tag{11.141}$$

We choose the eigenfunctions to be orthonormal

$$\int dx\, \chi_n^\dagger(x)\varphi_m(x) = \delta_{nm}. \tag{11.142}$$

Proposition:

- *The spectrum of $i\widehat{D}$—the eigenvalues λ_n—is not gauge invariant!*

Proof. Under a gauge transformation the operator changes

$$\begin{aligned} gi\widehat{D}(A^g)g^{-1} &= gi\gamma^\mu(\partial_\mu + A^g_\mu P_+)g^{-1} \\ &= i\gamma^\mu(\partial_\mu + A_\mu P_+) + ig\ \partial\!\!\!/ g^{-1} + i\ \partial\!\!\!/ g g^{-1} P_+ \\ &= i\widehat{D}(A) + ig\ \partial\!\!\!/ g^{-1} - ig\ \partial\!\!\!/ g^{-1} P_+ \\ &\neq i\widehat{D}(A) \end{aligned} \tag{11.143}$$

(note: $\partial\!\!\!/ g g^{-1} + g\ \partial\!\!\!/ g^{-1} = 0$ from $gg^{-1} = \mathbf{1}$). Q.E.D.

In contrast, we have:

Proposition:

- *The spectrum of the Dirac operator $i\ D\!\!\!\!/$—the eigenvalues λ_n^{Dirac}—is gauge invariant!*

Proof.

$$\begin{aligned} gi\ D\!\!\!\!/(A^g)g^{-1} &= gi\gamma^\mu(\partial_\mu + A^g_\mu)g^{-1} \\ &= i\gamma^\mu(\partial_\mu + A_\mu) + ig\ \partial\!\!\!/ g^{-1} - ig\ \partial\!\!\!/ g^{-1} \\ &= i\ D\!\!\!\!/(A). \end{aligned} \tag{11.144}$$

Consequently every eigenfunction ψ_n of $i\ D\!\!\!\!/(A)$

$$i\ D\!\!\!\!/(A)\psi_n = \lambda_n^{\text{Dirac}}\psi_n \tag{11.145}$$

has an associated eigenfunction $g^{-1}\psi_n$ of $i\ D\!\!\!\!/(A^g)$ with the same eigenvalue

$$i\ D\!\!\!\!/(A^g)g^{-1}\psi_n = \lambda_n^{\text{Dirac}} g^{-1}\psi_n. \qquad \text{Q.E.D.} \tag{11.146}$$

Theorem: The **classical action** remains invariant

$$S_{\text{AGG}} = \int dx\ \bar{\psi} i\widehat{D}\psi \equiv S'_{\text{AGG}} \tag{11.147}$$

under **gauge transformation on the positive chirality components**

$$\begin{aligned} A_\mu &\rightarrow A'_\mu = A_\mu + D_\mu v \\ \psi &\rightarrow \psi' = \psi - v\psi_+ \\ \bar{\psi} &\rightarrow \bar{\psi}' = \bar{\psi} + \bar{\psi}_- v, \end{aligned} \tag{11.148}$$

where v denotes the FP ghost and $\psi_+ = P_+\psi$, $\bar{\psi}_- = \bar{\psi}P_-$.

Proof.

$$\begin{aligned}
\bar{\psi}' i\widehat{D}(A')\psi' &= \bar{\psi}' i\gamma^\mu(\partial_\mu + A'_\mu P_+)\psi' \\
&= \bar{\psi} i\widehat{D}(A)\psi + \bar{\psi} i\gamma^\mu(\partial_\mu v + A_\mu v - vA_\mu)P_+\psi \\
&\quad + \bar{\psi} i\gamma^\mu(v\partial_\mu + vA_\mu)P_+\psi - \bar{\psi} i\gamma^\mu(\partial_\mu v + v\partial_\mu + A_\mu v)P_+\psi \\
&= \bar{\psi} i\widehat{D}(A)\psi. \qquad \text{Q.E.D.} \qquad (11.149)
\end{aligned}$$

11.5.2 Anomalous Jacobian

As we discussed in Chapter 5 it is the PI measure which changes under the above gauge transformation (11.148)

$$d\bar{\psi}d\psi \to d\bar{\psi}'d\psi' = J[v]d\bar{\psi}d\psi. \qquad (11.150)$$

The Jacobian $J[v]$ provides precisely the non-Abelian anomaly, here the consistent version (8.59) satisfying the Wess–Zumino consistency condition (8.82).

Path integral: The explicit calculation is a variant of Fujikawa's method described in Section 5.2. We expand the spinors ψ, ψ' in terms of the eigenfunctions φ_n, χ_n of the operator $i\widehat{D}$ (recall equations (11.141))

$$\begin{aligned}
\psi(x) &= \sum_n a_n\varphi_n(x) \\
\bar{\psi}(x) &= \sum_n \chi_m^\dagger(x)\bar{b}_m. \qquad (11.151)
\end{aligned}$$

The Grassmann elements a_n, $\bar{b}_m$ can also be written by the inner products

$$a_n = (\chi_n^\dagger, \psi), \quad \bar{b}_m = (\bar{\psi}, \varphi_m) \qquad (11.152)$$

since, on the other hand, we have

$$\begin{aligned}
\psi(x) &= \langle x|\psi\rangle \\
&= \sum_n \langle x|n\rangle \int dy\, \langle n|y\rangle\langle y|\psi\rangle \\
&= \sum_n \varphi_n(x)(\chi_n^\dagger, \psi). \qquad (11.153)
\end{aligned}$$

We notice that

$$\langle x|n\rangle = \varphi_n(x) \qquad \text{and} \qquad \langle n|y\rangle = \chi_n^\dagger(y) \qquad (11.154)$$

are *not* Hermitian conjugate due to the non-Hermiticity of $i\widehat{D}$.

The PI measure and action are re-expressed by

$$d\bar{\psi}d\psi = \prod_n d\bar{b}_n da_n \tag{11.155}$$

$$S_{\mathrm{AGG}} = \int dx\ \bar{\psi}i\widehat{D}\psi = \sum_n \lambda_n \bar{b}_n a_n. \tag{11.156}$$

(Note that the gauge noninvariance of the eigenvalues λ_n is compensated by the noninvariance of a_n, $\bar{b}_n$ to finally achieve a gauge invariant action S_{AGG}.) Then the generating functional is given by the following Grassmann integral

$$\begin{aligned} Z_{\mathrm{AGG}}[A_\mu] &= e^{-W_{\mathrm{AGG}}[A_\mu]} = \int d\bar{\psi}d\psi \exp\left[-\int dx\ \bar{\psi}i\widehat{D}\psi\right] \\ &= \int \prod_n d\bar{b}_n da_n \exp\left[-\sum_n \lambda_n \bar{b}_n a_n\right] \\ &= \prod_n \lambda_n = \det i\widehat{D}. \end{aligned} \tag{11.157}$$

This infinite product of eigenvalues makes sense—as usual—in terms of the ζ-function regularization (recall equation (5.175))

$$\det{}^R i\widehat{D} = \exp[-\zeta'_{i\widehat{D}}(0)]. \tag{11.158}$$

Jacobian: We calculate the Jacobian $J[v]$ à la Fujikawa. We study the Grassmann elements under the above restricted gauge transformation (11.148)

$$\begin{aligned} a'_n &= \sum_m C_{nm} a_m \\ \bar{b}'_m &= \sum_n D_{nm} \bar{b}_n, \end{aligned} \tag{11.159}$$

with

$$\begin{aligned} C_{nm} &= \delta_{nm} - \int dx\ \chi_n^\dagger(x) v(x) P_+ \varphi_m(x) \\ D_{nm} &= \delta_{nm} + \int dx\ \chi_n^\dagger(x) P_- v(x) \varphi_m(x), \end{aligned} \tag{11.160}$$

which leads to the **Jacobian**

$$\begin{aligned} J[v] &= [\det C \cdot \det D]^{-1} \\ &= \exp\left[\int dx \sum_n \chi_n^\dagger(x) v(x)(P_+ - P_-)\varphi_n(x)\right] \\ &= \exp\left[\int dx \sum_n \chi_n^\dagger(x) v(x)\gamma_5 \varphi_n(x)\right]. \end{aligned} \tag{11.161}$$

Regularization: We regularize the sum with a Gaussian cut-off

$$\sum_n \chi_n^\dagger v \gamma_5 \varphi_n = \lim_{M\to\infty} \sum_n \chi_n^\dagger v\gamma_5 \exp\left[-\frac{\lambda_n^2}{M^2}\right]\varphi_n \tag{11.162}$$

$$= \lim_{M\to\infty}\lim_{y\to x} \mathrm{Tr}\ v\gamma_5 \exp\left[-\frac{(i\widehat{D})^2}{M^2}\right]\delta(x-y).$$

Noting that we have in the γ_5-diagonal basis

$$(i\widehat{D})^2 = \begin{pmatrix} i\partial_- iD_+ & 0 \\ 0 & iD_+ i\partial_- \end{pmatrix} \tag{11.163}$$

so that the trace with respect to this chiral basis is

$$\mathrm{Tr}_{\mathbf{chiral}}(i\widehat{D})^2 = i^2(\partial_- D_+ + D_+\partial_-) = -(\not\partial \not{D} P_+ + \not{D}\not\partial P_-), \tag{11.164}$$

we rewrite the total trace Tr (11.162)

$$\begin{aligned} \mathrm{Tr}\ v\gamma_5 \exp\left[-\frac{(i\widehat{D})^2}{M^2}\right] &= \mathrm{Tr}\ v(P_+ - P_-)\exp\left[\frac{\partial_- D_+ + D_+\partial_-}{M^2}\right] \\ &= \mathrm{Tr}\ v\left(P_+ \exp\left[\frac{\not\partial\not{D}}{M^2}\right] - P_- \exp\left[\frac{\not{D}\not\partial}{M^2}\right]\right). \end{aligned} \tag{11.165}$$

As usual we evaluate the trace in the plane–wave basis

$$\lim_{M\to\infty}\lim_{y\to x}\mathrm{Tr}\ vP_+\exp\left[\frac{\not\partial\not{D}}{M^2}\right]\delta(x-y) \tag{11.166}$$

$$= \lim_{M\to\infty}\int \frac{d^4k}{(2\pi)^4}\mathrm{Tr}\ ve^{-ikx}P_+\exp\left[\frac{\not\partial\not{D}}{M^2}\right]e^{ikx}$$

$$= \lim_{M\to\infty} M^4 \int \frac{d^4k}{(2\pi)^4} e^{-k_\mu k_\mu}\mathrm{Tr}\ vP_+ \exp\left[\frac{\not\partial\not{D}}{M^2} + \frac{i(\not{k}\not{D} + \not\partial\not{k})}{M}\right]$$

where the plane waves have shifted the differential operators (recall equation (5.61)) and we have also rescaled the momentum $k_\mu \to Mk_\mu$; Tr now means the trace over the Dirac matrices and over the gauge group generators. We treat the second term of the trace (11.165) analogously.

Next we expand the exponential; we use the trace properties of the γ-matrices in our Euclidean convention (11.122), (11.123)

$$\begin{aligned} \mathrm{tr}\ \gamma_5 &= \mathrm{tr}\ \gamma_5\gamma^\mu\gamma^\nu = 0 \\ \mathrm{tr}\ \gamma_5\gamma^\mu\gamma^\nu\gamma^\alpha\gamma^\beta &= 4\varepsilon^{\mu\nu\alpha\beta} \end{aligned} \tag{11.167}$$

(with $\varepsilon^{1234} = \varepsilon^{1230} = 1$); then only terms proportional to $1/M^4$ survive.

We need the integrals

$$\begin{aligned} \int d^4k e^{-k_\mu k_\mu} &= \pi^2 \\ \int d^4k e^{-k_\mu k_\mu} k_\mu k_\nu &= \frac{\pi^2}{2} g_{\mu\nu} \\ \int d^4k e^{-k_\mu k_\mu} k_\mu k_\nu k_\alpha k_\beta &= \frac{\pi^2}{4}(g_{\mu\nu}g_{\alpha\beta} + g_{\mu\alpha}g_{\nu\beta} + g_{\mu\beta}g_{\nu\alpha}). \end{aligned} \tag{11.168}$$

In the expansion we keep only the negative parity terms, the terms proportional to $\varepsilon^{\mu\nu\alpha\beta}$. The positive parity terms can be absorbed by renormalization of the generating functional.

Anomaly: Then we find after a little calculation (for computational details see e.g. [Adam 1990]) the **non-Abelian anomaly** $v^a G^a[A_\mu]$

$$\begin{aligned} &\lim_{M\to\infty}\lim_{y\to x} \mathrm{Tr}\ v\gamma_5 \exp\left[-\frac{(i\widehat{D})^2}{M^2}\right]\delta(x-y) \\ &= \frac{1}{24\pi^2}\varepsilon^{\mu\nu\alpha\beta}\ \mathrm{tr}\ v\partial_\mu\left(A_\nu\partial_\alpha A_\beta + \frac{1}{2}A_\nu A_\alpha A_\beta\right) \\ &= -v^a G^a[A_\mu]. \end{aligned} \tag{11.169}$$

So the **Jacobian** $J[v]$ contains precisely the **anomaly** $G(v, A)$

$$\begin{aligned} J[v] &= \exp\left[\int dx \lim_{M\to\infty}\lim_{y\to x} \mathrm{Tr}\ v\gamma_5 \exp\left[-\frac{(i\widehat{D})^2}{M^2}\right]\delta(x-y)\right] \\ &= \exp\left[\frac{1}{24\pi^2}\int dx \varepsilon^{\mu\nu\alpha\beta}\ \mathrm{tr}\ v\partial_\mu\left(A_\nu\partial_\alpha A_\beta + \frac{1}{2}A_\nu A_\alpha A_\beta\right)\right] \\ &= \exp[-G(v, A)]. \end{aligned} \tag{11.170}$$

On the other hand, the Jacobian relates the generating functionals (where we can now use $Z[A_\mu]$ (11.126) instead of $Z_{\mathrm{AGG}}[A_\mu]$ (11.157) since both differ only by an overall constant which cancels)

$$Z[A_\mu^v] = J[v]Z[A_\mu], \tag{11.171}$$

which we can rewrite in terms of the 'quantum action' $W[A_\mu] = -\ln Z[A_\mu]$

$$\begin{aligned}\delta W[A_\mu] &= -\frac{1}{Z[A_\mu]}(Z[A_\mu^v] - Z[A_\mu]) \\ &= -J[v] + 1 \\ &= -\frac{1}{24\pi^2}\int \mathrm{tr}\; vd\left(AdA + \frac{1}{2}A^3\right) \\ &= G(v, A).\end{aligned} \tag{11.172}$$

So we produce the consistent, **non-Abelian anomaly** (note again, $\delta W \equiv \delta W_{\mathrm{AGG}}$)—Bardeen's result (6.79)—or in other terms the **anomalous Ward identity** defined by

$$\begin{aligned}\delta W[A_\mu] &= \int dx\; \delta A_\mu^b(x)\frac{\delta}{\delta A_\mu^b(x)}W[A_\mu] \\ &= -\int dx\; v^a D_\mu^{ba}\langle j^{\mu b}\rangle,\end{aligned} \tag{11.173}$$

with

$$\begin{aligned}\langle j^{\mu b}\rangle &= \frac{\delta}{\delta A_\mu^b}W[A_\mu] \\ &= \frac{\int d\bar{\psi}d\psi\; \bar{\psi}\gamma^\mu iT^b P_+\psi \exp\left[-\int dx\; \bar{\psi}iD_+\psi\right]}{\int d\bar{\psi}d\psi \exp\left[-\int dx\; \bar{\psi}iD_+\psi\right]}.\end{aligned} \tag{11.174}$$

The result (11.172) is also correct in Minkowski space, since there the current is defined without the factor i (see equation (4.46)). If we consider negative chirality fields the sign is reversed in result (11.172).

11.5.3 Alvarez-Gaumé and Ginsparg's index procedure

Atiyah and Singer [Atiyah, Singer 1984] and Alvarez, Singer and Zumino [Alvarez, Singer, Zumino 1984] (in case of gravitation) have discovered that the non-Abelian anomaly is related to a more refined index theorem—the **family index theorem** (see also [Sumitani 1984]). However, to establish this theorem one has to know the mathematical discipline of K-theory and the theory of index bundles. For a review of this field, see e.g. [Berline, Getzler, Vergne 1992], [Lawson, Michelsohn 1989], [Nash 1991], [Tröster 1994]. A 'physicist's approach' to determining the non-Abelian anomaly from an index theorem, avoiding these further mathematical techniques, has been proposed by Alvarez-Gaumé and Ginsparg [Alvarez-Gaumé, Ginsparg 1984, 1985]. They relate the anomaly in $2n$ dimensions to the AS index theorem in

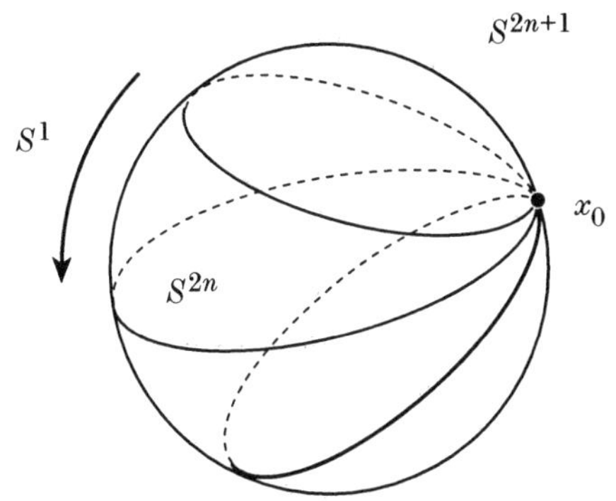

Fig. 11.3. Topological equivalence: $S^1 \times S^{2n} \sim S^{2n+1}$

a $(2n+2)$-dimensional space. We are going to explain this procedure—which somehow mimics the family index theorem.

Space of gauge potentials: We consider the $2n$-dimensional Euclidean space $\mathbf{R}^{2n}$ compactified to a sphere S^{2n}. We choose a semi-simple gauge group G, a simply connected compact Lie group; e.g. $SU(N)$ where the first homotopy group vanishes $\Pi_1(SU(N)) = 0$. We extend the space by working with a **one-parameter family of gauge elements**

$$g(\theta, x) : S^1 \times S^{2n} \to G \qquad \forall\, \theta \in S^1,\ x \in S^{2n} \tag{11.175}$$

with boundary condition

$$g(0, x_0) = g(2\pi, x_0) = \mathbf{1}. \tag{11.176}$$

Due to this boundary condition the space $S^1 \times S^{2n}$ looks topologically like the sphere S^{2n+1}, see Figure 11.3,

$$S^1 \times S^{2n} \sim S^{2n+1}. \tag{11.177}$$

So g is regarded as a map

$$g(\theta, x) : S^{2n+1} \to G \tag{11.178}$$

classified by the homotopy group (recall Section 2.2)

$$\Pi_{2n+1}(G) = \mathbf{Z}. \tag{11.179}$$

Specifically, we have $G = SU(N)$ with $N \geq n+1$.

Next we choose, without loss of generality, some reference gauge connec-

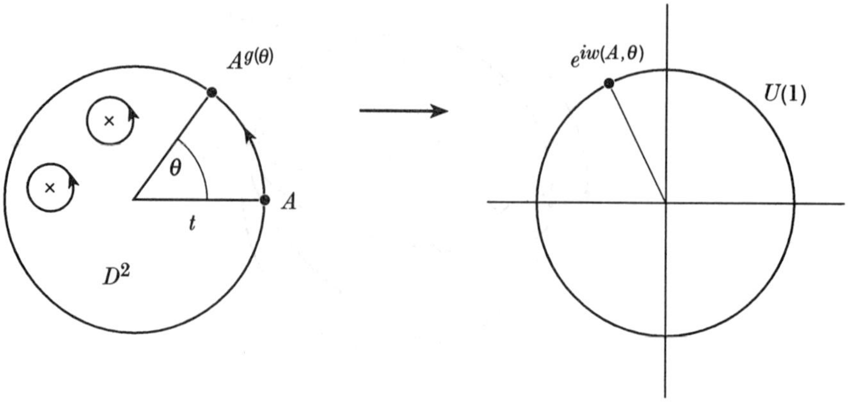

Fig. 11.4. Disk D^2: homotopy $A^{t,\theta}$, $\times \ldots$ points where the fermion determinant vanishes, $U(1)$ map defined by the determinant phase

tion A in the zero instanton sector (vanishing Pontrjagin index, recall Section 6.6) so that the Dirac operator $i\,\not{D}$ has *no* zero-modes and we consider the **one-parameter family of gauge potential transformations**

$$A^{g(\theta)} = g^{-1}(\theta, x)(A(x) + d)g(\theta, x). \tag{11.180}$$

The infinitesimal variation

$$A^{g(\theta)} \to A^{g(\theta+\delta\theta)} \tag{11.181}$$

involves an infinitesimal gauge transformation with the gauge element

$$v = g^{-1}\delta g = v_\alpha(\theta, x)d\theta^\alpha, \tag{11.182}$$

with

$$v_\alpha(\theta, x) = g^{-1}(\theta, x)\frac{\partial}{\partial\theta^\alpha}g(\theta, x) \tag{11.183}$$

and $\alpha = 1$ in our case. The element v (11.182) is nothing but the MC form on group space (and it is identified with the FP ghost, recall Section 8.1). Then we have for the infinitesimal gauge potential variation

$$\delta A^{g(\theta)} = A^{g(\theta+\delta\theta)} - A^{g(\theta)} = -\mathcal{D}v, \tag{11.184}$$

with the covariant derivative

$$\mathcal{D} = d + [A^{g(\theta)},\ ]. \tag{11.185}$$

The gauge potential $A^{g(\theta)}$—geometrically a connection—describes a circle in the space of gauge connections Sp $\mathcal{A}$. To construct a disk we introduce a homotopy. We consider the **two-parameter family of gauge potentials**—the **homotopy**

$$A^{t,\theta} = tA^{g(\theta)}, \qquad t \in [0,1] \tag{11.186}$$

interpolating between

$$A^{t=1,\theta} = A^{g(\theta)} \qquad \text{and} \qquad A^{t=0,\theta} = 0. \tag{11.187}$$

Note that here our assumption that A defines a trivial gauge potential is crucial, otherwise we cannot interpolate as in equation (11.186). Geometrically this homotopy is a 2-dimensional disk D^2 with boundary $\partial D^2 = S^1$ in Sp $\mathcal{A}$; see Figure 11.4.

Fermion determinant: Next we consider the **fermion determinant**

$$\det i\widehat{D}(A^{t,\theta}), \tag{11.188}$$

with the differential operator à la AGG

$$\widehat{D}(A^{t,\theta}) = \not\partial + \not A^{t,\theta} P_+. \tag{11.189}$$

This determinant is a complex functional of the gauge potential $A^{t,\theta}$ on the disk D^2. We may choose the reference gauge potential A such that the determinant does not vanish on the boundary $\partial D^2 = S^1$

$$\det i\widehat{D}(A^{t=1,\theta}) \neq 0. \tag{11.190}$$

Remembering that the real part of the determinant is proportional to the Dirac determinant, equation (11.139), and is thus gauge invariant, we have along the boundary

$$\begin{aligned} Z_{\text{AGG}}[A^{g(\theta)}] &= \exp\left[-W_{\text{AGG}}[A^{g(\theta)}]\right] \\ &= \det i\widehat{D}(A^{g(\theta)}) = (\det i\not D)^{1/2} e^{iw(A,\theta)}. \end{aligned} \tag{11.191}$$

Proposition:

- *It is only the imaginary part of the determinant which may receive an anomalous gauge variation!*

The determinant restricted to the boundary—the phase—defines a mapping, see Figure 11.4,

$$f(\theta) := e^{iw(A,\theta)}: \qquad \partial D^2 = S^1 \to U(1) = S^1, \tag{11.192}$$

which is characterized by the **winding number** (recall Section 2.2)

$$\begin{aligned} m &= \frac{1}{2\pi i}\int_0^{2\pi} d\theta\, f^{-1}(\theta)\frac{d}{d\theta}f(\theta) \\ &= \frac{1}{2\pi}\int_0^{2\pi} d\theta\, \frac{d}{d\theta}w(A,\theta). \end{aligned} \tag{11.193}$$

It counts the net winding of the phase of the determinant (11.188) as we move along the boundary of the disk.

On the other hand, the anomaly is the gauge variation of the 'quantum action'. Therefore we find

$$\begin{aligned} -G(v, A^{g(\theta)}) &= -\delta W[A^{g(\theta)}] \equiv -\delta W_{\text{AGG}}[A^{g(\theta)}] \\ &= i\delta w(A,\theta) = i d\theta \frac{d}{d\theta}w(A,\theta). \end{aligned} \tag{11.194}$$

Proposition:

- *The anomaly $G(v, A^{g(\theta)})$ is given by the local winding of the determinant phase!*

Strategy: The strategy according to Alvarez-Gaumé and Ginsparg is now the following:

i) We find a situation for which the winding number (11.193) is nonvanishing, $m \neq 0$, for *pure* topological reasons. Then the phase of the determinant must change under a gauge transformation and we have an anomaly.

ii) We find an appropriate $(2n+2)$-dimensional Dirac operator whose index is equal to this winding number. Then the index density, given by the AS index theorem, supplies an explicit expression for the non-Abelian anomaly.

Winding number: First we relate the winding number m (11.193) of the determinant phase to the local behaviour of $\det i\widehat{D}(A^{t,\theta})$ on the interior of the disk D^2. We note that we have chosen the Dirac operator $i\not{D}$ not to contain any zero-modes. Since we have (recall relation (11.191))

$$|\det i\widehat{D}(A^{g(\theta)})| = (\det i\not{D})^{1/2} \neq 0 \tag{11.195}$$

the operator $i\widehat{D}(A^{g(\theta)})$ also does not contain zero-modes. But the operator $i\widehat{D}(A^{t,\theta})$ defined on the interior of the disk D^2 may receive zero-modes because $A^{t,\theta}$ is *not* the gauge transformation of A. These zeros occur at isolated points. There the determinant—as a regularized product of eigenvalues—vanishes. Then the winding number of the phase of $\det i\widehat{D}(A^{t,\theta})$ around the boundary $\partial D^2 = S^1$ equals the sum of the local winding numbers of the phase around small circles enclosing the zeros, see Figure 11.4. This follows from a continuous deformation of the loop $\partial D^2 = S^1$ into small contours around the zeros. So we have

$$m = \sum_i m_i, \tag{11.196}$$

where $m_i = m(t_i, \theta_i)$ is the local winding number and t_i, θ_i denote the zero points in the disk D^2.

Now Alvarez-Gaumé and Ginsparg have shown via an 'adiabatic approximation' that these interior zeros of $\det i\widehat{D}(A^{t,\theta})$ correspond one-to-one to the zero-modes of a specific $(2n+2)$-dimensional Dirac operator $i\,\not{D}_{2n+2}$. The winding numbers at these zero points turn out to be $m_i = \pm 1$, the chirality of the $(2n+2)$-dimensional zero-modes. Therefore the **total winding number** m around ∂D^2 is equal to the difference of the positive and negative chirality zero-modes n_+, n_- of the operator $i\,\not{D}_{2n+2}$—this is precisely the **index**

$$m = n_+ - n_- = \text{index } i\,\not{D}_{2n+2}. \tag{11.197}$$

Recalling equation (11.193) we find the relation

$$\frac{1}{2\pi}\int_0^{2\pi} d\theta\, \frac{d}{d\theta} w(A,\theta) = \text{index } i\,\not{D}_{2n+2}. \tag{11.198}$$

Then, for $i\,\not{D}_{2n+2}$ given, the AS index theorem (11.114), (11.115) enables us to identify the anomalous local variation $\frac{d}{d\theta}w(A,\theta)$ with characteristic classes, up to a total derivative.

Construction of a Dirac operator in $(2n+2)$ **dimensions:** The bundle we consider is defined on the base manifold

$$S^2 \times S^{2n}, \tag{11.199}$$

where S^2 is a sphere in Sp $\mathcal{A}/\mathcal{G}$, the moduli space, and S^{2n} is the usual compactified space–time manifold. In order to construct a noncontractible sphere S^2 we actually take two disks in Sp $\mathcal{A}$ and glue them together at the boundaries. (We avoid such a complicating boundary term; compare with Figure 11.6.) The disk discussed before with parameters (t,θ) gives the upper patch

$$H^+ \times S^{2n} \tag{11.200}$$

and some trivial disk with parameters (s,θ) serves as the lower patch

$$H^- \times S^{2n}. \tag{11.201}$$

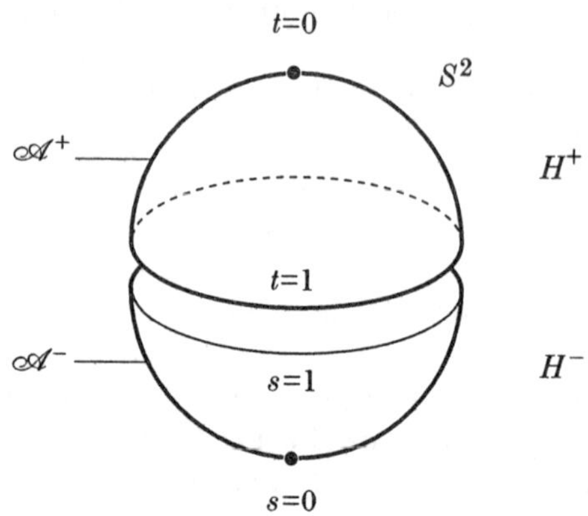

Fig. 11.5. The disks with parameters (t, θ) and (s, θ) are glued together at $t = s = 1$ to form a sphere S^2 in Sp $\mathcal{A}/\mathcal{G}$

The two patches are glued together to form a sphere; see Figure 11.5.

We choose as local gauge **connection 1-forms** in $(2n+2)$ dimensions

$$\begin{aligned} \mathcal{A}^+(t,\theta,x) &= tg^{-1}(A(x)+d+\delta)g \quad \text{on } H^+ \times S^{2n} \\ \mathcal{A}^-(s,\theta,x) &= A(x) \qquad\qquad\qquad \text{on } H^- \times S^{2n}, \end{aligned} \tag{11.202}$$

with $\delta = d\theta\, d/d\theta$. Specifically, we have

$$\mathcal{A}^+ = \mathcal{A}^+_a dx^a \quad , \quad \mathcal{A}^- = \mathcal{A}^-_a dx^a, \tag{11.203}$$

with $a = (\mu, t, \theta) = (1, \ldots, 2n, 2n+1, 2n+2)$ and

$$\begin{aligned} dx^{2n+1} &= dt \text{ or } ds \\ dx^{2n+2} &= d\theta \end{aligned} \tag{11.204}$$

and the components

$$\begin{aligned} &\mathcal{A}^+_\mu = A^{t,\theta}_\mu, \quad \mathcal{A}^+_t = 0, \quad \mathcal{A}^+_\theta = tg^{-1}\frac{d}{d\theta}g \\ &\mathcal{A}^-_\mu = A_\mu, \quad \mathcal{A}^-_t = 0, \quad \mathcal{A}^-_\theta = 0. \end{aligned} \tag{11.205}$$

On the overlap, on the 'equator' $(t = s = 1)$ the local connections fulfil the **compatibility condition** (recall our bundle discussion in Section 2.7)

$$\mathcal{A}^+ = g^{-1}(\theta,x)(\mathcal{A}^- + \widetilde{\Delta})g(\theta,x) \qquad \forall\,(\theta,x) \in S^1 \times S^{2n}. \tag{11.206}$$

The operator

$$\widetilde{\Delta} = d + \delta + \delta_t \tag{11.207}$$

means the **exterior derivative in** $(2n+2)$ **dimensions** with

$$\delta = d\theta \, \frac{d}{d\theta}, \qquad \delta_t = dt \, \frac{d}{dt} \quad \left(\text{or } \delta_t \to \delta_s = ds \, \frac{d}{ds} \right) \tag{11.208}$$

and the gauge element $g(\theta, x)$ represents the **transition function**. Note that here on the equator $(t = 1)$ we have

$$\widetilde{\Delta} \to \Delta = d + \delta \tag{11.209}$$

since $\delta_t \, g(\theta, x) = 0$. Therefore the pair

$$\mathcal{A} = \{\mathcal{A}^+, \mathcal{A}^-\} \tag{11.210}$$

defines a **global connection** on the principal bundle over the manifold $S^2 \times S^{2n}$ characterized by the transition function $g(\theta, x)$.

The **curvature 2-form** $\mathcal{F}$ in $(2n+2)$ dimensions is given by

$$\mathcal{F} = \widetilde{\Delta}\mathcal{A} + \mathcal{A}^2. \tag{11.211}$$

The local curvatures $\mathcal{F}^+$ and $\mathcal{F}^-$, defined on the upper and the lower patch, are related on the equator by the **compatibility condition**

$$\mathcal{F}^+ = g^{-1}(\theta, x)\mathcal{F}^- g(\theta, x) \qquad \forall \, (\theta, x) \in S^1 \times S^{2n}. \tag{11.212}$$

Finally we construct the **Dirac operator** in $(2n+2)$ dimensions by

$$i \, \not{D}_{2n+2} = i(\partial_a + \mathcal{A}_a)\Gamma^a. \tag{11.213}$$

$\mathcal{A}_a$ are the local gauge connections (11.205) and Γ^a are $(2n+2)$-dimensional Dirac matrices which are conveniently chosen in a Weyl-like basis

$$\begin{aligned} \Gamma^\mu &= \sigma_1 \otimes \gamma^\mu \\ \Gamma^{2n+1} &= \sigma_2 \otimes \mathbf{1} \\ \Gamma^{2n+2} &= \sigma_1 \otimes \gamma_5 \\ \Gamma_5 &= i^{n+1} \prod_{a=1}^{2n+2} \Gamma^a = \sigma_3 \otimes \mathbf{1}, \end{aligned} \tag{11.214}$$

with γ^μ and γ_5 being the usual Dirac matrices in $2n$-dimensional space.

Index theorem: We have prepared everything: the bundle, its connection and curvature, and the Dirac operator. Now we can apply the **AS index theorem** (11.114), (11.115) in $(2n+2)$ dimensions

$$\text{index } i\,\not{D}_{2n+2} = \frac{1}{(n+1)!}\left(\frac{i}{2\pi}\right)^{n+1}\int_{S^2\times S^{2n}} \text{tr } \mathcal{F}^{n+1}. \tag{11.215}$$

We show that the integral representation indeed gives the winding number we are interested in. Let us evaluate the integral. From our previous studies (recall e.g. Section 7.2) we know that tr $\mathcal{F}^{n+1}$ is closed, hence locally exact (Poincaré's lemma, Section 7.3). This means

$$\exists\, Q_{2n+1} \ldots \text{ Chern–Simons form}$$

such that

$$\text{tr } \mathcal{F}^{n+1} = \tilde{\Delta} Q_{2n+1}(\mathcal{A}, \mathcal{F}) \tag{11.216}$$

is valid on a local patch. Then the integral becomes

$$\begin{aligned}
&\int_{S^2\times S^{2n}} \text{tr } \mathcal{F}^{n+1} \\
&\quad = \int_{H^+\times S^{2n}} \tilde{\Delta} Q_{2n+1}(\mathcal{A}^+, \mathcal{F}^+) + \int_{H^-\times S^{2n}} \tilde{\Delta} Q_{2n+1}(\mathcal{A}^-, \mathcal{F}^-) \\
&\quad = \int_{S^1\times S^{2n}} [Q_{2n+1}(\mathcal{A}^+, \mathcal{F}^+)\big|_{t=1} - Q_{2n+1}(\mathcal{A}^-, \mathcal{F}^-)\big|_{s=1}].
\end{aligned} \tag{11.217}$$

In the last step we applied Stokes' theorem (2.128). So we obtain the difference of the Chern–Simons forms evaluated on the 'equator' $S^1 \times S^{2n}$ where $\mathcal{A}^+$ and $\mathcal{A}^-$, $\mathcal{F}^+$ and $\mathcal{F}^-$ are related by the gauge transformations (11.206), (11.212). We have already considered such a difference in Chapter 7 (recall equations (7.114)–(7.117)) and we find the result

$$Q_{2n+1}(\mathcal{A}^+, \mathcal{F}^+)\big|_{t=1} - Q_{2n+1}(\mathcal{A}^-, \mathcal{F}^-)\big|_{s=1} = Q_{2n+1}(g^{-1}\tilde{\Delta} g, 0) + \tilde{\Delta}\alpha_{2n}, \tag{11.218}$$

with

$$Q_{2n+1}(g^{-1}\tilde{\Delta} g, 0) = (-)^n \frac{n!(n+1)!}{(2n+1)!} \text{ tr } (g^{-1}\tilde{\Delta} g)^{2n+1}. \tag{11.219}$$

The second term α_{2n}—a local polynomial in the gauge connection—vanishes upon integration via Stokes' theorem. Then we arrive at the following **index formula**

$$\text{index } i\,\not{D}_{2n+2} = \frac{(-)^n n!}{(2n+1)!}\left(\frac{i}{2\pi}\right)^{n+1}\int_{S^1\times S^{2n}} \text{tr } (g^{-1}\tilde{\Delta} g)^{2n+1}. \tag{11.220}$$

Here on the 'equator' we have $\tilde{\Delta} \to \Delta$ and we emphasize again the topological invariance of the index which is determined *only* by the transition functions $g(\theta, x)$ of the bundle. As we know (recall our discussion in Section 6.6), this integral formula represents the winding number of the map $g(\theta, x)$ and counts the number of times the group G covers the sphere $S^{2n+1} \sim S^1 \times S^{2n}$ under the mapping; it is homotopically classified by $\Pi_{2n+1}(G) = \mathbf{Z}$. So the net winding of the determinant phase which equals the index of the Dirac operator (recall equation (11.198)) is traced back to the winding of $g(\theta, x)$.

Anomaly: Finally we determine the explicit expression of the non-Abelian anomaly from the index density given by theorem (11.215). We first observe that

$$\int_{S^1 \times S^{2n}} Q_{2n+1}(\mathcal{A}^-, \mathcal{F}^-)\Big|_{s=1} = 0 \tag{11.221}$$

since due to our chosen $\mathcal{A}^- = A(x)$, equation (11.205), there is no volume element $d\theta$ to integrate over the manifold $S^1 \times S^{2n}$. Then the index theorem (11.215) together with calculation (11.217) provides

$$\text{index } i\,\not{D}_{2n+2} = \frac{1}{(n+1)!}\left(\frac{i}{2\pi}\right)^{n+1} \int_{S^1 \times S^{2n}} Q_{2n+1}(\mathcal{A}^+, \mathcal{F}^+)\Big|_{t=1}, \tag{11.222}$$

where we rewrite $\mathcal{A}^+$, $\mathcal{F}^+$ in terms of $A^{g(\theta)}$, v (equations (11.180), (11.182))

$$\text{index } i\,\not{D}_{2n+2} = \frac{1}{(n+1)!}\left(\frac{i}{2\pi}\right)^{n+1} \int_{S^1 \times S^{2n}} Q_{2n+1}(A^{g(\theta)} + v, F(A^{g(\theta)})). \tag{11.223}$$

We expand the Chern–Simons form Q_{2n+1} in powers of v and recall that v is the only quantity that carries $d\theta$. Therefore only the linear term in v

$$Q^1_{2n}(v, A^{g(\theta)}, F(A^{g(\theta)})) \tag{11.224}$$

contributes to the integral over $S^1 \times S^{2n}$. Of course, this linear term (11.224) represents the chain term Q^1_{2n} of the SZ chain of descent equations (recall equations (9.77), (9.78)). Then we obtain

$$\text{index } i\,\not{D}_{2n+2} = \frac{1}{(n+1)!}\left(\frac{i}{2\pi}\right)^{n+1} \int_{S^1 \times S^{2n}} Q^1_{2n}(v, A^{g(\theta)}, F(A^{g(\theta)})). \tag{11.225}$$

Remembering now the relation: anomaly—local winding of the determinant phase—index density (equations (11.194), (11.198))

$$-G(v, A^{g(\theta)}) = i d\theta \frac{d}{d\theta} w(A, \theta) = i\delta w(A, \theta) \tag{11.226}$$

$$\frac{1}{2\pi}\int_{S^1}\delta w(A,\theta) = \text{index } i\,\not{D}_{2n+2} =: \int_{S^1}\text{Index } i\,\not{D}_{2n+2} \tag{11.227}$$

we find the **anomaly result**

$$\begin{aligned} -G(v,A^{g(\theta)}) &= 2\pi i \text{ Index } i\,\not{D}_{2n+2} \\ &= 2\pi i\frac{i^{n+1}}{(2\pi)^{n+1}(n+1)!}\int_{S^{2n}} Q^1_{2n}(v,A^{g(\theta)},F(A^{g(\theta)})). \end{aligned} \tag{11.228}$$

Finally, we may choose $\theta = 0$, $g(0,x) = \mathbf{1}$ and shift $n \to (n-1)$, then we have proved via the index theorem the following proposition which we have used frequently in the past.

Proposition:

- *The anomaly $G(v,A)$ is given by the chain term Q^1_{2n-2}*

$$G(v,A) = N\int_{S^{2n-2}} Q^1_{2n-2}(v,A,F), \tag{11.229}$$

with the normalization

$$N = -2\pi i\frac{i^n}{(2\pi)^n n!}. \tag{11.230}$$

For negative chirality fields we obtain an overall $(-)$ sign factor.

We already know the chain term Q^1_{2n-2} (see formula (9.122)),

$$Q^1_{2n-2}(v,A,F) = n(n-1)\int_0^1 dt(1-t)P(v,d(A,F_t^{n-2})), \tag{11.231}$$

with $F_t = tF + (t^2-t)A^2$. In 4 dimensions ($n = 3$) we find Bardeen's result

$$G(v,A) = -\frac{1}{24\pi^2}\int_{S^4}\text{tr } vd\left[AdA + \frac{1}{2}A^3\right]. \tag{11.232}$$

At the end of this chapter we want to summarize the several topological aspects which are important in our index discussion.

'Russian formula': Let us concentrate again on the 'equator' of the sphere S^2 ($t = s = 1$) in Sp $\mathcal{A}/\mathcal{G}$. For the connections we have

$$\begin{aligned} \mathcal{A}^+|_{t=1} &= A^{g(\theta)} + v \\ \mathcal{A}^-|_{s=1} &= A, \end{aligned} \tag{11.233}$$

with $v = g^{-1}\delta g$ the MC form, and for the curvatures we get

$$\begin{aligned}\mathcal{F}^+(\mathcal{A}^+)|_{t=1} &= \widetilde{\Delta}\mathcal{A}^+ + (\mathcal{A}^+)^2|_{t=1} \\ &= (d+\delta)(A^{g(\theta)} + v) + (A^{g(\theta)} + v)^2 \\ \mathcal{F}^-(\mathcal{A}^-)|_{s=1} &= \widetilde{\Delta}\mathcal{A}^- + (\mathcal{A}^-)^2|_{s=1} \\ &= dA + A^2 = F. \end{aligned} \tag{11.234}$$

Then the curvature compatibility condition (11.212) provides

$$\mathcal{F}^+(\mathcal{A}^+) = g^{-1}(\theta, x)\mathcal{F}^- g(\theta, x) = g^{-1}(\theta, x)Fg(\theta, x) \equiv F(A^{g(\theta)}), \tag{11.235}$$

with

$$F(A^{g(\theta)}) = dA^{g(\theta)} + (A^{g(\theta)})^2. \tag{11.236}$$

But the identity

$$\mathcal{F}^+(A^{g(\theta)} + v) \equiv F(A^{g(\theta)}) \tag{11.237}$$

and the version restricted to $\theta = 0$, $g(0, x) = \mathbf{1}$

$$\mathcal{F}^+(A + v) \equiv F(A) \tag{11.238}$$

we know already—it is the '**Russian formula**' found by Stora and Zumino in a quite different context. Equation (11.237) represents Zumino's identity (9.75) and equation (11.238) Stora's version (9.9). So in our fibre bundle construction the 'Russian formula' (11.237), (11.238)—which is the basic relation for the differential geometric derivation of the anomaly and of the whole chain of descent equations—emerges quite 'naturally' by the compatibility condition (11.212) of the curvature on the 'equator' $S^1 \times S^{2n}$. It represents the **Maurer–Cartan horizontality condition** of the curvature (see e.g. [Baulieu 1984], [Baulieu, Thierry-Mieg 1984], [Bonora, Cotta-Ramusino, Rinaldi, Stasheff 1987]).

Topology: The topological spaces we consider are

- Sp $\mathcal{A}\ldots$ The **affine space of all gauge connections**. This space is topologically trivial.
- $\mathcal{G}\ldots$ The **space of all pointed gauge transformation functions**

$$\mathcal{G} = \{g(x) : S^{2n} \to G | g(x_0) = \mathbf{1}\},$$

  where x_0 corresponds to ∞ in the uncompactified theory.
- Sp $\mathcal{A}/\mathcal{G}\ldots$ The **moduli space** obtained by identifying the points in Sp $\mathcal{A}$ which differ by a gauge transformation. This is the physical space. Generally, its topology is nontrivial.

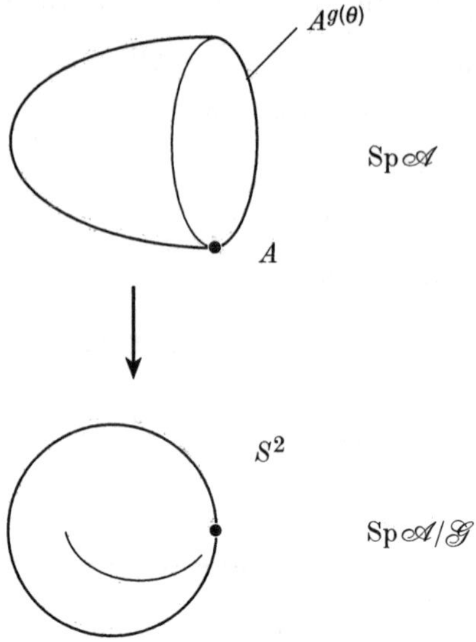

Fig. 11.6. The deformed disk $A^{t,\theta}$ in Sp $\mathcal{A}$ projected down to a sphere S^2 in Sp $\mathcal{A}/\mathcal{G}$

The topological origin of the non-Abelian anomaly is that the fermion determinant represents a section of a nontrivial line bundle over Sp $\mathcal{A}/\mathcal{G}$—the **determinant bundle**. In other words, the variation in the determinant phase along $\partial D^2 = S^1$ does not allow the determinant to be a single-valued functional over Sp $\mathcal{A}/\mathcal{G}$. This happens because the variation in the determinant phase under gauge transformations provides an obstruction to projecting the gauge orbits in Sp $\mathcal{A}$ down to the points in Sp $\mathcal{A}/\mathcal{G}$.

Specifically, we project the disk $D^2 = \{A^{t,\theta}\}$ in Sp $\mathcal{A}$ down to a sphere S^2 in Sp $\mathcal{A}/\mathcal{G}$. We continuously deform the disk D^2 and we identify the boundary $\partial D^2 = \{A^{g(\theta)}\}$ with a point in Sp $\mathcal{A}/\mathcal{G}$ obtaining a 2-dimensional sphere S^2 embedded in Sp $\mathcal{A}/\mathcal{G}$; see Figure 11.6. To show the nontriviality of the determinant bundle—the occurrence of an anomaly—we just have to consider the bundle over such a noncontractible sphere S^2 in Sp $\mathcal{A}/\mathcal{G}$.

Determinant bundle DET: So we construct a bundle with the fermion determinant $\det i\widehat{D}(A^{t,\theta})$. It assigns a complex number to each point of the sphere S^2 in Sp $\mathcal{A}/\mathcal{G}$ and defines a complex line bundle—the **determinant bundle DET**—over this two-parameter subspace S^2 of Sp $\mathcal{A}/\mathcal{G}$. A more abstract definition is given by Quillen [Quillen 1985] (for literature see e.g. [Mickelsson 1989]). The **structure group** is clearly $U(1)$. The **transition function** of DET—the phase function $e^{iw(A,\theta)}$ which represents the imaginary part of the 'quantum action'— is defined on the equator S^1 of the sphere S^2 and connects two fibre points, or two sections

$$\det i\widehat{D}(A^{g(\theta)}) = e^{iw(A,\theta)} \det i\widehat{D}(A).$$

Homotopy: In terms of homotopy groups (recall Section 2.2) we have the following situation. Complex line bundles, $U(1)$ bundles over S^{2n}, are classified by the homotopy group $\Pi_{2n-1}(U(1))$—the homotopy class of the transition functions on the 'equator' S^{2n-1}. This group is nonvanishing only for $n = 1$

$$\Pi_1(U(1)) = \mathbf{Z}. \tag{11.239}$$

The gauge elements $g(\theta, x)$ on the boundary of the disk $\partial D^2 = S^1$ describe a loop in $\mathcal{G}$ classified by

$$\Pi_1(\mathcal{G}) = \mathbf{Z}. \tag{11.240}$$

Since Sp $\mathcal{A}$ is contractible—which means topologically trivial $\Pi_n(\text{Sp } \mathcal{A}) = 0$ $\forall\, n$, specifically for $n = 2$—there exists the isomorphism

$$\Pi_2(\text{Sp } \mathcal{A}/\mathcal{G}) \simeq \Pi_1(\mathcal{G}). \tag{11.241}$$

Thus the 2-dimensional sphere S^2 in Sp $\mathcal{A}/\mathcal{G}$ is noncontractible iff the one-parameter family $g(\theta, x)$ (11.175) generates a nontrivial homotopy class (11.240) in $\mathcal{G}$ (compare with Figure 11.6). On the other hand, we obviously also have

$$\Pi_1(\mathcal{G}) \simeq \Pi_{2n+1}(G) \tag{11.242}$$

so that we finally find

$$\Pi_2(\text{Sp } \mathcal{A}/\mathcal{G}) \simeq \Pi_1(\mathcal{G}) \simeq \Pi_{2n+1}(G) \simeq \mathbf{Z}. \tag{11.243}$$

The gauge orbit space Sp $\mathcal{A}/\mathcal{G}$—the physical moduli space—has a noncontractible sphere S^2 if and only if the homotopy group $\Pi_{2n+1}(G)$ is nontrivial; for instance for $G = SU(N)$, $N \geq n+1$, with $N \geq 3$ in our case. Then the determinant bundle over this noncontractible sphere is nontrivial—an anomaly occurs (for more elaborate work see e.g. [Blau 1988, 1989]).

Résumé: We have reached our aim to determine the non-Abelian anomaly—the solution of the Wess–Zumino consistency condition—on pure topological grounds. The non-Abelian anomaly characterizes the nontrivial topology of the determinant bundle over the manifold $S^2 \times S^{2n}$. More precisely, the anomaly represents the local winding of the fermion determinant phase. This winding around the boundary of a disk in Sp $\mathcal{A}$—which is fabricated to a sphere S^2 in Sp $\mathcal{A}/\mathcal{G}$—corresponds to the index of a Dirac operator in $(2n + 2)$ dimensions. Then the index density given by the index theorem provides the explicit anomaly expression. An anomaly occurs whenever the homotopy group of the gauge group G is $\Pi_{2n+1}(G) = \mathbf{Z}$.

Note: The condition $\Pi_{2n+1}(G) = \mathbf{Z}$, however, is only a *sufficient condition* for the occurrence of an anomaly!

For example, in four dimensions a $U(1)$ gauge theory coupled to Weyl fermions gives $\Pi_5(U(1)) = 0$ but the theory is anomalous, which we know from perturbation theory. Such a situation is actually not puzzling. The determinant phase may have a local variation although the net winding number is vanishing when we integrate the phase over a complete loop—that means we consider the global topology.

To pick up this anomaly by topological techniques it seems we again need a more refined index version—the *local index theorems* as developed, for instance, by [Bismut 1991], [Bismut, Freed 1986a,b], [Knecht, Lazzarini, Stora 1991a,b] (a recent overview of index theorems and anomalies can be found in [Richter 1998], [Ekstrand 1999]).

12
Gravitation

In a field theory for elementary particles we also have to account for the effect of gravitation. It plays an important role, particularly for the anomalies in quantum field theory. Due to the basic principle of the equivalence of gravitation and inertia we can describe the effect of gravitation entirely by a curved space–time manifold. So we need to study Riemannian geometry. We regard gravitation as a gauge theory, where the gauge transformations are given by the coordinate- and by the Lorentz transformations. As in the case of YM fields, only the matter fields—the fermions—are quantized but not the gravitational field itself. In our presentation of gravitation we certainly rely on the classical books like [Sexl, Urbantke 1983], [Thirring 1992], [Weinberg 1972], [Misner, Thorne, Wheeler 1973]; additional literature close to our interest is [Nakahara 1990], [Visconti 1992], [Göckerler, Schücker 1987], [Birrell, Davies 1982], [d'Inverno 1992], [Wald 1984], [Sachs, Wu 1977], [Choquet-Bruhat, DeWitt-Morette 1982].

We want to explain the gravitational formalism to the extent that we specifically need for our description of the anomalies. So we begin with the Riemannian geometry (Section 12.1) where we introduce the metric, the Christoffel connection and curvature, and the general coordinate transformations. Next we consider the orthonormal tangent frame (Section 12.2) where we discuss the vielbein, the spin connection and the local Lorentz transformations. We derive the equations of motion from the action principle (Section 12.3), and we introduce the energy–momentum tensor here. We discuss the fermionic action in a separate section (Section 12.4) which is followed by the study of the invariances of the fermionic action, the Einstein-, the Lorentz- and the Weyl invariance (Section 12.5). Then we come to our main interest, the broken symmetries—the gravitational anomalies (Section 12.6). We shall distinguish between the Einstein-, Lorentz- and Weyl anomalies and we determine the corresponding consistency conditions. We shall find the equivalence between the Einstein- and Lorentz anomalies, and the type of covariant gravitational anomaly will be explained too. Our interest in gravitation is as in the gauge case. We introduce anticommuting Grassmann fields—ghosts—and the BRS algebra, then we are able to derive the SZ chain of descent equations (Section 12.7). Finally, we calculate the gravitational anomalies with help of the AS index theorems (Section 12.8), where we also include the mixed case of gauge and gravitational fields.

12.1 Riemannian geometry

A manifold can be endowed with a metric tensor in order to achieve a further structure. It generalizes the inner product of two vectors in flat Euclidean space to that in curved Riemannian space. Let us study the differential geometry and the analysis of tensors in this curved space.

12.1.1 Metric

Let M be a differentiable manifold, then the inner product of two vectors $X, Y \in T_p(M)$ is defined in the tangent space $T_p(M)$. We introduce the following metrics:

Riemannian metric: A **Riemannian metric** g is a tensor field of type $(0,2)$ on M with the subsequent properties at each point $p \in M$

$$\begin{aligned} &\text{i)} \qquad g_p(X,Y) = g_p(Y,X) \\ &\text{ii)} \qquad g_p(X,X) \geq 0, \qquad \text{equality only for } X = 0. \end{aligned} \tag{12.1}$$

So $g_p \equiv g|_p$, which means the tensor g evaluated at point p is a symmetric positive definite bilinear form.

The metric g_p is called a **pseudo-Riemannian metric** if we have

$$\text{iii)} \qquad g_p(X,Y) = 0 \qquad \forall X \in T_p(M) \qquad \Rightarrow Y = 0 \tag{12.2}$$

instead of property ii).

Let us recall our discussion in Section 2.3, the inner product of a vector $X \in T_p(M)$ with a dual vector, a 1-form $\omega \in T_p^*(M)$. It represents the map

$$(\ ,\ ) : T_p^*(M) \otimes T_p(M) \to \mathbf{R}, \tag{12.3}$$

where

$$(\omega, X) = \left(\omega_\mu dx^\mu, X^\nu \frac{\partial}{\partial x^\nu}\right) = \omega_\mu X^\mu \in \mathbf{R}. \tag{12.4}$$

When we have a metric we define the **inner product** between two vectors $X, Y \in T_p(M)$ by the map

$$g_p(X,Y) : T_p(M) \otimes T_p(M) \to \mathbf{R}. \tag{12.5}$$

Then we may define a linear map

$$g_p(X,\ ) : T_p(M) \to \mathbf{R} \tag{12.6}$$

by

$$Y \mapsto g_p(X, Y) \tag{12.7}$$

and we identify this map with a 1-form

$$g_p(X, \ ) \leftrightarrow \omega \in T_p^*(M). \tag{12.8}$$

On the other hand, the 1-form $\omega \in T_p^*(M)$ induces a vector $X \in T_p(M)$ by

$$(\omega, Y) = g_p(X, Y), \tag{12.9}$$

where X is unique due to property iii). So the metric g_p establishes an isomorphism between

$$T_p(M) \simeq T_p^*(M). \tag{12.10}$$

On a given chart (V, φ) of the manifold M with coordinates x^μ we can express the **metric tensor** $g \in T^0{}_2(M)$ as

$$g_p = g_{\mu\nu}(x) dx^\mu \otimes dx^\nu, \tag{12.11}$$

where

$$g_{\mu\nu}(x) = g_p\left(\frac{\partial}{\partial x^\mu}, \frac{\partial}{\partial x^\nu}\right). \tag{12.12}$$

We regard $g_{\mu\nu}$ as a matrix and $g^{\mu\nu}$ as its inverse according to

$$g_{\mu\nu} g^{\nu\sigma} = \delta_\mu{}^\sigma. \tag{12.13}$$

We denote the **determinant of the metric** by

$$g := \det g_{\mu\nu}. \tag{12.14}$$

Clearly, we have

$$\det g^{\mu\nu} = g^{-1}. \tag{12.15}$$

The isomorphism (12.10) between vectors and covectors is supplied by the metric

$$\omega_\mu = g_{\mu\nu} X^\nu, \qquad X^\mu = g^{\mu\nu}\omega_\nu \tag{12.16}$$

and

$$g_p(X, Y) = (\omega, Y) = \omega_\mu Y^\mu = g_{\mu\nu} X^\nu Y^\mu. \tag{12.17}$$

So we raise and lower the indices with the help of the metric tensor $g^{\mu\nu}$, $g_{\mu\nu}$. Generally, a mixed tensor is obtained analogously by

$$T^\mu{}_\nu = g^{\mu\sigma} T_{\sigma\nu}. \tag{12.18}$$

The **norm** $\|X\|$, or length, of a vector $X \in T_p(M)$ is defined by

$$\|X\|^2 = g_p(X,X) = g_{\mu\nu}X^\mu X^\nu. \tag{12.19}$$

We find the **infinitesimal distance squared**—the **line element**—by inserting the infinitesimal displacement $dx^\mu\, \partial/\partial x^\mu \in T_p(M)$ (here dx^μ is an infinitesimal vector component)

$$\begin{aligned} ds^2 &= g_p\left(dx^\mu \frac{\partial}{\partial x^\mu}, dx^\nu \frac{\partial}{\partial x^\nu}\right) = g_p\left(\frac{\partial}{\partial x^\mu}, \frac{\partial}{\partial x^\nu}\right) dx^\mu \cdot dx^\nu \\ &= g_{\mu\nu}(x) dx^\mu \cdot dx^\nu \end{aligned} \tag{12.20}$$

($dx^\mu \cdot dx^\nu$ actually denotes the symmetrized tensor product). This quantity is often called a metric in a sloppy way.

The quadratic form of $g_{\mu\nu}$ can be diagonalized. If $g_{\mu\nu}$ is Riemannian all diagonal terms are positive; in the case of a pseudo-Riemannian metric $g_{\mu\nu}$ may have positive and negative diagonal terms $g \to \text{diag}(-1,\ldots,1)$ with -1 occurring t times and 1 occurring s times. The pair (t,s) is called the **signature of the metric**. If $t=1$ we have a **Lorentz metric**. In this way we obtain

Riemannian metric	$\longrightarrow$	Euclidean metric $\delta = \text{diag}\,(1,1,\ldots,1)$
Lorentz metric	$\longrightarrow$	Minkowski metric $\eta = \text{diag}\,(-1,1,\ldots,1)$

(Note that here we choose a convention for $\eta_{\mu\nu}$ which differs in sign from the previous use.) Let M be a differentiable manifold with a Riemannian, a pseudo-Riemannian or a Lorentz metric g; then the pair (M,g) is called a Riemannian, a pseudo-Riemannian or a Lorentz manifold. Consider a Lorentz manifold; then the vectors $X \in T_p(M)$ are classified by

$g(X,X) > 0$	X is **spacelike**
$g(X,X) = 0$	X is **lightlike** $\quad (X \neq 0)$
$g(X,X) < 0$	X is **timelike**.

Induced metric: We begin with some definitions.

Let $f: M \to N$ be a smooth map with $\dim M \leq \dim N$, then

i) f is an **immersion** of M into N if the tangent map $f_*: T_p(M) \to T_{f(p)}(N)$ (recall Section 2.6.1) is an injection (one-to-one).

ii) f is an **embedding** if f is an injection and an immersion. The image $f(M)$ is called a **submanifold** of N.

Now, if the map f is an embedding of M into N the pullback **induces the metric** on M

$$g_M = f^* g_N \tag{12.21}$$

or in coordinates

$$g_{M\mu\nu}(x) = g_{N\alpha\beta}(f(x)) \frac{\partial y^\alpha}{\partial x^\mu} \frac{\partial y^\beta}{\partial x^\nu}, \tag{12.22}$$

where x and y are the coordinates of p and $f(p)$.

Examples: Let us consider the unit sphere S^2 embedded into the 3-dimensional space with Euclidean metric $(\mathbf{R}^3, \delta)$. Let (θ, φ) be the polar coordinates of S^2; we define the **embedding** f as the inclusion

$$f : (\theta, \varphi) \mapsto (y_1 = \sin\theta \cos\varphi, y_2 = \sin\theta \sin\varphi, y_3 = \cos\theta). \tag{12.23}$$

Then we obtain for the **induced metric**

$$\begin{aligned} g_{S^2} &= g_{\mu\nu}\, dx^\mu \otimes dx^\nu \\ &= \delta_{\alpha\beta} \frac{\partial y^\alpha}{\partial x^\mu} \frac{\partial y^\beta}{\partial x^\nu} dx^\mu \otimes dx^\nu \\ &= d\theta \otimes d\theta + \sin^2\theta\, d\varphi \otimes d\varphi. \end{aligned} \tag{12.24}$$

If we embed the circle S^1 into the plane $(\mathbf{R}^2, \delta)$ by

$$f : (r, \varphi) \mapsto (y_1 = r\cos\varphi, y_2 = r\sin\varphi) \tag{12.25}$$

we find for the metric of the plane in polar coordinates

$$g = dr \otimes dr + r^2 d\varphi \otimes d\varphi \tag{12.26}$$

and for the induced metric on the circle S^1 (with constant radius $r = L$)

$$g_{S^1} = L^2 d\varphi \otimes d\varphi. \tag{12.27}$$

Isometry: Spaces of constant curvature like the sphere admit a special symmetry group—the isometry group.

Let (M, g) be a (pseudo) Riemannian manifold, the diffeomorphism $f : M \to M$ is an **isometry** if it preserves the metric

$$f^* g_{f(p)} = g_p, \tag{12.28}$$

which means

$$g_{f(p)}(f_* X, f_* Y) = g_p(X, Y) \tag{12.29}$$

for the vectors $X, Y \in T_p(M)$. In coordinate language—x is the coordinate of p and y of $f(p)$—we have

$$g_{\alpha\beta}(f(x))\frac{\partial y^\alpha}{\partial x^\mu}\frac{\partial y^\beta}{\partial x^\nu} = g_{\mu\nu}(x). \tag{12.30}$$

Isometries form a group—the **isometry group**—preserving the length of vectors. For instance, in $\mathbf{R}^m$ we have the isometry group

$$f : x \mapsto Ax + b, \tag{12.31}$$

with $A \in SO(m)$, $b \in \mathbf{R}^m$.

Conformal transformation: Let (M, g) be a (pseudo) Riemannian manifold; the diffeomorphism $f : M \to M$ is called a **conformal transformation** if it preserves the metric up to a scale

$$f^* g_{f(p)} = e^{2\sigma(p)} g_p, \tag{12.32}$$

that is

$$g_{f(p)}(f_* X, f_* Y) = e^{2\sigma(p)} g_p(X, Y) \tag{12.33}$$

for $X, Y \in T_p(M)$ and $\sigma(p) \in \mathcal{F}(M)$ (the set of smooth functions). In terms of coordinates we write

$$g_{\alpha\beta}(f(x))\frac{\partial y^\alpha}{\partial x^\mu}\frac{\partial y^\beta}{\partial x^\nu} = e^{2\sigma(x)} g_{\mu\nu}(x). \tag{12.34}$$

The conformal transformations on (M, g) also form a group—the **conformal group**. They preserve the angles between vectors, but not the lengths.

Let g, g' be two metrics on M; if we have

$$g'_p = e^{2\sigma(p)} g_p \tag{12.35}$$

we say that g'_p is **conformally related** to g_p and the transition $g_p \to g'_p$ is named **Weyl rescaling**. The Weyl rescalings form the **Weyl group**.

If there is an atlas where we find for each chart (V, φ)

$$g_{\mu\nu}(p) = e^{2\sigma(p)} \eta_{\mu\nu}, \tag{12.36}$$

with $p \in V$, the manifold (M, g) is called **locally conformally flat**.

Theorem: All spaces with constant curvature are conformally flat!

Example: Let us consider the unit sphere S^2 with the standard metric

$$ds^2 = d\theta^2 + \sin^2\theta\, d\varphi^2. \tag{12.37}$$

Then the map

$$f : (\theta, \varphi) \mapsto (u, v) \tag{12.38}$$

defined by

$$u = \log \left| \tan \frac{\theta}{2} \right|, \qquad v = \varphi \tag{12.39}$$

generates a conformally flat metric. We calculate

$$\begin{aligned} du &= \frac{du}{d\theta} d\theta = \frac{d}{d\theta} \log \left| \tan \frac{\theta}{2} \right| d\theta \\ &= \frac{1}{2 \sin \frac{\theta}{2} \cos \frac{\theta}{2}} d\theta = \frac{1}{\sin \theta} d\theta \end{aligned} \tag{12.40}$$

so that

$$ds^2 = \sin^2 \theta \left(\frac{d\theta^2}{\sin^2 \theta} + d\varphi^2 \right) = \sin^2 \theta \left(du^2 + dv^2 \right) \tag{12.41}$$

is indeed conformally flat.

Theorem: Any 2-dimensional Riemannian manifold is conformally flat!

Proof. Let (x, y) be the local coordinates of M with metric

$$ds^2 = g_{xx} dx^2 + 2 g_{xy} dx dy + g_{yy} dy^2. \tag{12.42}$$

The determinant of the metric is given by

$$g = g_{xx} g_{yy} - g_{xy}^2. \tag{12.43}$$

We rewrite the metric (12.42) in the following way

$$\begin{aligned} ds^2 &= g_{xx} dx^2 + \frac{\sqrt{g_{xx}}(g_{xy} + i\sqrt{g})}{\sqrt{g_{xx}}} dx dy \\ &\quad + \frac{\sqrt{g_{xx}}(g_{xy} - i\sqrt{g})}{\sqrt{g_{xx}}} dx dy + \frac{(g_{xy} + i\sqrt{g})(g_{xy} - i\sqrt{g})}{g_{xx}} dy^2 \\ &= \left(\sqrt{g_{xx}} dx + \frac{g_{xy} + i\sqrt{g}}{\sqrt{g_{xx}}} dy \right) \left(\sqrt{g_{xx}} dx + \frac{g_{xy} - i\sqrt{g}}{\sqrt{g_{xx}}} dy \right). \end{aligned} \tag{12.44}$$

Now the theory of differential equations tells us that there exists an integrating factor $\lambda(x, y) = \lambda_1(x, y) + i\lambda_2(x, y)$ such that

$$\begin{aligned} \lambda \left(\sqrt{g_{xx}} dx + \frac{g_{xy} + i\sqrt{g}}{\sqrt{g_{xx}}} dy \right) &= du + i dv \\ \lambda^* \left(\sqrt{g_{xx}} dx + \frac{g_{xy} - i\sqrt{g}}{\sqrt{g_{xx}}} dy \right) &= du - i dv. \end{aligned} \tag{12.45}$$

Defining

$$\frac{1}{\lambda\lambda^*} = |\lambda|^{-2} =: e^{2\sigma} \tag{12.46}$$

we get for the metric

$$ds^2 = \frac{1}{\lambda\lambda^*}(du^2 + dv^2) = e^{2\sigma}(du^2 + dv^2) \tag{12.47}$$

which is indeed conformally flat. Q.E.D.

12.1.2 Equivalence principle

The basis of gravitation, or general relativity, is the principle of the equivalence of gravitation and inertia. It rests upon the equality of gravitational and inertial mass as demonstrated experimentally by Eötvös and Dicke. It was Einstein who concluded that we do not detect an external static homogeneous gravitational field in a freely falling elevator. The reason is that the observer and the elevator itself react to the gravitational field with the same acceleration.

Equivalence principle:

- *At every space–time point in an arbitrary gravitational field it is possible to choose a locally inertial system such that within a sufficiently small region the laws of nature are determined by special relativity.*

If there is no gravitational field we find an inertial system in the whole space; the line-element is of the form

$$ds^2 = \eta_{\mu\nu} dx^\mu \cdot dx^\nu = -dt^2 + d\vec{x}^2. \tag{12.48}$$

If there exists a gravitational field, then the line-element can be chosen in the above form only in some small region; in a neighbourhood the system will be accelerated and the line-element is determined by

$$ds^2 = g_{\mu\nu}(x)\, dx^\mu \cdot dx^\nu, \tag{12.49}$$

where the tensor $g_{\mu\nu}(x)$ describes the gravitational field.

The mathematical technique to implement the equivalence principle is the analysis of tensors, which we will study next. We want to construct physical equations that are invariant under general coordinate transformations. So we must know how the quantities appearing in these equations behave under these transformations.

Coordinate transformations: Assuming a general coordinate transformation $x \to x'(x)$ the following quantities transform as

scalar Φ

$$\Phi'(x') = \Phi(x) \tag{12.50}$$

contravariant vector V^μ

$$V'^\mu(x') = \frac{\partial x'^\mu}{\partial x^\nu} V^\nu(x) \tag{12.51}$$

covariant vector V_μ

$$V'_\mu(x') = V_\nu(x) \frac{\partial x^\nu}{\partial x'^\mu} \tag{12.52}$$

mixed tensor $\mathcal{T}^{\mu\nu}{}_\sigma$

$$\mathcal{T}'^{\mu\nu}{}_\sigma(x') = \frac{\partial x'^\mu}{\partial x^\alpha} \frac{\partial x'^\nu}{\partial x^\beta} \mathcal{T}^{\alpha\beta}{}_\gamma(x) \frac{\partial x^\gamma}{\partial x'^\sigma} \tag{12.53}$$

tensor density $t^{\mu\nu}$

$$t'^{\mu\nu}(x') = \left| \frac{\partial x'}{\partial x} \right|^W \frac{\partial x'^\mu}{\partial x^\alpha} \frac{\partial x'^\nu}{\partial x^\beta} t^{\alpha\beta}(x), \tag{12.54}$$

where $\left|\frac{\partial x'}{\partial x}\right|$ represents the Jacobian of the coordinate transformation $x \to x'$ and W is the **weight** of the density.

Considering, for instance, the metric tensor

$$g'_{\mu\nu}(x') = g_{\alpha\beta}(x) \frac{\partial x^\alpha}{\partial x'^\mu} \frac{\partial x^\beta}{\partial x'^\nu} \tag{12.55}$$

we find for the determinant

$$g' = \det g'_{\mu\nu} = \left| \frac{\partial x}{\partial x'} \right|^2 \det g_{\mu\nu} = \left| \frac{\partial x'}{\partial x} \right|^{-2} g, \tag{12.56}$$

which is a scalar density of weight $W = -2$. Hence we can re-express a tensor density $t^{\mu\nu}$ by a tensor $\mathcal{T}^{\mu\nu}$ and an appropriate determinant factor

$$t^{\mu\nu} = |g|^{-W/2}\, \mathcal{T}^{\mu\nu}. \tag{12.57}$$

The quantity

$$\begin{aligned} |g'|^{W/2}\, t'^{\mu\nu} &= \left| \frac{\partial x'}{\partial x} \right|^{-W} \left| \frac{\partial x'}{\partial x} \right|^{W} \frac{\partial x'^\mu}{\partial x^\alpha} \frac{\partial x'^\nu}{\partial x^\beta} |g|^{W/2}\, t^{\alpha\beta} \\ &= \frac{\partial x'^\mu}{\partial x^\alpha} \frac{\partial x'^\nu}{\partial x^\beta} |g|^{W/2}\, t^{\alpha\beta} \end{aligned} \tag{12.58}$$

indeed transforms as a tensor.

The volume element occurring in the integration transforms as

$$d^4x' = \left|\frac{\partial x'}{\partial x}\right| d^4x. \tag{12.59}$$

Hence the element

$$\sqrt{|g|}\, d^4x \tag{12.60}$$

—named the **invariant volume element**—remains invariant under coordinate transformations

$$\sqrt{|g'|}\, d^4x' = \left|\frac{\partial x'}{\partial x}\right|^{-1} \left|\frac{\partial x'}{\partial x}\right| \sqrt{|g|}\, d^4x = \sqrt{|g|}\, d^4x. \tag{12.61}$$

12.1.3 Christoffel connection and curvature

In Riemannian geometry we also introduce a connection and curvature.

Christoffel connection: For a Riemannian manifold M with metric $g_{\mu\nu}$ there exists a *unique* **connection 1-form**—the **Christoffel connection**—

$$\Gamma^\alpha{}_\beta = \Gamma^\alpha{}_{\mu\beta} dx^\mu, \tag{12.62}$$

where $\Gamma^\alpha{}_{\mu\beta}$ denotes the familiar **Christoffel symbol**. It is determined by the metric tensor via formula

$$\Gamma^\alpha{}_{\mu\beta} = \frac{1}{2} g^{\alpha\lambda}(-\partial_\lambda g_{\mu\beta} + \partial_\mu g_{\beta\lambda} + \partial_\beta g_{\lambda\mu}) \tag{12.63}$$

and it is symmetric in the lower indices

$$\Gamma^\alpha{}_{\mu\beta} = \Gamma^\alpha{}_{\beta\mu} \tag{12.64}$$

(we shall demonstrate that it transforms as a connection, theorem (12.82), later on).

For example, in 2-dimensional Euclidean space $(\mathbf{R}^2, \delta)$ all components of the Christoffel symbol $\Gamma^\alpha{}_{\mu\beta}$ clearly vanish if we refer to Cartesian coordinates. But if we consider polar coordinates the metric is given by equation (12.26); hence

$$g_{\mu\nu} = \begin{pmatrix} 1 & 0 \\ 0 & r^2 \end{pmatrix}. \tag{12.65}$$

Then formula (12.63) provides the Christoffel symbols explicitly

$$\begin{aligned}
\Gamma^{\varphi}{}_{\varphi r} = \Gamma^{\varphi}{}_{r\varphi} &= \frac{1}{2} g^{\varphi\varphi}(-\partial_{\varphi} g_{r\varphi} + \partial_r g_{\varphi\varphi} + \partial_{\varphi} g_{\varphi r}) \\
&= \frac{1}{2}\frac{1}{r^2} 2r = \frac{1}{r} \\
\Gamma^{r}{}_{\varphi\varphi} &= -r
\end{aligned} \tag{12.66}$$

and all other components vanish

$$\Gamma^{r}{}_{rr} = \Gamma^{r}{}_{r\varphi} = \Gamma^{r}{}_{\varphi r} = \Gamma^{\varphi}{}_{rr} = \Gamma^{\varphi}{}_{\varphi\varphi} = 0. \tag{12.67}$$

Analogously, when embedding the unit sphere S^2 into Euclidean space $(\mathbf{R}^3, \delta)$ we have the induced metric (recall equation (12.24))

$$g_{\mu\nu} = \begin{pmatrix} 1 & 0 \\ 0 & \sin^2\theta \end{pmatrix} \tag{12.68}$$

and we calculate from formula (12.63)

$$\begin{aligned}
\Gamma^{\theta}{}_{\varphi\varphi} &= -\sin\theta\cos\theta \\
\Gamma^{\varphi}{}_{\theta\varphi} &= \Gamma^{\varphi}{}_{\varphi\theta} = \cot\theta;
\end{aligned} \tag{12.69}$$

all other components vanish

$$\Gamma^{\theta}{}_{\theta\theta} = \Gamma^{\theta}{}_{\theta\varphi} = \Gamma^{\theta}{}_{\varphi\theta} = \Gamma^{\varphi}{}_{\theta\theta} = \Gamma^{\varphi}{}_{\varphi\varphi} = 0. \tag{12.70}$$

Covariant derivative: We define the **covariant derivative** as we did in gauge theories

$$\nabla = d + [\Gamma, \ ], \tag{12.71}$$

with

$$\nabla = \nabla_{\mu} dx^{\mu}, \qquad d = \partial_{\mu} dx^{\mu}, \qquad \Gamma = \Gamma_{\mu} dx^{\mu} \tag{12.72}$$

and Γ is the matrix notation of $\Gamma^{\alpha}{}_{\beta}$. Applied to a tensor-valued p-form

$$\mathcal{T} \to \mathcal{T}^{\alpha\ldots}_{\beta\ldots} = \frac{1}{p!}\mathcal{T}^{\alpha\ldots}_{\beta\ldots,\mu_1\ldots\mu_p}(x) dx^{\mu_1}\ldots dx^{\mu_p} \tag{12.73}$$

we have

$$\nabla\mathcal{T} = d\mathcal{T} + [\Gamma, \mathcal{T}], \tag{12.74}$$

where the commutator means (analogous to definition (2.353))

$$\begin{aligned}
[\Gamma, \mathcal{T}] \quad \to \quad & [\Gamma, \mathcal{T}]^{\alpha\ldots}_{\beta\ldots} = \\
= \quad & \Gamma^{\alpha}{}_{\nu}\mathcal{T}^{\nu\ldots}_{\beta\ldots} + \text{ all upper indices} \\
& - (-)^p \mathcal{T}^{\alpha\ldots}_{\nu\ldots}\Gamma^{\nu}{}_{\beta} - (-)^p \text{ all lower indices}
\end{aligned} \tag{12.75}$$

(the factor $(-)^p$ stems from interchanging the 1-form Γ with the p-form $\mathcal{T}$).

Definition: A connection is called a **Riemannian connection** if

i) *the metric is covariantly constant*

$$\nabla_\alpha g_{\mu\nu} = 0 \tag{12.76}$$

ii) *the connection has zero torsion*

$$T^\alpha = \nabla dx^\alpha = 0. \tag{12.77}$$

A Riemannian connection is sometimes also called the **Levi–Civita connection** in the literature.

Torsion: Let us consider the torsion introduced above. The **torsion 2-form**

$$T^\alpha = \frac{1}{2} T^\alpha{}_{\mu\nu} dx^\mu dx^\nu \tag{12.78}$$

is defined by

$$\begin{aligned} T^\alpha &= \nabla dx^\alpha = ddx^\alpha + [\Gamma, dx]^\alpha \\ &= \Gamma^\alpha{}_\nu dx^\nu = \Gamma^\alpha{}_{\mu\nu} dx^\mu dx^\nu. \end{aligned} \tag{12.79}$$

So the torsion-free condition implies

$$T^\alpha{}_{\mu\nu} = \Gamma^\alpha{}_{\mu\nu} - \Gamma^\alpha{}_{\nu\mu} = 0; \tag{12.80}$$

this is the symmetry of the Christoffel symbol

$$\Gamma^\alpha{}_{\mu\nu} = \Gamma^\alpha{}_{\nu\mu}. \tag{12.81}$$

Note: Here dx^α is thought of as the component of a vectorial 1-form

$$dx^\alpha \otimes \frac{\partial}{\partial x^\alpha} = \delta^\alpha{}_\beta dx^\beta \otimes \frac{\partial}{\partial x^\alpha}.$$

The index α is a space–time index, therefore we can contract the indices as in equation (12.79) and we get a 2-form. For this reason we find a concept such as torsion in gravitation but, in contrast, not in YM gauge theories.

Theorem: Conditions i) and ii) determine the Christoffel connection Γ uniquely in terms of the metric $g_{\mu\nu}$ and its derivative $\partial_\lambda g_{\mu\nu}$ via formula

$$\Gamma^\alpha{}_{\mu\beta} = \frac{1}{2} g^{\alpha\lambda}(-\partial_\lambda g_{\mu\beta} + \partial_\mu g_{\beta\lambda} + \partial_\beta g_{\lambda\mu}). \tag{12.82}$$

Proof. Using the constancy of the metric

$$\begin{aligned} \nabla g_{\mu\nu} &= dg_{\mu\nu} + [\Gamma, g]_{\mu\nu} \\ &= dg_{\mu\nu} - g_{\alpha\nu}\Gamma^{\alpha}{}_{\mu} - g_{\mu\alpha}\Gamma^{\alpha}{}_{\nu} = 0 \end{aligned} \tag{12.83}$$

we have for the components

$$\nabla_{\beta} g_{\mu\nu} = \partial_{\beta} g_{\mu\nu} - g_{\alpha\nu}\Gamma^{\alpha}{}_{\beta\mu} - g_{\mu\alpha}\Gamma^{\alpha}{}_{\beta\nu} = 0. \tag{12.84}$$

Adding and subtracting cyclic permutations, respecting the symmetry of the Christoffel symbols, we find

$$\partial_{\beta} g_{\mu\nu} - \partial_{\mu} g_{\nu\beta} + \partial_{\nu} g_{\beta\mu} = 2 g_{\mu\alpha}\Gamma^{\alpha}{}_{\nu\beta}, \tag{12.85}$$

which is

$$\Gamma^{\alpha}{}_{\nu\beta} = \frac{1}{2} g^{\alpha\lambda}(-\partial_{\lambda} g_{\nu\beta} + \partial_{\nu} g_{\beta\lambda} + \partial_{\beta} g_{\lambda\nu}). \qquad \text{Q.E.D.} \tag{12.86}$$

Curvature: The **curvature 2-form** R associated with the connection is given by the Cartan structure equation

$$R^{\alpha}{}_{\beta} = d\Gamma^{\alpha}{}_{\beta} + \Gamma^{\alpha}{}_{\lambda}\Gamma^{\lambda}{}_{\beta} \tag{12.87}$$

or briefly in matrix notation

$$R = d\Gamma + \Gamma^2. \tag{12.88}$$

On the other hand, we write for a 2-form

$$R^{\alpha}{}_{\beta} = \frac{1}{2} R^{\alpha}{}_{\beta\mu\nu} dx^{\mu} dx^{\nu}, \tag{12.89}$$

where $R^{\alpha}{}_{\beta\mu\nu}$ denotes the **Riemann tensor**. Then equation (12.87) corresponds to

$$R^{\alpha}{}_{\beta\mu\nu} = \partial_{\mu}\Gamma^{\alpha}{}_{\nu\beta} - \partial_{\nu}\Gamma^{\alpha}{}_{\mu\beta} + \Gamma^{\alpha}{}_{\mu\lambda}\Gamma^{\lambda}{}_{\nu\beta} - \Gamma^{\alpha}{}_{\nu\lambda}\Gamma^{\lambda}{}_{\mu\beta}. \tag{12.90}$$

The Riemann tensor is antisymmetric in the following indices

$$\begin{aligned} R^{\alpha}{}_{\beta\mu\nu} &= -R^{\alpha}{}_{\beta\nu\mu} \\ R^{\alpha}{}_{\beta\mu\nu} &= -R_{\beta}{}^{\alpha}{}_{\mu\nu} \end{aligned} \tag{12.91}$$

and satisfies the cyclicities—the **Bianchi identities**

$$\begin{aligned} R^{\alpha}{}_{\beta\mu\nu} + R^{\alpha}{}_{\mu\nu\beta} + R^{\alpha}{}_{\nu\beta\mu} &= 0 \\ \nabla_{\sigma} R^{\alpha}{}_{\beta\mu\nu} + \nabla_{\mu} R^{\alpha}{}_{\beta\nu\sigma} + \nabla_{\nu} R^{\alpha}{}_{\beta\sigma\mu} &= 0. \end{aligned} \tag{12.92}$$

We rewrite the last equation with differential forms

$$\nabla R = 0. \tag{12.93}$$

We emphasize that $R^\alpha{}_\beta$ represents a matrix valued 2-form and it *must not be confused* with the **Ricci tensor**

$$\mathcal{R}_{\beta\nu} = R^\alpha{}_{\beta\alpha\nu}. \tag{12.94}$$

The Ricci tensor is a symmetric (in β and ν) 0-form which we need rather rarely. Its contraction gives the **Ricci scalar**

$$\mathcal{R} = g^{\beta\nu}\mathcal{R}_{\beta\nu}. \tag{12.95}$$

Note: These contraction operations are not possible in YM gauge theories since there we have mixed indices—space–time and gauge group indices—whereas in gravitation there are only space–time indices.

The Riemann curvature tensor is also determined by the commutator of the covariant derivatives analogously to the gauge case (recall theorems (3.302), (3.308) and our discussion of differential forms in Section 6.2).

Theorem:

$$[\nabla_\mu, \nabla_\nu]^\alpha{}_\beta = R^\alpha{}_{\beta\mu\nu}. \tag{12.96}$$

Applying the commutator to vectors we obtain

$$\begin{aligned}
[\nabla_\mu, \nabla_\nu]A^\alpha &= [\nabla_\mu, \nabla_\nu]^\alpha{}_\beta A^\beta = R^\alpha{}_{\beta\mu\nu}A^\beta \\
[\nabla_\mu, \nabla_\nu]A_\beta &= [\nabla_\mu, \nabla_\nu]_\beta{}^\alpha A_\alpha = R_\beta{}^\alpha{}_{\mu\nu}A_\alpha \\
&= -A_\alpha R^\alpha{}_{\beta\mu\nu}
\end{aligned} \tag{12.97}$$

and to tensors we get

$$\begin{aligned}
[\nabla_\mu, \nabla_\nu]\mathcal{T}_{\alpha\beta} &= [\nabla_\mu, \nabla_\nu]_\alpha{}^\lambda \mathcal{T}_{\lambda\beta} + [\nabla_\mu, \nabla_\nu]_\beta{}^\lambda \mathcal{T}_{\alpha\lambda} \\
&= -\mathcal{T}_{\lambda\beta}R^\lambda{}_{\alpha\mu\nu} - \mathcal{T}_{\alpha\lambda}R^\lambda{}_{\beta\mu\nu}.
\end{aligned} \tag{12.98}$$

So applying the commutator to tensors gives:

Theorem:

$$\begin{aligned}
[\nabla_\mu, \nabla_\nu] &= [R_{\mu\nu}, \;] \\
[\nabla_\mu, \nabla_\nu]\mathcal{T} &= [R_{\mu\nu}, \mathcal{T}]
\end{aligned} \tag{12.99}$$

or in components

$$\begin{aligned}
[\nabla_\mu, \nabla_\nu]^{\alpha\ldots}_{\beta\ldots} &= [R_{\mu\nu},\ ]^{\alpha\ldots}_{\beta\ldots} \\
[\nabla_\mu, \nabla_\nu]\mathcal{T}^{\alpha\ldots}_{\beta\ldots} &= R^\alpha{}_{\lambda\mu\nu}\mathcal{T}^{\lambda\ldots}_{\beta\ldots} + \text{ all upper indices} \\
&\quad - \mathcal{T}^{\alpha\ldots}_{\lambda\ldots} R^\lambda{}_{\beta\mu\nu} - \text{ all lower indices.}
\end{aligned} \tag{12.100}$$

Within differential forms we have:

Theorem:

$$\nabla^2 = [R,\ ]. \tag{12.101}$$

Of course, we can quickly verify relation (12.101) by direct calculation (recall the gauge case, equation (6.56)).

Remark: However, in the case of a nonvanishing torsion

$$T^\alpha{}_{\mu\nu} = \Gamma^\alpha{}_{\mu\nu} - \Gamma^\alpha{}_{\nu\mu} \neq 0 \tag{12.102}$$

we would get

$$[\nabla_\mu, \nabla_\nu]A^\alpha = R^\alpha{}_{\beta\mu\nu}A^\beta + (\Gamma^\lambda{}_{\mu\nu} - \Gamma^\lambda{}_{\nu\mu})\nabla_\lambda A^\alpha. \tag{12.103}$$

12.1.4 Variations and derivatives

Variations: Later on, when we study the action principle, we need to know the variations of fields and tensors when the metric is varied

$$g_{\mu\nu}(x) \to g_{\mu\nu}(x) + \delta g_{\mu\nu}(x). \tag{12.104}$$

Here we consider the change of the quantities $g^{\mu\nu}$, g and $\mathcal{R}_{\beta\nu}$. Let us start with the relation

$$g^{\mu\nu}g_{\nu\lambda} = \delta^\mu{}_\lambda. \tag{12.105}$$

Let δ be some variation, then we have

$$\delta g^{\mu\nu}g_{\nu\lambda} + g^{\mu\nu}\delta g_{\nu\lambda} = 0 \tag{12.106}$$

yielding

$$\delta g^{\mu\nu} = -g^{\mu\sigma}\delta g_{\sigma\lambda}g^{\lambda\nu}. \tag{12.107}$$

For the determinant we vary the formula

$$g = \det g_{\mu\nu} = e^{\operatorname{tr}\ln g_{\mu\nu}}, \tag{12.108}$$

which gives

$$\delta g = g g^{\mu\nu}\delta g_{\mu\nu} \tag{12.109}$$

$$\delta\sqrt{|g|} \;=\; \frac{1}{2}\sqrt{|g|}\, g^{\mu\nu}\delta g_{\mu\nu}. \tag{12.110}$$

We use formula (12.63) to reformulate the **contraction of the Christoffel symbol**

$$\begin{aligned} \Gamma^{\alpha}{}_{\mu\alpha} &= \frac{1}{2}g^{\alpha\lambda}(-\partial_\lambda g_{\mu\alpha} + \partial_\mu g_{\alpha\lambda} + \partial_\alpha g_{\lambda\mu}) \\ &= \frac{1}{2}g^{\alpha\lambda}\partial_\mu g_{\alpha\lambda}. \end{aligned} \tag{12.111}$$

Choosing as a variation the ordinary derivative $\delta \to \partial_\mu$ we find

$$\Gamma^{\alpha}{}_{\alpha\mu} = \Gamma^{\alpha}{}_{\mu\alpha} = \frac{1}{\sqrt{|g|}}\,\partial_\mu\sqrt{|g|}. \tag{12.112}$$

Theorem: Palatini identity

$$\delta\mathcal{R}_{\mu\nu} = \nabla_\alpha\delta\Gamma^{\alpha}{}_{\nu\mu} - \nabla_\nu\delta\Gamma^{\alpha}{}_{\alpha\mu}. \tag{12.113}$$

(Note that $\delta\Gamma^{\alpha}{}_{\nu\mu}$ is a tensor; whereas the analogy in the YM gauge case δA^a is a vector.)

Proof.

$$\begin{aligned} \mathcal{R}_{\mu\nu} &= R^{\alpha}{}_{\mu\alpha\nu} = \partial_\alpha\Gamma^{\alpha}{}_{\nu\mu} - \partial_\nu\Gamma^{\alpha}{}_{\alpha\mu} \\ &\quad + \Gamma^{\alpha}{}_{\alpha\lambda}\Gamma^{\lambda}{}_{\nu\mu} - \Gamma^{\alpha}{}_{\nu\lambda}\Gamma^{\lambda}{}_{\alpha\mu} \end{aligned} \tag{12.114}$$

hence

$$\begin{aligned} \delta\mathcal{R}_{\mu\nu} &= \partial_\alpha\delta\Gamma^{\alpha}{}_{\nu\mu} - \partial_\nu\delta\Gamma^{\alpha}{}_{\alpha\mu} \\ &\quad + \delta\Gamma^{\alpha}{}_{\alpha\lambda}\Gamma^{\lambda}{}_{\nu\mu} + \Gamma^{\alpha}{}_{\alpha\lambda}\delta\Gamma^{\lambda}{}_{\nu\mu} \\ &\quad - \delta\Gamma^{\alpha}{}_{\nu\lambda}\Gamma^{\lambda}{}_{\alpha\mu} - \Gamma^{\alpha}{}_{\nu\lambda}\delta\Gamma^{\lambda}{}_{\alpha\mu}. \end{aligned} \tag{12.115}$$

On the other hand, we calculate the expressions

$$\begin{aligned} \nabla_\alpha\delta\Gamma^{\alpha}{}_{\nu\mu} &= \partial_\alpha\delta\Gamma^{\alpha}{}_{\nu\mu} + [\Gamma_\alpha, \delta\Gamma]^{\alpha}{}_{\nu\mu} \\ &= \partial_\alpha\delta\Gamma^{\alpha}{}_{\nu\mu} + \Gamma^{\alpha}{}_{\alpha\lambda}\delta\Gamma^{\lambda}{}_{\nu\mu} \\ &\quad - \delta\Gamma^{\alpha}{}_{\lambda\mu}\Gamma^{\lambda}{}_{\alpha\nu} - \delta\Gamma^{\alpha}{}_{\nu\lambda}\Gamma^{\lambda}{}_{\alpha\mu} \end{aligned} \tag{12.116}$$

$$\begin{aligned} \nabla_\nu\delta\Gamma^{\alpha}{}_{\alpha\mu} &= \partial_\nu\delta\Gamma^{\alpha}{}_{\alpha\mu} + \Gamma^{\alpha}{}_{\nu\lambda}\delta\Gamma^{\lambda}{}_{\alpha\mu} \\ &\quad - \delta\Gamma^{\alpha}{}_{\lambda\mu}\Gamma^{\lambda}{}_{\nu\alpha} - \delta\Gamma^{\alpha}{}_{\alpha\lambda}\Gamma^{\lambda}{}_{\nu\mu} \end{aligned} \tag{12.117}$$

and we find

$$\delta\mathcal{R}_{\mu\nu} = \nabla_\alpha\delta\Gamma^{\alpha}{}_{\nu\mu} - \nabla_\nu\delta\Gamma^{\alpha}{}_{\alpha\mu}. \qquad \text{Q.E.D.} \tag{12.118}$$

Derivatives: There are important cases where the covariant derivative simplifies to an ordinary derivative.

i) Take some **scalar** Φ; then we clearly have

$$\nabla_\mu \Phi = \partial_\mu \Phi \tag{12.119}$$

since $[\Gamma_\mu, \Phi] = 0$.

ii) For a **covariant vector** V_ν we get

$$\nabla_\mu V_\nu = \partial_\mu V_\nu + [\Gamma_\mu, V]_\nu = \partial_\mu V_\nu - V_\lambda \Gamma^\lambda{}_{\mu\nu} \tag{12.120}$$

so that the difference is

$$\nabla_\mu V_\nu - \nabla_\nu V_\mu = \partial_\mu V_\nu - \partial_\nu V_\mu . \tag{12.121}$$

iii) For a **contravariant vector** V^μ we observe the following important property

$$\nabla_\mu V^\mu = \partial_\mu V^\mu + [\Gamma_\mu, V]^\mu = \partial_\mu V^\mu + \Gamma^\mu{}_{\mu\lambda} V^\lambda . \tag{12.122}$$

Inserting the Christoffel contraction (12.112) we obtain

$$\begin{aligned} \nabla_\mu V^\mu &= \partial_\mu V^\mu + \frac{1}{\sqrt{|g|}} \, \partial_\lambda \sqrt{|g|} \, V^\lambda \\ &= \frac{1}{\sqrt{|g|}} \, \partial_\mu (\sqrt{|g|} \, V^\mu) \end{aligned} \tag{12.123}$$

or

$$\sqrt{|g|} \nabla_\mu V^\mu = \partial_\mu (\sqrt{|g|} \, V^\mu). \tag{12.124}$$

As a consequence we find a **covariant version of the Gauss theorem**

$$\int d^4x \sqrt{|g|} \, \nabla_\mu V^\mu = \int d^4x \partial_\mu (\sqrt{|g|} \, V^\mu) = \int df_\mu \sqrt{|g|} \, V^\mu \to 0 \tag{12.125}$$

if $V^\mu \to 0$ at the boundary. Note the appearence of the determinant $\sqrt{|g|}$ to achieve an invariant volume element.

iv) For a **contravariant tensor** we analogously have

$$\begin{aligned} \nabla_\mu T^{\mu\nu} &= \partial_\mu T^{\mu\nu} + [\Gamma_\mu, T]^{\mu\nu} \\ &= \partial_\mu T^{\mu\nu} + \Gamma^\mu{}_{\mu\lambda} T^{\lambda\nu} + \Gamma^\nu{}_{\mu\lambda} T^{\mu\lambda} \\ &= \frac{1}{\sqrt{|g|}} \, \partial_\mu (\sqrt{|g|} T^{\mu\nu}) + \Gamma^\nu{}_{\mu\lambda} T^{\mu\lambda} . \end{aligned} \tag{12.126}$$

If the tensor is antisymmetric

$$\mathcal{T}^{\mu\lambda} \to \mathcal{T}^{\mu\lambda}_{\text{asym}} = -\mathcal{T}^{\lambda\mu}_{\text{asym}} \tag{12.127}$$

the last term vanishes (since $\Gamma^\nu{}_{\mu\lambda}$ is symmetric) and we get

$$\nabla_\mu \mathcal{T}^{\mu\nu}_{\text{asym}} = \frac{1}{\sqrt{|g|}}\,\partial_\mu(\sqrt{|g|}\,\mathcal{T}^{\mu\nu}_{\text{asym}}). \tag{12.128}$$

v) A **covariant tensor** gives

$$\begin{aligned}\nabla_\sigma \mathcal{T}_{\mu\nu} &= \partial_\sigma \mathcal{T}_{\mu\nu} + [\Gamma_\sigma, \mathcal{T}]_{\mu\nu} \\ &= \partial_\sigma \mathcal{T}_{\mu\nu} - \mathcal{T}_{\lambda\nu}\Gamma^\lambda{}_{\sigma\mu} - \mathcal{T}_{\mu\lambda}\Gamma^\lambda{}_{\sigma\nu}.\end{aligned} \tag{12.129}$$

Again, if the tensor is antisymmetric all Γ-terms cancel in the sum of cyclic permutations

$$\nabla_\sigma \mathcal{T}^{\text{asym}}_{\mu\nu} + \nabla_\mu \mathcal{T}^{\text{asym}}_{\nu\sigma} + \nabla_\nu \mathcal{T}^{\text{asym}}_{\sigma\mu} = \partial_\sigma \mathcal{T}^{\text{asym}}_{\mu\nu} + \partial_\mu \mathcal{T}^{\text{asym}}_{\nu\sigma} + \partial_\nu \mathcal{T}^{\text{asym}}_{\sigma\mu}. \tag{12.130}$$

vi) For a **mixed tensor** we find

$$\begin{aligned}\nabla_\mu \mathcal{T}^{\mu\nu}{}_\nu &= \partial_\mu \mathcal{T}^{\mu\nu}{}_\nu + [\Gamma_\mu, \mathcal{T}]^{\mu\nu}{}_\nu \\ &= \partial_\mu \mathcal{T}^{\mu\nu}{}_\nu + \Gamma^\mu{}_{\mu\lambda}\mathcal{T}^{\lambda\nu}{}_\nu + \Gamma^\nu{}_{\mu\lambda}\mathcal{T}^{\mu\lambda}{}_\nu - \mathcal{T}^{\mu\nu}{}_\lambda\Gamma^\lambda{}_{\mu\nu} \\ &= \frac{1}{\sqrt{|g|}}\,\partial_\mu(\sqrt{|g|}\,\mathcal{T}^{\mu\nu}{}_\nu) + \Gamma^\nu{}_{\mu\lambda}\mathcal{T}^{\mu\lambda}{}_\nu - \mathcal{T}^{\mu\nu}{}_\lambda\Gamma^\lambda{}_{\mu\nu}\end{aligned} \tag{12.131}$$

$$\nabla_\nu \mathcal{T}^{\mu\nu}{}_\mu = \frac{1}{\sqrt{|g|}}\,\partial_\nu(\sqrt{|g|}\,\mathcal{T}^{\mu\nu}{}_\mu) + \Gamma^\mu{}_{\nu\lambda}\mathcal{T}^{\lambda\nu}{}_\mu - \mathcal{T}^{\mu\nu}{}_\lambda\Gamma^\lambda{}_{\nu\mu}. \tag{12.132}$$

In the difference all Γ-terms cancel since the Γ's are symmetric

$$\sqrt{|g|}\,\nabla_\mu \mathcal{T}^{\mu\nu}{}_\nu - \sqrt{|g|}\,\nabla_\nu \mathcal{T}^{\mu\nu}{}_\mu = \partial_\mu(\sqrt{|g|}\,\mathcal{T}^{\mu\nu}{}_\nu) - \partial_\nu(\sqrt{|g|}\,\mathcal{T}^{\mu\nu}{}_\mu). \tag{12.133}$$

This is an ordinary derivative and vanishes upon integration by virtue of Gauss' theorem (12.125) if we assume $\mathcal{T} \to 0$ at the boundary.

12.1.5 General coordinate transformations

We have to study more precisely the transformation laws of tensor quantities. We consider a general coordinate transformation $x \to x'(x)$; then a **tensor-valued p-form transforms** as

$$\mathcal{T}(x) \stackrel{x\to x'}{\longrightarrow} \mathcal{T}'(x') = \Lambda^{-1}(x)\mathcal{T}(x)\Lambda(x). \tag{12.134}$$

The **coordinate transformation matrix**

$$(\Lambda^{-1})^{\mu}{}_{\nu}(x) = \frac{\partial x'^{\mu}}{\partial x^{\nu}} \tag{12.135}$$

acts on *all* upper coordinate indices of the tensor-valued p-form and the matrix

$$\Lambda^{\nu}{}_{\mu}(x) = \frac{\partial x^{\nu}}{\partial x'^{\mu}} \tag{12.136}$$

acts on *all* lower indices. Clearly, we have

$$(\Lambda^{-1})^{\mu}{}_{\nu}\Lambda^{\nu}{}_{\sigma} = \frac{\partial x'^{\mu}}{\partial x^{\nu}}\frac{\partial x^{\nu}}{\partial x'^{\sigma}} = \delta^{\mu}{}_{\sigma}. \tag{12.137}$$

In components we write

$$\begin{aligned} \mathcal{T}'(x') &\rightarrow \frac{1}{p!}\mathcal{T}'^{\alpha\ldots}_{\beta\ldots,\mu_1\ldots\mu_p}(x')dx'^{\mu_1}\ldots dx'^{\mu_p} \\ &= (\Lambda^{-1})^{\alpha}{}_{\bar{\alpha}}\ldots\frac{1}{p!}\mathcal{T}^{\bar{\alpha}\ldots}_{\bar{\beta}\ldots,\bar{\mu}_1\ldots\bar{\mu}_p}(x)\Lambda^{\bar{\beta}}{}_{\beta}\ldots \\ &\quad\cdot\Lambda^{\bar{\mu}_1}{}_{\mu_1}\ldots\Lambda^{\bar{\mu}_p}{}_{\mu_p}(\Lambda^{-1})^{\mu_1}{}_{\nu_1}\ldots(\Lambda^{-1})^{\mu_p}{}_{\nu_p}dx^{\nu_1}\ldots dx^{\nu_p} \\ &= (\Lambda^{-1})^{\alpha}{}_{\bar{\alpha}}(x)\ldots\frac{1}{p!}\mathcal{T}^{\bar{\alpha}\ldots}_{\bar{\beta}\ldots,\nu_1\ldots\nu_p}(x)dx^{\nu_1}\ldots dx^{\nu_p}\Lambda^{\bar{\beta}}{}_{\beta}(x)\ldots. \end{aligned} \tag{12.138}$$

The **Christoffel connection transforms** as

$$\Gamma(x) \stackrel{x\rightarrow x'}{\longrightarrow} \Gamma'(x') = \Lambda^{-1}(x)[\Gamma(x) + d]\Lambda(x) \tag{12.139}$$

(here d acts only on Λ), which follows either from formula (12.63) or from the requirement that $\nabla\mathcal{T}$ transforms as a tensor.

The **curvature transforms** again, tensor-like

$$R(x) \stackrel{x\rightarrow x'}{\longrightarrow} R'(x') = \Lambda^{-1}(x)R(x)\Lambda(x). \tag{12.140}$$

Diffeomorphism group: The group of coordinate reparametrizations—the diffeomorphism group—actually allows locally for two representations:

i) **passive coordinate transformation:** all tensor-valued forms change according to

$$\mathcal{T}(x) \stackrel{x\rightarrow x'}{\longrightarrow} \mathcal{T}'(x')$$

and the integration measure changes too

$$dV(x) \stackrel{x\rightarrow x'}{\longrightarrow} dV(x').$$

ii) **active coordinate transformation:** all tensor-valued forms vary as

$$\mathcal{T}(x) \to \mathcal{T}'(x).$$

So the forms are shifted back to the point x and the integration measure $dV(x)$ remains unchanged.

We work with an **infinitesimal coordinate transformation**

$$x^\alpha \to x'^\alpha = x^\alpha - \xi^\alpha(x). \tag{12.141}$$

Then the transformation matrices are infinitesimal as well

$$\begin{aligned} (\Lambda^{-1})^\alpha{}_\beta(x) = \frac{\partial x'^\alpha}{\partial x^\beta} = \frac{\partial x^\alpha}{\partial x^\beta} - \frac{\partial \xi^\alpha(x)}{\partial x^\beta} &= \delta^\alpha{}_\beta - (v_\xi)^\alpha{}_\beta \\ \Lambda^\alpha{}_\beta(x) = \frac{\partial x^\alpha}{\partial x'^\beta} = \left(\frac{1}{\partial x'/\partial x}\right)^\alpha{}_\beta &= \delta^\alpha{}_\beta + (v_\xi)^\alpha{}_\beta, \end{aligned} \tag{12.142}$$

where

$$(v_\xi)^\alpha{}_\beta = \frac{\partial \xi^\alpha(x)}{\partial x^\beta} = \partial_\beta \xi^\alpha. \tag{12.143}$$

In matrix notation we write

$$\begin{aligned} \Lambda^{-1}(x) &= \mathbf{1} - v_\xi(x) \\ \Lambda(x) &= \mathbf{1} + v_\xi(x), \end{aligned} \tag{12.144}$$

with the **infinitesimal coordinate transformation tensor**

$$v_\xi = \partial \xi. \tag{12.145}$$

Passive coordinate transformation: We find the infinitesimal passive change in the following way

$$\begin{aligned} \mathcal{T}(x) \overset{x\to x'}{\longrightarrow} \mathcal{T}'(x') &= \Lambda^{-1}(x)\mathcal{T}(x)\Lambda(x) \\ &\mathrel{\widehat{=}} \mathcal{T}(x) + \delta^{c'}_{v_\xi}\mathcal{T}(x). \end{aligned} \tag{12.146}$$

The index c stands for coordinate transformation and the prime for passive. Inserting expressions (12.144)

$$\begin{aligned} \mathcal{T}'(x') &= (\mathbf{1} - v_\xi(x))\mathcal{T}(x)(\mathbf{1} + v_\xi(x)) \\ &= \mathcal{T}(x) + [\mathcal{T}, v_\xi](x) \end{aligned} \tag{12.147}$$

(the commutator contains contractions of the tensor indices only) we get the **infinitesimal passive coordinate transformation of a tensor form**

$$\delta^{c'}_{v_\xi}\mathcal{T}(x) = \mathcal{T}'(x') - \mathcal{T}(x) = [\mathcal{T}, v_\xi](x). \tag{12.148}$$

Specifically, for the **curvature tensor 2-form** we have

$$\delta^{c'}_{v_\xi} R(x) = R'(x') - R(x) = [R, v_\xi](x) \tag{12.149}$$

and for the **Christoffel connection 1-form**

$$\begin{aligned}\delta^{c'}_{v_\xi} \Gamma(x) &= \Gamma'(x') - \Gamma(x) = \nabla v_\xi(x) \\ &= dv_\xi(x) + [\Gamma, v_\xi](x).\end{aligned} \tag{12.150}$$

Lie derivative: The active change, on the other hand, is closely related to the Lie derivative which we have introduced in Section 2.6. Here we need the Lie derivative along the flow of the vector field ξ^α given by the infinitesimal coordinate change (12.141). We can take over the formula (2.356), derived in Section 2.6.5, the **Lie derivative along ξ of a tensor-valued p-form**

$$\begin{aligned}\mathcal{L}_\xi \mathcal{T}^{\alpha\dots}_{\beta\dots} = & [\xi^\nu \partial_\nu \mathcal{T}^{\alpha\dots}_{\beta\dots,\mu_1\dots\mu_p}(x) \\ & - \partial_\nu \xi^\alpha \mathcal{T}^{\nu\dots}_{\beta\dots,\mu_1\dots\mu_p}(x) - \text{ all upper indices} \\ & + \mathcal{T}^{\alpha\dots}_{\nu\dots,\mu_1\dots\mu_p}(x)\partial_\beta \xi^\nu + \text{ all lower indices} \\ & + \mathcal{T}^{\alpha\dots}_{\beta\dots,\nu\dots\mu_p}(x)\partial_{\mu_1}\xi^\nu + \text{ all form indices}] \\ & \cdot \frac{1}{p!}\, dx^{\mu_1}\dots dx^{\mu_p}.\end{aligned} \tag{12.151}$$

In matrix notation we write

$$\begin{aligned}\mathcal{L}_\xi \mathcal{T} &= \xi \cdot \partial\, \mathcal{T} - [\partial\xi, \mathcal{T}] \\ \mathcal{L}_\xi &= \xi \cdot \partial - [\partial\xi, \ ],\end{aligned} \tag{12.152}$$

where here the commutator definition *also* includes the contraction of the form indices!

In terms of the interior product, however, we have the formula (recall equation (2.359))

$$\begin{aligned}\mathcal{L}_\xi \mathcal{T} &= (i_\xi d + d i_\xi)\mathcal{T} - [v_\xi, \mathcal{T}] \\ \mathcal{L}_\xi &= (i_\xi d + d i_\xi) - [v_\xi, \ ],\end{aligned} \tag{12.153}$$

where now the commutator contains *only* the contraction of the upper and lower tensor indices as in definition (2.353) (remember that $\xi \cdot \partial$ together with the contraction of the form indices gives precisely the interior product term, equation (2.358)).

Let us collect the necessary **properties of the Lie derivative**

i) $[\mathcal{L}_\xi, i_\xi] = 0$

ii) $[\mathcal{L}_\xi, d] = 0$

$$\begin{array}{ll} \text{iii)} & [\mathcal{L}_{\xi_1}, i_{\xi_2}] = i_{[\xi_1,\xi_2]} \\ \text{iv)} & [\mathcal{L}_{\xi_1}, \mathcal{L}_{\xi_2}] = \mathcal{L}_{[\xi_1,\xi_2]}, \end{array} \tag{12.154}$$

where the **Lie bracket** is defined by (recall equations (2.340), (2.341))

$$\begin{aligned} [\xi_1, \xi_2] &= [\xi_1, \xi_2]^\nu \partial_\nu \\ [\xi_1, \xi_2]^\nu &= \xi_1^\mu \partial_\mu \xi_2^\nu - \xi_2^\mu \partial_\mu \xi_1^\nu . \end{aligned} \tag{12.155}$$

The Lie derivative depends only on the structure of the manifold and *not* on the connection. For a Riemannian connection, which is torsion-free, we can replace the ordinary by the covariant derivative in formula (12.152)

$$\mathcal{L}_\xi = \xi \cdot \nabla - [\nabla \xi, \;] \tag{12.156}$$

since

$$\begin{aligned} \xi \cdot \nabla - [\nabla \xi, \;] &= \xi^\nu \partial_\nu + \xi^\lambda [\Gamma^\alpha{}_{\lambda\beta}, \;] - [\partial_\beta \xi^\alpha + \Gamma^\alpha{}_{\beta\lambda}\xi^\lambda, \;] \\ &= \xi \cdot \partial - [\partial_\beta \xi^\alpha, \;] = \mathcal{L}_\xi \end{aligned} \tag{12.157}$$

(recall that torsion-freedom implies symmetric Christoffel symbols).

Note: However, in the case of a nonvanishing torsion

$$T^\alpha{}_{\lambda\beta} = \Gamma^\alpha{}_{\lambda\beta} - \Gamma^\alpha{}_{\beta\lambda} \neq 0 \tag{12.158}$$

we find instead

$$\mathcal{L}_\xi = \xi \cdot \nabla - [\nabla_\beta \xi^\alpha + T^\alpha{}_{\lambda\beta}\xi^\lambda, \;], \tag{12.159}$$

which can be quickly verified

$$\begin{aligned} \xi \cdot \nabla - [\nabla_\beta \xi^\alpha + T^\alpha{}_{\lambda\beta}\xi^\lambda, \;] &= \xi \cdot \partial + \xi^\lambda [\Gamma^\alpha{}_{\lambda\beta}, \;] - [\partial_\beta \xi^\alpha, \;] \\ &\quad - [\Gamma^\alpha{}_{\beta\lambda}\xi^\lambda, \;] - [(\Gamma^\alpha{}_{\lambda\beta} - \Gamma^\alpha{}_{\beta\lambda})\xi^\lambda, \;] \\ &= \xi \cdot \partial - [\partial_\beta \xi^\alpha, \;] = \mathcal{L}_\xi . \end{aligned} \tag{12.160}$$

Active coordinate transformations: Now we turn to the active change of a tensor field. We perform a Taylor expansion to first order in ξ

$$\begin{aligned} T^{\alpha\ldots}_{\beta\ldots,\mu\ldots}(x) &\stackrel{x\to x'=x-\xi}{\longrightarrow} T'^{\alpha\ldots}_{\beta\ldots,\mu\ldots}(x-\xi) \\ &= T'^{\alpha\ldots}_{\beta\ldots,\mu\ldots}(x) - \xi^\lambda \partial_\lambda T^{\alpha\ldots}_{\beta\ldots,\mu\ldots}(x). \end{aligned} \tag{12.161}$$

We subtract the initial field

$$\begin{aligned} &T'^{\alpha\ldots}_{\beta\ldots,\mu\ldots}(x') - T^{\alpha\ldots}_{\beta\ldots,\mu\ldots}(x) \\ &= T'^{\alpha\ldots}_{\beta\ldots,\mu\ldots}(x) - T^{\alpha\ldots}_{\beta\ldots,\mu\ldots}(x) - \xi^\lambda \partial_\lambda T^{\alpha\ldots}_{\beta\ldots,\mu\ldots}(x) \end{aligned} \tag{12.162}$$

and insert for the l.h.s. the commutator result (12.148). In this way we find the **infinitesimal active coordinate transformation of a tensor field**

$$\begin{aligned}\delta^c_\xi \mathcal{T}^{\alpha\ldots}_{\beta\ldots,\mu\ldots}(x) &= \mathcal{T}'^{\alpha\ldots}_{\beta\ldots,\mu\ldots}(x) - \mathcal{T}^{\alpha\ldots}_{\beta\ldots,\mu\ldots}(x) \\ &= \xi^\lambda \partial_\lambda \mathcal{T}^{\alpha\ldots}_{\beta\ldots,\mu\ldots}(x) - [\partial\xi, \mathcal{T}(x)]^{\alpha\ldots}_{\beta\ldots,\mu\ldots} \\ &= \mathcal{L}_\xi \mathcal{T}^{\alpha\ldots}_{\beta\ldots,\mu\ldots}(x). \end{aligned} \tag{12.163}$$

Here the commutator also includes the contraction of the form indices $\mu\ldots$. If we multiply equation (12.163) by the wedge product $\frac{1}{p!}dx^{\mu_1}\ldots dx^{\mu_p}$ we obtain the **change of a tensor-valued p-form**

$$\delta^c_\xi \mathcal{T}(x) = \mathcal{T}'(x) - \mathcal{T}(x) = \mathcal{L}_\xi \mathcal{T}(x) \tag{12.164}$$

(written in matrix notation) and using formula (12.153) further we get

$$\delta^c_\xi \mathcal{T}(x) = (i_\xi d + d i_\xi)\mathcal{T}(x) - [v_\xi, \mathcal{T}](x). \tag{12.165}$$

Specifically, the **change of the curvature 2-form** is

$$\begin{aligned}\delta^c_\xi R(x) &= R'(x) - R(x) = \mathcal{L}_\xi R(x) \\ &= (i_\xi d + d i_\xi)R(x) - [v_\xi, R](x);\end{aligned} \tag{12.166}$$

the **change of the metric** is

$$\begin{aligned}\delta^c_\xi g_{\mu\nu}(x) &= g'_{\mu\nu}(x) - g_{\mu\nu}(x) = \mathcal{L}_\xi g_{\mu\nu}(x) \\ &= \nabla_\mu \xi_\nu + \nabla_\nu \xi_\mu.\end{aligned} \tag{12.167}$$

Let us calculate result (12.167) explicitly. Using the covariant form (12.156) and the covariant constancy (12.76) of the metric we get

$$\begin{aligned}\delta^c_\xi g_{\mu\nu}(x) &= \mathcal{L}_\xi g_{\mu\nu}(x) = \xi\cdot\nabla g_{\mu\nu} - [\nabla\xi, g]_{\mu\nu} \\ &= -[\nabla\xi, g]_{\mu\nu} = g_{\lambda\nu}\nabla_\mu \xi^\lambda + g_{\mu\lambda}\nabla_\nu\xi^\lambda \\ &= \nabla_\mu\xi_\nu + \nabla_\nu\xi_\mu. \qquad \text{Q.E.D.}\end{aligned} \tag{12.168}$$

Expanding the Christoffel connection instead of a tensor field

$$\Gamma'^{\alpha}{}_{\mu\beta}(x') - \Gamma^{\alpha}{}_{\mu\beta}(x) = \Gamma'^{\alpha}{}_{\mu\beta}(x) - \Gamma^{\alpha}{}_{\mu\beta}(x) - \xi^\lambda\partial_\lambda\Gamma^{\alpha}{}_{\mu\beta}(x) \tag{12.169}$$

we find analogously the **infinitesimal active coordinate transformation of the Christoffel connection**

$$\begin{aligned}\delta^c_\xi \Gamma^{\alpha}{}_{\mu\beta}(x) &= \Gamma'^{\alpha}{}_{\mu\beta}(x) - \Gamma^{\alpha}{}_{\mu\beta}(x) \\ &= \mathcal{L}_\xi \Gamma^{\alpha}{}_{\mu\beta}(x) + \partial_\mu (v_\xi)^\alpha{}_\beta.\end{aligned} \tag{12.170}$$

The multiplication with dx^μ gives the **transformation of the Christoffel connection 1-form** (written in matrix notation)

$$\delta_\xi^c \Gamma(x) = \Gamma'(x) - \Gamma(x) = \mathcal{L}_\xi \Gamma(x) + dv_\xi. \tag{12.171}$$

So the connection picks up an additional term dv_ξ. Re-expressing the Lie derivative in terms of formula (12.153) gives

$$\begin{aligned}\delta_\xi^c \Gamma(x) &= (i_\xi d + di_\xi)\Gamma(x) + [\Gamma, v_\xi](x) + dv_\xi \\ &= (i_\xi d + di_\xi)\Gamma(x) + \nabla v_\xi.\end{aligned} \tag{12.172}$$

These active coordinate transformations are also named **Einstein transformations.**

Let us eventually collect the passive coordinate transformations

$$\begin{aligned}&\delta_{v_\xi}^{c'} \mathcal{T} = [\mathcal{T}, v_\xi] && \text{tensor-valued } p\text{-form} \\ &\delta_{v_\xi}^{c'} \Gamma = \nabla v_\xi && \text{Christoffel connection 1-form} \\ &\delta_{v_\xi}^{c'} \omega = 0 && \text{scalar-valued } p\text{-form}\end{aligned} \tag{12.173}$$

and compare with the active ones

$$\begin{aligned}\delta_\xi^c \mathcal{T} &= (i_\xi d + di_\xi)\mathcal{T} + \delta_{v_\xi}^{c'} \mathcal{T} \\ \delta_\xi^c \Gamma &= (i_\xi d + di_\xi)\Gamma + \delta_{v_\xi}^{c'} \Gamma \\ \delta_\xi^c \omega &= (i_\xi d + di_\xi)\omega,\end{aligned} \tag{12.174}$$

then we find for the difference

$$\delta_\xi^c - \delta_{v_\xi}^{c'} = i_\xi d + di_\xi. \tag{12.175}$$

Killing vector: Let $\xi = \xi^\mu \partial/\partial x^\mu \in T_p(M)$ be a vector field on a Riemannian manifold (M, g) and let the infinitesimal displacement $\varepsilon\xi$ generate an isometry—the **isometric map**

$$f : x^\mu \to x^\mu + \varepsilon \xi^\mu. \tag{12.176}$$

This means that the metric is preserved (recall equation (12.30))

$$g_{\alpha\beta}(x + \varepsilon\xi) \frac{\partial(x^\alpha + \varepsilon\xi^\alpha)}{\partial x^\mu} \frac{\partial(x^\beta + \varepsilon\xi^\beta)}{\partial x^\nu} = g_{\mu\nu}(x). \tag{12.177}$$

The expansion to order ε gives

$$\begin{aligned}(g_{\alpha\beta}(x) + \varepsilon\xi^\lambda \partial_\lambda g_{\alpha\beta}(x))(\delta^\alpha{}_\mu + \varepsilon\partial_\mu \xi^\alpha)(\delta^\beta{}_\nu + \varepsilon\partial_\nu \xi^\beta) &= g_{\mu\nu}(x) \\ g_{\mu\nu}(x) + \varepsilon(\xi^\lambda \partial_\lambda g_{\mu\nu}(x) + g_{\alpha\nu}(x)\partial_\mu \xi^\alpha + g_{\mu\beta}(x)\partial_\nu \xi^\beta) &= g_{\mu\nu}(x),\end{aligned}$$

which implies the **Killing equation**

$$\xi^\lambda \partial_\lambda g_{\mu\nu}(x) + g_{\alpha\nu}(x)\partial_\mu \xi^\alpha + g_{\mu\beta}(x)\partial_\nu \xi^\beta = 0. \tag{12.178}$$

The solution ξ is called the **Killing vector**. It plays an important role in the study of the symmetries of manifolds.

The Killing equation can be rewritten in terms of the Lie derivative (recall equation (12.168))

$$\mathcal{L}_\xi g_{\mu\nu}(x) = \nabla_\mu \xi_\nu + \nabla_\nu \xi_\mu = 0; \tag{12.179}$$

more explicitly

$$\partial_\mu \xi_\nu + \partial_\nu \xi_\mu - 2\xi_\lambda \Gamma^\lambda{}_{\mu\nu} = 0. \tag{12.180}$$

For example, let us consider the familiar Minkowski space $(\mathbf{R}^4, \eta)$, which is flat, hence all Christoffel connections $\Gamma^\lambda{}_{\mu\nu}$ vanish. Then we get for the Killing equation

$$\partial_\mu \xi_\nu + \partial_\nu \xi_\mu = 0, \tag{12.181}$$

with the general solution

$$\xi_\mu = a_{\mu\nu} x^\nu + b_\mu, \qquad a_{\mu\nu} = -a_{\nu\mu}. \tag{12.182}$$

This is precisely the infinitesimal transformation of the Poincaré group containing 10 free parameters ($m(m+1)/2$ in m dimensions).

Finally, if we allow the map f (12.176) to be **conformal** instead of isometric then the metric is conserved only up to a scale (recall equation (12.34))

$$g_{\alpha\beta}(x+\varepsilon\xi)\frac{\partial(x^\alpha + \varepsilon\xi^\alpha)}{\partial x^\mu}\frac{\partial(x^\beta + \varepsilon\xi^\beta)}{\partial x^\nu} = e^{2\sigma(x)} g_{\mu\nu}(x) \tag{12.183}$$

and the **conformal Killing vector** ξ is determined by the equation

$$\mathcal{L}_\xi \, g_{\mu\nu} = 2\sigma g_{\mu\nu}. \tag{12.184}$$

Consider, for instance, the Euclidean space $(\mathbf{R}^m, \delta)$ with x^μ the coordinates, then the **dilatation vector**

$$\xi = x^\mu \, \frac{\partial}{\partial x^\mu} \tag{12.185}$$

represents a conformal Killing vector since we have

$$\begin{aligned} \mathcal{L}_\xi \, \delta_{\mu\nu} &= \xi \cdot \partial \, \delta_{\mu\nu} - [\partial\xi, \delta]_{\mu\nu} \\ &= \delta_{\lambda\nu}\partial_\mu x^\lambda + \delta_{\mu\lambda}\partial_\nu x^\lambda \\ &= 2\delta_{\mu\nu}. \end{aligned} \tag{12.186}$$

12.2 Tangent frame

Later on we also introduce matter fields, where we are mainly interested in fermions which are described by spinor fields. The spinor, however, does not 'know' how to transform under a general coordinate transformation—the diffeomorphism group. The mathematical reason is that the $GL(m, \mathbf{R})$ group has no spinor representations. (By a spinor representation we mean a *finite* dimensional double-valued representation, for an *infinite* double-valued representation, see [Ne'eman, Šijački 1985].) So how can we incorporate spinors into general relativity? Spinors obey Lorentz transformations, hence we must work with the **group of local frame rotations** $SO(m)$ or $SO(1, m-1)$—the **Lorentz group**. This change from diffeomorphisms to Lorentz transformations can be achieved by the vielbein.

12.2.1 Vielbein

For spinors we need a tangent frame, and we establish a tangent frame which is orthonormal everywhere in the whole space. Due to the equivalence principle we can find at every point x_0 a set of coordinates $\xi^a_{x_0}$, $a = 1, \dots, m = \dim M$, which are locally inertial at x_0. Then the **metric in any noninertial system** is given by

$$g_{\mu\nu}(x) = \eta_{ab} e^a{}_\mu(x) e^b{}_\nu(x), \tag{12.187}$$

where

$$e^a{}_\mu(x) = \partial_\mu \xi^a_{x_0}(x). \tag{12.188}$$

The Minkowski metric η_{ab} can also be replaced by the Euclidean metric δ_{ab}. Of course, at a point different from x_0 we have to choose a new, different set of coordinates—unless space–time is flat—in order to achieve $e^a{}_\mu$ be orthonormal. The quantity $e^a{}_\mu$ is named **vielbein—zweibein** in two dimensions, **vierbein** in 4 dimensions. The Greek letter index μ refers to the original coordinates; the Latin letter index a refers to the new, locally inertial ones and is the index of the orthonormal tangent frame.

Introducing the **vielbein 1-form**

$$e^a(x) = e^a{}_\mu(x) dx^\mu \in T^*_p(M) \tag{12.189}$$

the metric can be rewritten by

$$g_p = g_{\mu\nu}(x) dx^\mu \otimes dx^\nu = \eta_{ab} e^a(x) \otimes e^b(x). \tag{12.190}$$

The vielbein $e^a{}_\mu(x)$ is a hybrid object. Under general coordinate transformations $x \to x'$ it behaves like a covariant vector

$$e^a{}_\mu(x) \overset{x \to x'}{\longrightarrow} e'^a{}_\mu(x') = \frac{\partial x^\nu}{\partial x'^\mu} e^a{}_\nu(x) \tag{12.191}$$

but under local frame rotations it transforms as

$$e^a{}_\mu(x) \to e'^a{}_\mu(x) = (L^{-1}(x))^a{}_b e^b{}_\mu(x). \tag{12.192}$$

The matrix $L^a{}_b(x) \in SO(1, m-1)$ or $SO(m)$ represents the **local Lorentz transformation** satisfying

$$\eta_{ab} L^a{}_c(x) L^b{}_d(x) = \eta_{cd}. \tag{12.193}$$

From the metric (12.187) we see that the vielbein $e^a{}_\mu$—the orthonormal tangent frame—is defined only up to local frame rotations; the rotation (12.192) satisfies the metric (12.187) as well.

Definition: Denoting the **inverse vielbein** by $E \equiv e^{-1}$, it is defined by

$$E_a{}^\mu(x) = \eta_{ab} g^{\mu\nu}(x) e^b{}_\nu(x). \tag{12.194}$$

It has the following **coordinate** and **Lorentz transformations**

$$E_a{}^\mu(x) \quad \to \quad E'_a{}^\mu(x') = \frac{\partial x'^\mu}{\partial x^\nu} E_a{}^\nu(x) \tag{12.195}$$

$$E_a{}^\mu(x) \quad \to \quad E'_a{}^\mu(x) = E_b{}^\mu(x) L^b{}_a(x) \tag{12.196}$$

and the **vector field**

$$E_a(x) = E_a{}^\mu(x) \frac{\partial}{\partial x^\mu} \in T_p(M) \tag{12.197}$$

is the **dual** quantity to the 1-form e^a (12.189). The vielbein and its inverse satisfy the equations

$$e^a \cdot E_b = e^a{}_\mu E_b{}^\mu = \delta^a{}_b \tag{12.198}$$

$$E_a{}^\mu e^a{}_\nu = \delta^\mu{}_\nu \tag{12.199}$$

$$\eta_{ab} = g_{\mu\nu}(x) E_a{}^\mu(x) E_b{}^\nu(x). \tag{12.200}$$

Considering the **vielbein determinant**

$$e := |\det e^a{}_\mu| \tag{12.201}$$

we find from relation (12.187)

$$g = \det g_{\mu\nu} = \pm(\det e^a{}_\mu)^2, \tag{12.202}$$

with (+) for a Euclidean metric and (−) for a Minkowski metric; hence

$$e = \sqrt{|g|}. \tag{12.203}$$

Returning to the oriented **invariant volume element** for a manifold M we then have

$$\begin{aligned} dV_M &= \sqrt{|g|}\, d^m x = \sqrt{|g|}\, dx^1 \ldots dx^m \\ &= e\, dx^1 \ldots dx^m = e^1 \ldots e^m, \end{aligned} \tag{12.204}$$

where clearly the dots denote a wedge product.

Variations: For a general **variation of the vielbein** we get (analogous to the metric equations (12.107)–(12.110))

$$\delta e^a{}_\mu = -e^a{}_\nu \delta E_b{}^\nu e^b{}_\mu \tag{12.205}$$

$$\delta e = e E_a{}^\mu \delta e^a{}_\mu, \qquad \frac{\delta e}{\delta e^a{}_\mu} = e E_a{}^\mu \tag{12.206}$$

$$\delta e = \frac{1}{2} \frac{1}{\sqrt{|g|}}\, \delta g = \frac{1}{2}\, e\, g^{\mu\nu} \delta g_{\mu\nu}, \tag{12.207}$$

which implies for the **contracted Christoffel symbol** (recall equation (12.112))

$$\Gamma^\alpha{}_{\alpha\mu} = \Gamma^\alpha{}_{\mu\alpha} = \frac{1}{e}\, \partial_\mu e. \tag{12.208}$$

In particular, we need the **infinitesimal active coordinate transformation of the vielbein** (we thereby lose the orthonormality)

$$\begin{aligned} \delta^c_\xi e^a{}_\mu(x) &= e'^a{}_\mu(x) - e^a{}_\mu(x) = \mathcal{L}_\xi e^a{}_\mu(x) \\ &= \xi^\nu \partial_\nu e^a{}_\mu(x) + e^a{}_\nu(x) \partial_\mu \xi^\nu \end{aligned} \tag{12.209}$$

or replacing $\partial \to \nabla$ in $\mathcal{L}_\xi$

$$\delta^c_\xi e^a{}_\mu(x) = \xi^\nu \nabla_\nu e^a{}_\mu(x) + e^a{}_\nu(x) \nabla_\mu \xi^\nu. \tag{12.210}$$

For the vielbein determinant we then find

$$\delta^c_\xi e = e E_a{}^\mu \delta^c_\xi e^a{}_\mu = \xi^\nu \partial_\nu e + e \partial_\nu \xi^\nu. \tag{12.211}$$

We shall use all these formulae frequently in our anomaly calculations later on.

Reference system: Now, we can take a contravariant vector field $V^\mu(x)$ and use the vielbein $e^a{}_\mu(x)$ to refer its components to a coordinate system which is locally inertial at x

$$V^a = e^a{}_\mu(x) V^\mu(x). \tag{12.212}$$

Analogously we can refer covariant vectors $V_\mu(x)$ to a locally inertial system and, generally, also tensor fields

$$\begin{aligned} V_b &= V_\mu(x) E^\mu{}_b(x) & (12.213)\\ \mathcal{T}^a{}_b &= e^a{}_\mu(x) \mathcal{T}^\mu{}_\nu(x) E_b{}^\nu(x). & (12.214) \end{aligned}$$

So we use the vielbein $e^a{}_\mu$ and its inverse $E_a{}^\mu$ to refer a tensor quantity from a coordinate basis to an orthonormal tangent frame and vice versa.

Note: Actually the coordinate system (CS) is also a tangent frame, but it is *not* orthonormal in the whole space (unless space–time is flat). In contrast, what we call the tangent frame (TF) is the frame which *is* an orthonormal tangent frame in the *whole* space.

12.2.2 Spin connection and curvature

Spin connection: Certainly, we can also define the connection in reference to the tangent frame e^a—in fact, when including spinor fields into gravity we have to. Then we call the connnection: **spin connection 1-form**

$$\begin{aligned} \omega^a{}_b = \omega^a{}_{bc} e^c = \omega^a{}_{b\mu} dx^\mu \quad &\in \quad \text{Lie } SO(m) \otimes \Lambda^1 \quad \text{or} \\ &\in \quad \text{Lie } SO(1, m-1) \otimes \Lambda^1. \qquad (12.215) \end{aligned}$$

This is the connection which is contained in the Dirac operator, in the fermionic action—as we shall see in Section 12.4—and which interacts with the fermions, the spin 1/2 particles. In the tangent frame the gravitational formalism appears quite analogous to the YM gauge case.

Covariant derivative: The **covariant derivative** is defined by

$$D = d + [\omega, \ ], \qquad D = D_c e^c = D_\mu dx^\mu \qquad (12.216)$$

and acts on tensor-valued p-forms which are referred to the tangent frame.

Curvature: The **curvature 2-form** R associated with the spin connection ω is, again, given by the Cartan structure equation

$$\begin{aligned} R = d\omega + \omega^2 & \qquad \text{matrix notation} \\ R^a{}_b = d\omega^a{}_b + \omega^a{}_c \omega^c{}_b & \qquad \text{component notation.} \qquad (12.217) \end{aligned}$$

On the other hand, as a 2-form the curvature is expressed by

$$R^a{}_b = \frac{1}{2} R^a{}_{bcd} e^c e^d, \qquad (12.218)$$

where $e^c e^d$ denotes the wedge product and $R^a{}_{bcd}$ the Riemann tensor referred to the tangent frame. Clearly, with the help of the vielbeins we can change to the coordinate basis

$$\begin{aligned} R^a{}_b &= \frac{1}{2} R^a{}_{bcd} e^c{}_\mu e^d{}_\nu dx^\mu dx^\nu \\ &= \frac{1}{2} R^a{}_{b\mu\nu} dx^\mu dx^\nu \end{aligned} \tag{12.219}$$

and

$$E_a{}^\alpha R^a{}_{bcd} e^b{}_\beta e^c{}_\mu e^d{}_\nu = R^\alpha{}_{\beta\mu\nu}. \tag{12.220}$$

As we discussed before, the curvature is also determined by the commutator of the covariant derivatives.

Theorem:

$$[D_\mu, D_\nu]^a{}_b = R^a{}_{b\mu\nu} \tag{12.221}$$

$$[D_\mu, D_\nu]^a{}_b = [R_{\mu\nu}, \;]^a{}_b \quad \text{for tangent frame tensors} \tag{12.222}$$

$$D^2 = [R, \;] \quad \text{within differential forms.} \tag{12.223}$$

The **Bianchi identity** can be written as in the YM case

$$DR = 0. \tag{12.224}$$

Torsion: The **torsion 2-form** is defined by

$$\begin{aligned} T = De &= de + \omega e && \text{matrix notation} \\ T^a = (De)^a &= de^a + \omega^a{}_b e^b && \text{component notation.} \end{aligned} \tag{12.225}$$

Applying d gives

$$\begin{aligned} dT &= dDe = d^2 e + d\omega e - \omega de \\ &= Re - \omega T \end{aligned} \tag{12.226}$$

which is called the **torsion consistency condition**.

Again, we use the vielbeins to establish the bridge to the coordinate basis

$$\begin{aligned} T^a &= \frac{1}{2} T^a{}_{bc} e^b e^c = \frac{1}{2} T^a{}_{bc} e^b{}_\mu e^c{}_\nu dx^\mu dx^\nu \\ &= \frac{1}{2} T^a{}_{\mu\nu} dx^\mu dx^\nu \end{aligned} \tag{12.227}$$

and

$$\begin{aligned} T^a{}_{\mu\nu} &= T^a{}_{bc} e^b{}_\mu e^c{}_\nu \\ T^\alpha{}_{\mu\nu} &= E_a{}^\alpha T^a{}_{bc} e^b{}_\mu e^c{}_\nu . \end{aligned} \tag{12.228}$$

Definition: As in the coordinate basis the connection is called the **Riemannian connection** if

i) *the metric is covariantly constant*

$$D_c \eta_{ab} = 0 \Longleftrightarrow \omega_{ab} = -\omega_{ba} \tag{12.229}$$

ii) *the connection has zero torsion*

$$T^a = 0. \tag{12.230}$$

Theorem: The Riemann conditions i) and ii) determine the spin connection ω uniquely in terms of the vielbein e^a and its derivative de^a via the formula

$$\omega_{abc} = \frac{1}{2}(\xi_{abc} + \xi_{bca} - \xi_{cab}), \tag{12.231}$$

with

$$\xi^a{}_{bc} = de^a(E_b, E_c). \tag{12.232}$$

Proof. First, we notice that the covariantly constant metric

$$\begin{aligned} D\eta_{ab} = d\eta_{ab} + [\omega, \eta]_{ab} &= 0 \\ \eta_{cb}\omega^c{}_a + \eta_{ac}\omega^c{}_b &= 0 \end{aligned} \tag{12.233}$$

implies the antisymmetry of the spin connection matrix

$$\omega_{ab} = -\omega_{ba}, \quad \text{where} \quad \omega_{ab} = \eta_{ac}\omega^c{}_b, \tag{12.234}$$

which is equivalent to ω_{ab} being an element of the Lie algebra Lie $SO(m)$.

Next, we define a quantity $\xi^a{}_{bc}$ by

$$de^a =: \frac{1}{2}\, \xi^a{}_{bc} e^b e^c, \tag{12.235}$$

then the torsion-free condition gives

$$\begin{aligned} T^a = de^a + \omega^a{}_c e^c &= 0 \\ \frac{1}{2}\, \xi^a{}_{bc} e^b e^c + \omega^a{}_{cb} e^b e^c &= 0; \end{aligned} \tag{12.236}$$

hence

$$\xi_{abc} = \omega_{abc} - \omega_{acb}. \tag{12.237}$$

We raise and lower the tangent frame indices with the metric tensor η_{ab}. Next we add and subtract the cyclic permutations of equation (12.237) and we find the explicit expression for the spin connection

$$\omega_{abc} = \frac{1}{2}(\xi_{abc} + \xi_{bca} - \xi_{cab}). \tag{12.238}$$

We now rewrite the quantity $\xi^a{}_{bc}$, defined via equation (12.235). We evaluate the wedge product $e^b e^c$ at the vector fields $E_{\bar{b}}$, $E_{\bar{c}}$ (analogous to equation (2.111) of Section 2.4)

$$\begin{aligned} e^b e^c(E_{\bar{b}}, E_{\bar{c}}) &= (e^b \otimes e^c - e^c \otimes e^b)(E_{\bar{b}}, E_{\bar{c}}) \\ &= \delta^b{}_{\bar{b}}\delta^c{}_{\bar{c}} - \delta^c{}_{\bar{b}}\delta^b{}_{\bar{c}}; \end{aligned} \tag{12.239}$$

then we get

$$de^a(E_{\bar{b}}, E_{\bar{c}}) = \frac{1}{2}\, \xi^a{}_{bc} e^b e^c(E_{\bar{b}}, E_{\bar{c}}) = \xi^a{}_{\bar{b}\bar{c}}. \tag{12.240}$$

Note that $\xi^a{}_{bc}$ is antisymmetric in bc. So finally in terms of components we have

$$\begin{aligned} \xi^a{}_{bc} &= \frac{\partial}{\partial x^\mu}\, e^a{}_\nu dx^\mu dx^\nu \left(E_b{}^\alpha \frac{\partial}{\partial x^\alpha}, E_c{}^\beta \frac{\partial}{\partial x^\beta} \right) \\ &= \partial_\mu e^a{}_\nu (E_b{}^\mu E_c{}^\nu - E_b{}^\nu E_c{}^\mu) \\ &= (\partial_\mu e^a{}_\nu - \partial_\nu e^a{}_\mu) E_b{}^\mu E_c{}^\nu. \qquad \text{Q.E.D.} \end{aligned} \tag{12.241}$$

Remark: The zero torsion implies by virtue of condition (12.226)

$$Re = 0 \tag{12.242}$$

which is the **first Bianchi identity** of the curvature tensor

$$R^a{}_{bcd} + R^a{}_{cdb} + R^a{}_{dbc} = 0. \tag{12.243}$$

12.2.3 Examples in 2 dimensions

Now let us illustrate the concepts introduced on the 2-dimensional sphere.

Metric: Generally, the **metric** is expressed by

$$g_p = g_{\mu\nu}(x) dx^\mu \otimes dx^\nu = \eta_{ab}\, e^a(x) \otimes e^b(x); \tag{12.244}$$

here, on the sphere S^2 with radius L we have the standard metric (12.24)

$$\begin{aligned} g_{S^2} &= L^2 d\theta \otimes d\theta + L^2 \sin^2\theta d\varphi \otimes d\varphi \\ &= e^1 \otimes e^1 + e^2 \otimes e^2 \end{aligned} \tag{12.245}$$

or, written as a **line-element**

$$\begin{aligned} ds^2 &= L^2 d\theta^2 + L^2 \sin^2\theta d\varphi^2 \\ &= (e^1)^2 + (e^2)^2. \end{aligned} \tag{12.246}$$

Zweibein: So the **zweibein 1-form** is

$$e^a = e^a{}_\varphi d\varphi + e^a{}_\theta d\theta, \qquad a = 1,2, \tag{12.247}$$

with

$$e^1 = Ld\theta, \qquad e^2 = L\sin\theta d\varphi, \tag{12.248}$$

and the **zweibeins** are

$$\begin{aligned} e^1{}_\varphi &= 0 \qquad & e^2{}_\varphi &= L\sin\theta \\ e^1{}_\theta &= L \qquad & e^2{}_\theta &= 0. \end{aligned} \tag{12.249}$$

The exterior derivative as a 1-form is

$$d = \frac{\partial}{\partial\varphi}\, d\varphi + \frac{\partial}{\partial\theta}\, d\theta, \tag{12.250}$$

so that we get

$$de^1 = 0 \qquad de^2 = L\cos\theta\ d\theta d\varphi. \tag{12.251}$$

The **inverse zweibein**

$$E_a = E_a{}^\varphi \frac{\partial}{\partial\varphi} + E_a{}^\theta \frac{\partial}{\partial\theta} \tag{12.252}$$

has to satisfy the system of equations

$$e^a{}_\varphi E_b{}^\varphi + e^a{}_\theta E_b{}^\theta = \delta^a{}_b, \tag{12.253}$$

implying the components

$$\begin{aligned} E_1{}^\varphi &= 0 \qquad & E_2{}^\varphi &= \frac{1}{L\sin\theta} \\ E_1{}^\theta &= \frac{1}{L} \qquad & E_2{}^\theta &= 0 \end{aligned} \tag{12.254}$$

and the vectors

$$E_1 = \frac{1}{L}\frac{\partial}{\partial\theta} \qquad E_2 = \frac{1}{L\sin\theta}\frac{\partial}{\partial\varphi}. \tag{12.255}$$

Spin connection: The **spin connection 1-form** $\omega^a{}_b = \omega^a{}_{bc}e^c$ is given by formula (12.231), (12.232).

Let us first calculate the quantities

$$\begin{aligned} \xi^1{}_{bc} &= de^1(E_b, E_c) = 0 \\ \xi^2{}_{bc} &= de^2(E_b, E_c) = L\cos\theta \; d\theta d\varphi(E_b, E_c). \end{aligned} \tag{12.256}$$

Generally we have (analogous to equation (2.111) of Section 2.4)

$$d\theta d\varphi(E_a, E_b) = E_a{}^\theta E_b{}^\varphi - E_b{}^\theta E_a{}^\varphi \tag{12.257}$$

and particularly

$$d\theta d\varphi(E_1, E_2) = -d\theta d\varphi(E_2, E_1) = \frac{1}{L^2 \sin\theta}, \tag{12.258}$$

so that we get

$$\xi^2{}_{12} = -\xi^2{}_{21} = \frac{\cos\theta}{L\sin\theta}. \tag{12.259}$$

Then formula $\omega_{abc} = (\xi_{abc} + \xi_{bca} - \xi_{cab})/2$ provides the (nonvanishing) **spin connection components**

$$-\omega^1{}_{22} = \omega^2{}_{12} = \xi^2{}_{12} = \frac{\cos\theta}{L\sin\theta} \tag{12.260}$$

and the **spin connection 1-form**

$$-\omega^1{}_2 = \omega^2{}_1 = \omega^2{}_{12} e^2 = \cos\theta d\varphi. \tag{12.261}$$

Curvature: For the **curvature 2-form** given by the spin connection

$$R^a{}_b = d\omega^a{}_b + \omega^a{}_c \omega^c{}_b \tag{12.262}$$

we find specifically in 2 dimensions

$$\begin{aligned} -R^2{}_1 &= R^1{}_2 = d\omega^1{}_2 = \sin\theta d\theta d\varphi \equiv \frac{1}{L^2}\, e^1 e^2 \\ R^1{}_1 &= R^2{}_2 = 0, \end{aligned} \tag{12.263}$$

so that

$$R^a{}_b = \frac{1}{L^2}\, e^a e^b. \tag{12.264}$$

So the **Riemann tensor** itself (recall $R^1{}_2 = R^1{}_{212} e^1 e^2$)

$$-R^1{}_{221} = R^1{}_{212} = \frac{1}{L^2} \tag{12.265}$$

expresses the constant curvature of the sphere S^2.

Note: Of course, in this simple 2-dimensional example we can deduce the spin connection even more quickly from the zero torsion condition (12.230) (which in fact implies the formulae we used)

$$T^a = (De)^a = de^a + \omega^a{}_b e^b = 0. \tag{12.266}$$

From the antisymmetry of ω_{ab} we know that

$$\omega_{11} = \omega_{22} = 0, \tag{12.267}$$

then we get for $a = 2$

$$\begin{aligned} de^2 + \omega^2{}_1 e^1 &= 0 \\ \omega^2{}_1 d\theta &= \cos\theta \; d\varphi d\theta \\ \omega^2{}_1 &= \cos\theta d\varphi. \end{aligned} \tag{12.268}$$

S^n-sphere: If we generalize the above calculations to the n-dimensional sphere S^n using stereographic projection coordinates, we find for the infinitesimal **line-element**

$$ds^2 = \frac{\sum_{a=1}^n (dx^a)^2}{\left[1 + \left(\frac{r}{2L}\right)^2\right]^2} = \sum_{a=1}^n (e^a)^2, \tag{12.269}$$

with

$$r^2 = \sum_{a=1}^n (x^a)^2, \tag{12.270}$$

and for the **vielbein**

$$e^a = \frac{dx^a}{1 + \left(\frac{r}{2L}\right)^2}. \tag{12.271}$$

The invariant volume element is

$$dV_{S^n} = e^1 \dots e^n = e \, dx^1 \dots dx^n = \frac{d^n x}{\left[1 + \left(\frac{r}{2L}\right)^2\right]^n} \tag{12.272}$$

giving the 'volume' over S^n

$$V(S^n) = \frac{2\pi^{(n+1)/2}}{\Gamma\left(\frac{n+1}{2}\right)} L^n. \tag{12.273}$$

The **spin connection 1-form** turns out to be

$$\omega^a{}_b = \frac{x^a dx^b - x^b dx^a}{2L^2\left[1 + \left(\frac{r}{2L}\right)^2\right]}, \tag{12.274}$$

supplying the **curvature**

$$R^a{}_b = \frac{1}{L^2}\, e^a e^b, \qquad R_{abcd} = \frac{1}{L^2}(\delta_{ac}\delta_{bd} - \delta_{ad}\delta_{bc}) \tag{12.275}$$

and the Ricci quantities

$$\mathcal{R}_{bd} = R^a{}_{bad} = \frac{n-1}{L^2}\,\delta_{bd}, \qquad \mathcal{R} = \mathcal{R}^b{}_b = \frac{n(n-1)}{L^2}. \tag{12.276}$$

$\mathbf{R}^2$-plane: Finally, let us consider the $\mathbf{R}^2$-plane; we express the **line-element** by polar coordinates

$$ds^2 = dr^2 + r^2 d\varphi^2 = (e^1)^2 + (e^2)^2. \tag{12.277}$$

Then the **zweibeins**

$$e^a = e^a{}_r dr + e^a{}_\varphi d\varphi \tag{12.278}$$

follow as

$$\begin{aligned} e^1 &= dr & \qquad e^2 &= r d\varphi \\ e^1{}_r &= 1 & e^2{}_r &= 0 \\ e^1{}_\varphi &= 0 & e^2{}_\varphi &= r. \end{aligned} \tag{12.279}$$

The exterior derivative

$$d = \frac{\partial}{\partial r}\, dr + \frac{\partial}{\partial \varphi}\, d\varphi \tag{12.280}$$

gives

$$de^1 = 0, \qquad de^2 = dr d\varphi, \tag{12.281}$$

so that we find the **connection 1-form** from the zero torsion condition

$$\begin{aligned} T^2 = (De)^2 = de^2 + \omega^2{}_1 e^1 &= 0 \\ \omega^2{}_1 dr &= d\varphi\, dr; \end{aligned}$$

hence

$$-\omega^1{}_2 = \omega^2{}_1 = d\varphi \equiv \frac{e^2}{r}. \tag{12.282}$$

The **curvature** clearly vanishes

$$-R^1{}_2 = R^2{}_1 = d\omega^2{}_1 = 0 \tag{12.283}$$

as it should be on a plane.

12.2.4 Relation between Christoffel- and spin connection

For a Riemannian manifold with conditions (12.76), (12.77) or (12.229), (12.230) there exists *one unique connection* which is given by the Christoffel connection Γ if we refer to a coordinate system (CS) or by the spin connection ω if we refer to the tangent frame (TF). The associated curvature is then determined by

$$R(\Gamma) = d\Gamma + \Gamma^2 \qquad \text{in CS} \tag{12.284}$$

$$R(\omega) = d\omega + \omega^2 \qquad \text{in TF.} \tag{12.285}$$

Again, we emphasize that $R^\alpha{}_\beta(\Gamma)$ and $R^a{}_b(\omega)$ describe one and the same geometrical object—namely the *curvature*—which is either referred to CS or to TF. With the help of the vielbeins we can shift the curvature—in general any tensor—from its CS reference to the TF reference and vice versa

$$R^\alpha{}_\beta = E_a{}^\alpha R^a{}_b e^b{}_\beta. \tag{12.286}$$

In order to ensure this relation for the curvature, the connections Γ and ω must transform into each other in a definite way. Alternatively, the connections Γ and ω depend on the metric $g_{\mu\nu}$ and on the vielbein $e^a{}_\mu$ respectively (recall formulae (12.63) and (12.231), (12.232)), $g_{\mu\nu}$ and $e^a{}_\mu$ are closely related to each other, consequently Γ and ω must be related too.

Let us find this relation. First, we notice that

$$\nabla_\alpha (g^{\mu\nu} e^a{}_\mu e^b{}_\nu) = 0 \tag{12.287}$$

$$(\nabla_\alpha g^{\mu\nu}) e^a{}_\mu e^b{}_\nu + g^{\mu\nu} (\nabla_\alpha e^a{}_\mu) e^b{}_\nu + g^{\mu\nu} e^a{}_\mu \nabla_\alpha e^b{}_\nu = 0$$

(recall the covariant constancy of the metric); hence

$$E^{b\mu} \nabla_\alpha e^a{}_\mu + E^{a\mu} \nabla_\alpha e^b{}_\mu = 0. \tag{12.288}$$

Secondly, by using

$$de^a = \nabla e^a = (\nabla e^a{}_\mu) E_c{}^\mu e^c \tag{12.289}$$

we rewrite the torsion

$$\begin{aligned} T^a &= de^a + \omega^a{}_c e^c \\ &= \frac{1}{2} [E_c{}^\mu \nabla_b e^a{}_\mu - E_b{}^\mu \nabla_c e^a{}_\mu + \omega^a{}_{cb} - \omega^a{}_{bc}] e^b e^c. \end{aligned} \tag{12.290}$$

The zero torsion condition then gives

$$E_c{}^\mu \nabla_b e^a{}_\mu - E_b{}^\mu \nabla_c e^a{}_\mu + \omega^a{}_{cb} - \omega^a{}_{bc} = 0 \tag{12.291}$$

to which we subtract and add cyclic permutations. Using identity (12.288) and the antisymmetry of ω_{ab} we obtain

$$\omega^a{}_{bc} = E^{a\mu}\nabla_c e_{b\mu} = -E_b{}^\mu \nabla_c e^a{}_\mu. \tag{12.292}$$

Noting that

$$E^{a\mu}\nabla e_{b\mu} = g^{\mu\nu} e^a{}_\nu \nabla(g_{\mu\alpha}E_b{}^\alpha) = e^a{}_\mu \nabla E_b{}^\mu \tag{12.293}$$

we find the desired relation:

Theorem:

$$\omega^a{}_b = e^a{}_\mu \nabla E_b{}^\mu = e^a{}_\mu dE_b{}^\mu + e^a{}_\mu \Gamma^\mu{}_\nu E_b{}^\nu \tag{12.294}$$

$$\omega^a{}_b = -E_b{}^\mu \nabla e^a{}_\mu = -E_b{}^\mu de^a{}_\mu + E_b{}^\mu e^a{}_\nu \Gamma^\nu{}_\mu. \tag{12.295}$$

Here the covariant derivative ∇ acts only on the vielbein.

Next we turn to the covariant derivative

$$\begin{aligned} D^a{}_b &= \delta^a{}_b d + \omega^a{}_b \\ &= e^a{}_\mu \delta^\mu{}_\nu E_b{}^\nu d + e^a{}_\mu \delta^\mu{}_\nu dE_b{}^\nu + e^a{}_\mu \Gamma^\mu{}_\nu E_b{}^\nu \\ &= e^a{}_\mu (\delta^\mu{}_\nu d + \Gamma^\mu{}_\nu) E_b{}^\nu \end{aligned} \tag{12.296}$$

and we find:

Theorem:

$$D^a{}_b = e^a{}_\mu \nabla^\mu{}_\nu E_b{}^\nu = e^a{}_\mu \nabla E_b{}^\mu. \tag{12.297}$$

Here the derivative also acts further to the right.

Note: As emphasized before, there is only one covariant derivative. If we refer to a CS it is ∇ containing the Christoffel connection Γ, if we refer to the TF it is D containing the spin connection ω, and both can be transformed into each other via the vielbeins.

For instance, if we apply the derivative to a vector we have

$$\begin{aligned} DV^a &= (DV)^a = D^a{}_b V^b = e^a{}_\mu \nabla^\mu{}_\nu E_b{}^\nu V^b \\ &= e^a{}_\mu \nabla^\mu{}_\nu V^\nu = e^a{}_\mu \nabla V^\mu. \end{aligned} \tag{12.298}$$

To reverse the above relations we calculate

$$E_a{}^\alpha \omega^a{}_b e^b{}_\beta = E_a{}^\alpha e^a{}_\mu dE_b{}^\mu e^b{}_\beta + E_a{}^\alpha e^a{}_\mu \Gamma^\mu{}_\nu E_b{}^\nu e^b{}_\beta; \tag{12.299}$$

we use

$$dE_b{}^\mu e^b{}_\beta + E_b{}^\mu de^b{}_\beta = 0 \tag{12.300}$$

following from $E\, e = \mathbf{1}$ and we obtain the relation:

	coordinate system	tangent frame
basis	$\{dx^\mu\}, \left\{\dfrac{\partial}{\partial x^\mu}\right\}$	$\{e^a\}, \{E_a\}$
metric	$g_{\mu\nu}(x)$, $V_\mu = g_{\mu\nu}V^\nu$	η_{ab}, $V_a = \eta_{ab}V^b$
connection	$\Gamma = \Gamma_\mu dx^\mu$	$\omega = \omega_c e^c$
covariant derivative	$\nabla = d + [\Gamma, \ ]$	$D = d + [\omega, \ ]$
curvature	$R = d\Gamma + \Gamma^2$	$R = d\omega + \omega^2$
torsion	$T^\alpha = \nabla dx^\alpha$	$T^a = De^a$
tensor	$T^{\alpha\ldots}_{\beta\ldots}$	$T^{a\ldots}_{b\ldots}$

Table 12.1.

transformations	
coordinate system	tangent frame
$dx^\mu = E_a{}^\mu(x)e^a(x)$	$e^a(x) = e^a{}_\mu(x)dx^\mu$
$\dfrac{\partial}{\partial x^\mu} = e^a{}_\mu(x)E_a(x)$	$E_a(x) = E_a{}^\mu(x)\dfrac{\partial}{\partial x^\mu}$
$g_{\mu\nu}(x) = \eta_{ab}e^a{}_\mu(x)e^b{}_\nu(x)$	$\eta_{ab} = g_{\mu\nu}(x)E_a{}^\mu(x)E_b{}^\nu(x)$
$\Gamma = E(De)$	$\omega = e(\nabla E)$
$\nabla = EDe$	$D = e\nabla E$
$R^\alpha{}_\beta = E_a{}^\alpha R^a{}_b e^b{}_\beta$	$R^a{}_b = e^a{}_\alpha R^\alpha{}_\beta E_b{}^\beta$
$T^\alpha = E_a{}^\alpha T^a$	$T^a = e^a{}_\alpha T^\alpha$
$T^{\alpha\ldots}_{\beta\ldots} = E_a{}^\alpha \ldots T^{a\ldots}_{b\ldots} e^b{}_\beta \ldots$	$T^{a\ldots}_{b\ldots} = e^a{}_\alpha \ldots T^{\alpha\ldots}_{\beta\ldots} E_b{}^\beta \ldots$

Table 12.2.

Theorem:

$$\Gamma^\alpha{}_\beta = E_a{}^\alpha de^a{}_\beta + E_a{}^\alpha \omega^a{}_b e^b{}_\beta = E_a{}^\alpha De^a{}_\beta \tag{12.301}$$

$$\nabla^\alpha{}_\beta = E_a{}^\alpha De^a{}_\beta. \tag{12.302}$$

Again, in equation (12.301) the derivative acts only on the vielbein, whereas in equation (12.302) the derivative acts further to the right.

Let us summarize in our practical matrix notation:

Theorem:

$$\begin{aligned}
\omega &= e(\nabla E) = edE + e\Gamma E \\
\Gamma &= E(De) = Ede + E\omega e \\
D &= e\nabla E = e(d+\Gamma)E \\
\nabla &= EDe = E(d+\omega)e.
\end{aligned} \tag{12.303}$$

Finally, we collect the discussed quantities for the two kinds of reference—CS or TF—in Tables 12.1 and 12.2. What we observe is the following. We may regard $\Gamma^\alpha{}_\beta$ as a **gauge transformation** of $\omega^a{}_b$ with the vielbein $e^a{}_\beta \in GL(m, \mathbf{R})$ as a **gauge element**. Similarly $\nabla^\alpha{}_\beta$ represents a $GL(m, \mathbf{R})$ gauge transformation of $D^a{}_b$, the curvature $R^\alpha{}_\beta$ a transformation of $R^a{}_b$, the torsion T^α a transformation of T^a, generally a tensor $\mathcal{T}^{\alpha\dots}_{\beta\dots}$ a gauge transformation of $\mathcal{T}^{a\dots}_{b\dots}$.

12.2.5 Local Lorentz transformations

As we have discussed already, the vielbein is determined only up to a frame rotation—a local Lorentz transformation $L^a{}_b(x) \in SO(1, m-1)$ or $SO(m)$

$$\begin{aligned}
e &\to L^{-1}e \\
E &\to EL.
\end{aligned} \tag{12.304}$$

Under such a rotation the following **Lorentz transformations** hold

$$\begin{aligned}
\omega &\to L^{-1}(\omega + d)L \\
D &\to L^{-1}DL \\
R &\to L^{-1}RL \\
T &\to L^{-1}T \\
\mathcal{T} &\to L^{-1}\mathcal{T}L,
\end{aligned} \tag{12.305}$$

where we use the convenient matrix notation; it is understood that L^{-1} acts on *all* upper indices and L on *all* lower indices. All equations follow when using expression (12.294) as a definition for the spin connection. These transformations are quite analogous to the gauge transformations in YM gauge theory—the Lorentz transformation $L(x)$ plays the role of the gauge group element $g(x)$—and we regard the equations (12.305) as **gauge transformations**, the Lorentz group as a $SO(1, m-1)$ or $SO(m)$ **gauge group.**

Mainly we consider an **infinitesimal Lorentz transformation**

$$L(x) = \mathbf{1} + \alpha(x) \qquad\qquad L^{-1}(x) = \mathbf{1} - \alpha(x)$$

$$L^a{}_b(x) = \delta^a{}_b + \alpha^a{}_b(x), \tag{12.306}$$

then the 'rotation angle', the matrix α_{ab}, is antisymmetric

$$\alpha_{ab} = -\alpha_{ba}. \tag{12.307}$$

This is a consequence of the definition of the Lorentz transformation

$$\eta_{ab} L^a{}_c L^b{}_d = \eta_{cd}$$

$$\eta_{ab}(\delta^a{}_c + \alpha^a{}_c)(\delta^b{}_d + \alpha^b{}_d) = \eta_{cd};$$

hence

$$\eta_{ab}\alpha^a{}_c\delta^b{}_d + \eta_{ab}\delta^a{}_c\alpha^b{}_d = 0$$

$$\alpha_{dc} + \alpha_{cd} = 0. \tag{12.308}$$

For our purpose—the study of anomalies—we work with the **infinitesimal Lorentz transformations**

$$\begin{aligned} \delta^L_\alpha e &= -\alpha e \\ \delta^L_\alpha E &= E\alpha \\ \delta^L_\alpha \omega &= D\alpha \\ \delta^L_\alpha R &= [R, \alpha] \\ \delta^L_\alpha T &= -\alpha T. \end{aligned} \tag{12.309}$$

Again, we regard these equations as **gauge transformations** with the **gauge element** $\alpha(x)$ analogous to the YM case.

12.3 Action principle

In order to derive the dynamical equations for a physical system we use the **principle of least action**. It means that the action remains stationary with respect to small variations of a field. This view provides the important bridge between symmetry and conservation law.

12.3.1 Einstein–Hilbert action

The theory of general relativity describes the dynamics of geometry. The corresponding action—the **Einstein–Hilbert action**—is given by

$$S_{\text{EH}} = \frac{1}{2\kappa}\int d^m x\, \sqrt{|g|}\, \mathcal{R}, \tag{12.310}$$

where $(c = 1)$

$$\kappa = 8\pi G, \tag{12.311}$$

G denotes the gravitational constant, and $\mathcal{R}$ is the Ricci scalar (12.95). The factor $1/2\kappa$ is chosen to reproduce the correct Newtonian limit in the presence of matter. Clearly S_{EH} is a scalar under general coordinate transformations.

We vary the metric arbitrarily, where the variation should vanish at the boundary of the integration domain (including infinity)

$$g_{\mu\nu}(x) \to g_{\mu\nu}(x) + \delta g_{\mu\nu}(x), \qquad \delta g_{\mu\nu}(x) \overset{|x|\to\text{boundary}}{\longrightarrow} 0. \tag{12.312}$$

We consider the metric $g_{\mu\nu}(x)$ as an external gravitational field. Then we apply the **action principle**

$$\delta S_{\text{EH}} = 0 \tag{12.313}$$

to gain the equations of motion, the Einstein equations. We have to vary

$$\begin{aligned} \delta(\sqrt{|g|}\,\mathcal{R}) &= \delta(\sqrt{|g|}\,g^{\mu\nu}\mathcal{R}_{\mu\nu}) \qquad (12.314) \\ &= \delta\sqrt{|g|}\,g^{\mu\nu}\mathcal{R}_{\mu\nu} + \sqrt{|g|}\,\delta g^{\mu\nu}\mathcal{R}_{\mu\nu} + \sqrt{|g|}\,g^{\mu\nu}\delta\mathcal{R}_{\mu\nu}. \end{aligned}$$

For the variation of the Ricci tensor we use Palatini's identity (12.113); multiplication with the metric gives

$$\sqrt{|g|}\,g^{\mu\nu}\delta\mathcal{R}_{\mu\nu} = \sqrt{|g|}\,\nabla_\alpha(g^{\mu\nu}\delta\Gamma^\alpha{}_{\nu\mu}) - \sqrt{|g|}\,\nabla_\nu(g^{\mu\nu}\delta\Gamma^\alpha{}_{\alpha\mu}), \tag{12.315}$$

which we convert into an ordinary derivative via formula (12.133)

$$\sqrt{|g|}\,g^{\mu\nu}\delta\mathcal{R}_{\mu\nu} = \partial_\alpha(\sqrt{|g|}\,g^{\mu\nu}\delta\Gamma^\alpha{}_{\nu\mu}) - \partial_\nu(\sqrt{|g|}\,g^{\mu\nu}\delta\Gamma^\alpha{}_{\alpha\mu}). \tag{12.316}$$

This expression vanishes upon integration by virtue of Gauss' theorem since $\delta\Gamma^\alpha_{\nu\mu} \to 0$ at the boundary (that means we also postulate that $\delta\partial_\lambda g_{\mu\nu} \to 0$ at the boundary, besides $\delta g_{\mu\nu} \to 0$).

For the remaining variations we use formulae (12.107), (12.110), we find

$$\delta(\sqrt{|g|}\,\mathcal{R}) = \sqrt{|g|}\left(\frac{1}{2}g^{\mu\nu}\mathcal{R} - \mathcal{R}^{\mu\nu}\right)\delta g_{\mu\nu}, \tag{12.317}$$

and from the action principle

$$\delta S_{\text{EH}} = \frac{1}{2\kappa}\int d^m x\sqrt{|g|}\left(\frac{1}{2}g^{\mu\nu}\mathcal{R} - \mathcal{R}^{\mu\nu}\right)\delta g_{\mu\nu} = 0 \tag{12.318}$$

we obtain the **Einstein equation without matter**

$$\mathcal{R}^{\mu\nu} - \frac{1}{2}g^{\mu\nu}\mathcal{R} = 0. \tag{12.319}$$

The symmetric tensor

$$G^{\mu\nu} := \mathcal{R}^{\mu\nu} - \frac{1}{2}g^{\mu\nu}\mathcal{R} \tag{12.320}$$

is usually called the **Einstein tensor**. So without matter, for a free gravitational field we just have

$$G^{\mu\nu} = 0. \tag{12.321}$$

12.3.2 Energy–momentum tensor

Next we include matter by considering the action

$$S_M = \int d^m x \sqrt{|g|}\, \mathcal{L}_M, \tag{12.322}$$

where $\mathcal{L}_M$ is the Lagrange density of matter—scalars, fermions, gauge fields—as discussed in Chapter 3.

The change of the matter density under a variation $\delta g_{\mu\nu}$ defines the **energy–momentum tensor** $T^{\mu\nu}$ of the matter system

$$\begin{aligned} \delta S_M &= \frac{1}{2} \int d^m x \sqrt{|g|}\, T^{\mu\nu} \delta g_{\mu\nu} \\ \frac{\delta S_M}{\delta g_{\mu\nu}} &= \frac{1}{2} \sqrt{|g|}\, T^{\mu\nu}. \end{aligned} \tag{12.323}$$

The following **properties** hold:

i) $T^{\mu\nu} = T^{\nu\mu}$ **symmetry** (12.324)

ii) $\nabla_\mu T^{\mu\nu} = 0$ **covariant conservation.** (12.325)

Property i) follows from the symmetry of $\delta g_{\mu\nu}$; we prove property ii) later on.

Now we consider a gravitational field which is coupled to matter; the corresponding action is

$$S = S_{\mathrm{EH}} + S_M \tag{12.326}$$

and the action principle

$$\delta S = \frac{1}{2\kappa} \int d^m x \sqrt{|g|} \left(\frac{1}{2} g^{\mu\nu} \mathcal{R} - \mathcal{R}^{\mu\nu} + \kappa T^{\mu\nu} \right) \delta g_{\mu\nu} = 0 \tag{12.327}$$

provides the **Einstein equation with matter**

$$\begin{aligned} \mathcal{R}^{\mu\nu} - \frac{1}{2} g^{\mu\nu} \mathcal{R} &= \kappa T^{\mu\nu} \\ G^{\mu\nu} &= \kappa T^{\mu\nu}. \end{aligned} \tag{12.328}$$

So energy and momentum influence the structure of space–time represented by the Einstein tensor.

Sometimes an additional constant—the **cosmological constant** Λ—is included in the action without destroying the invariance of the action

$$S_{\mathrm{EH}} = \frac{1}{2\kappa} \int d^m x \sqrt{|g|} (\mathcal{R} + \Lambda). \tag{12.329}$$

But we do not pursue this case further.

Proof. Now we prove the important property of the covariant conservation of the energy–momentum tensor (equation (12.325)). We start with the Bianchi identity (12.92) and respect the constancy of the metric (12.76)

$$\nabla_\sigma R_{\alpha\beta\mu\nu} + \nabla_\mu R_{\alpha\beta\nu\sigma} + \nabla_\nu R_{\alpha\beta\sigma\mu} = 0. \tag{12.330}$$

We contract the indices α and μ which yields the Ricci tensors

$$\nabla_\sigma \mathcal{R}_{\beta\nu} - \nabla_\nu \mathcal{R}_{\beta\sigma} - \nabla_\alpha R_\beta{}^\alpha{}_{\nu\sigma} = 0,$$

contracting again between β and ν gives

$$\begin{aligned} \nabla_\sigma \mathcal{R} - \nabla_\nu \mathcal{R}^\nu{}_\sigma - \nabla_\alpha \mathcal{R}^\alpha{}_\sigma &= 0 \\ \nabla_\mu \left(\mathcal{R}^{\mu\nu} - \frac{1}{2} g^{\mu\nu} \mathcal{R} \right) &= 0 \\ \nabla_\mu T^{\mu\nu} &= 0. \qquad \text{Q.E.D.} \end{aligned} \tag{12.331}$$

In the last step we used the Einstein equation (12.328).

Examples:

i) *Scalar fields*
Let us first consider scalar fields given by the action

$$S_{\text{scalar}} = -\frac{1}{2} \int d^4 x \sqrt{|g|} [g^{\mu\nu} \partial_\mu \phi \partial_\nu \phi + m^2 \phi^2]. \tag{12.332}$$

The variation gives (remember formulae (12.107), (12.110))

$$\begin{aligned} \delta S_{\text{scalar}} &= -\frac{1}{2} \int d^4 x \, \delta \sqrt{|g|} [g^{\mu\nu} \partial_\mu \phi \partial_\nu \phi + m^2 \phi^2] \\ &\quad - \frac{1}{2} \int d^4 x \sqrt{|g|} \, \delta g^{\mu\nu} \partial_\mu \phi \partial_\nu \phi \\ &= \frac{1}{2} \int d^4 x \sqrt{|g|} [\partial^\mu \phi \partial^\nu \phi \\ &\quad - \frac{1}{2} g^{\mu\nu} (g^{\alpha\beta} \partial_\alpha \phi \partial_\beta \phi + m^2 \phi^2)] \delta g_{\mu\nu} \end{aligned} \tag{12.333}$$

so that we find for the energy–momentum tensor

$$T^{\mu\nu} = \partial^\mu \phi \partial^\nu \phi - \frac{1}{2} g^{\mu\nu} (g^{\alpha\beta} \partial_\alpha \phi \partial_\beta \phi + m^2 \phi^2). \tag{12.334}$$

ii) *Photon fields*
Varying the photon action (recall equation (3.252)) provides

$$\begin{aligned} S_{\mathrm{ED}} &= -\frac{1}{4}\int d^4x\, \sqrt{|g|}\, g^{\mu\alpha} g^{\nu\beta} F_{\mu\nu} F_{\alpha\beta} \\ \delta S_{\mathrm{ED}} &= \frac{1}{2}\int d^4x\, \sqrt{|g|}\left[F^{\mu\alpha} F^{\nu}{}_{\alpha} - \frac{1}{4} g^{\mu\nu} F_{\alpha\beta} F^{\alpha\beta}\right]\delta g_{\mu\nu} \\ T^{\mu\nu} &= F^{\mu\alpha} F^{\nu}{}_{\alpha} - \frac{1}{4} g^{\mu\nu} F_{\alpha\beta} F^{\alpha\beta}. \end{aligned} \tag{12.335}$$

Vielbein formalism: When working with fermions we have to express the action in terms of the vielbein $e^a{}_\mu$. Then we vary the vielbein

$$e^a{}_\mu(x) \to e^a{}_\mu(x) + \delta e^a{}_\mu(x), \qquad \delta e^a{}_\mu(x) \overset{|x|\to\text{boundary}}{\longrightarrow} 0 \tag{12.336}$$

and the **energy–momentum tensor** of the fermionic system is defined by

$$\begin{aligned} \delta S_M &= \int d^m x\, e T^\mu{}_a \delta e^a{}_\mu \\ \frac{\delta S_M}{\delta e^a{}_\mu} &= e T^\mu{}_a. \end{aligned} \tag{12.337}$$

We extract the vielbein determinant $e = |\det e^a{}_\mu|$ to achieve a true tensor quantity instead of a tensor density. The tensor $T^\mu{}_a$ is a quantity which represents a contravariant coordinate vector and a covariant Lorentz vector. It is related to the previously introduced **symmetric tensor** $T^{\mu\nu}$ by

$$T^{\mu\nu} = \frac{1}{2}(T^\mu{}_a E^{a\nu} + T^\nu{}_a E^{a\mu}). \tag{12.338}$$

Alternatively, we could also vary the inverse vielbein

$$E_a{}^\nu(x) \to E_a{}^\nu(x) + \delta E_a{}^\nu(x), \qquad \delta E_a{}^\nu(x) \overset{|x|\to\text{boundary}}{\longrightarrow} 0, \tag{12.339}$$

then the **energy–momentum tensor** is defined by

$$\begin{aligned} \delta S_M &= -\int d^m x\, e T^a{}_\nu \delta E_a{}^\nu \\ \frac{\delta S_M}{\delta E_a{}^\nu} &= -e\, T^a{}_\nu, \end{aligned} \tag{12.340}$$

in accordance with definition (12.337).

12.4 Fermionic action

In order to describe the anomalies we need the fermionic action. Fermions are represented by spinor fields which are subjected to the Lorentz group (and not to the diffeomorphism group). So we must refer the Dirac operator in the action to the tangent frame, which we achieve with help of the vielbein.

According to the equivalence principle special relativity is valid in a locally inertial system and it does not matter which inertial frame we choose at a point x. Therefore the action must be invariant under local frame rotations—Lorentz transformations—and clearly also under a general coordinate change.

Covariant derivative: To construct such an invariant action we need a covariant derivative of a spinor which transforms locally like a Lorentz-vector spinor (like a tensor product of a Lorentz vector and a spinor). This we achieve via the vielbein

$$D_a\,\psi = E_a{}^{\mu}\,D_\mu\psi. \tag{12.341}$$

Under a Lorentz transformation $L(x)$ the **spinor transforms** as

$$\begin{aligned}\psi(x) &\rightarrow \rho^{-1}(L(x))\psi(x)\\ \bar{\psi}(x) &\rightarrow \bar{\psi}(x)\,\rho(L(x)),\end{aligned} \tag{12.342}$$

where $\rho(L(x))$ denotes the **spinor representation** of $L(x)$. For an infinitesimal Lorentz transformation (12.306), $L = \mathbf{1} + \alpha$, we have

$$\begin{aligned}\rho(L(x)) &= \mathbf{1} + \frac{1}{2}\alpha_{ab}(x)\sigma^{ab}\\ \rho^{-1}(L(x)) &= \mathbf{1} - \frac{1}{2}\alpha_{ab}(x)\sigma^{ab}.\end{aligned} \tag{12.343}$$

The matrices $\frac{1}{2}\sigma^{ab}$ are called the **generators for the spinor representation of the Lorentz group**. A matrix σ^{ab} is antisymmetric and can be represented by

$$\sigma^{ab} = \frac{1}{4}[\gamma^a,\gamma^b], \qquad \sigma^{ab} = -\sigma^{ba}. \tag{12.344}$$

It fulfils the frequently used relation

$$[\sigma^{ab},\sigma^{cd}] = \eta^{ad}\sigma^{bc} - \eta^{ac}\sigma^{bd} + \eta^{bc}\sigma^{ad} - \eta^{bd}\sigma^{ac}. \tag{12.345}$$

We construct the **covariant derivative** as

$$D_\mu = \partial_\mu + \omega_\mu \tag{12.346}$$

and specify the additional connection term ω_μ later on. For the moment we just require the **transformation property**

$$\omega_\mu \to \rho^{-1}\omega_\mu\rho + \rho^{-1}\partial_\mu\rho; \tag{12.347}$$

then the covariant derivative (12.346) satisfies the desired transformation (we use $\partial_\mu\rho^{-1}\rho + \rho^{-1}\partial_\mu\rho = 0$ following from $\rho^{-1}\rho = \mathbf{1}$).

$$\begin{aligned} E_a{}^\mu D_\mu\psi &\to E_b{}^\mu L^b{}_a(\partial_\mu + \rho^{-1}\omega_\mu\rho + \rho^{-1}\partial_\mu\rho)\rho^{-1}\psi \\ &= E_b{}^\mu L^b{}_a\rho^{-1}(\partial_\mu + \omega_\mu)\psi \\ &= E_b{}^\mu L^b{}_a\rho^{-1}(L)D_\mu\psi. \end{aligned} \tag{12.348}$$

In the spinor formalism, where we refer to TF, we have a constant metric η_{ab} and constant Dirac matrices γ^a. Whereas referring to CS, where the metric $g_{\mu\nu}(x)$ is x-dependent, we get x-dependent Dirac matrices given by

$$\gamma^\mu(x) = E_a{}^\mu(x)\gamma^a. \tag{12.349}$$

Then we also have

$$\{\gamma^\mu(x), \gamma^\nu(x)\} = 2g^{\mu\nu}(x), \qquad \{\gamma^a, \gamma^b\} = 2\eta^{ab}. \tag{12.350}$$

Fermionic action: Defining the Dirac operator

$$\not{D} = \gamma^\mu(x)D_\mu = E_a{}^\mu(x)\gamma^a D_\mu \tag{12.351}$$

the **fermionic action** is given by

$$\begin{aligned} S_\psi &= \int d^m x\, e\, \mathcal{L}_\psi = \int d^m x\, e\, \bar{\psi} i \not{D}\, \psi \\ &= \int d^m x\, e\, \bar{\psi}\, i E_a{}^\mu\gamma^a D_\mu\psi. \end{aligned} \tag{12.352}$$

Note: Whenever we work with spinors (in this section and in the following ones) we choose for convenience the metric signature $(+1, -1, -1, -1)$ as in the gauge case. Therefore we have in the Lagrangian the operator $i\not{D}$ which is also Hermitian for a Euclidean signature, and the Euclidean signature is finally used in the index and anomaly calculations.

The action (12.352) is invariant under *both* the general coordinate transformations and the Lorentz transformations. $E_a{}^\mu(x)D_\mu$ describes a coordinate scalar and $\bar{\psi}(x)$, $\psi(x)$ are coordinate scalars too. A Lorentz transformation, on the other hand, leaves the Lagrangian invariant

$$\begin{aligned}\bar\psi i E_a{}^\mu \gamma^a D_\mu \psi &\to \bar\psi \rho(L) i E_b{}^\mu L^b{}_a (L^{-1})^a{}_c \gamma^c \rho^{-1}(L) D_\mu \psi \\ &= \bar\psi i E_b{}^\mu \gamma^b D_\mu \psi. \end{aligned} \tag{12.353}$$

Spin connection: Now we determine the explicit form of ω_μ. Under an infinitesimal Lorentz transformation it changes as

$$\begin{aligned}\omega_\mu &\to \left(\mathbf{1} - \frac{1}{2}\alpha_{ab}\sigma^{ab}\right)\omega_\mu\left(\mathbf{1} + \frac{1}{2}\alpha_{ab}\sigma^{ab}\right) \\ &\quad + \left(\mathbf{1} - \frac{1}{2}\alpha_{ab}\sigma^{ab}\right)\partial_\mu\left(\mathbf{1} + \frac{1}{2}\alpha_{ab}\sigma^{ab}\right) \\ &= \omega_\mu + \partial_\mu \alpha_{ab}\frac{\sigma^{ab}}{2} + \alpha_{ab}\left[\omega_\mu, \frac{\sigma^{ab}}{2}\right] \\ &= \omega_\mu + D_\mu \alpha \end{aligned} \tag{12.354}$$

when defining

$$\alpha(x) := \alpha_{ab}(x)\frac{\sigma^{ab}}{2}. \tag{12.355}$$

So we get

$$\delta^L_\alpha \omega_\mu = D_\mu \alpha. \tag{12.356}$$

This clearly reminds us of the transformation behaviour of the spin connection which we have introduced in the previous sections (e.g. in Section 12.2.5). Indeed, we identify ω_μ of equation (12.346) with the spin connection given by equation (12.294)

$$\omega_\mu \equiv \omega^a{}_{b\mu}\frac{\sigma_a{}^b}{2} = e^a{}_\nu \nabla_\mu E_b{}^\nu \frac{\sigma_a{}^b}{2}. \tag{12.357}$$

Then expression (12.357) satisfies the transformation property (12.354). To show the converse we have

$$\omega_{ab\mu}\frac{\sigma^{ab}}{2} \to \omega_{ab\mu}\frac{\sigma^{ab}}{2} + \partial_\mu\alpha_{ab}\frac{\sigma^{ab}}{2} + \frac{1}{4}\alpha_{cd}\omega_{ab\mu}[\sigma^{ab}, \sigma^{cd}]; \tag{12.358}$$

for the last term we use the commutator formula (12.345), then we recover precisely the spin connection transformation law (12.309)

$$\begin{aligned}\omega_{ab\mu} &\to \omega_{ab\mu} + \partial_\mu \alpha_{ab} + (\omega_{ac\mu}\alpha^c{}_b - \alpha_{ac}\omega^c{}_{b\mu}) \\ &= \omega_{ab\mu} + D_\mu \alpha_{ab}. \qquad \text{Q.E.D.}\end{aligned} \tag{12.359}$$

Equivalent actions: In the literature one can also find other types of fermionic action which are, however, equivalent to our expression (12.352) (see e.g. [van Nieuwenhuizen 1988], [Alvarez-Gaumé, Ginsparg 1985], [Alvarez-Gaumé, Witten 1983]). Since these types are convenient for our anomaly discussions we will show their equivalence.

Let us first note some identities which are very useful.

Recall the vielbein $e^a{}_\mu$; it represents a tensorfield—the unit tensor in a mixed tangent frame and coordinate basis—and *not* 4 covectorfields. When applying the covariant derivative one certainly has to contract *all* the indices (which we indicate by the subscript *tot*); the result then is:

Theorem: Vielbein condition

$$D_\mu^{\rm tot} e^a{}_\nu = \partial_\mu e^a{}_\nu + \omega^a{}_{b\mu} e^b{}_\nu - e^a{}_\lambda \Gamma^\lambda{}_{\mu\nu} = 0. \tag{12.360}$$

Proof. The condition (12.360) holds due to the equivalence of the Christoffel- and spin connection, equation (12.301), which guarantees that

$$\begin{aligned} D_\mu^{\rm tot} e^a{}_\nu &= \partial_\mu e^a{}_\nu + \omega^a{}_{b\mu} e^b{}_\nu - e^a{}_\lambda \Gamma^\lambda{}_{\mu\nu} \\ &= \partial_\mu e^a{}_\nu + \omega^a{}_{b\mu} e^b{}_\nu - e^a{}_\lambda E_b{}^\lambda \partial_\mu e^b{}_\nu - e^a{}_\lambda E_b{}^\lambda \omega^b{}_{c\mu} e^c{}_\nu \\ &= 0. \qquad \text{Q.E.D.} \end{aligned} \tag{12.361}$$

Generally, the **covariant derivative for a hybrid frame–coordinate tensor** is defined by

$$\begin{aligned} D_\mu^{\rm tot} T^{a\alpha}{}_\beta &= \partial_\mu T^{a\alpha}{}_\beta + [\omega_\mu, T^\alpha{}_\beta]^a + [\Gamma_\mu, T^a]^\alpha{}_\beta \\ &= \partial_\mu T^{a\alpha}{}_\beta + \omega^a{}_{b\mu} T^{b\alpha}{}_\beta \\ &\quad + \Gamma^\alpha{}_{\mu\lambda} T^{a\lambda}{}_\beta - T^{a\alpha}{}_\lambda \Gamma^\lambda{}_{\mu\beta}. \end{aligned} \tag{12.362}$$

Hence we obtain for the inverse vielbein the condition:

Theorem: Inverse vielbein condition

$$D_\mu^{\rm tot} E_a{}^\nu = \partial_\mu E_a{}^\nu - E_b{}^\nu \omega^b{}_{a\mu} + \Gamma^\nu{}_{\mu\lambda} E_a{}^\lambda = 0, \tag{12.363}$$

specifically, for $\mu = \nu$

$$D_\mu^{\rm tot} E_a{}^\mu = \partial_\mu E_a{}^\mu + \Gamma^\mu{}_{\mu\lambda} E_a{}^\lambda - E_b{}^\mu \omega^b{}_{a\mu} = 0. \tag{12.364}$$

Note that condition (12.360) or (12.363) is a result and not an independent condition.

Using furthermore the formula (12.208) for condition (12.364)

$$D_\mu^{\rm tot} E_a{}^\mu = \frac{1}{e} \partial_\mu (e E_a{}^\mu) - E_b{}^\mu \omega^b{}_{a\mu} = 0 \tag{12.365}$$

we find the relation

$$\partial_\mu(eE_a{}^\mu) = eE_b{}^\mu \omega^b{}_{a\mu}. \tag{12.366}$$

Finally, what we also need are the following commutation relations

$$[\sigma^{ab}, \gamma^c] = \gamma^a \eta^{bc} - \gamma^b \eta^{ac} \tag{12.367}$$

$$\{\gamma^a, \sigma^{bc}\} = \begin{cases} \gamma^a\gamma^b\gamma^c & \text{for } a,b,c \text{ different} \\ 0 & \text{otherwise.} \end{cases} \tag{12.368}$$

Lagrangian reformulation I: Now we are prepared to reformulate the Lagrangian (12.352)

$$\begin{aligned} e\mathcal{L}_\psi &= e\bar{\psi} i E_a{}^\mu \gamma^a D_\mu \psi \\ &= eE_a{}^\mu i\bar{\psi}\gamma^a \partial_\mu \psi + \frac{1}{2} eE_a{}^\mu \omega_{bc\mu} i\bar{\psi}\gamma^a \sigma^{bc}\psi. \end{aligned} \tag{12.369}$$

We split the second term into

$$\gamma^a \sigma^{bc} = \frac{1}{2}[\gamma^a, \sigma^{bc}] + \frac{1}{2}\{\gamma^a, \sigma^{bc}\}; \tag{12.370}$$

we use the commutator formula (12.367) and obtain

$$\begin{aligned} e\mathcal{L}_\psi &= eE_a{}^\mu i\bar{\psi}\gamma^a \partial_\mu \psi + \frac{1}{2} eE_b{}^\mu \omega^b{}_{a\mu} i\bar{\psi}\gamma^a \psi \\ &\quad + \frac{1}{2} eE_a{}^\mu \omega_{bc\mu} i\bar{\psi}\, \frac{1}{2}\{\gamma^a, \sigma^{bc}\}\psi. \end{aligned} \tag{12.371}$$

Next we replace the derivative in the Lagrangian

$$\partial_\mu \to \frac{1}{2}(\partial_\mu - \overleftarrow{\partial_\mu}) \tag{12.372}$$

which does not change the equations of motion. In this case, however, the derivative $\overleftarrow{\partial_\mu}$ (acting to the left) acts on the product $eE_a{}^\mu \bar{\psi}$. Therefore we get a term which is

$$-\frac{1}{2}\partial_\mu(eE_a{}^\mu) i\bar{\psi}\gamma^a \psi = -\frac{1}{2} eE_b{}^\mu \omega^b{}_{a\mu} i\bar{\psi}\gamma^a \psi \tag{12.373}$$

when using formula (12.366). This term cancels the second term in equation (12.371) precisely and we are left with the Lagrangian

$$\begin{aligned} e\mathcal{L}_\psi &= \frac{1}{2} eE_a{}^\mu \bar{\psi} i\gamma^a \overleftrightarrow{\partial_\mu}\, \psi \\ &\quad + \frac{1}{2} eE_a{}^\mu \omega_{bc\mu} i\bar{\psi}\, \frac{1}{2}\{\gamma^a, \sigma^{bc}\}\psi. \end{aligned} \tag{12.374}$$

Here $\overleftrightarrow{\partial}_\mu$ is defined by

$$\overleftrightarrow{\partial}_\mu = \partial_\mu - \overleftarrow{\partial}_\mu \tag{12.375}$$

and acts *only* on the spinors ψ, $\bar{\psi}$. Defining a **left-acting covariant derivative**

$$\overleftarrow{D}_\mu = \overleftarrow{\partial}_\mu - \omega_\mu, \tag{12.376}$$

we finally find the **Lagrangian**

$$e\mathcal{L}_\psi = \frac{1}{2}\, e\bar{\psi} i(\gamma^\mu D_\mu - \overleftarrow{D}_\mu \gamma^\mu)\psi. \tag{12.377}$$

On the other hand, a quick way to arrive at expression (12.377) is to use the following theorem.

Theorem:

$$[D_\mu, \gamma_\nu] = 0. \tag{12.378}$$

Proof. We calculate

$$D_\mu \gamma_\nu = \gamma_a \partial_\mu e^a{}_\nu(x) + \gamma_\nu \partial_\mu + \omega_\mu \gamma_\nu - \gamma_\lambda \Gamma^\lambda{}_{\mu\nu} \tag{12.379}$$

$$[D_\mu, \gamma_\nu] = \gamma_a \partial_\mu e^a{}_\nu + \frac{1}{2}\omega_{ab\mu}[\sigma^{ab}, \gamma_\nu] - \gamma_\lambda \Gamma^\lambda{}_{\mu\nu}; \tag{12.380}$$

we insert the equivalence of the connections, equation (12.301), and the commutator result (12.367), then we find

$$\begin{aligned} [D_\mu, \gamma_\nu] &= \gamma_a \partial_\mu e^a{}_\nu + \omega_{ab\mu}\gamma^a e^b{}_\nu \\ &\quad - \gamma_a \partial_\mu e^a{}_\nu - \gamma_a \omega^a{}_{b\mu} e^b{}_\nu = 0. \quad \text{Q.E.D.} \end{aligned} \tag{12.381}$$

Using the commutator identity (12.378) now we decompose the Lagrangian (12.352), we integrate by parts and recover result (12.377)

$$\begin{aligned} e\mathcal{L}_\psi &= e\bar{\psi} i\gamma^\mu D_\mu \psi = \frac{1}{2} e\bar{\psi} i(\gamma^\mu D_\mu + D_\mu \gamma^\mu)\psi \\ &= \frac{1}{2} e\bar{\psi} i(\gamma^\mu D_\mu - \overleftarrow{D}_\mu \gamma^\mu)\psi. \end{aligned} \tag{12.382}$$

Lagrangian reformulation II: Another possibility of reformulating Lagrangian (12.352) is the following. We can substitute

$$D_\mu \to \frac{1}{2}\, \overleftrightarrow{D}_\mu, \tag{12.383}$$

with

$$\overleftrightarrow{D}_\mu = D_\mu - \overleftarrow{D}_\mu = \overleftrightarrow{\partial}_\mu + 2\omega_\mu \tag{12.384}$$

already at the beginning so that we have the **Lagrangian**

$$e\mathcal{L}_\psi = \frac{1}{2} e\bar{\psi} i \overleftrightarrow{\not{D}} \psi. \tag{12.385}$$

But then the Dirac operator

$$\overleftrightarrow{\not{D}} = \gamma^\mu(x) \overleftrightarrow{D}_\mu = E_a{}^\mu(x)\gamma^a \overleftrightarrow{D}_\mu \tag{12.386}$$

acts left on *both*, on the spinor $\bar{\psi}$ and on $eE_a{}^\mu$. Again, the term $\partial_\mu(eE_a{}^\mu)$ cancels with a commutator term leading to expression (12.374).

Remark: Considering expression (12.374) in two dimensions $a = 1, 2$ (where the γ-matrices change into Pauli matrices, recall Section 4.5) we notice that

$$\begin{aligned} \sigma^{12} &\sim [\sigma_1, \sigma_2] \sim \sigma_3 \\ \{\gamma^a, \sigma^{12}\} &\to \{\sigma_a, \sigma_3\} = 0, \end{aligned}$$

the anticommutator term vanishes and we find the important result:

Proposition:

- *In two dimensions there is no spin connection term in the fermionic action!*

Résumé: Finally we consider massless fermions in a given chirality $P_+ = \frac{1}{2}(1+\gamma_5)$. We can construct fermionic actions which are invariant under Lorentz transformations *and* general coordinate transformations. These actions are equivalent to each other and are given by

$$\begin{aligned} S_\psi &= \int d^m x \, e \, \bar{\psi} \, i \, \not{D} P_+ \psi \\ &= \int d^m x \, \frac{1}{2} \, e \, \bar{\psi} \, i \, \overleftrightarrow{\not{D}} \, P_+ \psi \\ &= \int d^m x \, \frac{1}{2} \, e \, \bar{\psi} \, i(\gamma^\mu D_\mu - \overleftarrow{D}_\mu \, \gamma^\mu) P_+ \psi. \end{aligned} \tag{12.387}$$

12.5 Invariances

Now we want to study the classical symmetries of the fermion action, and we choose the following expression

$$S_\psi = \int d^m x \, \frac{1}{2} \, e \, \bar{\psi} \, i(\gamma^\mu D_\mu - \overleftarrow{D}_\mu \, \gamma^\mu) P_+ \psi. \tag{12.388}$$

Calculating the energy–momentum tensor

$$e\,T^{\mu}{}_{a} = \frac{\delta S_{\psi}}{\delta e^{a}{}_{\mu}} = \frac{1}{2}\, e\,\bar{\psi}\, i(\gamma_a D^{\mu} - \overleftarrow{D^{\mu}}\,\gamma_a)P_{+}\psi \tag{12.389}$$

we find for the **symmetric energy momentum tensor**

$$T^{\mu\nu} = \frac{1}{4}\,\bar{\psi}\, i(\gamma^{\mu} D^{\nu} - \overleftarrow{D^{\nu}}\,\gamma^{\mu})P_{+}\psi + \mu \leftrightarrow \nu. \tag{12.390}$$

The antisymmetric part vanishes as we shall see. In the derivation of expression (12.390) we have used the equations of motion (as a consequence the term proportional to $\delta e/\delta e^{a}{}_{\mu}$ vanishes)

$$\begin{aligned} \frac{\delta S_{\psi}}{\delta \bar{\psi}} = 0 \quad &\Longrightarrow \quad \not{D} P_{+}\psi = 0 \\ \frac{\delta S_{\psi}}{\delta \psi} = 0 \quad &\Longrightarrow \quad \bar{\psi}\,\overleftarrow{\not{D}}\,P_{+} = 0, \end{aligned} \tag{12.391}$$

only then does a general variation of the action

$$\delta S_{\psi} = \int d^{m}x \left[\frac{\delta S_{\psi}}{\delta \psi}\,\delta\psi + \frac{\delta S_{\psi}}{\delta \bar{\psi}}\,\delta\bar{\psi} + \frac{\delta S_{\psi}}{\delta e^{a}{}_{\mu}}\,\delta e^{a}{}_{\mu} \right] \tag{12.392}$$

agree with the definition (12.337) of the energy–momentum tensor (since the first two terms drop).

12.5.1 Lorentz invariance

The fermionic action is invariant under **infinitesimal local Lorentz transformations**

$$\begin{aligned} \delta^{L}_{\alpha}\, e^{a}{}_{\mu} &= -\alpha^{a}{}_{b}\, e^{b}{}_{\mu} \\ \delta^{L}_{\alpha}\, E_{a}{}^{\mu} &= E_{b}{}^{\mu}\,\alpha^{b}{}_{a} \\ \delta^{L}_{\alpha}\, e &= 0 \\ \delta^{L}_{\alpha}\, \omega^{a}{}_{b\mu} &= D_{\mu}\alpha^{a}{}_{b} = \partial_{\mu}\alpha^{a}{}_{b} + \omega^{a}{}_{c\mu}\alpha^{c}{}_{b} - \alpha^{a}{}_{c}\omega^{c}{}_{b\mu} \\ \delta^{L}_{\alpha}\, \psi &= -\frac{1}{2}\,\alpha_{ab}\sigma^{ab}\psi \\ \delta^{L}_{\alpha}\, \bar{\psi} &= \frac{1}{2}\,\alpha_{ab}\bar{\psi}\sigma^{ab}. \end{aligned} \tag{12.393}$$

The Lorentz-like variation of the action (12.388) is then

$$\delta^{L}_{\alpha}\, S_{\psi} = \int d^{m}x\; e\,T^{\mu}{}_{a}\,\delta^{L}_{\alpha}\, e^{a}{}_{\mu} = -\int d^{m}x\; e\,T^{\mu}{}_{a}\,\alpha^{a}{}_{b}\, e^{b}{}_{\mu} \tag{12.394}$$

supplying the following proposition.

Proposition:

- *The Lorentz variation of the action is given by*

$$\delta^L_\alpha S_\psi = \int d^m x \, e \, \alpha_{ab} \, T^{ab}. \tag{12.395}$$

Theorem: The **energy–momentum tensor is symmetric!**

$$\delta^L_\alpha S_\psi = 0 \Longleftrightarrow T^{ab} = T^{ba} \tag{12.396}$$

(since $\alpha_{ab} = -\alpha_{ba}$). Clearly, if T^{ab} is symmetric, $T^{\mu\nu}$ is symmetric too.

12.5.2 Einstein invariance

The fermionic action (12.388) is also invariant under **Einstein transformations**

$$\begin{aligned}
\delta^c_\xi \, e^a{}_\mu &= \xi^\nu \partial_\nu e^a{}_\mu + e^a{}_\nu \partial_\mu \xi^\nu = \xi^\nu \nabla_\nu e^a{}_\mu + e^a{}_\nu \nabla_\mu \xi^\nu \\
\delta^c_\xi \, E_a{}^\mu &= \xi^\nu \partial_\nu E_a{}^\mu - \partial_\nu \xi^\mu E_a{}^\nu = \xi^\nu \nabla_\nu E_a{}^\mu - \nabla_\nu \xi^\mu E_a{}^\nu \\
\delta^c_\xi \, e &= \xi^\nu \partial_\nu e + e \partial_\nu \xi^\nu \\
\delta^c_\xi \, \omega^a{}_{b\mu} &= \xi^\nu \partial_\nu \omega^a{}_{b\mu} + \omega^a{}_{b\nu} \partial_\mu \xi^\nu \\
\delta^c_\xi \, \psi &= \xi^\nu \partial_\nu \psi \\
\delta^c_\xi \, \bar{\psi} &= \xi^\nu \partial_\nu \bar{\psi}.
\end{aligned} \tag{12.397}$$

(Geometrically, we consider the trivial fibre-bundle case.) Then the Einstein-like variation of the action becomes

$$\begin{aligned}
\delta^c_\xi \, S_\psi &= \int d^m x \, e \, T^\mu{}_a \, \delta^c_\xi \, e^a{}_\mu \\
&= \int d^m x (e^a{}_\nu \nabla_\mu \xi^\nu + \xi^\nu \nabla_\nu e^a{}_\mu) e T^\mu{}_a,
\end{aligned} \tag{12.398}$$

which we are going to calculate further. We integrate the first term by parts

$$e^a{}_\nu \nabla_\mu \xi^\nu e T^\mu{}_a = e \nabla_\mu \xi^\nu T^\mu{}_\nu \longrightarrow -e \xi^\nu \nabla_\mu T^\mu{}_\nu \tag{12.399}$$

and for the second term we use the connection formula (12.295)

$$\xi^\nu \nabla_\nu e^a{}_\mu e T^\mu{}_a = \xi^\nu E_b{}^\mu \nabla_\nu e^a{}_\mu e T^b{}_a = \xi^\nu \omega_{ab\nu} e T^{ab}. \tag{12.400}$$

Altogether we obtain:

Proposition:

- *The Einstein variation of the action is given by*

$$\delta_\xi^c S_\psi = -\int d^m x e \xi^\nu (\nabla_\mu T^\mu{}_\nu - \omega_{ab\nu} T^{ab}). \qquad (12.401)$$

The second term, however, vanishes here since, due to Lorentz invariance, the tensor T^{ab} is symmetric whereas $\omega_{ab\nu}$ is antisymmetric. So we find:

Theorem: The **energy–momentum tensor is covariantly conserved!**

$$\delta_\xi^c\, S_\psi = 0 \text{ and } \delta_\alpha^L\, S_\psi = 0 \Longrightarrow \nabla_\mu T^{\mu\nu} = 0. \qquad (12.402)$$

(Notice that we need both Einstein- *and* Lorentz invariance.)

Alternative approach: We also arrive at the above conclusions by splitting the vielbein variation into a symmetric and an antisymmetric part.

Lemma:

$$\delta e^a{}_\mu = \frac{1}{2}\, \delta g_{\mu\nu} E^{a\nu} - \alpha^a{}_b e^b{}_\mu, \qquad (12.403)$$

with

$$\alpha^a{}_b = \frac{1}{2}(E^{a\nu}\delta e_{b\nu} - E_b{}^\nu \delta e^a{}_\nu). \qquad (12.404)$$

We insert the splitting (12.403) into the variation of the action

$$\begin{aligned} \delta S_\psi &= \int d^m x\, e\, T^\mu{}_a\, \delta e^a{}_\mu \\ &= \frac{1}{2}\int d^m x \sqrt{|g|}\, T^{\mu\nu}\delta g_{\mu\nu} + \int d^m x\, e\, \alpha_{ab} T^{ab}; \end{aligned} \qquad (12.405)$$

due to Lorentz invariance the tensor T^{ab} is symmetric whereas α_{ab} is antisymmetric; that implies the vanishing of the second term and we are left with

$$\delta S_\psi = \frac{1}{2}\int d^m x \sqrt{|g|}\, T^{\mu\nu}\delta g_{\mu\nu}. \qquad (12.406)$$

Thus due to Lorentz symmetry we can choose the action to depend only on $g_{\mu\nu}$. The Einstein-like variation gives

$$\begin{aligned} \delta_\xi^c\, S_\psi &= \frac{1}{2}\int d^m x \sqrt{|g|}\, T^{\mu\nu}\delta_\xi^c g_{\mu\nu} \\ &= \frac{1}{2}\int d^m x \sqrt{|g|}\, T^{\mu\nu}(\nabla_\mu \xi_\nu + \nabla_\nu \xi_\mu). \end{aligned} \qquad (12.407)$$

Integrating by parts, using again the symmetry of $T^{\mu\nu}$ we finally find:

Proposition:

- *Due to Lorentz symmetry the Einstein variation of the action is given by*

$$\delta_\xi^c\, S_\psi = -\int d^m x \sqrt{|g|}\; \xi_\nu \nabla_\mu T^{\mu\nu}. \tag{12.408}$$

This is clearly in accordance with result (12.401).

Note: If we choose, on the other hand, a Lorentz-like variation $\delta \to \delta_\alpha^L$ in equation (12.405) the first term vanishes and we recover result (12.395). In this case the $\alpha^a{}_b$ of equation (12.404) represents precisely the frame 'rotation angle' (12.306).

12.5.3 Weyl invariance

Finally we discuss the conformal or Weyl transformations (recall Section 12.1.1). Let us consider the **Weyl rescaling** of the vielbein

$$\begin{aligned} e^a{}_\mu{}'(x) &= \rho(x) e^a{}_\mu(x) \\ E_a{}^{\mu\prime}(x) &= \rho^{-1}(x) E_a{}^\mu(x) \\ e' &= \rho^m(x)\, e, \end{aligned} \tag{12.409}$$

which alters the metric by

$$\begin{aligned} g'_{\mu\nu}(x) &= \rho^2(x)\rho_{\mu\nu}(x) \\ g' &= \rho^{2m}(x)\, g, \end{aligned} \tag{12.410}$$

where

$$\rho(x) = e^{\sigma(x)}. \tag{12.411}$$

Spin connection and Dirac operator **transform conformally** according to

$$\begin{aligned} \omega'_\mu(x) &= \rho^{-r}(x)\omega_\mu(x)\rho^r(x) + \rho^{-r}(x)\partial_\mu \rho^r(x) \\ D'_\mu(x) &= \rho^{-r}(x) D_\mu(x) \rho^r(x), \end{aligned} \tag{12.412}$$

together with the Fermi fields

$$\begin{aligned} \psi'(x) &= \rho^{-r}(x)\psi(x) \\ \bar\psi'(x) &= \rho^{-r}(x)\bar\psi(x), \end{aligned} \tag{12.413}$$

where

$$r = \frac{m-1}{2}, \qquad m = \dim M. \tag{12.414}$$

These scaling transformations are determined by the required scaling invariance of the Lagrangian.

Choosing the **transformation parameter infinitesimal**

$$\rho(x) = 1 + \sigma(x) \tag{12.415}$$

we have the **Weyl transformations**

$$\begin{aligned}
\delta_\sigma^W \, e^a{}_\mu &= \sigma \, e^a{}_\mu \\
\delta_\sigma^W \, E_a{}^\mu &= -\sigma \, E_a{}^\mu \\
\delta_\sigma^W \, e &= m \, \sigma \, e \\
\delta_\sigma^W \, \omega^a{}_{b\mu} &= \partial_\nu \sigma (e^a{}_\mu E_b{}^\nu - e_{b\mu} E^{a\nu}) \\
\delta_\sigma^W \, \psi &= -r \, \sigma \, \psi \\
\delta_\sigma^W \, \bar{\psi} &= -r \, \sigma \, \bar{\psi}.
\end{aligned} \tag{12.416}$$

Note: We want to emphasize that the Weyl transformation of the connection is already determined by the rescaling of the vielbein *independent* of the construction of the Lagrangian. The reason is that the connection is uniquely determined by the vielbein and its derivative. Remember formulae (12.238), (12.241), a Weyl variation gives (we keep only the derivative $\partial_\mu \sigma$-terms, all σ-terms cancel)

$$\begin{aligned}
\delta_\sigma^W \, \omega_{ab\mu} &= \delta_\sigma^W (\omega_{abc} e^c{}_\mu) \\
&= \frac{1}{2} (\partial_\lambda \sigma e_{a\mu} E_b{}^\lambda - \partial_\mu \sigma e_{a\lambda} E_b{}^\lambda + \partial_\mu \sigma e_{b\nu} E_a{}^\nu \\
&\quad - \partial_\nu \sigma e_{b\mu} E_a{}^\nu - \partial_\lambda \sigma e_{b\mu} E_a{}^\lambda + \partial_\nu \sigma e_{a\mu} E_b{}^\nu) \\
&= \partial_\nu \sigma (e_{a\mu} E_b{}^\nu - e_{b\mu} E_a{}^\nu). \qquad \text{Q.E.D.}
\end{aligned} \tag{12.417}$$

So it turns out that our Lagrangian (12.388), constructed to be Lorentz- and Einstein-invariant, is Weyl-invariant in addition.

Finally, what does the Weyl symmetry imply for the energy–momentum tensor?

Proposition:

- *The Weyl variation of the action is given by*

$$\delta_\sigma^W \, S_\psi = \int d^m x \, e \, T^\mu{}_a \delta_\sigma^W \, e^a{}_\mu = \int d^m x \, e \, \sigma \, T^\mu{}_\mu. \tag{12.418}$$

Theorem: The **energy–momentum tensor is traceless!**

$$\delta_\sigma^W \, S_\psi = 0 \Longleftrightarrow T^\mu{}_\mu = 0. \tag{12.419}$$

12.6 Gravitational anomalies

Now we come to our main interest, the anomalies. We consider the quantum case—the 'quantum action'. Then the invariances—the Einstein-, Lorentz-, Weyl invariance—which we have discussed previously might be broken. In this case we speak of gravitational anomalies.

12.6.1 Einstein-, Lorentz-, Weyl anomaly

In the quantized theory we work with the generating functional for the Green functions. We quantize only the Fermi fields; the gravitational field represented by the metric or by the vielbein we still consider as an external (nonquantized) field. Then we have for the generating functional

$$Z[e^a{}_\mu] = e^{-W[e^a{}_\mu]} = \int d\bar{\psi} d\psi \exp\left[-\int dx\; e\mathcal{L}_\psi\right], \tag{12.420}$$

where the Lagrangian $e\mathcal{L}_\psi$ is given by equation (12.388). The variation of the 'quantum action'

$$\delta W[e^a{}_\mu] = \int dx\; \delta e^a{}_\mu(x) \frac{\delta}{\delta e^a{}_\mu(x)} W[e^a{}_\mu] \tag{12.421}$$

provides the **energy–momentum tensor**

$$\begin{aligned} \delta W[e^a{}_\mu] &= \int dx\; e\langle T^\mu{}_a\rangle \delta e^a{}_\mu \\ \frac{\delta W[e^a{}_\mu]}{\delta e^a{}_\mu} &= e\langle T^\mu{}_a\rangle \end{aligned} \tag{12.422}$$

analogous to the classical case (12.337). The bracket $\langle\ \rangle$ denotes the mean value with respect to the path integral. Choosing in particular

$$\begin{aligned} \delta \longrightarrow \delta^L_\alpha \quad & \text{Lorentz transformation} \\ \delta^c_\xi \quad & \text{Einstein transformation} \\ \delta^W_\sigma \quad & \text{Weyl transformation} \end{aligned}$$

and inserting the corresponding transformations of the vielbein we obtain the several variations of the 'quantum action'. These variations might not vanish as in the classical case, then we have anomalies.

Proposition: Gravitational anomalies

- **Lorentz anomaly**

$$\delta^L_\alpha W[e^a{}_\mu] = \int dx\; e\alpha_{ab}\langle T^{ab}\rangle =: G^L(\alpha). \tag{12.423}$$

- **Einstein anomaly**

$$\delta^c_\xi W[e^a{}_\mu] = -\int dx\; e\xi^\nu(\nabla_\mu\langle T^\mu{}_\nu\rangle - \omega_{ab\nu}\langle T^{ab}\rangle) =: G^E(\xi).. \tag{12.424}$$

- **Weyl anomaly**

$$\delta^W_\sigma W[e^a{}_\mu] = \int dx\; e\sigma\langle T^\mu{}_\mu\rangle =: G^W(\sigma). \tag{12.425}$$

Note: Now $\langle T^{ab}\rangle$ means the **antisymmetric part** of the (total) **energy–momentum tensor** $\langle T^{\mu a}\rangle$

$$\langle T^{ab}\rangle = \frac{1}{2}(e^a{}_\mu\langle T^{\mu b}\rangle - e^b{}_\mu\langle T^{\mu a}\rangle). \tag{12.426}$$

In equation (12.423) only the antisymmetric $\langle T^{ab}\rangle$ contributes, the symmetric part of the energy–momentum tensor also exists but does not contribute here.

Physical significance: The gravitational anomalies mean the following:

- The **Lorentz anomaly** is equivalent to the existence of an antisymmetric part of the energy–momentum tensor.

$$\langle T^{ab}\rangle = -\langle T^{ba}\rangle. \tag{12.427}$$

- The **Einstein anomaly** involves the nonconservation of the energy–momentum tensor

$$\nabla_\mu\langle T^{\mu\nu}\rangle \neq 0. \tag{12.428}$$

- The **Weyl anomaly** expresses the nonvanishing tensor trace

$$\langle T^\mu{}_\mu\rangle \neq 0 \tag{12.429}$$

  (and is sometimes called **trace anomaly** for this reason).

Remark:

- Even in the absence of an Einstein anomaly, $G^E(\xi) = 0$, the energy–momentum tensor is *not* covariantly conserved!

This nonconservation is a consequence of the Lorentz anomaly, implying the existence of an antisymmetric part $\langle T^{ab}\rangle$, and does *not* signal the breakdown of general coordinate invariance. Physically, it means that there is an energy flow into the degrees of freedom which specify the direction of the vielbeins—the metric is not the only external field that the system interacts with [Leutwyler 1985b, 1986a].

Whereas the Weyl anomaly has been found already in the seventies [Capper, Duff 1974], [Deser, Duff, Isham 1976], [Coleman, Jackiw 1971] (for a review see [Duff 1994], [Deser 1996]) the study of the Einstein- and Lorentz anomaly started with the work of Alvarez-Gaumé and Witten [Alvarez-Gaumé, Witten 1983] in the eighties.

Pure Einstein anomaly: However, it is possible to make the energy–momentum tensor symmetric [Alvarez-Gaumé, Ginsparg 1985] so that Lorentz symmetry is preserved—no Lorentz anomaly occurs—then the second term in equation (12.424) vanishes and we are left with a **pure Einstein anomaly**

$$\delta^c_\xi W = -\int dx\; e\xi_\nu \nabla_\mu \langle T^{\mu\nu}\rangle = G^E(\xi). \tag{12.430}$$

- Then the nonconservation of energy and momentum is equivalent to the breaking of coordinate invariance!

In this case the 'quantum action' depends only on $g_{\mu\nu}$ (recall our discussion in Section 12.5.2) and we have

$$\begin{aligned} \delta^c_\xi W[g_{\mu\nu}] &= \int dx\; \delta^c_\xi g_{\mu\nu} \frac{\delta}{\delta g_{\mu\nu}} W[g_{\mu\nu}] \\ &= \frac{1}{2}\int dx\; \sqrt{|g|}\; \langle T^{\mu\nu}\rangle \delta^c_\xi g_{\mu\nu}, \end{aligned} \tag{12.431}$$

with

$$\frac{\delta W[g_{\mu\nu}]}{\delta g_{\mu\nu}} = \frac{1}{2}\sqrt{|g|}\; \langle T^{\mu\nu}\rangle. \tag{12.432}$$

The insertion of the metric transformation (12.167) leads directly to the anomaly equation (12.430).

Résumé: The anomalies in gravitation are the analogues of the anomaly in Yang–Mills gauge theory as discussed in Chapter 8. Particularly, the pure Einstein anomaly (12.430), which represents the nonconservation of the energy–momentum tensor, is the precise analogue to the gauge anomaly, which expresses the nonconservation of the gauge current (compare with equation (8.58) of Section 8.2).

As in the gauge case there exist different methods of evaluating the gravitational anomalies. The anomalies are given by a local polynomial in the connection and curvature. This polynomial was first found within perturbation theory. The authors [Alvarez-Gaumé, Witten 1983], [van Nieuwenhuizen 1988], [Langouche 1984] have calculated Feynman diagrams where the external gravitational field—the metric (or vielbein) in a linearized version—couples to a fermion loop via the energy–momentum tensor. Another way to find the gravitational anomalies is the heat kernel method of [Leutwyler 1985b], [Leutwyler, Mallik 1986] or Fujikawa's path integral approach [Fujikawa 1983], [Fujikawa, Tomiya, Yasuda 1985]. Finally, one can determine the anomalies algebraically and by topological techniques; we want to present this in the forthcoming sections.

12.6.2 Consistency conditions

The gravitational anomalies have to satisfy certain consistency conditions analogous to the gauge case. We neglect the Weyl anomaly for the moment. What we need first are the commutation relations for the generators of the infinitesimal symmetry transformations which are defined on the space of local functionals of e, ω or Γ. For convenience we use a matrix notation, e.g. $e \equiv e^a{}_\mu$ denotes the vielbein here (and not the determinant as before).

The generator of a variation acting on some functional $U[e,\omega]$ or $U[e,\Gamma]$ is represented by

$$\delta = \int dx \left[\delta e \frac{\delta}{\delta e} + \delta\omega \frac{\delta}{\delta\omega}\right] \tag{12.433}$$

or

$$\delta = \int dx \left[\delta e \frac{\delta}{\delta e} + \delta\Gamma \frac{\delta}{\delta\Gamma}\right]. \tag{12.434}$$

Considering the Lorentz- and Einstein transformations we have (recall equations (12.309), (12.171), (12.172), (12.209))

$$\delta^L_\alpha e = -\alpha e \qquad \delta^L_\alpha \omega = D\alpha \tag{12.435}$$

and

$$\begin{aligned} \delta^c_\xi e &= \mathcal{L}_\xi e & \delta^c_\xi \Gamma &= \mathcal{L}_\xi \Gamma + dv_\xi \\ &= i_\xi de + ev_\xi & &= (i_\xi d + di_\xi)\Gamma + \nabla v_\xi. \end{aligned} \tag{12.436}$$

Then we obtain for the **generator of a Lorentz transformation**

$$\delta^L_\alpha = \int dx \left[-\alpha e \frac{\delta}{\delta e} + D\alpha \frac{\delta}{\delta\omega}\right] \tag{12.437}$$

and for the **generator of an Einstein transformation**

$$\delta^c_\xi = \int dx \left[(i_\xi de + ev_\xi)\frac{\delta}{\delta e} + ((i_\xi d + di_\xi)\Gamma + \nabla v_\xi)\frac{\delta}{\delta\Gamma}\right]. \tag{12.438}$$

They satisfy the following commutation relations [Bardeen, Zumino 1984],

[Stora 1984, 1986], [Langouche, Schücker, Stora 1984], [Alvarez-Gaumé, Ginsparg 1985]. Here we consider only the case of trivial fibre bundles; we discuss the formalism for nontrivial bundles at the BRS level later on (Section 12.7.3).

Theorem: Commutation relations

$$\begin{aligned} [\delta^L_{\alpha_1}, \delta^L_{\alpha_2}] &= \delta^L_{[\alpha_1,\alpha_2]} \\ [\delta^c_{\xi_1}, \delta^c_{\xi_2}] &= \delta^c_{[\xi_2,\xi_1]} \\ [\delta^L_{\alpha}, \delta^c_{\xi}] &= \delta^L_{\xi\cdot\partial\alpha}. \end{aligned} \tag{12.439}$$

These operators are sometimes also called **Ward operators**. The Lorentz part behaves analogously to the gauge case (remember the proof (8.9)–(8.14)); we demonstrate the correctness of the Einstein part below. The Lorentz- and Einstein generator δ^L_α and δ^c_ξ do not commute but give $\delta^L_{\xi\cdot\partial\alpha}$ due to the $i_\xi d = \xi \cdot \partial$ terms in equation (12.438); we show the correctness again below.

Now, applying these commutation relations to the 'quantum action' together with

$$\begin{aligned} &\delta^L_\alpha W[e,\omega] = G^L(\alpha,\omega) && \text{Lorentz anomaly} \\ &\delta^c_\xi W[e,\Gamma] = G^E(v_\xi,\Gamma) && \text{Einstein anomaly} \end{aligned}$$

(analogously to the gauge case of Section 8.2) we find:

Theorem: Consistency conditions

$$\begin{aligned} \delta^L_{\alpha_1} G^L(\alpha_2,\omega) - \delta^L_{\alpha_2} G^L(\alpha_1,\omega) &= G^L([\alpha_1,\alpha_2],\omega) \\ \delta^c_{\xi_1} G^E(v_{\xi_2},\Gamma) - \delta^c_{\xi_2} G^E(v_{\xi_1},\Gamma) &= G^E(v_{[\xi_2,\xi_1]},\Gamma) \\ \delta^L_\alpha G^E(v_\xi,\Gamma) - \delta^c_\xi G^L(\alpha,\omega) &= G^L(\xi\cdot\partial\alpha,\omega). \end{aligned} \tag{12.440}$$

These equations determine the structure of the anomalies—the **consistent gravitational anomalies**—in terms of the connection and curvature. We have to take the normalization, however, from somewhere else, e.g. from perturbation theory, from the path integral method or from the index theorem. Of course, there also exist the **trivial solutions**

$$\begin{aligned} G^L_{\text{triv}}(\alpha,\omega) &= \delta^L_\alpha \widehat{G}[\omega] \\ G^E_{\text{triv}}(v_\xi,\Gamma) &= \delta^c_\xi \widehat{G}[\Gamma] \end{aligned} \tag{12.441}$$

where $\widehat{G}[\omega]$ or $\widehat{G}[\Gamma]$ represents a *local* polynomial in the fields ω or Γ and their derivatives.

On the other hand, we could also work with the passive coordinate transformations. There we have (recall equations (12.148), (12.150))

$$\delta^{c'}_{v_\xi}\, e = e\, v_\xi \qquad \delta^{c'}_{v_\xi}\, \Gamma = \nabla\, v_\xi. \tag{12.442}$$

Then the **generator for a passive coordinate transformation** is

$$\delta^{c'}_{v_\xi} = \int dx \left[e v_\xi \frac{\delta}{\delta e} + \nabla v_\xi \frac{\delta}{\delta \Gamma} \right] \tag{12.443}$$

and satisfies, analogous to the gauge case, the relations:

Theorem: Commutation relations

$$\begin{aligned} [\delta^{c'}_{v_{\xi_1}}, \delta^{c'}_{v_{\xi_2}}] &= \delta^{c'}_{[v_{\xi_1}, v_{\xi_2}]} \\ [\delta^{c'}_{v_\xi}, \delta^{L}_{\alpha}] &= 0. \end{aligned} \tag{12.444}$$

The passive generator of the diffeomorphism group certainly leads to the same Einstein anomaly as the active one [Alvarez-Gaumé, Ginsparg 1985], [Bardeen, Zumino 1984]

$$\delta^{c'}_{v_\xi} W[e, \Gamma] = G^E(v_\xi, \Gamma) \equiv \int_M \operatorname{tr} v_\xi G^E[\Gamma]. \tag{12.445}$$

The reason is that the difference of the generators—when acting on functionals—yields a total derivative (recall equation (12.175)) which vanishes under suitable boundary conditions

$$\begin{aligned} (\delta^c_\xi - \delta^{c'}_{v_\xi}) U[e, \Gamma] &= \int_{M_m} (i_\xi d + d i_\xi) Q_m[e, \Gamma] \\ &= \int_{M_m} d(i_\xi Q_m) = \int_{\partial M_m} i_\xi Q_m \to 0. \end{aligned} \tag{12.446}$$

(Q_m is a form of maximal degree m.)

Using the commutation relations (12.444) gives:

Theorem: Consistency conditions

$$\begin{aligned} \delta^{c'}_{v_{\xi_1}} G^E(v_{\xi_2}, \Gamma) - \delta^{c'}_{v_{\xi_2}} G^E(v_{\xi_1}, \Gamma) &= G^E([v_{\xi_1}, v_{\xi_2}], \Gamma) \\ \delta^{c'}_{v_\xi} G^L(\alpha, \omega) - \delta^L_\alpha G^E(v_\xi, \Gamma) &= 0. \end{aligned} \tag{12.447}$$

The second equation, however, is trivially satisfied since

$$\delta^{c'}_{v_\xi} \omega = \delta^L_\alpha \Gamma = 0. \tag{12.448}$$

Clearly, the commutation relations of the active and passive diffeomorphism generators are not independent of each other.

Proposition:

- *The commutation relations of the active diffeomorphism generators* (12.439) *are equivalent to the passive ones* (12.444) *and as a consequence the active consistency conditions* (12.440) *are equivalent to the passive ones* (12.447)*!*

Proof. Using formula (12.175) we calculate

$$\begin{aligned}\delta^c_{\xi_1}\delta^c_{\xi_2}W[e,\Gamma] &= \delta^c_{\xi_1}G^E(v_{\xi_2},\Gamma)\\ &= \int_M \operatorname{tr}\, v_{\xi_2} di_{\xi_1}G^E[\Gamma] + \delta^{c'}_{v_{\xi_1}}G^E(v_{\xi_2},\Gamma)\end{aligned} \qquad (12.449)$$

(note that v_{ξ_2} is a 0-form and $G^E(\Gamma)$ a form of maximal degree). Then by using the 'passive' relation (12.447) we get for the commutator

$$\begin{aligned}&[\delta^c_{\xi_1},\delta^c_{\xi_2}]W[e,\Gamma]\\ &= \int_M \operatorname{tr}\,(-i_{\xi_1}dv_{\xi_2} + i_{\xi_2}dv_{\xi_1} + [v_{\xi_1},v_{\xi_2}])G^E[\Gamma].\end{aligned} \qquad (12.450)$$

We evaluate the bracket (recall $v_\xi = \partial\xi$)

$$\begin{aligned}&i_{\xi_2}d(v_{\xi_1})^\alpha{}_\beta - i_{\xi_1}d(v_{\xi_2})^\alpha{}_\beta + [v_{\xi_1},v_{\xi_2}]^\alpha{}_\beta =\\ &= \xi_2^\nu\partial_\nu\partial_\beta\xi_1^\alpha - \xi_1^\nu\partial_\nu\partial_\beta\xi_2^\alpha + \partial_\nu\xi_1^\alpha\partial_\beta\xi_2^\nu - \partial_\nu\xi_2^\alpha\partial_\beta\xi_1^\nu\\ &= \partial_\beta[\xi_2,\xi_1]^\alpha\\ &= \left(v_{[\xi_2,\xi_1]}\right)^\alpha{}_\beta\end{aligned} \qquad (12.451)$$

and we recover the desired 'active' commutation relation (12.439)

$$\begin{aligned}[\delta^c_{\xi_1},\delta^c_{\xi_2}]W[e,\Gamma] &= \int_M \operatorname{tr}\, v_{[\xi_2,\xi_1]}G^E[\Gamma]\\ &= \delta^c_{[\xi_2,\xi_1]}W[e,\Gamma].\end{aligned} \qquad (12.452)$$

For the mixed relations we first calculate the 'active' relation (12.440) (remember the generators act on functionals)

$$\begin{aligned}&\delta^L_\alpha G^E(v_\xi,\Gamma) - \delta^c_\xi G^L(\alpha,\omega)\\ &= -\int_M \operatorname{tr}\,\alpha di_\xi G^L[\omega] = \int_M \operatorname{tr}\, i_\xi d\alpha G^L[\omega]\\ &= G^L(\xi\cdot\partial\alpha,\omega)\end{aligned} \qquad (12.453)$$

and then we use this result to show the 'passive' relation (12.447)

$$\delta^L_\alpha G^E(v_\xi,\Gamma) - \delta^{c'}_{v_\xi} G^L(\alpha,\omega) \tag{12.454}$$
$$= -\int_M \text{tr } \alpha di_\xi G^L[\omega] + \int_M \text{tr } \alpha di_\xi G^L[\omega] = 0. \qquad \text{Q.E.D.}$$

12.6.3 Equivalence of Einstein- and Lorentz anomaly

We have introduced two kinds of gravitational anomaly—the Einstein- and the Lorentz anomaly—depending on whether we perform a general coordinate transformation on the 'quantum action' or a Lorentz transformation (again, we disregard the Weyl anomaly). These two anomalies are, however, not independent of each other. In fact, Bardeen and Zumino [Bardeen, Zumino 1984] discovered the following feature:

Proposition:

- *There exists a local, nonpolynomial, counterterm S for the 'quantum action' W that shifts the gravitational anomaly from pure Einstein-type to Lorentz-type and vice versa!*

So the Einstein- and Lorentz anomalies appear as two different aspects of one and the same anomalous phenomenon.

Let us choose the 'quantum action' $W[g,\Gamma]$ such that it depends only on the metric g and the connection Γ (this is always possible). Then the action is Lorentz invariant and we get a pure Einstein anomaly

$$\delta^c_\xi W[g,\Gamma] = G^E(v_\xi,\Gamma) = \int_{M_{2n}} \text{tr } v_\xi G^E[\Gamma] =: -\int_{M_{2n}} Q^1_{2n}(v_\xi,\Gamma). \tag{12.455}$$

Here we introduce a polynomial in v_ξ and Γ—the chain term $Q^1_{2n}(v_\xi,\Gamma)$—analogous to the gauge case; we shall discuss its significance in the chain of descent equations later on (Section 12.7). For convenience we further use the $GL(2n)$ matrix notation for the vielbein, $e \equiv e^a{}_\mu$, and we introduce the following t-parameter family—the **homotopy**

$$e_t = \mathbf{1} + t(e - \mathbf{1}) \qquad t \in I = [0,1], \tag{12.456}$$

with

$$e_{t=1} = e \qquad e_{t=0} = \mathbf{1}. \tag{12.457}$$

Bardeen–Zumino counterterm: The **BZ counterterm** can be constructed as [Bardeen, Zumino 1984]

$$S = \int_{M_{2n}\times I} Q_{2n+1}(\omega^{e_t} + v_t), \tag{12.458}$$

where

$$\begin{aligned} \omega^{e_t} &= e_t^{-1}(\omega + d)e_t \\ v_t &= e_t^{-1}\delta_t e_t = e_t^{-1}\frac{\partial}{\partial t}e_t dt \end{aligned} \tag{12.459}$$

and Q_{2n+1} means, as usual, the Chern–Simons form. The **homotopic connection** ω^{e_t} interpolates between the spin connection, for $t = 0$,

$$\omega^{e_{t=0}} = \omega \tag{12.460}$$

and the Christoffel connection, for $t = 1$,

$$\omega^{e_{t=1}} = \omega^e = e^{-1}(\omega + d)e = \Gamma. \tag{12.461}$$

Γ is considered as a $GL(2n)$ gauge transformation of ω. So S is a local functional—it contains a finite number of vielbeins e, de and connections ω, $d\omega$—but it is nonpolynomial.

Next we define a **homotopic metric** (again, in matrix notation)

$$g_t = e_t^T\ \eta\ e_t, \tag{12.462}$$

with

$$g_{t=1} = e^T\ \eta\ e = g \qquad g_{t=0} = \eta. \tag{12.463}$$

For a variation with respect to the parameter t we find

$$\begin{aligned} \delta_t g_t &= \delta_t e_t^T \eta e_t + e_t^T \eta \delta_t e_t \\ &= v_t^T g_t + g_t v_t, \end{aligned} \tag{12.464}$$

with

$$v_t^T = (e_t^{-1}\delta_t e_t)^T = \delta_t e_t^T (e_t^{-1})^T; \tag{12.465}$$

furthermore

$$\delta_t \omega^{e_t} = D_t v_t = dv_t + [\omega^{e_t}, v_t], \tag{12.466}$$

$$\delta_t W[g_t, \omega^{e_t}] = -\int_{M_{2n}} Q^1_{2n}(v_t, \omega^{e_t}). \tag{12.467}$$

To evaluate the counterterm (12.458) we expand the Chern–Simons form in powers of v_t

$$Q_{2n+1}(\omega^{e_t} + v_t) = Q^0_{2n+1}(\omega^{e_t}) + Q^1_{2n}(v_t, \omega^{e_t}) + \ldots, \tag{12.468}$$

where only the linear term Q^1_{2n} contributes to the integral (here $v_t^2 = 0$ and the series stops). Then we obtain

$$
\begin{aligned}
S &= \int_{M_{2n}\times I} Q_{2n+1}(\omega^{e_t}+v_t) = \int_0^1 \int_{M_{2n}} Q^1_{2n}(v_t,\omega^{e_t}) \\
&= -\int_0^1 \delta_t W[g_t,\omega^{e_t}] = W[\eta,\omega] - W[g,\Gamma] \qquad (12.469)
\end{aligned}
$$

and we find that S satisfies the variations

$$
\begin{aligned}
\delta^c_\xi S &= -\delta^c_\xi W[g,\Gamma] = -G^E(v_\xi \Gamma) = \int_{M_{2n}} Q^1_{2n}(v_\xi,\Gamma) \\
\delta^L_\alpha S &= \delta^L_\alpha W[\eta,\omega] = G^L(\alpha,\omega) = -\int_{M_{2n}} Q^1_{2n}(\alpha,\omega). \qquad (12.470)
\end{aligned}
$$

Anomaly shift: So we can use the counterterm S to shift the anomaly! For instance, adding S to W

$$W \to W + S \qquad (12.471)$$

shifts the anomaly from Einstein-type to Lorentz-type

$$
\begin{aligned}
\delta^c_\xi(W+S) &= 0 \\
\delta^L_\alpha(W+S) &= G^L(\alpha,\omega). \qquad (12.472)
\end{aligned}
$$

Choosing, on the other hand, the 'quantum action' such that only the Lorentz anomaly occurs (which is possible)

$$
\begin{aligned}
\delta^c_\xi W &= 0 \\
\delta^L_\alpha W &= G^L(\alpha,\omega), \qquad (12.473)
\end{aligned}
$$

then the subtraction

$$W \to W - S \qquad (12.474)$$

shifts the anomaly back to pure Einstein-type

$$
\begin{aligned}
\delta^c_\xi(W-S) &= G^E(v_\xi,\Gamma) \\
\delta^L_\alpha(W-S) &= 0. \qquad (12.475)
\end{aligned}
$$

Physically, this means that we have a choice of the energy–momentum tensor. We can choose between a symmetric tensor which is not covariantly conserved and a conserved tensor which is not symmetric!

Remark: Calculating, finally, the gravitational anomalies by regularizing the determinant of the Weyl operator

$$\delta W = -\delta \ln \det i \not{D} P_+,$$

with the help of the heat kernel method Leutwyler [Leutwyler 1985b], [Leutwyler, Mallik 1986] found out that there exists *only* the Lorentz anomaly—coordinate invariance remains intact! To shift the anomaly now from Lorentz-type to Einstein-type involves the nonpolynomial BZ counterterm. This term, however, lies *outside* the class of counterterms which occur in the regularization procedure!

12.6.4 Covariant gravitational anomaly

The gravitational anomalies discussed so far obey the consistency conditions and are called consistent anomalies for this reason. However, the corresponding energy–momentum tensor $\langle T^{\mu\nu}\rangle$ does not transform covariantly under Einstein transformations. But Bardeen and Zumino [Bardeen, Zumino 1984] discovered that—as in the gauge case, which we have studied in Chapter 10—one can also construct a covariantly transforming energy–momentum tensor leading to a covariant gravitational anomaly.

Definition: We introduce a general **variation**

$$\delta_\varphi\, g_{\mu\nu}(x) = \varphi_{\mu\nu}(x) \tag{12.476}$$

and we regard this variation as a derivative in the **space of all metrics** Sp $\mathcal{M} = \{g_{\mu\nu}(x)\}$.

The variation δ_φ is a derivative along an arbitrary (but fixed) tensor field $\varphi_{\mu\nu}$

$$\delta_\varphi = i_\varphi \delta = \int dx\ \varphi_{\mu\nu}(x)\frac{\delta}{\delta g_{\mu\nu}(x)}, \tag{12.477}$$

where δ denotes the **exterior derivative** in Sp $\mathcal{M}$. So δ_φ is an operator which gives a metric $g_{\mu\nu}$ an arbitrary (but fixed), symmetric increment $\varphi_{\mu\nu}$.

Einstein transformation law: In order to derive the transformation law for the energy–momentum tensor we proceed analogously to the gauge case of Chapter 10. We apply the following commutator to the 'quantum action'

$$[\delta^c_\xi, \delta_\varphi]W[g] = (\delta^c_\xi\delta_\varphi - \delta_\varphi\delta^c_\xi)W[g], \tag{12.478}$$

where we have chosen $W[g]$ to depend on g; on the one hand we obtain

$$= \int dx\ \varphi_{\mu\nu}(x)\delta^c_\xi\frac{\delta}{\delta g_{\mu\nu}(x)}W[g] - \delta_\varphi G^E(v_\xi, \Gamma) \tag{12.479}$$

and on the other

$$= -\int dx\ \delta^c_\xi\varphi_{\mu\nu}(x)\frac{\delta}{\delta g_{\mu\nu}(x)}W[g]. \tag{12.480}$$

Equating these results gives

$$\int dx[\delta_\xi^c \varphi_{\mu\nu}(x) + \varphi_{\mu\nu}(x)\delta_\xi^c] \frac{\delta}{\delta g_{\mu\nu}} W[g] = \delta_\varphi G^E(v_\xi, \Gamma). \tag{12.481}$$

Inserting the definition of the energy–momentum tensor (12.432)

$$\frac{\delta}{\delta g_{\mu\nu}} W[g] = \frac{1}{2}\sqrt{|g|}\,\langle T^{\mu\nu}\rangle \tag{12.482}$$

provides the transformation law for that tensor.

Theorem: Einstein transformation law

$$\delta_\xi^c \int dx \sqrt{|g|}\, \varphi_{\mu\nu} \langle T^{\mu\nu}\rangle = 2\delta_\varphi G^E(v_\xi, \Gamma). \tag{12.483}$$

Of course, we could also start by allowing $\varphi_{\mu\nu}$ to transform covariantly as a tensor, $\delta_\xi^c\, \varphi_{\mu\nu} = \mathcal{L}_\xi\, \varphi_{\mu\nu}$; then the commutator vanishes, $[\delta_\xi^c, \delta_\varphi] = 0$, and the application to $W[g]$ provides result (12.483) immediately.

What does the transformation law (12.483) mean? If there were no anomaly, the left-hand side would be Einstein transformation invariant. Since $\varphi_{\mu\nu}$ transforms covariantly as a tensor, $\langle T^{\mu\nu}\rangle$ must transform covariantly too.

- Hence in the presence of an anomaly—Einstein anomaly—the covariant transformation law of the energy–momentum tensor is broken!

Bardeen–Zumino polynomial: Bardeen and Zumino [Bardeen, Zumino 1984] found that, analogous to the gauge case (recall Chapter 10), one can construct a new energy–momentum tensor which transforms covariantly.

Proposition:

- *There exists a local polynomial tensor—the* **BZ polynomial** $\mathcal{P}^{\mu\nu}$*—with transformation property*

$$\delta_\xi^c \int dx \sqrt{|g|}\, \varphi_{\mu\nu} \mathcal{P}^{\mu\nu} = -2\delta_\varphi\, G^E(v_\xi, \Gamma), \tag{12.484}$$

so that the **covariant energy–momentum tensor**

$$\langle \widetilde{T}^{\mu\nu}\rangle = \langle T^{\mu\nu}\rangle + \mathcal{P}^{\mu\nu} \tag{12.485}$$

transforms covariantly, which means the **Einstein invariance** *of*

$$\delta_\xi^c \int dx \sqrt{|g|}\, \varphi_{\mu\nu}\, \langle \widetilde{T}^{\mu\nu}\rangle = 0. \tag{12.486}$$

Definition: Then the **covariant gravitational anomaly** is defined by

$$\widetilde{G}^E(\xi) = -\int dx\sqrt{|g|}\,\xi_\nu\nabla_\mu\langle\widetilde{T}^{\mu\nu}\rangle. \tag{12.487}$$

Formulae: The explicit expression for the BZ polynomial $\mathcal{P}^{\mu\nu}$ can be found by solving the transformation property (12.484) for $\mathcal{P}^{\mu\nu}$. This is possible since the consistent anomaly is known, say, by the chain term Q^1_{2n-2} (see Section 12.7). Then we can evaluate the covariant gravitational anomaly, the procedure works analogously to the YM case, which we explained in Chapter 10. We just have to replace $A \to \Gamma$, $F \to R$ and $v \to V_\xi$, and we can read off the final result from formula (10.46). In this way we find Bardeen–Zumino's formula in $(2n-2)$ dimensions:

Covariant gravitational anomaly

$$\widetilde{G}^E(V_\xi,\Gamma) = -n\int_{M_{2n-2}} P(V_\xi, R^{n-1}), \tag{12.488}$$

where V_ξ means

$$(V_\xi)^\alpha{}_\beta = \nabla_\beta\xi^\alpha = \partial_\beta\xi^\alpha + \Gamma^\alpha{}_{\beta\lambda}\xi^\lambda. \tag{12.489}$$

$P(V_\xi, R^{n-1})$ is an invariant polynomial of V_ξ and R^{n-1} as defined in Section 7.1. We have now included the normalization of the invariant polynomial into P and it is given by the index theorem (see Section 12.8). Note that the leading term is a factor n larger than in the consistent case. Actually, the covariant anomaly depends on R, but let us keep Γ in the argument. As in the gauge case the covariant anomaly does not fulfil the Stora–Zumino chain of descent equations but is part of more complicated algebraic relations (see e.g. [Ader, Gieres, Noirot 1991]).

12.7 BRS algebra and descent equations

We can simplify the consistency conditions described in Section 12.6.2 considerably by introducing anticommuting Grassmann variables—ghosts—and the BRS operator, as we did in the YM gauge case in Chapter 8.

12.7.1 BRS algebra

We regard the following quantities as anticommuting Grassmann fields:

$\xi^\alpha(x)$ ... **Einstein ghost**

$\alpha_{ab}(x)$ ... **Lorentz ghost**

$\sigma(x)$ ... **Weyl ghost.**

Accordingly the infinitesimal transformation operators become BRS-like

$$\begin{array}{lll} \delta^c_\xi & \longrightarrow & s_E \\ \delta^L_\alpha & \longrightarrow & s_L \\ \delta_\sigma & \longrightarrow & s_W . \end{array}$$

These operators act on the vielbein as the usual transformations (12.397), (12.393), (12.416) (we consider for the moment only the trivial bundle case)

$$\begin{aligned} s_E\, e^a{}_\mu &= \xi \cdot \partial\, e^a{}_\mu + e^a{}_\nu \partial_\mu \xi^\nu \\ s_L\, e^a{}_\mu &= -\alpha^a{}_b\, e^b{}_\mu \\ s_W\, e^a{}_\mu &= \sigma\, e^a{}_\mu . \end{aligned} \tag{12.490}$$

Defining the **total BRS operator** now by

$$s = s_E + s_L + s_W \tag{12.491}$$

we get

$$s\, e^a{}_\mu = \xi \cdot \partial\, e^a{}_\mu + e^a{}_\nu \partial_\mu \xi^\nu - \alpha^a{}_b e^b{}_\mu + \sigma\, e^a{}_\mu . \tag{12.492}$$

We require the **BRS property:**

$$s \text{ antiderivative and } s^2 = 0. \tag{12.493}$$

This means, on any local functional F depending on e, ξ, α, σ, we have

$$s^2 F[e, \xi, \alpha, \sigma] = 0 \tag{12.494}$$

and specifically

$$s^2\, e^a{}_\mu = 0. \tag{12.495}$$

From condition (12.495) follow the **BRS transformations for the ghosts**

$$\begin{aligned} s\, \xi^\alpha &= \xi \cdot \partial\, \xi^\alpha \\ s\, \alpha^a{}_b &= \xi \cdot \partial\, \alpha^a{}_b - \alpha^a{}_c \alpha^c{}_b \\ s\, \sigma &= \xi \cdot \partial\, \sigma . \end{aligned} \tag{12.496}$$

Consequently we obtain

$$\begin{aligned} s(v_\xi)^\alpha{}_\beta &= s\, \partial_\beta \xi^\alpha = \partial_\beta (\xi^\nu \partial_\nu \xi^\alpha) \\ &= \xi \cdot \partial (v_\xi)^\alpha{}_\beta - (v_\xi)^\alpha{}_\nu (v_\xi)^\nu{}_\beta \end{aligned} \tag{12.497}$$

or

$$s\, v_\xi = \xi \cdot \partial\, v_\xi - v_\xi^2 . \tag{12.498}$$

Proposition: Total gravitational anomaly

- *The BRS transformed 'quantum action' $W[e]$ represents the total gravitational anomaly*

$$\begin{aligned} s\,W[e] &= G^{\text{grav}}_{\text{tot}}(e,\xi,\alpha,\sigma) \\ (s_E+s_L+s_W)W[e] &= G^E(e,\xi)+G^L(e,\alpha)+G^W(e,\sigma). \end{aligned} \tag{12.499}$$

In terms of functional derivatives we write

$$G^{\text{grav}}_{\text{tot}} = s\,W[e] = \int dx\; se^a{}_\mu \frac{\delta}{\delta e^a{}_\mu(x)} W[e] \tag{12.500}$$

with $s\,e^a{}_\mu$ given by equations (12.490). The nilpotency (12.493) implies:

Theorem: **Consistency condition** for the gravitational anomaly

$$s\,G^{\text{grav}}_{\text{tot}}(e,\xi,\alpha,\sigma) = 0. \tag{12.501}$$

This compact equation is equivalent to the set of conditions discussed in Section 12.6.2. So the search for an explicit anomaly expression is reduced to the solution of a cohomology problem. This means we are seeking local functionals in the vielbein and in the ghosts which are linear in the ghosts and closed under s but not exact. Hence the **trivial solution** is

$$G^{\text{grav}}_{\text{triv}} = s\,\widehat{G}[e], \tag{12.502}$$

where $\widehat{G}[e]$ represents a local functional.

Lie derivative: As in the gauge case the condition (12.501) is part of a whole chain of equations. To derive this chain we first need the appropriate BRS algebra on the basic variables. We find it convenient to express the BRS transformations in terms of the Lie derivative (12.152) along ξ

$$\mathcal{L}_\xi = \xi\cdot\partial - [\partial\xi,\ ]. \tag{12.503}$$

But now $\xi = \xi^\alpha\,\partial_\alpha$ represents the **ghost vector field** of diffeomorphisms. Rewriting the Lie derivative we find

$$\mathcal{L}_\xi = \widehat{\mathcal{L}}_\xi - [v_\xi,\ ], \tag{12.504}$$

where here the commutator contracts only tensor indices. The **Lie derivative operator** acting on forms

$$\widehat{\mathcal{L}}_\xi = i_\xi d - d i_\xi \tag{12.505}$$

now contains a minus sign due to the anticommutativity of the ghost field. We demonstrate this quickly for the example

$$\begin{aligned}\widehat{\mathcal{L}}_\xi \, e^a &= (i_\xi d - di_\xi)e^a{}_\mu dx^\mu \\ &= \xi^\nu \partial_\nu e^a{}_\mu dx^\mu - \xi^\nu \partial_\mu e^a{}_\nu dx^\mu + \partial_\mu \xi^\nu e^a{}_\nu dx^\mu + \xi^\nu \partial_\mu e^a{}_\nu dx^\mu \\ &= (\xi^\nu \partial_\nu e^a{}_\mu + \partial_\mu \xi^\nu e^a{}_\nu) dx^\mu = \mathcal{L}_\xi e^a \end{aligned} \tag{12.506}$$

(recall that ξ^ν and dx^μ anticommute). The ghost ξ makes i_ξ act as a derivative and $\widehat{\mathcal{L}}_\xi$ as an antiderivative!

BRS algebra: Now we establish the BRS algebra. Again, we neglect the Weyl transformations for simplicity. (The Weyl anomaly was first discovered by [Coleman, Jackiw 1971], [Crewther 1972] and [Capper, Duff 1974]. Its inclusion in the descent equations can be found e.g. in [Ebner, Heid, Lopes Cardoso 1987].)

Let us recall next the effect of the transformation operators described in Sections 12.1.5 and 12.2.5. From formulae (12.166), (12.172) and (12.309) we can read off the following equations when considering the fields ξ and α as anticommuting ghosts

$$\begin{aligned} s_E\Gamma &\leftarrow \delta^c_\xi \Gamma = (i_\xi d - di_\xi)\Gamma - \nabla v_\xi \\ s_E R(\Gamma) &\leftarrow \delta^c_\xi R(\Gamma) = (i_\xi d - di_\xi)R(\Gamma) + [R(\Gamma), v_\xi] \\ s_E\omega &\leftarrow \delta^c_\xi \omega = (i_\xi d - di_\xi)\omega \\ s_E R(\omega) &\leftarrow \delta^c_\xi R(\omega) = (i_\xi d - di_\xi)R(\omega) \\ s_L\omega &\leftarrow \delta^L_\alpha \omega = -D\alpha \\ s_L R(\omega) &\leftarrow \delta^L_\alpha R(\omega) = [R(\omega), \alpha] \\ s_L\Gamma &\leftarrow \delta^L_\alpha \Gamma = 0. \end{aligned} \tag{12.507}$$

For the **BRS operator** that we now use

$$s = s_E + s_L, \tag{12.508}$$

with the transformation properties (12.496) and (12.498), we find:

BRS algebra

i) *for the coordinate system*

$$\begin{aligned} s\, v_\xi &= \widehat{\mathcal{L}}_\xi v_\xi - v_\xi^2 \\ s\, \Gamma &= \widehat{\mathcal{L}}_\xi \Gamma - \nabla v_\xi \\ s\, R &= \widehat{\mathcal{L}}_\xi R + [R, v_\xi], \end{aligned} \tag{12.509}$$

where $R = d\Gamma + \Gamma^2$,

ii) *for the tangent frame*

$$\begin{aligned} s\,\alpha &= \widehat{\mathcal{L}}_\xi \alpha - \alpha^2 \\ s\,\omega &= \widehat{\mathcal{L}}_\xi \omega - D\alpha \\ s\,R &= \widehat{\mathcal{L}}_\xi R + [R, \alpha], \end{aligned} \tag{12.510}$$

where $R = d\omega + \omega^2$.

In addition we know that for the ghost vector field $\xi = \xi^\alpha \partial_\alpha$ we have

$$\mathcal{L}_\xi\, \xi = [\xi, \xi], \tag{12.511}$$

where $[\xi, \xi]$ means the **graded Lie bracket** of the ghosts

$$[\xi, \xi] = [\xi, \xi]^\alpha \partial_\alpha = 2\xi^\nu \partial_\nu \xi^\alpha \partial_\alpha. \tag{12.512}$$

So we can rewrite the BRS transformation (12.496) for the ghost ξ by

$$s\,\xi = \frac{1}{2}\mathcal{L}_\xi\, \xi = \frac{1}{2}[\xi, \xi]. \tag{12.513}$$

Finally, for the vielbein we find from equations (12.490)

$$s\,e^a = \widehat{\mathcal{L}}_\xi\, e^a - \alpha^a{}_b\, e^b. \tag{12.514}$$

Shifted BRS algebra: As suggested by Ader, Gieres and Noirot [Ader, Gieres, Noirot 1991] the two sets of equations (12.509) and (12.510) can be simplified. For that we define a **shifted BRS operator**

$$s_\xi := s - \widehat{\mathcal{L}}_\xi, \qquad s_\xi^2 = 0 \tag{12.515}$$

(introduced by [Ne'eman, Takasugi, Thierry-Mieg 1980], [Baulieu, Bellon 1986], [Baulieu, Thierry-Mieg 1984]).

This new BRS operator s_ξ removes the explicit dependence on ξ and decouples the diffeomorphism from the Lorentz group. In terms of s_ξ we thus get the following algebra:

Shifted BRS algebra

i) *for the coordinate system*

$$\begin{aligned} s_\xi\, v_\xi &= -v_\xi^2 \\ s_\xi\, \Gamma &= -\nabla v_\xi \\ s_\xi\, R(\Gamma) &= [R(\Gamma), v_\xi], \end{aligned} \tag{12.516}$$

ii) *for the tangent frame*

$$\begin{aligned} s_\xi\, \alpha &= -\alpha^2 \\ s_\xi\, \omega &= -D\alpha \\ s_\xi\, R(\omega) &= [R(\omega), \alpha]. \end{aligned} \tag{12.517}$$

Additionally we have

$$s_\xi\, \xi = -\frac{1}{2}[\xi, \xi] \tag{12.518}$$

$$s_\xi\, e^a = -\alpha^a{}_b\, e^b. \tag{12.519}$$

12.7.2 Descent equations

The shifted BRS algebra, which we have now derived, is of the same form as the BRS algebra in YM gauge theory. Therefore we may proceed completely analogously to the gauge case (Chapter 9) in order to find the chain of descent equations. The following is valid for gravitation with a trivial fibre-bundle structure; we discuss the generalization to nontrivial bundles subsequently.

First, we note the validity of the following familiar identity (due to the shifted BRS algebra (12.516), (12.517)).

Theorem: 'Russian formula'

$$\begin{aligned} \widehat{R}(\widehat{\Gamma}) &= \Delta\widehat{\Gamma} + \widehat{\Gamma}^2 \equiv d\Gamma + \Gamma^2 = R(\Gamma) \\ \widehat{R}(\widehat{\omega}) &= \Delta\widehat{\omega} + \widehat{\omega}^2 \equiv d\omega + \omega^2 = R(\omega), \end{aligned} \tag{12.520}$$

where

$$\widehat{\Gamma} = \Gamma + v_\xi, \qquad \widehat{\omega} = \omega + \alpha, \qquad \Delta = d + s_\xi \tag{12.521}$$

and

$$\Delta^2 = 0 \quad \text{by virtue of } d\widehat{\mathcal{L}}_\xi + \widehat{\mathcal{L}}_\xi d = 0. \tag{12.522}$$

Next we start from a symmetric invariant polynomial of degree n (as defined in Section 7.1) from which we know that it is locally exact—we have the **'transgression'**

$$P(R^n) = dQ_{2n-1}(\Gamma, R). \tag{12.523}$$

The term Q_{2n-1} represents the **Chern–Simons form** given by formula (7.45). This 'transgression' is also valid for the shifted fields—**'shifted transgression'**

$$P(\widehat{R}^n) = \Delta Q_{2n-1}(\widehat{\Gamma}, \widehat{R}) \tag{12.524}$$

by virtue of the Bianchi identity

$$\widehat{\Delta}\widehat{R} = \Delta R + [\widehat{\Gamma}, \widehat{R}] = 0. \tag{12.525}$$

Then the 'Russian formula' (12.520) implies the identity

$$\Delta Q_{2n-1}(\Gamma + v_\xi, R) = dQ_{2n-1}(\Gamma, R). \tag{12.526}$$

Expanding the Chern–Simons form in powers of the ghost v_ξ

$$Q_{2n-1}(\Gamma + v_\xi, R) = \sum_{k=0}^{2n-1} Q^k_{2n-1-k}(v_\xi, \Gamma, R); \tag{12.527}$$

comparing terms of the same order in equation (12.526) leads to:

Stora–Zumino chain of descent equations

$$\begin{aligned} P(R^n) - dQ^0_{2n-1} &= 0 \\ s_\xi\, Q^k_{2n-1-k} + dQ^{k+1}_{2n-2-k} &= 0 \\ s_\xi\, Q^{2n-1}_0 &= 0 \end{aligned} \tag{12.528}$$

and $k = 0, 1, \ldots, 2n-2$.

Identification: When we integrate the equations over a boundaryless manifold we get for $k = 1$

$$s_\xi \int Q^1_{2n-2}(v_\xi, \Gamma, R) = 0 \tag{12.529}$$

(the d-term vanishes via Stokes theorem). Since we also have (after integration)

$$\widehat{\mathcal{L}}_\xi Q^1_{2n-2} = i_\xi dQ^1_{2n-2} - di_\xi Q^1_{2n-2} \to 0 \tag{12.530}$$

(Q^1_{2n-2} is a form of maximal degree and $di_\xi Q^1_{2n-2}$ is exact) we recover the consistency condition (12.501) for the gravitational anomaly

$$s \int Q^1_{2n-2}(v_\xi, \Gamma, R) = 0. \tag{12.531}$$

So we can identify the pure Einstein anomaly with the chain term Q^1_{2n-2}.

On the other hand, the descent equations are also valid when we refer to the tangent frame. Then we have the fields $\alpha, \omega, R(\omega)$ and we can identify the Lorentz anomaly with the term Q^1_{2n-2}.

Proposition:

- *The chain term* Q^1_{2n-2} *represents either the pure Einstein anomaly*

$$G^E(v_\xi, \Gamma) \equiv -\int Q^1_{2n-2}(v_\xi, \Gamma, R) \tag{12.532}$$

or the Lorentz anomaly

$$G^L(\alpha, \omega) \equiv -\int Q^1_{2n-2}(\alpha, \omega, R). \tag{12.533}$$

Here the normalization is included in Q^1_{2n-2}.

Finally, we also want to refer to other approaches of the cohomology set-up for gravitational anomalies [Baulieu, Thierry-Mieg 1984], [Bonora, Pasti, Tonin 1986], [Bandelloni 1986] and to extensions of the cohomology formalism to gravity *with* torsion [Moritsch, Schweda, Sorella 1994], [Moritsch, Schweda, Sommer 1995], [Moritsch 1994].

Résumé: We start with one and the same invariant polynomial $P(R^n)$. Since the curvature is given either by $R = d\Gamma + \Gamma^2$ or by $R = d\omega + \omega^2$ we may consider the polynomial either as a function of the $GL(2n)$ curvature or of the $SO(2n)$ curvature. This leads to a first cocycle for the term $Q^1_{2n-2}(v, \Gamma, R)$ or $Q^1_{2n-2}(\alpha, \omega, R)$. Both terms represent the anomaly—the Einstein or the Lorentz anomaly—and satisfy the same chain. Generally, we obtain the chain terms in version $Q^k_{2n-1-k}(v_\xi, \Gamma, R)$ or $Q^k_{2n-1-k}(\alpha, \omega, R)$. We also know the explicit expressions; they are determined, for instance, by the formulae (9.120) and (9.136) derived in Section 9.5. For example:

Chern–Simons form

$$Q^0_{2n-1}(\Gamma, R) = n \int_0^1 dt\ P(\Gamma, R_t^{n-1}), \quad R_t = tR + (t^2 - t)\Gamma^2 \tag{12.534}$$

$$Q^0_{2n-1}(\omega, R) = n \int_0^1 dt\ P(\omega, R_t^{n-1}), \quad R_t = tR + (t^2 - t)\omega^2. \tag{12.535}$$

Anomaly term

$$Q^1_{2n-2}(v_\xi, \Gamma, R) = n(n-1) \int_0^1 dt(1-t)P(v_\xi, d(\Gamma, R_t^{n-2})) \tag{12.536}$$

$$Q^1_{2n-2}(\alpha, \omega, R) = n(n-1) \int_0^1 dt(1-t)P(\alpha, d(\omega, R_t^{n-2})). \tag{12.537}$$

Choosing $n = 2$ specifically we obtain for the **Chern–Simons form**

$$
\begin{aligned}
Q_3^0(\Gamma, R) &\sim \operatorname{tr}\left(\Gamma d\Gamma + \frac{2}{3}\Gamma^3\right) \\
Q_3^0(\omega, R) &\sim \operatorname{tr}\left(\omega d\omega + \frac{2}{3}\omega^3\right)
\end{aligned} \tag{12.538}
$$

and for the **2-dimensional anomaly term**

$$
\begin{aligned}
Q_2^1(v_\xi, \Gamma) &\sim \operatorname{tr}\, v_\xi d\Gamma \\
Q_2^1(\alpha, \omega) &\sim \operatorname{tr}\, \alpha d\omega.
\end{aligned} \tag{12.539}
$$

Here the invariant polynomial $P(R^n)$ contains a normalization constant which we determine later on (Section 12.8).

Remark: Finally, we also observe that the polynomial vanishes

$$
P(R^{2k+1}) = 0 \tag{12.540}
$$

($k = 0, 1, 2, \ldots$) for odd products of R. This is due to the antisymmetry of $R^\alpha{}_\beta$ or $R^a{}_b$. So n must be even, $n = 2k+2$. Consequently the gravitational anomaly determined by Q^1_{2n-2} can occur in only $2n-2 = 4k+2 = 2, 6, 10, \ldots$ dimensions!

12.7.3 BRS for nontrivial fibre bundles

The chain of descent equations that we derived previously is valid for gravitation—considered as a $SO(m)$ or $GL(m, \mathbf{R})$ gauge theory—with a trivial bundle structure. But as in the YM gauge case (Section 9.2) the whole formalism can be extended to nontrivial fibre bundles. Again, this generalization has been elaborated by Stora [Stora 1986] and his collaborators [Langouche, Schücker, Stora 1984], a work we want to follow now.

To lift the diffeomorphism of the base into the bundle an additional fixed background connection ω_0 is required (see Stora's GIFT lectures [Stora 1986]). Then the BRS algebra is introduced in the following way:

BRS algebra with $s^2 = 0$

$$
\begin{aligned}
se &= -\alpha e + \mathcal{L}^0_\xi e \\
s\omega &= -D\alpha + i_\xi R(\omega) - D i_\xi(\omega - \omega_0) \\
s\alpha &= -\alpha^2 + \mathcal{L}^0_\xi \alpha - \frac{1}{2} i_\xi i_\xi R(\omega_0) \\
s\xi &= \frac{1}{2}[\xi, \xi],
\end{aligned} \tag{12.541}
$$

where the **covariant Lie derivative operator** acting on forms is defined by

$$\mathcal{L}^0_\xi = i_\xi D(\omega_0) - D(\omega_0) i_\xi \tag{12.542}$$

and

$$D(\omega_0) = d + [\omega_0, \ ]. \tag{12.543}$$

The background connection is not transformed

$$s\omega_0 = 0. \tag{12.544}$$

Note: Clearly, for $\omega_0 = 0$—the case of a trivial bundle—we recover the BRS algebra given before (equations (12.510), (12.513), (12.514)), where we need the identity

$$s_E\omega = \widehat{\mathcal{L}}_\xi\omega \equiv i_\xi R(\omega) - Di_\xi\omega. \tag{12.545}$$

As before, the **gravitational anomaly** G^{Grav} is determined by the **consistency condition**

$$sG^{\mathrm{grav}}(\xi, \alpha) = 0. \tag{12.546}$$

It is convenient to change (mix) the ghosts $\alpha \to \widetilde{\alpha}$, where

$$\widetilde{\alpha} := \alpha + i_\xi(\omega - \omega_0). \tag{12.547}$$

Then we obtain a new set of algebra which does not contain ω_0 any more (the BRS algebra (12.541) corresponds to a Lie algebra which seems to depend on ω_0; however, one finds that actually it does not depend on ω_0):

BRS algebra

$$\begin{aligned} se &= -\widetilde{\alpha}e + (i_\xi D - Di_\xi)e \\ s\omega &= -D\widetilde{\alpha} + i_\xi R(\omega) \\ s\widetilde{\alpha} &= -\widetilde{\alpha}^2 + \frac{1}{2} i_\xi i_\xi R(\omega) \\ s\xi &= \frac{1}{2}[\xi, \xi]. \end{aligned} \tag{12.548}$$

If we now perform a 'Stora shift'

$$\begin{aligned} \omega &\to \widetilde{\omega} = \omega + \widetilde{\alpha} \\ d &\to \Delta = d + s, \end{aligned} \tag{12.549}$$

we find the following identity of the curvatures.

Theorem: 'Russian formula'

$$\widetilde{R}(\widetilde{\omega}) \equiv e^{i_\xi} R(\omega) = (\mathbf{1} + i_\xi + \frac{1}{2} i_\xi i_\xi) R(\omega), \tag{12.550}$$

where

$$\widetilde{R}(\widetilde{\omega}) = \Delta\widetilde{\omega} + \widetilde{\omega}^2. \tag{12.551}$$

To derive the chain of descent equations we have to consider the 'transgression' for the shifted fields which is valid for nontrivial fibre bundles (recall equations (7.28), (7.29) of Section 7.2).

Theorem: 'Shifted transgression'

$$P(\widetilde{R}^n(\widetilde{\omega})) - P(R^n(\omega_0)) = \Delta Q_{2n-1}(\widetilde{\omega}, \omega_0) \tag{12.552}$$

$$Q_{2n-1}(\widetilde{\omega}, \omega_0) = n \int_0^1 dt P(\widetilde{\omega} - \omega_0, \widetilde{R}_t^{n-1}), \tag{12.553}$$

with

$$\widetilde{R}_t = \widetilde{R}(\widetilde{\omega}_t) = \Delta\widetilde{\omega}_t + \widetilde{\omega}_t^2, \quad \widetilde{\omega}_t = \omega_0 + t(\widetilde{\omega} - \omega_0). \tag{12.554}$$

The homotopic curvature can be re-expressed in the following way

$$\widetilde{R}_t = R_t + t i_\xi R + \frac{1}{2} t i_\xi i_\xi R + (t^2 - t)([\widetilde{\alpha}, \omega - \omega_0] + \widetilde{\alpha}^2), \tag{12.555}$$

with

$$R_t \equiv R(\omega_t) = d\omega_t + \omega_t^2, \quad \omega_t = \omega_0 + t(\omega - \omega_0). \tag{12.556}$$

Applying the 'Russian formula' to the 'shifted transgression'

$$e^{i_\xi} P(R^n(\omega)) - P(R^n(\omega_0)) = \Delta Q_{2n-1}(\widetilde{\omega}, \omega_0), \tag{12.557}$$

expanding both sides in powers of the ghosts

$$\begin{aligned} &(\mathbf{1} + i_\xi + \frac{1}{2} i_\xi i_\xi + \cdots) P(R^n(\omega)) - P(R^n(\omega_0)) \\ &= \quad (d + s)(Q^0_{2n-1} + Q^1_{2n-2} + Q^2_{2n-3} + \cdots), \end{aligned} \tag{12.558}$$

and comparing terms of the same order, we obtain:

Chain of descent equations

$$P(R^n(\omega)) - P(R^n(\omega_0)) = dQ^0_{2n-1} \tag{12.559}$$

$$sQ^0_{2n-1} + dQ^1_{2n-2} = i_\xi P(R^n(\omega)) \tag{12.560}$$

$$sQ^1_{2n-2} + dQ^2_{2n-3} = \frac{1}{2} i_\xi i_\xi P(R^n(\omega)) \qquad (12.561)$$
$$\cdots \qquad \cdots$$

Since we have

$$i_\xi i_\xi P(R^n(\omega)) = 0 \quad \text{in } (2n-2) \text{ dimensions}$$

we find that Q^1_{2n-2} satisfies the consistency condition (12.546).

Proposition:

- *The chain term* Q^1_{2n-2} *represents the gravitational anomaly*

$$G^{\text{grav}}(\xi, \alpha) = -\int_{M_{2n-2}} Q^1_{2n-2}. \qquad (12.562)$$

The normalization is included in Q^1_{2n-2}, respectively in the invariant polynomial $P(R^n)$.

Formula: The chain terms can be solved explicitly; the result is, for example [Langouche, Schücker, Stora 1984]:

Anomaly term

$$\begin{aligned} Q^1_{2n-2} &= n \int_0^1 dt P(\widetilde{\alpha}, R_t^{n-1}) \\ &+ \; n(n-1) \int_0^1 dt P(\omega - \omega_0, (t^2 - t)[\widetilde{\alpha}, \omega - \omega_0], R_t^{n-2}) \\ &+ \; n(n-1) \int_0^1 dt P(\omega - \omega_0, t i_\xi R, R_t^{n-2}). \end{aligned} \qquad (12.563)$$

Of course, we may choose $\omega_0 = 0$ only for a trivial bundle and we recover the formulae discussed before.

12.8 Index theorem for gravitation

As in the gauge case the gravitational anomalies can be completely determined by topological techniques. These rest upon the family index theorem of Atiyah and Singer [Atiyah, Singer 1984] and, specifically for gravitation, on the theorem of Alvarez, Singer and Zumino [Alvarez, Singer, Zumino 1984], and, furthermore, on the local index theorems of Bismut [Bismut 1991], [Bismut, Freed 1986a,b], [Knecht, Lazzarini, Stora 1991a,b]. We, however, again follow the procedure of Alvarez-Gaumé and Ginsparg [Alvarez-Gaumé, Ginsparg 1984] described in Section 11.5.3. The analogy to the gauge case is very close.

12.8.1 Alvarez-Gaumé and Ginsparg's index procedure

Diffeomorphisms: In gravitation we work with the **topological spaces**:

Sp $\mathcal{M}\ldots$	**space of all metrics** Sp $\mathcal{M} = \{g_{\mu\nu}\}$ on the manifold M_{2n}.
$\mathcal{D}\ldots$	**space of all pointed diffeomorphisms** $\mathcal{D} = \{x'(x) \vert x'(x_0) = x_0,\, dx'(x_0) = \mathbf{1}\}$.
Sp $\mathcal{M}/\mathcal{D}\ldots$	**moduli space** equivalence class of all metrics up to diffeomorphism.

These spaces play the role of the spaces Sp $\mathcal{A}$, $\mathcal{G}$, Sp $\mathcal{A}/\mathcal{G}$ in the case of gauge theories.

Alvarez-Gaumé and Ginsparg relate the anomaly in $2n$ dimensions to the Atiyah–Singer index theorem in $(2n+2)$ dimensions. The extra 2 dimensions, parametrized by (t,θ), describe 2-dimensional families of metrics on M_{2n}—2-dimensional, compact submanifolds in Sp $\mathcal{M}$. These are projected down to a sphere S^2 in Sp $\mathcal{M}/\mathcal{D}$ (the topologically nontrivial space), where the metrics—which represent the upper and lower hemisphere in Sp $\mathcal{M}/\mathcal{D}$—are related to each other along the 'equator' of the sphere by a loop in the diffeomorphism group $\mathcal{D}$. The anomaly represents an obstruction to this projection. To show the occurrence of the anomaly the determinant bundle is considered over such a noncontractible sphere S^2 in Sp $\mathcal{M}/\mathcal{D}$.

So we introduce a **one-parameter family of diffeomorphisms**

$$x \to x'(\theta, x), \tag{12.564}$$

giving rise to a loop in the diffeomorphisms. Then we get the coordinate transformation matrix

$$(\Lambda^{-1})^{\mu}{}_{\nu}(\theta, x) = \frac{\partial x'^{\mu}(\theta, x)}{\partial x^{\nu}}, \tag{12.565}$$

the exterior derivative in the additional θ-variable

$$\delta = d\theta\, \frac{\partial}{\partial \theta}, \tag{12.566}$$

and the Maurer–Cartan form in group space

$$v_{\xi}(\theta, x) = \Lambda^{-1}(\theta, x)\delta\Lambda(\theta, x). \tag{12.567}$$

The Christoffel connection Γ and the curvature $R = d\Gamma + \Gamma^2$ transform as

$$\Gamma(\theta, x) = \Lambda^{-1}(\theta, x)[\Gamma(x) + d]\Lambda(\theta, x)$$
$$R(\theta, x) = \Lambda^{-1}(\theta, x)R(x)\Lambda(\theta, x). \tag{12.568}$$

The anomaly occurs by considering the generating functional which represents the fermion determinant.

Proposition:

- *The noninvariance of the generating functional, the fermion determinant, under a general coordinate transformation occurs in a phase variation as we move along $\mathcal{D}$. The local winding of the determinant phase generates the anomaly!*

Procedure:

i) We find a topological situation where the winding number of the phase is nonvanishing.
ii) We construct a $(2n+2)$-dimensional Dirac operator—by adding as dimensions the variables θ and t—whose index is equal to this winding number. Then the index density supplies the gravitational anomaly.

The whole procedure is nothing but a repetition of the gauge case, described extensively in Section 11.5.3, so that we may take over from formula (11.228) the final result:

Proposition: Einstein anomaly

- *The Einstein anomaly is given by*

$$-G^E(v_\xi, \Gamma(\theta)) = 2\pi i \text{ Index } i\, \not{D}_{2n+2}, \tag{12.569}$$

where

$$\text{index } i\, \not{D}_{2n+2} = \int_{S^1} \text{Index } i\, \not{D}_{2n+2}. \tag{12.570}$$

Index theorem: Now the index theorem enters, stating that the index can be expressed as an integral over certain characteristic classes [Atiyah, Singer 1968a,b,c, 1971a,b, 1984] (for further literature we refer to the books [Gilkey 1984], [Berline, Getzler, Vergne 1992], [Lawson, Michelsohn 1989], [Booss, Bleecker 1985] and [Nash 1991]). In our case of the Dirac operator (12.351) which contains the spin connection $\omega = \omega_{ab}\, \sigma^{ab}/2$ the characteristic class is given by the Dirac genus of the compact manifold M.

Definition: The **Dirac genus** is defined by

$$\widehat{A}(M) = \prod_a \frac{x_a/2}{\sinh x_a/2}. \tag{12.571}$$

The quantities x_a denote the skew eigenvalues of the curvature 2-form

$$\frac{R_{ab}}{2\pi} = \begin{pmatrix} 0 & x_1 & & & \\ -x_1 & 0 & & & \\ & & \ddots & & \\ & & & 0 & x_n \\ & & & -x_n & 0 \end{pmatrix}, \tag{12.572}$$

which we consider as a matrix in the Lie algebra of $SO(2n)$. Each x_a expresses a 2-form. The Dirac genus $\widehat{A}(M)$ can be expanded to arbitrary order in the curvature R. Rewriting the results of AGG [Alvarez-Gaumé, Ginsparg 1985] we have

$$\begin{aligned} \widehat{A}(M) &= \mathbf{1} + \frac{1}{(4\pi)^2}\frac{1}{12}\,\mathrm{tr}\,R^2 + \frac{1}{(4\pi)^4}\left[\frac{1}{288}(\mathrm{tr}\,R^2)^2 + \frac{1}{360}\,\mathrm{tr}\,R^4\right] \\ &+ \frac{1}{(4\pi)^6}\left[\frac{1}{10368}(\mathrm{tr}\,R^2)^3 + \frac{1}{4320}\,\mathrm{tr}\,R^2\,\mathrm{tr}\,R^4 + \frac{1}{5670}\,\mathrm{tr}\,R^6\right] \\ &+ \frac{1}{(4\pi)^8}\left[\frac{1}{497664}(\mathrm{tr}\,R^2)^4 + \frac{1}{103680}(\mathrm{tr}\,R^2)^2\,\mathrm{tr}\,R^4\right. \\ &\left. + \frac{1}{68040}\,\mathrm{tr}\,R^2\,\mathrm{tr}\,R^6 + \frac{1}{259200}(\mathrm{tr}\,R^4)^2 + \frac{1}{75600}\,\mathrm{tr}\,R^8\right] \\ &+ \dots \end{aligned} \tag{12.573}$$

So the Dirac genus represents a sum of invariant polynomials in the curvature 2-form R to a given finite order depending on the dimension of the manifold.

Theorem: Atiyah–Singer index theorem

$$\text{index } D_+ = \int_{M_{2n}} \widehat{A}(M). \tag{12.574}$$

$D_+ = i\,\not{D}P_+$ denotes the Weyl operator and $\not{D} = \not{\partial} + \not{\omega}$ the Dirac operator (12.351) for gravitation. Clearly, we pick up only the n-th term of $\widehat{A}(M)$ proportional to the volume $2n$-form.

Note: Up to now we have not worried about the specific form of the manifold. However, choosing for gravitation, for instance, the sphere S^{2n} with radius L the Riemannian curvature (recall equation (12.275)) implies vanishing traces

$$R^a{}_b = \frac{1}{L^2}\, e^a\, e^b \quad \Rightarrow \quad \mathrm{tr}\; R^{2n} = 0 \tag{12.575}$$

and consequently

$$\widehat{A}(S^{2n}) = \mathbf{1}. \tag{12.576}$$

So in this case we achieve only a trivial result.

Anomaly: We apply the AS index theorem now to the Dirac operator in $(2n+2)$-dimensions chosen on a Weyl-like basis (recall equations (11.213), (11.214)). The $(2n+2)$-dimensional manifold M_{2n+2} constructed à la AGG is a product manifold of a 2-dimensional sphere S^2 in the variables (θ, t) with the usual compact manifold M_{2n} of space–time

$$M_{2n+2} = S^2 \times M_{2n}. \tag{12.577}$$

So we can express the index (12.574) by

$$2\pi i \cdot \text{ index } i\,\not{D}_{2n+2} = 2\pi i \int_{S^2 \times M_{2n}} \widehat{A}(M) \equiv \int_{S^2 \times M_{2n}} P(R^{n+1}(\theta)), \tag{12.578}$$

which picks up a certain invariant polynomial $P(R^{n+1})$ to order $(n+1)$. Proceeding now as usual, we rewrite the polynomial locally as the exterior derivative (in $(2n+2)$ dimensions) of the Chern–Simons form

$$P(R^{n+1}(\theta)) = (d + \delta + \delta_t) Q_{2n+1}. \tag{12.579}$$

We apply Stokes' theorem then we get the Chern–Simons form over the 'equator' $S^1 \times M_{2n}$

$$2\pi i \cdot \text{ index } i\,\not{D}_{2n+2} = \int_{S^1 \times M_{2n}} Q_{2n+1}(\Gamma(\theta) + v_\xi, R(\theta)). \tag{12.580}$$

We expand in powers of v_ξ—the only quantity that carries $d\theta$—and we gain the SZ chain term Q^1_{2n}—the linear term in v_ξ—as the sole contribution

$$2\pi i \cdot \text{ index } i\,\not{D}_{2n+2} = \int_{S^1 \times M_{2n}} Q^1_{2n}(v_\xi, \Gamma(\theta), R(\theta)). \tag{12.581}$$

Choosing finally $\theta = 0$ we find from equations (12.569), (12.570), (12.581):

Proposition: Einstein anomaly

- *The pure Einstein anomaly is given by the chain term* Q^1_{2n}

$$G^E(v_\xi, \Gamma) = -\int_{M_{2n}} Q^1_{2n}(v, \Gamma, R). \tag{12.582}$$

This is in accordance with our differential-geometric chain result (12.532) (by shifting $n \to n-1$).

Now we can calculate the chain term Q^1_{2n} *inclusive* of the normalization from the invariant polynomial $P(R^{n+1})$ (recall equation (12.578)).

Normalization:

- *The invariant polynomial—and consequently the chain term* Q^1_{2n}*—is determined completely by the Dirac genus*

$$P(R^{n+1}) = 2\pi i\, \widehat{A}(M_{2n})\Big|_{2n+2}. \tag{12.583}$$

The subscript $2n+2$ means that for a $2n$-dimensional theory we have to start with an invariant polynomial on a $(2n+2)$-dimensional manifold. But the indices of the curvature $R^\mu{}_\nu$ or $R^a{}_b$ still act only on the $2n$-dimensional space–time manifold M_{2n}. It is the form part of R which carries the extra dimensions.

Lorentz transformations: On the other hand, we also consider Lorentz transformations, then we refer to the tangent frame and we work with the **topological spaces**:

Sp $\mathcal{B}\ldots$	**space of all tangent frames** Sp $\mathcal{B} = \{e^a\}$.
$\mathcal{L}\ldots$	**group of all local Lorentz transformations**
Sp $\mathcal{B}/\mathcal{L}\ldots$	**moduli space.**

More precisely, Sp $\mathcal{B}$ means the space of all sections of the tangent frame bundle, and $\mathcal{L}$ means the space of all maps of the base manifold into the Lorentz group. Actually we work on the space Sp $\mathcal{M}$ × Sp $\mathcal{B}$ / $\mathcal{D} \times \mathcal{L}$.

We introduce the **one-parameter family of local Lorentz transformations**

$$L(\theta, x) \tag{12.584}$$

and the MC form in group space

$$\alpha(\theta, x) = L^{-1}(\theta, x)\delta L(\theta, x). \tag{12.585}$$

The spin connection ω and the curvature $R(\omega) = d\omega + \omega^2$ transform as

$$\begin{aligned} \omega(\theta, x) &= L^{-1}(\theta, x)[\omega(x) + d]L(\theta, x) \\ R(\theta, x) &= L^{-1}(\theta, x)R(x)L(\theta, x). \end{aligned} \tag{12.586}$$

To find the anomaly we proceed as described above, we just have to replace $v_\xi \to \alpha$, $\Gamma \to \omega$ and we obtain the final result.

Proposition: Lorentz anomaly

- *The Lorentz anomaly is given by the chain term* Q^1_{2n}

$$G^L(\alpha, \omega) = -\int_{M_{2n}} Q^1_{2n}(\alpha, \omega, R). \tag{12.587}$$

Again, the chain term Q^1_{2n} is now correctly normalized via equation (12.583).

12.8.2 Examples in two dimensions

Let us illustrate the 2-dimensional case, $n = 1$, which is easy to handle. We have to begin with the invariant polynomial (with the correct normalization); it is determined by equation (12.583) and expansion (12.573):

Invariant polynomial

$$P(R^2) = 2\pi i \, \widehat{A}(M_2)\Big|_{\text{dim}=4} = \frac{i}{96\pi} \text{ tr } R^2. \tag{12.588}$$

Then the chain term Q^1_2 following from the polynomial (12.588) is given by:

Chain term

$$Q^1_2(v_\xi, \Gamma) = \frac{i}{96\pi} \text{ tr } v_\xi d\Gamma, \quad Q^1_2(\alpha, \omega) = \frac{i}{96\pi} \text{ tr } \alpha d\omega. \tag{12.589}$$

For the anomalies we find the expressions:

Pure Einstein anomaly

$$\begin{aligned} G^E(v_\xi, \Gamma) &= -\int d^2x \; e \; \xi_\nu \; \nabla_\mu \; \langle T^{\mu\nu} \rangle \\ &= -\int_{M_2} Q^1_2(v_\xi, \Gamma) \\ &= \frac{-i}{96\pi} \int_{M_2} \text{tr } v_\xi d\Gamma. \end{aligned} \tag{12.590}$$

Lorentz anomaly

$$\begin{aligned} G^L(\alpha,\omega) &= \int d^2x\, e\, \alpha_{ab}\, \langle T^{ab}\rangle \\ &= -\int_{M_2} Q_2^1(\alpha,\omega) \\ &= \frac{-i}{96\pi}\int_{M_2} \mathrm{tr}\ \alpha\omega. \end{aligned} \tag{12.591}$$

As discussed in Chapter 5 the factor $(-i)$ is removed in Minkowski space.

What does this mean for the energy–momentum tensor?

Conventions: Specifically in 2 dimensions with a Minkowski signature we use the following conventions here:

ε-tensor referred to TF

$$\begin{aligned} &\varepsilon^{01} = -\varepsilon^{10} = 1 && \varepsilon_{01} = -\varepsilon_{10} = -1 \\ &\varepsilon^{ac}\varepsilon_{cb} = \delta^a{}_b && \varepsilon^{ab}\varepsilon_{ab} = -2. \\ &\varepsilon_{ab} = \eta_{ac}\eta_{bd}\varepsilon^{cd} \end{aligned} \tag{12.592}$$

ε-tensor referred to CS , $e = \det e^a{}_\mu$

$$\begin{aligned} &\bar\varepsilon^{\mu\nu} = \frac{1}{e}\,\varepsilon^{\mu\nu} && \bar\varepsilon_{\mu\nu} = e\,\varepsilon_{\mu\nu} \\ &\bar\varepsilon^{01} = -\bar\varepsilon^{10} = \frac{1}{e} && \bar\varepsilon_{01} = -\bar\varepsilon_{10} = -e \\ &\bar\varepsilon^{\mu\lambda}\,\bar\varepsilon_{\lambda\nu} = \delta^\mu{}_\nu && \bar\varepsilon^{\mu\nu}\,\bar\varepsilon_{\mu\nu} = -2 \\ &\bar\varepsilon^{\mu\nu} = E_a{}^\mu E_b{}^\nu \varepsilon^{ab} && \bar\varepsilon_{\mu\nu} = g_{\mu\rho}g_{\nu\sigma}\bar\varepsilon^{\rho\sigma} = e^a{}_\mu e^b{}_\nu \varepsilon_{ab}. \end{aligned} \tag{12.593}$$

Formulae: With these notations we derive the following useful formulae which we will need below

$$\begin{aligned} &R_{\alpha\beta\mu\nu} = -\frac{1}{2}\,\bar\varepsilon_{\alpha\beta}\bar\varepsilon_{\mu\nu}\mathcal{R} && e^a{}_\mu e^b{}_\nu - e^a{}_\nu e^b{}_\mu = \varepsilon^{ab}\bar\varepsilon_{\mu\nu} \\ &R_{abcd} = -\frac{1}{2}\,\varepsilon_{ab}\varepsilon_{cd}\mathcal{R} && dx^\mu dx^\nu = \varepsilon^{\mu\nu} dx^0 dx^1 = \varepsilon^{\mu\nu} d^2x \\ &R^{ab}{}_{\mu\nu} = -\frac{1}{2}\,\varepsilon^{ab}\bar\varepsilon_{\mu\nu}\mathcal{R} && \bar\varepsilon^{\mu\nu}\varepsilon^{ab}\partial_\mu\omega_{ab\nu} = \mathcal{R}, \end{aligned} \tag{12.594}$$

where $\mathcal{R}$ is the Ricci scalar (12.95)

$$\mathcal{R} = R^{\mu\nu}{}_{\mu\nu} = R^{ab}{}_{ab} = 2R^{01}{}_{01}. \tag{12.595}$$

Lorentz anomaly: Let us consider the case of a Lorentz anomaly (12.591). In components we have

$$G^L(\alpha, \omega) = \int d^2xe\alpha_{ab}\langle T^{ab}\rangle = \frac{1}{96\pi}\int d^2x\ \alpha^a{}_b\partial_\mu\omega^b{}_{a\nu}\varepsilon^{\mu\nu}. \tag{12.596}$$

Proposition:

- *The Lorentz anomaly is equivalent to the existence of an antisymmetric energy–momentum tensor of the following form*

$$\langle T^{ab}\rangle = \frac{1}{96\pi}\ \bar{\varepsilon}^{\mu\nu}\partial_\mu\omega^{ba}{}_\nu = \frac{1}{192\pi}\ \varepsilon^{ab}\mathcal{R}. \tag{12.597}$$

We have shifted the anomaly into the Lorentz part which means that there is no anomaly in the diffeomorphisms, $G^E(\xi) = 0$, no Einstein anomaly. However, this does not mean a conservation of the energy–momentum tensor—as discussed already in Section 12.6.1—but implies (recall formula (12.424))

$$\nabla_\mu\ \langle T^\mu{}_\nu\rangle = \omega_{ab\nu}\ \langle T^{ab}\rangle. \tag{12.598}$$

Inserting result (12.597) we get:

Proposition:

- *The Lorentz anomaly also corresponds to the following nonconservation of the energy–momentum tensor*

$$\nabla_\mu\ \langle T^\mu{}_\nu\rangle = \frac{1}{192\pi}\ \varepsilon^{ab}\omega_{ab\nu}\mathcal{R}. \tag{12.599}$$

We observe that $\langle T^{\mu\nu}\rangle$ does not transform in a covariant way.

Weyl anomaly: For completeness we also quote the result of the Weyl anomaly without going into the details of the calculation. These can be found e.g. in [Ebner, Heid, Lopes Cardoso 1987]

$$\begin{aligned} G^W(\sigma) &= \int d^2x\ e\ \sigma\ \langle T^\mu{}_\mu\rangle \\ &= \frac{1}{48\pi}\int d^2x\ e\ \sigma\left(\mathcal{R} - \frac{1}{2}\ \varepsilon^{ab}\nabla_\mu\omega_{ab}{}^\mu\right) \end{aligned} \tag{12.600}$$

and consequently we have:

Proposition:

- *The Weyl anomaly expresses the following nonvanishing tensor trace*

$$\langle T^{\mu}{}_{\mu}\rangle = \frac{1}{48\pi}\left(\mathcal{R} - \frac{1}{2}\,\varepsilon^{ab}\nabla_{\mu}\omega_{ab}{}^{\mu}\right). \tag{12.601}$$

Covariant gravitational anomaly: As discussed in Section 12.6.4 we can find a covariantly transforming energy–momentum tensor

$$\langle \widetilde{T}^{\mu\nu}\rangle = \langle T^{\mu\nu}\rangle + \mathcal{P}^{\mu\nu} \tag{12.602}$$

by adding the local BZ polynomial $\mathcal{P}^{\mu\nu}$. Then the covariant gravitational anomaly is determined by $\langle \widetilde{T}^{\mu\nu}\rangle$ and can be calculated explicitly via the formula (12.488) of Bardeen and Zumino ($n = 2$). Again, the correctly normalized invariant polynomial is given by equation (12.588) and we obtain:

Covariant gravitational anomaly

$$\begin{aligned}
\widetilde{G}^{E}(V_{\xi}, R) &= -\int d^2x\, e\, \xi_{\nu}\, \nabla_{\mu}\, \langle \widetilde{T}^{\mu\nu}\rangle \\
&= -2\int_{M_2} P(V_{\xi}, R) \\
&= \frac{-i}{48\pi}\int_{M_2} \mathrm{tr}\; V_{\xi}R.
\end{aligned} \tag{12.603}$$

In components we have (after partial integration and removing $(-i)$ in Minkowski space)

$$\widetilde{G}^{E}(V_{\xi}, R) = \frac{-1}{96\pi}\int d^2x\; \xi^{\beta}\nabla_{\alpha}R^{\alpha}{}_{\beta\mu\nu}\varepsilon^{\mu\nu} \tag{12.604}$$

so that we find for the covariant energy–momentum tensor

$$\nabla_{\mu}\, \langle \widetilde{T}^{\mu}{}_{\beta}\rangle = \frac{1}{96\pi}\, \bar{\varepsilon}^{\mu\nu}\nabla_{\alpha}R^{\alpha}{}_{\beta\mu\nu}. \tag{12.605}$$

Re-expressing in terms of the Ricci scalar $\mathcal{R}$ (formula (12.594)) gives:

Proposition:

- *The covariant gravitational anomaly represents the following nonconservation of the covariant energy–momentum tensor*

$$\nabla_{\mu}\, \langle \widetilde{T}^{\mu\nu}\rangle = \frac{1}{96\pi}\, \bar{\varepsilon}^{\mu\nu}\nabla_{\mu}\mathcal{R} = \frac{1}{96\pi}\, \bar{\varepsilon}^{\mu\nu}\partial_{\mu}\mathcal{R}. \tag{12.606}$$

Of course, the BZ polynomial has to satisfy

$$\begin{aligned} \nabla_\mu \mathcal{P}^{\mu\nu} &= \nabla_\mu \langle \widetilde{T}^{\mu\nu} \rangle - \nabla_\mu \langle T^{\mu\nu} \rangle \\ &= \frac{1}{96\pi} \left(\bar{\varepsilon}^{\mu\nu} \nabla_\mu \mathcal{R} - \frac{1}{2}\, \varepsilon^{ab} \omega_{ab}{}^{\nu} \mathcal{R} \right) \end{aligned} \tag{12.607}$$

and can be found to be of the following form:

BZ polynomial

$$\mathcal{P}^{\mu\nu} = -\frac{1}{192\pi} \left[\bar{\varepsilon}^{\mu\alpha} \bar{\varepsilon}^{\nu\beta} \varepsilon^{ab} (\nabla_\alpha \omega_{ab\beta} + \nabla_\beta \omega_{ab\alpha}) - \bar{\varepsilon}^{\mu\nu} \mathcal{R} \right]. \tag{12.608}$$

(Relation (12.607) can be easily checked, all we need is the identity $\nabla_\mu \bar{\varepsilon}^{\alpha\beta} = 0$.)

We notice that the trace of the polynomial is

$$\mathcal{P}^{\mu}{}_{\mu} = \frac{1}{96\pi}\, \varepsilon^{ab} \nabla_\mu \omega_{ab}{}^{\mu} \tag{12.609}$$

and cancels precisely the second term in the trace (12.601) so that we get for the trace of the covariant energy–momentum tensor—the **covariant trace anomaly**

$$\langle \widetilde{T}^{\mu}{}_{\mu} \rangle = \frac{1}{48\pi}\, \mathcal{R}. \tag{12.610}$$

We have calculated all the results in this section within differential geometry and topology; they are in agreement with the results of [Leutwyler 1985b], [Leutwyler, Mallik 1986], [Langouche 1984], [Ebner, Heid, Lopes Cardoso 1987] and [Alvarez-Gaumé, Witten 1983] which are achieved by quite different methods, and last but not least they agree with the recent calculation [Bertlmann, Kohlprath 1999, 2000] within the dispersion relation approach described in Section 4.4.

12.8.3 Mixed anomalies

Finally, we want to consider a fermionic theory with gravitational *and* Yang–Mills gauge fields. Then we use the Dirac operator

$$\not{D} = E_a{}^{\mu} \gamma^a (\partial_\mu + A_\mu + \omega_\mu), \tag{12.611}$$

with the YM field $A_\mu = A^i_\mu T^i$ and the spin connection $\omega_\mu = \omega_{ab\mu} \sigma^{ab}/2$. The corresponding quantum field theory may generate gauge-, gravitational- and mixed anomalies. The mixed anomalies arise in perturbation theory from a fermion loop which couples to gravitational *and* YM fields.

Again, these anomalies can be calculated via the index theorem [Alvarez-Gaumé, Witten 1983], [Alvarez, Singer, Zumino 1984]. In this case we use the following index theorem (see e.g. [Gilkey 1984], [Berline, Getzler, Vergne 1992], [Lawson, Michelsohn 1989]).

Theorem: Atiyah–Singer index theorem

$$\text{index } D_+ = \int_{M_{2n}} \text{ch }(F)\, \widehat{A}(M), \tag{12.612}$$

where $D_+ = i\, \not{D} P_+$ and $\not{D} = \not{\partial} + \not{A} + \not{\omega}$.

The expression ch (F) $\widehat{A}(M)$ means the wedge product of the Chern character (11.112) with the Dirac genus (12.571), (12.573) in a given order corresponding to the dimension of the manifold. Wedging the expansions (11.112) and (12.573) we gain the explicit invariant polynomials in several dimensions

$$\begin{aligned}
\dim = 2 \qquad & \text{index } D_+ = \frac{i}{2\pi} \int_{M_2} \text{tr } F \\
\dim = 4 \qquad & \text{index } D_+ = \frac{1}{(2\pi)^2} \int_{M_4} \left[\frac{-1}{2} \text{ tr } F^2 + \frac{r}{48} \text{ tr } R^2 \right] \\
\dim = 6 \qquad & \text{index } D_+ = \frac{1}{(2\pi)^3} \int_{M_6} \left[\frac{-i}{6} \text{ tr } F^3 + \frac{i}{48} \text{ tr } F \text{ tr } R^2 \right] \\
\dim = 8 \qquad & \text{index } D_+ = \frac{1}{(2\pi)^4} \int_{M_8} \left[\frac{1}{24} \text{ tr } F^4 - \frac{1}{96} \text{ tr } F^2 \text{ tr } R^2 \right. \\
& \qquad \left. + \frac{r}{4608} (\text{tr } R^2)^2 + \frac{r}{5760} \text{ tr } R^4 \right],
\end{aligned} \tag{12.613}$$

where r denotes the dimension of the gauge group representation.

We observe that mixed terms involving both F and R occur only for dimensions higher than or equal to 6 and consequently there exist mixed anomaly contributions only in these dimensions. We can calculate the anomalies from the above invariant polynomials in the previously desribed way. For further literature we refer to [Alvarez-Gaumé, Witten 1983], [Alvarez-Gaumé, Ginsparg 1985], [Minkowski 1988].

12.8.4 Axial gravitational anomaly

At the end of the book we return to the ABJ anomaly of Chapter 4. The ABJ anomaly, or axial anomaly, is described by the triangle diagram which contains one axial current and two vector currents that couple to gauge fields. Replacing the gauge fields by gravitational fields Delbourgo and Salam [Delbourgo, Salam 1972], Eguchi and Freund [Eguchi, Freund 1976] found that the fermion triangle with one axial current and two energy–momentum tensors also generates an **axial gravitational anomaly**. We do not consider the perturbative calculation here but instead we determine this anomaly via the index theorem, which we are now familiar with.

As we discussed in Chapters 5 and 11 the axial anomaly—the anomalous divergence of the axial current—occurs in the Jacobian of the path integral for the generating functional of the Green functions. This Jacobian contains, after suitable regularization, an index density that leads to the following relation (in Euclidean space, recall Theorem (11.20)).

Theorem: Anomaly–index relation

$$\mathcal{A} = d * j^5 = 2i \cdot \text{ Index } D_+. \tag{12.614}$$

The index density, on the other hand, is given by the AS index theorem (12.574)

$$\text{index } D_+ = \int_{M_{2n}} \text{Index } D_+ = \int_{M_{2n}} \widehat{A}(M). \tag{12.615}$$

(We do not compactify space–time to a sphere S^{2n}, where $\widehat{A}(S^{2n}) = 1$.) Now we use the expansion of the Dirac genus, equation (12.573), and we find:

Proposition: Axial gravitational anomaly

- *The axial anomaly in the presence of gravitational fields is given by*

$$\mathcal{A}^{\text{grav}} = d * j^5 = 2i\, \widehat{A}(M_{2n})\Big|_{n=2} = -i\, \frac{-1}{96\pi^2}\, \text{tr } R^2. \tag{12.616}$$

As usual, the factor $(-i)$ is removed in Minkowski space. In components we get

$$\mathcal{A}^{\text{grav}} = \partial^\mu j^5_\mu = -i\, \frac{-1}{384\pi^2}\, \varepsilon^{\mu\nu\rho\sigma} R^\alpha{}_{\beta\mu\nu} R^\beta{}_{\alpha\rho\sigma}. \tag{12.617}$$

Its physical meaning is as in the gauge case; the anomaly signals the breakdown of the external (axial) symmetry in the presence of gravitational fields.

Bibliography

Abdalla, E., Abdalla, M.C.B., Rothe, K.D. (1991). *Two-dimensional quantum field theory*. World Scientific, Singapore.

Abud, M., Ader, J.-P., Gieres, F. (1990). *Nucl. Phys. B* **339**, 687.

Actor, A. (1979). *Rev. Mod. Phys.* **51**, 461.

Adam, C. (1990). Funktionalanalytische und geometrische Aspekte von Anomalien. Diploma thesis at the University of Vienna.

Adam, C., Bertlmann, R.A., Hofer, P. (1992). *Z. Phys. C* **56**, 123.

—— (1993). *La Rivista del Nuovo Cimento*, Vol. **16**, N.8, p. 1–52.

Ader, J.-P., Gieres, F., Noirot, Y. (1991). *Phys. Lett.* **256B**, 401.

Adler, S.L. (1969). *Phys. Rev.* **177**, 2426.

—— (1970a). Perturbation theory anomalies, in: *Lectures on elementary particles and quantum field theory.* 1970 Brandeis University Summer Institute in Theoretical Physics, S. Deser, M. Grisaru and H. Pendleton (eds.), Vol.I, p.1, MIT Press, Cambridge.

—— (1970b). π^0-Decay, in: *High-energy physics and nuclear structure*, S. Devons (ed.), p. 647, Plenum Press, New York.

Adler, S.L., Bardeen, W.A. (1969). *Phys. Rev.* **182**, 1517.

Aharonov, Y., Bohm, D. (1959). *Phys. Rev.* **115**, 485.

Aitchison, I.J.R., Hey, A.J.G. (1989). *Gauge theories in particle physics* (2nd edn). Adam Hilger, Bristol.

Altarelli, G., Ross, G.G. (1988). *Phys. Lett. B* **212**, 391.

Altarelli, G., Stirling, W.J. (1989). *Particle World* **1**, 40.

Altarelli, G., Nason, P., Ridolfi, G. (1994). *Phys. Lett. B* **320**, 152.

Alvarez, O., Singer, I.M., Zumino, B. (1984). *Commun. Math. Phys.* **96**, 409.

Alvarez-Gaumé, L., Baulieu, L. (1983). *Nucl. Phys. B* **212**, 255.

Alvarez-Gaumé, L., Witten, E. (1983). *Nucl. Phys. B* **234**, 269.

Alvarez-Gaumé, L., Ginsparg, P. (1984). *Nucl. Phys. B* **243**, 449.

Alvarez-Gaumé, L., Ginsparg, P. (1985). *Ann. Phys. (NY)* **161**, 423.

Ambjørn, J., Greensite, J., Peterson, C. (1983). *Nucl. Phys. B* **221**, 381.

Andrianov, A., Bonora, L. (1984a). *Nucl. Phys. B* **233**, 232.

—— (1984b). *Nucl. Phys. B* **233**, 247.

Ansourian, M.M. (1977). *Phys. Lett.* **70B**, 301.

Atiyah, M.F., Singer, I.M. (1968a). *Ann. Math.* **87**, 485.

—— (1968b). *Ann. Math.* **87**, 531.

—— (1968c). *Ann. Math.* **87**, 546.

—— (1971a). *Ann. Math.* **93**, 119

—— (1971b). *Ann. Math.* **93**, 139.

—— (1984). *Proc. Natl. Acad. Sci. USA*, Vol. **81**, 2597.

Atiyah, M.F., Bott, R., Patodi, V.K. (1973). *Inventiones Math.* **19**, 279; errata: *Inventiones Math.* **28**, 277.

Atiyah, M.F., Patodi, V.K., Singer, I.M. (1976). *Math. Proc. Cambridge Philos. Soc.* **79**, 71.

Babelon, O., Viallet, C.M. (1979). *Phys. Lett.* **85B**, 246.

—— (1981). *Commun. Math. Phys.* **81**, 515.

Baier, R., Pilon, E. (1991). *Z. Phys. C* **52**, 339.

Balachandran, A.P., Marmo, G., Nair, V.P., Trahern, C.G. (1982). *Phys. Rev. D* **25**, 2713.

Ball, R.D. (1989). *Phys. Rep.* **182**, 1.

Bamberg, P., Sternberg, S. (1990). *A course in mathematics for students of physics.* Vol. **2**, Cambridge University Press.

Bandelloni, G. (1986). *J. Math. Phys.* **27**, 1128.

Banerjee, H., Banerjee, R., Mitra, P. (1986). *Z. Phys. C* **32**, 445.

Bao, D., Nair, V.P. (1985). *Commun. Math. Phys.* **101**, 437.

Bardeen, W.A. (1969). *Phys. Rev.* **184**, 1848.

Bardeen, W.A., Zumino, B. (1984). *Nucl. Phys. B* **244**, 421.

Baulieu, L. (1984). *Nucl. Phys. B* **241**, 557.

—— (1985). *Phys. Rep.* **128C**, 1.

Baulieu, L., Bellon, M. (1986). *Nucl. Phys. B* **266**, 75.

Baulieu, L., Thierry-Mieg, J. (1982). *Nucl. Phys. B* **197**, 477.

—— (1984). *Phys. Lett.* **144B**, 221.

Becchi, C., Rouet, A., Stora, R. (1974). *Phys. Lett.* **52B**, 344.

—— (1975). *Commun. Math. Phys.* **42**, 127.

—— (1976a). Renormalizable models with broken symmetries, in: *Renormalization theory*, G. Velo and A.S. Wightman (eds.), p. 299, D. Reidel Publishing Company, Dordrecht.

—— (1976b). Gauge field models, in: *Renormalization theory*, G. Velo and A.S. Wightman (eds.), p. 269, D. Reidel Publishing Company, Dordrecht.

—— (1976c). *Ann. Phys. (NY)* **98**, 287.

Belavin, A.A., Polyakov, A.M., Schwartz, A.S., Tyupkin, Y.S. (1975). *Phys. Lett.* **59B**, 85.

Bell, J.S. (1987). *Speakable and unspeakable in quantum mechanics*, Cambridge University Press.

Bell, J.S., Bertlmann, R.A. (1980). *Z. Phys. C* **4**, 11.

—— (1981a). *Nucl. Phys. B* **177**, 218.

—— (1981b). *Nucl. Phys. B* **187**, 285.

—— (1983). *Nucl. Phys. B* **227**, 435.

—— (1984). *Phys. Lett.* **137B**, 107.

Bell, J.S., Jackiw, R. (1969). *Nuovo Cimento A* **60**, 47.

Berline, N., Getzler, E., Vergne, M. (1992). *Heat kernels and Dirac operators. Grundlęhren der mathematischen Wissenschaften* **298**, Springer-Verlag, Berlin.

Bertlmann, R.A. (1982). *Nucl. Phys. B* **204**, 387.

—— (1986). *Phys. Rep.* **134**, 279.

—— (1992). Magic moments with John Bell, in: *Bell's theorem and the foundations of modern physics*, A. van der Merwe, F. Selleri and G. Tarozzi (eds.), p. 31, World Scientific, Singapore.

Bertlmann, R.A., Kohlprath, E. (1999). *Gravitational anomalies, Schwinger terms and dispersion relations*. Preprint University of Vienna, UWThPh-1999-63.

—— (2000). *Gravitational anomalies in a dispersive approach*. Preprint University of Vienna, UWThPh-2000-10, arXiv:hep-th/0003172, to appear in *Phys. Lett. B*.

Bertlmann, R.A., Launer, G., de Rafael, E. (1985). *Nucl. Phys. B* **250**, 61.

Bertlmann, R.A., Dominguez, C.A., Loewe, M., Perrottet, M., de Rafael, E. (1988). *Z. Phys. C* **39**, 231.

Bijnens, J., Ecker, G., Pich, A. (1992). *Phys. Lett. B* **286**, 341.

Birrell, N.D., Davies, P.C.W. (1982). *Quantum fields in curved space*. Cambridge University Press.

Bismut, J.-M. (1991). *Superconnexions, indice local des familles, déterminant de la cohomologie et métriques de Quillen*. Preprint Université de Paris-Sud, Mathématiques, Orsay, No. 91–04.

Bismut, J.-M., Freed, D.S. (1986a). *Commun. Math. Phys.* **106**, 159.

—— (1986b). *Commun. Math. Phys.* **107**, 103.

Blau, M. (1988). *Phys. Lett.* **209**, 503.

—— (1989). *J. Math. Phys.* **30**, 2226.

Bonora, L., Cotta-Ramusino, P. (1983). *Commun. Math. Phys.* **87**, 589.

—— (1986). *Phys. Rev. D* **33**, 3055.

Bonora, L., Bregola, M., Pasti, P. (1985). *Phys. Rev. D* **31**, 2665.

Bonora, L., Pasti, P., Tonin, M. (1986). *J. Math. Phys.* **27**, 2259.

Bonora, L., Cotta-Ramusino, P., Rinaldi, M., Stasheff, J. (1987). *Commun. Math. Phys.* **112**, 237.

Booss, B., Bleecker, D.D. (1985). *Topology and analysis. Universitext*, Springer-Verlag, New York.

Bott, R., Tu, L.W. (1982). *Differential forms in algebraic topology, Graduate texts in mathematics* **82**, Springer-Verlag, New York.

Bouchiat, C., Iliopolous, J., Meyer, Ph. (1972). *Phys. Lett.* **38B**, 519.

Brandt, F., Dragon, N., Kreuzer, M. (1989). *Phys. Lett.* **231B**, 263.

—— (1990a). *Nucl. Phys. B* **332**, 224.

—— (1990b). *Nucl. Phys. B* **332**, 250.

—— (1990c). *Nucl. Phys. B* **340**, 187.

Brown, L.S., Carlitz, R.D., Lee, C. (1977). *Phys. Rev. D* **16**, 417.

Capper, D.M., Duff, M.J. (1974). *Nuovo Cimento A* **23**, 173.

Caprini, I., Micu, L., Visinescu, M. (1992). *Z. Phys. C* **56**, 225.

Chambers, R.G. (1960). *Phys. Rev. Lett.* **5**, 3.

Cheng, T.-P., Li, L.-F. (1988). *Gauge theory of elementary particle physics*, Clarendon Press, Oxford.

Choquet-Bruhat, Y., DeWitt-Morette, C. with Dillard-Bleick, M. (1982). *Analysis, manifolds and physics*, North Holland, Amsterdam.

Coleman, S., Jackiw, R. (1971). *Ann. Phys. (NY)* **67**, 552.

Coquereaux, R., Jadczyk, A. (1988). *Riemannian geometry, fibre bundles, Kaluza–Klein theories and all that.* World Scientific, Singapore.

Cotta-Ramusino, P., Reina, C. (1984). *J. Geom. Phys.* Vol. **1**, 121.

Crewther, R. (1972). *Phys. Rev. Lett.* **28**, 1421.

Curtis, W.D., Miller, F.R. (1985). *Differential manifolds and theoretical physics.* Academic Press, San Diego.

Cutkosky, R.E. (1960). *J. Math. Phys.* **1**, 429.

Daniel, M., Viallet, C.M. (1980). *Rev. Mod. Phys.* **52**, 175.

Delbourgo, R., Salam, A. (1972). *Phys. Lett.* **40B**, 381.

Delbourgo, R., Thompson, G. (1985). *Phys. Rev. D* **32**, 3300.

Deser, S. (1996). *Helv. Phys. Acta* **69**, 570, arXiv:hep-th/9609138.

Deser, S., Duff, M.J., Isham, C.J. (1976). *Nucl. Phys. B* **111**, 45.

Deser, S., Jackiw, R., Templeton, S. (1984). *Ann. Phys. (NY)* **140**, 372.

d'Inverno, R. (1992). *Introducing Einstein's relativity*. Clarendon Press, Oxford.

Dirac, P.A.M. (1931). *Proc. Roy. Soc. A* **133**, 60.

—— (1933). *Physikalische Zeitschrift der Sowjetunion* **3**, 64.

—— (1958). *The principles of quantum mechanics* (4th edn). Clarendon Press, Oxford.

Dittrich, W., Reuter, M. (1986). *Selected topics in gauge theories. Lecture notes in physics* **244**, Springer-Verlag, Berlin.

Dolgov, A.D., Zakharov, V.I. (1971). *Nucl. Phys. B* **27**, 525.
Dubois-Violette, M., Talon, M., Viallet, C.M. (1985a). *Phys. Lett. B* **158**, 231.
—— (1985b). *Commun. Math. Phys.* **102**, 105.
Duff, M.J. (1994). *Class Quant. Grav.* **11**, 1387.
Dunne, G.V., Jackiw, R., Trugenberger, C.A. (1989). *Ann. Phys. (NY)* **194**, 197.
Ebner, M., Heid, R., Lopes Cardoso, G. (1987). *Z. Phys. C* **37**, 85.
Ecker, G. (1982). *Acta Phys. Austr.*, Suppl. XXIV, 3.
Ecker, G., Neufeld, H., Pich, A. (1992). *Phys. Lett. B* **278**, 337.
—— (1994). *Nucl. Phys. B* **413**, 321.
Eguchi, T., Freund, P.G.O. (1976). *Phys. Rev. Lett.* **37**, 1251.
Eguchi, T., Gilkey, P.B., Hanson, A.J. (1980). *Phys. Rep.* **66**, 213.
Einhorn, M.B., Jones, D.R.T. (1984). *Phys. Rev. D* **29**, 331.
Ekstrand, C. (1999). *Schwinger terms from external field problems.* Ph.D. thesis at the Royal Institute of Technology, Stockholm.
Faddeev, L.D. (1984). *Phys. Lett.* **145B**, 81.
Faddeev, L.D., Popov, V.N. (1967). *Phys. Lett.* **25B**, 29.
Faddeev, L.D., Shatashvili, S.L. (1986). *Phys. Lett.* **167B**, 225.
Faddeev, L.D., Slavnov, A.A. (1980). *Gauge fields, introduction to quantum theory*, Benjamin-Cummings, Reading, Massachusetts.
Falqui, G., Reina, C. (1985). *Commun. Math. Phys.* **102**, 503.
Felsager, B. (1981). *Geometry, particles and fields* (4th edn). Odense University Press, Copenhagen.
Feynman, R.P. (1948). *Rev. Mod. Phys.* **20**, 267.
Feynman, R.P., Hibbs, A.R. (1965). *Quantum mechanics and path integrals.* McGraw-Hill, New York.
Flanders, H. (1963). *Differential forms with applications to physical sciences.* Academic Press, New York.
Frankel, T. (1997). *The geometry of physics, an introduction.* Cambridge University Press.
Fritzsch, H. (1991). *Phys. Lett. B* **256**, 75.
Fujikawa, K. (1979). *Phys. Rev. Lett.* **42**, 1195.
—— (1980). *Phys. Rev. D* **21**, 2848.
—— (1983). *Nucl. Phys. B* **226**, 437.
—— (1984). *Phys. Rev. D* **29**, 285.
—— (1985). *Phys. Rev. D* **31**, 341.
—— (1989). *Aspects of the BRS symmetry and anomalies.* RITP Hiroshima University preprint RRK 89–8.
Fujikawa, K., Tomiya, M., Yasuda, O. (1985). *Z. Phys. C* **28**, 289.

Fukuda, H., Miyamoto, Y. (1949). *Prog. Theor. Phys.* **4**, 347.

Gamboa Saraví, R.E., Muschietti, M.A., Schaposnik, F.A., Solomin, J.E. (1983). *Commun. Math. Phys.* **89**, 363.

—— (1984a). *Phys. Lett.* **138B**, 145.

—— (1984b). *Ann. Phys. (NY)* **157**, 360.

Gasser, J., Leutwyler, H. (1984). *Ann. Phys. (NY)* **158**, 142.

Gell-Mann, M., Goldberger, M., Thirring, W. (1954). *Phys. Rev.* **95**, 1612, Abstract ibid p.654.

Georgi, H., Glashow, S.L. (1972). *Phys. Rev. D* **6**, 429.

Gieres, F. (1988). *Geometry of supersymmetric gauge theories. Lecture notes in physics* **302**, Springer-Verlag, Berlin.

Gilkey, P.B. (1984). *Invariance theory, the heat equation, and the Atiyah–Singer index theorem. Mathematics lecture series* **11**, Publish or Perish, Wilmington, Delaware.

Gipson, J.M. (1986). *Phys. Rev. D* **33**, 1061.

Göckeler, M., Schücker, T. (1987). *Differential geometry, gauge theories, and gravity*. Cambridge University Press.

Green, M.B., Schwarz, J.H. (1984). *Phys. Lett.* **149B**, 117.

Green, M.B., Schwarz, J.H., West, P.C. (1985). *Nucl. Phys. B* **254**, 327.

Greub, W., Halperin, S., Vanstone, R. (1972). *Connections, curvature and cohomology*, Vol. I, II, III. Academic Press, San Diego.

Gribov, V.N. (1978). *Nucl. Phys. B* **139**, 1.

Gross, D., Jackiw, R. (1972). *Phys. Rev. D* **6**, 477.

Grosse, H. (1988). *Models in statistical physics and quantum field theory. Trieste notes in physics*, Springer-Verlag, Berlin.

Grossmann, B. (1985). *Phys. Lett.* **152B**, 93.

Harada, K., Tsutsui, I. (1987). *Phys. Lett.* **183B**, 311.

Hawking, S.W. (1977). *Commun. Math. Phys.* **55**, 133.

Hirzebruch, F. (1978). *Topological methods in algebraic geometry. Grundlehren der mathematischen Wissenschaften* **131**. Springer-Verlag, Berlin.

Höhne, U. (1990). Das QCD Vakuum. Diploma thesis at the University of Vienna.

't Hooft, G. (1976a). *Phys. Rev. D* **14**, 3432.

—— (1976b). *Phys. Rev. Lett.* **37**, 8.

't Hooft, G., Veltman, M., (1972). *Nucl. Phys. B* **44**, 189.

Hořejší, J. (1985). *Phys. Rev. D* **32**, 1029.

—— (1986). *J. Phys. G: Nucl. Phys.* **12**, L7–L12.

—— (1992a). *Czech. J. Phys.* **42**, 241.

—— (1992b). *Czech. J. Phys.* **42**, 345.

—— (1994). *Introduction to electroweak unification—standard model from tree unitarity*. World Scientific, Singapore.

Hořejší, J., Novotný, J., Zavialov, O.I. (1988). *Phys. Lett. B* **213**, 173.

Huang, K. (1982). *Quarks, leptons and gauge fields*. World Scientific, Singapore.

Inomata, A. (1986). A possible test of the topological effect in quantum mechanics, in: *Ann. of NY Acad. of Sciences* **480**, p. 217, ed. D.M. Greenberger.

Isham, C.J. (1989). *Modern differential geometry for physicists*. World Scientific, Singapore.

Itzykson, C., Zuber, J.-B. (1980). *Quantum field theory*. McGraw-Hill, New York.

Jackiw, R. (1977). *Rev. Mod. Phys.* **49**, 681.

—— (1985a). *Phys. Rev. Lett.* **54**, 159, 2380.

—— (1985b). *Phys. Lett.* **154B**, 303.

—— (1985c). *Comments Nucl. Part. Phys.* Vol. **15**, No. 3, p. 99.

—— (1985d). Field theoretic investigations in current algebra, Topological investigations of quantized gauge theories, in: *Current algebra and anomalies*, S.B. Treiman, R. Jackiw, B. Zumino and E. Witten (eds.), p. 81, and p. 211, World Scientific, Singapore.

—— (1991). *Europhysics News*, Vol. **22**, No. 4, p. 76.

—— (1995). *Diverse topics in theoretical and mathematical physics*. World Scientific, Singapore.

Jackiw, R., Rajaraman, R. (1985). *Phys. Rev. Lett.* **54**, 1219.

Jackiw, R., Rebbi, C. (1976). *Phys. Rev. D* **14**, 517.

—— (1977). *Phys. Rev. D* **16**, 1052.

Jo, S.-G. (1985). *Phys. Lett.* **163B**, 353.

Johnson, K. (1963). *Phys. Lett.* **5**, 253.

Kastler, D., Stora, R. (1986a). *Jour. Geom. Phys.* **3**, 437.

—— (1986b). *Jour. Geom. Phys.* **3**, 483.

Kelnhofer, G. (1991). *Z. Phys.* **52**, 89.

Kiskis, J. (1977). *Phys. Rev. D* **15**, 2329.

Knecht, M., Lazzarini, S., Stora, R. (1991a). *Phys. Lett. B* **262**, 25.

—— (1991b). *Phys. Lett. B* **273**, 63.

Kobayashi, S., Nomizu, K. (1963). *Foundations of differential geometry*, Vol. I. Wiley-Interscience, New York.

—— (1969). *Foundations of differential geometry*, Vol. II. Wiley-Interscience, New York.

Kogut, J.B., Susskind, L. (1975). *Phys. Rev. D* **11**, 1477.

Korthals Altes, C.P., Perrottet, M. (1972). *Phys. Lett.* **39B**, 546.

Krasnikov, N.V. (1984). *JETP Lett.* **40**, 1170.

—— (1985a). *Nuovo Cimento* **95A**, 325.

—— (1985b). *Nuovo Cimento* **89A**, 308.

—— (1985c). *JETP Lett.* **41**, 586.

Kummer, W. (1970). *Acta Phys. Austr.* Suppl. VII, 567.

—— (1976). *Acta Phys. Austr.* Suppl. XV, 423.

—— (1990). Gauge-independence of anomalies, in: *Fields and particles*, H. Mitter and W. Schweiger (eds.), p. 231, Springer-Verlag, Berlin.

Langouche, F., (1984). *Phys. Lett.* **148B**, 93.

Langouche, F., Schücker, T., Stora, R. (1984). *Phys. Lett.* **145B**, 342.

Lawson, H.B., Michelsohn, M.-L. (1989). *Spin geometry*. Princeton University Press.

Lee, T.D., Yang, C.N. (1962). *Phys. Rev.* **128**, 885.

Lehmann, H., Symanzik, K., Zimmermann, W. (1955). *Nuovo Cimento* **1**, 205.

Leibbrandt, G. (1990). Unified gauge formalism at two loops, in: *Physical and nonstandard gauges*, P. Gaigg, W. Kummer and M. Schweda (eds.), p. 87, *Lecture notes in physics* **361**, Springer-Verlag, Berlin.

Leinaas, J.M., Olaussen, K. (1982). *Phys. Lett.* **108B**, 199.

Leutwyler, H. (1985a). *Phys. Lett.* **152**, 78.

—— (1985b). *Phys. Lett.* **153B**, 65; **155B**, 469 (E).

—— (1986a). *Helv. Phys. Acta* **59**, 201.

—— (1986b). On the determinant of the Weyl operator, in: *Quantum field theory and quantum statistics. Essays in honour of the sixtieth birthday of E.S. Fradkin*, Vol. 2, I.A. Batalin, C.J. Isham and G.A. Vilkovisky (eds.), p. 327, Adam Hilger, Bristol.

—— (1990). *Helv. Phys. Acta* **63**, 660.

Leutwyler, H., Mallik, S. (1986). *Z. Phys. C* **33**, 205.

Levy, D. (1987). *Nucl. Phys. B* **282**, 367.

Lott, J., Rajaraman, R. (1985). *Phys. Lett.* **165B**, 321.

Löwenstein, J., Swieca, J. (1971). *Ann. Phys. (NY)* **68**, 172.

Lucha, W., Schöberl, F.F. (1996). *Pfadintegrale in Quantenmechanik und Quantenfeldtheorie*. Spektrum Akademischer Verlag, Heidelberg.

Lurié, D. (1968). *Particles and fields*. Wiley-Interscience, New York.

Manton, N.S. (1985). *Ann. Phys. (NY)* **159**, 220.

Mañes, J., Stora, R., Zumino, B. (1985). *Commun. Math. Phys.* **102**, 157.

Mickelsson, J. (1985a). *Phys. Rev. Lett.* **54**, 2379.

—— (1985b). *Commun. Math. Phys.* **97**, 361.

—— (1989). *Current algebras and groups. Plenum monographs in nonlinear physics*, Plenum Press, New York.

Minkowski, P. (1988). *Phys. Lett.* **201B**, 328.

Misner, C.W., Thorne, K.S., Wheeler, J.A. (1973). *Gravitation*. Freeman, San Francisco.

Möllenstedt, G., Bayh, W. (1962). *Phys. Bl.* **18**, 299.

Moritsch, O. (1994). On the algebraic structure of gravity with torsion including Weyl symmetry. Thesis, Technical University of Vienna.

Moritsch, O., Schweda, M., Sommer, T. (1995). *Class. Quantum Grav.* **12**, 2059.

Moritsch, O., Schweda, M., Sorella, S.P. (1994). *Class. Quantum Grav.* **11**, 1225.

Morozov, A.Yu. (1986). *Sov. Phys. Usp.* **29** (11), 993.

Nakahara, M. (1990). *Geometry, topology and physics*. Adam Hilger, Bristol.

Narison, S., Shore, G.M., Veneziano, G. (1993). *Nucl. Phys. B* **391**, 69.

Nash, C. (1991). *Differential topology and quantum field theory*. Academic Press, London.

Nash, C., Sen, S. (1983). *Topology and geometry for physicists*. Academic Press, London.

Ne'eman, Y., Šijački, Dj. (1985). *Phys. Lett.* **157B**, 275.

Ne'eman, Y., Takasugi, E., Thierry-Mieg, J. (1980). *Phys. Rev. D* **22**, 2371.

Nielsen, N.K., Schroer, B. (1977). *Nucl. Phys. B* **127**, 493.

Nielsen, N.K., Römer, H., Schroer, B. (1977). *Phys. Lett.* **70B**, 445.

Nielson, H.B., Ninomiya, M. (1983). *Phys. Lett. B* **130**, 389.

—— (1991). *Int. Journal Mod. Phys. A*, Vol. **6**, 2913.

van Nieuwenhuizen, P., (1988). *Anomalies in quantum field theory: cancellation of anomalies in $d = 10$ supergravity. Leuven notes in mathematical and theoretical physics*, Vol. **3**, Leuven University Press.

Pauli, W., Villars, F. (1949). *Rev. Mod. Phys.* **21**, 434.

Peshkin, M., Tonomura, A. (1989). *The Aharonov–Bohm effect. Lecture notes in physics* **340**, Springer-Verlag, Berlin.

Pietschmann, H. (1983). *Weak interactions, formulae, results and derivations*. Springer-Verlag, Berlin.

Piguet, O., Sorella, S.P. (1992). *Nucl. Phys. B* **381**, 373.

—— (1993). *Nucl. Phys. B* **395**, 661.

Pokorski, S. (1987). *Gauge field theories*. Cambridge University Press.

Quillen, D. (1985). *Funct. Anal. Appl.* **19**, 31.

Quirós, M., de Urries, F.J., Hoyos, J., Mazón, M.L., Rodriguez, E. (1981). *J. Math. Phys.* **22**, 1767.

Rajaraman, R. (1982). *Introduction to solitons and instantons in quantum field theory*. North-Holland, Amsterdam.

—— (1985a). *Phys. Lett.* **154B**, 305.

—— (1985b). *Phys. Lett.* **162B**, 148.

Reuter, M. (1985). *Phys. Rev. D* **31**, 1374.

Richter, K. (1998). *Gravitational anomalies and the families index theorem.* Diploma thesis at the University of Vienna.

Römer, H. (1979). *Phys. Lett.* **83B**, 172.

—— (1981). *Phys. Lett.* **101B**, 55.

Rupp, C. (1998). *Berry phase, Schwinger term, and anomalies in quantum field theory.* Diploma thesis at the University of Vienna.

Ryder, L.H. (1988). *Quantum field theory.* Cambridge University Press.

Sachs, R.K., Wu, H. (1977). *General relativity for mathematicians. Graduate texts in mathematics* **48**, Springer-Verlag, New York.

Sakurai, J.J. (1985). *Modern quantum mechanics.* Benjamin-Cummings, Menlo Park.

Schwinger, J. (1951). *Phys. Rev.* **82**, 664.

—— (1962). *Phys. Rev.* **128**, 2425.

—— (1969). *Particles, sources and fields.* Vol. I, II, Addison-Wesley, Reading.

Seeley, R.T. (1967). *Amer. Math. Soc. Proc. Symp. Pure Math.* **10**, 288.

Sexl, R.U., Urbantke, H.K. (1983). *Gravitation und Kosmologie.* Bibliographisches Institut, Mannheim.

Shifman, M.A. (1991). *Phys. Rep.* **209**, 341.

—— (1994). *Instantons in gauge theories.* World Scientific, Singapore.

Shore, G.M., Veneziano, G. (1993). *Mod. Phys. Lett. A* **8**, 373.

Shuryak, E.V. (1988). *QCD vacuum, hadrons and superdense matter.* World Scientific, Singapore.

Singer, I.M. (1978). *Commun. Math. Phys.* **60**, 7.

Singer, I.M., Thorpe, J.A. (1967). *Lecture notes on elementary topology and geometry. Undergraduate texts in mathematics*, Springer–Verlag, New York.

Sorella, S.P. (1993). *Commun. Math. Phys.* **157**, 231.

Sorella, S.P., Tătaru, L. (1994). *Phys. Lett. B* **324**, 351.

Steinberger, J. (1949). *Phys. Rev.* **76**, 1180.

Stora, R. (1977). Continuum gauge theories, in: *New developments in quantum field theory and statistical mechanics.* 1976 Cargèse Lectures, M. Lévy and P. Mitter (eds.), p. 201, Plenum Press, New York.

—— (1984). Algebraic structure and topological origin of anomalies, in: *Recent progress in gauge theories.* 1983 Cargèse Lectures, H. Lehmann (ed.), NATO ASI series, Plenum Press, New York.

—— (1986). Algebraic structure of chiral anomalies, in: *New perspectives in quantum field theory.* 1985 GIFT lectures, Jaca, Spain, J. Abad, M. Asorey and A. Cruz (eds.), World Scientific, Singapore.

—— (1991). *Private communications.*
Sumitani, T. (1984). *J. Phys. A: Math. Gen.* **17**, L811.
Sutherland, D.G. (1967). *Nucl. Phys. B* **2**, 433.
Takahashi, Y. (1957). *Nuovo Cimento* **6**, 370.
Thierry-Mieg, J. (1980). *J. Math. Phys.* **21**, 2834.
Thirring, W. (1992). *A course in mathematical physics*, Vol. I, II, Springer-Verlag, New York.
Tonomura, A., Matsuda, T., Suzuki, R., Fukuhara, A., Osakabe, N., Umezaki, H., Endo, J., Shinagawa, K., Sugita, Y. and Fujiwara, H. (1982). *Phys. Rev. Lett.* **48**, 1443.
Tourrenc, P. (1977). *Phys. Rev. D* **16**, 3421.
Trautman, A. (1984). *Differential geometry for physicists.* Bibliopolis, Napoli.
Tröster, A. (1994). Nonabelian anomalies and the Atiyah–Singer index theorem. Diploma thesis at the University of Vienna.
Tsutsui, I. (1989). *Phys. Lett.* **229B**, 51.
Tyutin, I.V. (1975). *Gauge invariance in field theory and statistical mechanics.* Lebedev preprint FIAN, No. 39.
Veltman, M. (1967). *Proc. Roy. Soc. A* **301**, 107.
Visconti, A. (1992). *Introductory differential geometry for physicists.* World Scientific, Singapore.
Wald, R.M. (1984). *General relativity*. University of Chicago Press.
Ward, J.C. (1950). *Phys. Rev.* **78**, 1824.
Warner, F.W. (1983). *Foundations of differential manifolds and Lie groups. Graduate texts in mathematics* **94**, Springer-Verlag, New York.
Weinberg, S. (1972). *Gravitation and cosmology: principles and applications of the general theory of relativity*. Wiley, New York.
Wess, J., Zumino, B. (1971). *Phys. Lett.* **37B**, 95.
Witten, E. (1989a). *Commun. Math. Phys.* **121**, 351.
—— (1989b). *Nucl. Phys. B* **311**, 46.
Wu, T.T., Yang, N.C. (1975). *Phys. Rev. D* **12**, 3845.
Wu, Y.-S., Zee, A. (1985). *Phys. Lett.* **152B**, 98.
Yang, N.C., Mills, R.L. (1954). *Phys. Rev.* **96**, 191.
Zhang, Y.-Z. (1987). *Phys. Lett. B* **189**, 149.
Zumino, B. (1984). Chiral anomalies and differential geometry, in: *Relativity, groups and topology II*. 1983 Les Houches Lectures, B.S. DeWitt and R. Stora (eds.), p. 1291, North-Holland, Amsterdam.
Zumino, B. (1985a). *Nucl. Phys. B* **253**, 477.
—— (1985b). Anomalies, cocycles and Schwinger terms, in: *Symposium on anomalies, geometry and topology*, W.A. Bardeen and A.R. White (eds.), p. 111, World Scientific, Singapore.
Zumino, B., Wu, Y.S., Zee, A. (1984). *Nucl. Phys. B* **239**, 477.

Index